GREEN ALGAE

GREEN ALGAE

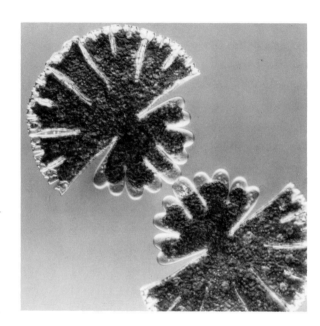

STRUCTURE,
REPRODUCTION
AND EVOLUTION
IN SELECTED GENERA

JEREMY D. PICKETT-HEAPS
University of Colorado

Sinauer Associates, Inc., Publishers
Sunderland, Massachusetts

This book is dedicated to the memory of my beloved wife Chi, who thought up the project and got me started on it, and who was tragically killed in a flying accident on the 5th of June, 1971.

First Edition

GREEN ALGAE:
Structure, Reproduction and Evolution
in Selected Genera

Copyright © 1975 by Sinauer Associates, Inc.

Printed in U.S.A.

Library of Congress Catalog Card Number: 74-24363

ISBN 0-87893-652-1

THE COVER

The photomicrograph on the cover depicts cell division in *M. denticulata.* The parent alga has split into two halves *(upper left and lower right)* each of which is generating a new semicell *(center)*. Magnification, about × 6,500.

CONTENTS

Abbreviations Used
on Figures   vii

1. Introduction   I

2. The Volvocales   7

2.1. Chlamydomonas   20
   2.1a. Cell division   40
   2.1b. Sexual reproduction   54

2.2. Volvox   59
   2.2a. Physiology and genetics of reproduction in V. carteri and V. aureus   61
   2.2b. Cellular and colonial structure   63
   2.2c. Structure of reproductive cells   65

2.3. Tetraspora   65
   2.3a. Cell division   66

3. The Chlorococcales   69

3.1. Chlorococcum   72

3.2. Chlorella   74
   3.2a. Cell division   74

3.3. Kirchneriella   82
   3.3a. Cell division   82

3.4. Ankistrodesmus   88
   3.4a. Cell division   90

3.5. Tetraedron   90
   3.5a. Cell division   92

3.6. Scenedesmus   96
   3.6a. Cell division and colony formation   106

3.7. Hydrodictyon   112

3.7a. Vegetative reproduction   116
3.7b. Sexual (and asexual) reproduction involving polyhedra   130

3.8. Pediastrum   140
   3.8a. Cell division and colony formation   144

3.9. Sorastrum   150
   3.9a. Cell division and colony formation   150

4. The Ulotrichales   167

4.1. Stichococcus   170
   4.1a. Cell division   172

4.2. Klebsormidium   174
   4.2a. Cell division   178
   4.2b. Zoosporogenesis   182

4.3. Coleochaete   186
   4.3a. Cell division   188
   4.3b. The hair cell   190
   4.3c. The zoospore   192

4.4. Ulothrix   194
   4.4a. Cell division   196

4.5. Microspora   198
   4.5a. Cell division   200

4.6. Stigeoclonium   204
   4.6a. Cell division   204
   4.6b. The zoospore   206

4.7. Ulva   208
   4.7a. Cell division   212
   4.7b. Sexual reproduction   214

4.8. Fritschiella   218
   4.8a. Cell division   218

5. The Oedogoniales 219

5.1. Oedogonium 220
   5.1a. Vegetative structure 220
   5.1b. Cell division 220
   5.1c. The zoospore cycle 242
   5.1d. Sexual reproduction 266

5.2. Bulbochaete 282
   5.2a. Vegetative structure 282
   5.2b. Cell division 284
   5.2c. Hair cell formation 288
   5.2d. The zoospore 296
   5.2e. Sexual reproduction 298

6. The Conjugales (Zygnematales) 357

6.1. Saccoderm desmids 358
   6.1a. Netrium 358

6.2. The Zygnemoideae 366
   6.2a. Spirogyra 368
   6.2b. Zygnema 382
   6.2c. Mougeotia 384

6.3. Unconstricted placoderm desmids 386
   6.3a. Closterium 386

6.4. Constricted placoderm desmids 412
   6.4a. Micrasterias 420
   6.4b. Cosmarium 442
   6.4c. Sexual reproduction in constricted placoderm desmids 452

7. The Charales 467

7.1. Vegetative cells 474
7.2. Cell division 478
7.3. Rhizoids 482
7.4. Formation of the antheridia 488
7.5. Spermatogenesis 494
7.6. Oogenesis 504

8. Evolution and Cell Morphology 513

8.1. The evolution of cytoplasmic systems: general conclusions 514
8.2. Phylogenetic significance of motile cell morphology 524
8.3. Cytomorphologic indications of phylogenetic affinities 536
   8.3a. The volvocine lines of evolution 536
   8.3b. The siphonous lines of evolution 542
   8.3c. The ulotrichalean lines of evolution 546
   8.3d. The archetypal Green Algal cell: chlamydomonad or prasinophyte? 562
8.4. A comparison of phylogenies 564
   8.4a. Phylogeny of the flagellar apparatus 565
   8.4b. The volvocine lines of evolution 568
   8.4c. The siphonous lines of evolution 569
   8.4d. The ulotrichalean lines of evolution and the origin of higher land plants 569

Appendices
I. Stereo-Electron Microscopy 577
II. Films on Green Algae 583
III. On the Use of Field Specimens 585

Bibliography 587

Acknowledgments 590

Index 601

ABBREVIATIONS USED ON FIGURES

Many abbreviations used on the figures are explained in the captions. The following summary may, however, be useful.

a	adhesive mucilage
as	androspore
ax	axillary shoot
az	azygote
b	basal cell
bb	basal bodies
bm	basal mucilage
br	branch
c	chloroplast
ca	coenobial adhesive
cb	choanoid body
cf	cleavage furrow
ch	chromosome
cn	centrioles
cp	cell plate
cr	corona
ct	cleavage microtubules
cv	contractile vacuole
cw	coenobial wall
d	dome
dc	division cap
dfc	downward-facing cap
dm	dwarf male
dr	division ring
e	endoplasmic reticulum
ec	empty (parental) cell
es	eyespot
ew	end wall
exw	expanding wall

f	flagellum
fb	flagellar base
fh	flagellar hair
fi	flagellar insertion
ft	fertilization tubule
g	Golgi bodies
h	hair cell
hl	hyaline layer
i	interzone (of spindle)
in	internode
is	isthmus
k	kinetochore
l	lipid body
m	mitochondrion
ma	mastigonemes
mb	manubrium
mc	microtubule center
ml	mucilage
mls	multilayered structure
mp	mucilage pores
mt	manchette microtubules
n	nucleus
nc	nucleolus
nd	node
ne	nuclear envelope
nr	nucleolar remnants
o	oogonium
ow	older wall
p	peroxisome
pc	primary capitulum
pd	plastid
pe	perinuclear envelope

pf	polar fenestra
ph	phycoplast
pl	pole
plw	papilla wall
pm	plasmalemma
pv	papilla vacuole
pw	parental wall
py	pyrenoid
rh	rhizoplast
rs	(wall) rupture site
rt	rootlet templates
s	septum
sc	secondary capitulum
sf	spermatogenous filaments
sfc	suffultory cell
sh	shield cell
sl	statoliths
sp	spines
st	starch
t	microtubules
tb	microtubular band
tc	tiered cap
ts	trilaminar sheath
ufc	upward-facing cap
v	vacuole
vc	vegetative cell
ve	vacuolar envelope
vg	vierergruppe
vs	vesicles
w	(cell) wall
wl	warty layer
xw	cross wall
z	zygote

1

INTRODUCTION

It is impossible to judge the most opportune time to write a book such as this one. Inevitably, the coverage of the subject matter in the following pages will be incomplete. Since I wish to document the structure of algal cells as they pass through various stages in their life cycles at the light and electron microscopical levels in reasonably complete detail, many interesting and representative Green Algae* cannot be mentioned, as

*I shall break with botanical tradition slightly and capitalize the terms "Green," "Brown," and "Red" Algae, in order to stress to the nonbotanist that these adjectives define large divisions in the plant kingdom. This procedure seems logical

there is little or no suitable, coherent information available on them. However, I believe that algae in general represent a fascinating and vast range of organisms whose potential for the study of fundamental processes of life has been largely unexploited, except in such isolated cases as *Acetabularia*. The study of algae is not yet one of the more glamorous disciplines in modern biology and seldom generates much enthusiasm in students. This situation does not seem about to alter and contrasts with the increasing concern over cancer, for example, which attracts ever more money and scientists toward a field whose rather narrowly defined objectives have remained elusive in spite of the tremendous expenditures of human energy and money in their pursuit (leading one of my envious colleagues to remark that these days, more scientists are living off cancer than dying from it!). Many biologists now realize that it may be more relevant, and ultimately more profitable, to attempt to understand how cells actually function than to begin by trying to understand what has gone wrong in cellular systems whose intricacies we have hardly begun to comprehend. I believe the algae will prove useful in attaining a general understanding of the relation between biological structure and function. This book, incomplete though it is, hopefully supports my contention that these organisms should command more interest from biologists because they provide excellent material for the deeper investigation of diverse biological phenomena.

It is a cliché, but one easily forgotten, that algae comprise an extremely important group of organisms. They constitute the primary source of food in the sea upon which all other marine life depends, and it has been estimated that marine phytoplankton fix about 10^{10} tons of carbon a year, probably more than the amount fixed by all terrestial plants. In freshwater habitats, too, their significance is often profound—as even detergent manufacturers have been recently obliged to acknowledge. Yet exposure of biology students to the study of algae is often cursory at best. An extreme example I can cite concerns one freshman year biology course within whose 100-plus lecture schedule the algae have been dis-

missed in eight minutes and the fungi in five. One concludes that many biologists regard the plant kingdom with disinterest and any plant smaller than a pea seedling with disdain. Even botanists sometimes seem to suffer a distressing narrowness of vision; some recent, excellent texts on the structure of "plant" cells have restricted their coverage entirely (or almost entirely) to higher plants (e.g., Clowes and Juniper, 1968; Ledbetter and Porter, 1970; Frey-Wyssling and Muhlethaler, 1965).

The dominance of work on animal and other highly evolved cells in formulating theories and dogmas in cell biology is hardly surprising, but it is an unbalanced approach, quite dangerous and beset with pitfalls. For example, some specific cases: histones have been proposed as a class of proteins perhaps fundamentally important in understanding chromosome structure, gene expression, etc., and yet the proponents of such ideas seldom acknowledge the existence of some eucaryotic dinoflagellates which apparently manage very well without needing histones (Leadbeater and Dodge, 1967; Kubai and Ris, 1969). Plant cytologists frequently reiterate their belief that microtubules are important in the formation of fibrillar cell walls, and yet rarely recall that many more primitive cells (e.g., desmids), including fungi, may elaborate highly complex fibrillar walls without involving microtubules.

Most biology textbooks in describing cell division invoke—or if they are more cautious, merely imply—the existence of various vital functions for centrioles during mitosis. Yet discerning students and teachers, who perforce must swallow these dogmas, cannot help being mystified by the ability of plant cells to divide perfectly well without centrioles. Moreover, centrioles often appear de novo in many plants and other organisms—inexplicable behavior in organelles which many biologists believe to be autonomous and self-replicating and which are often said to contain their own nucleic acid. (The presence of DNA in the centriole still remains to be established; the evidence supporting its presence is usually grossly overrated.) A great deal of the fashionable speculation concerning the nature and function of the centriole is at last being relegated to the archives of cytology, but not before it had acquired wide and apparently uncritical acceptance from scientists who, in the process, often demonstrated an unscientific intoler-

_____

to me, particularly since their more correct Latin names (Chlorophyceae, Phaeophyceae, Rhodophyceae) are always capitalized.

ance of mounting evidence contrary to their own beliefs (not, one sadly notes, an unusual state of affairs). Yet anyone who has studied higher plants, algae, fungi, and protozoa in detail knows how difficult it has been to reconcile the dogma surrounding the functioning of the centriole with experimental observations. It should be apparent, therefore, that a cell biologist who formulates general theories concerning the nature and function of cell constituents runs the risk of being severely embarrassed in the future if he restricts his outlook and neglects to consider what other, more diverse organisms may have to tell us.

I invite attention to a new, challenging aspect of cell biology—the phylogeny of cell structures. At both ends of the scale of living matter, evolution can to some extent explain diversity, whether it be in amino acid sequences in proteins or in whole organisms themselves. Increasing diversity in many ultrastructural systems is also becoming documented, and I have already tried to indicate how such diversity in spindle structures, for example, may be explained in terms of the evolution of the mitotic apparatus (Pickett-Heaps, 1969a, 1972a). An important justification for relating cell division to cell phylogeny can be made on the grounds that the intricate mechanisms involved in reproduction are of supreme importance to the viability of the organism, both in the short and in the long term. Cell division, morphologically and physiologically, must be extremely reliable and preferably rapid to allow the organism to profit when possible by its discovery of a favorable environment. Considered in its entirety, cell division must be a most conservative process, very stable genetically. Yet within the basic mitotic and cytokinetic mechanisms utilized by the Green Algae, there has obviously been room for profitable variation. Considering how long the algae (including the Green Algae) have been in existence (Klein and Cronquist, 1967, suggest that all other eukaryotic cells were derived from these organisms), an accumulation of some nonlethal and apparently advantageous variations in the basic cell division systems is not surprising. We can perhaps draw a reasonably close analogy with the structure of fundamentally important and extremely primitive proteins (e.g., cytochromes) in which changes to certain functionally vital portions of the amino acid sequence cannot be tolerated and therefore do not occur and are not found in ex-

tant organisms, while in other parts of the molecule, variation can be tolerated to a greater or lesser extent and under differing restrictions (e.g., some substitutions of one amino acid for another are relatively unrestricted, while other substitutions may be acceptable only if the old amino acid and its replacement have similar properties). The transmission of nonlethal changes in the amino acid sequence of cytochrome C provides a 1.2-billion-year record of molecular evolution which can be closely related to the genealogy of the organism concerned (Dickerson, 1972). Thus, cytochrome C from man, for example, has much more in common with cytochrome C from a whale than it does with cytochrome C from a dogfish or a wheat seedling. Family trees can be constructed entirely from similarities and dissimilarities in amino acid sequences in this protein, and they "agree remarkably well with those obtained from classical morphology" (Dickerson, 1972). Furthermore, vital proteins such as cytochrome C are at present evolving much more slowly than less important proteins, for the obvious reason that changes to a vital cell component are more likely to be lethal to the organism than changes to a nonvital component. These considerations are almost certainly valid too when considering evolution of the elaborate processes of mitosis and cytokinesis, and we should expect, therefore, such cytoplasmic systems to be essentially conservative while allowing some relatively nonimportant variations. Thus, the centriole is almost certainly unimportant in mitosis as such (Pickett-Heaps, 1969a, 1972a), and hence its absence from the spindle is not significant in terms of spindle function; its absence, however, may be highly significant (e.g., in the Red Algae) when considering another aspect of the cell's attributes—namely, its ability to form flagella.

I believe that this new aspect of cell biology—the phylogeny of cell structures—can be easily investigated in the algae, since we can discern in many classes of them the evolutionary development of complex cells or colonial forms from simpler ancestors represented by present-day organisms (e.g., the volvocine line of evolution from *Chlamydomonas* to *Volvox*; Chapters 2 and 8). One could reasonably expect comparison of the ultrastructure of such organisms to reveal some evidence of a concurrent evolution of intracellular systems. My discussion of phylogeny in this book (Chapter 8) is based mainly upon ultrastructural and morphological comparisons of cells. In this regard, it is significant that Hibberd and Leedale (1971) have suggested the establishment of a new class, the Eustigmatophyceae, formerly in the Xanthophyceae, largely upon the highly characteristic and unusual ultrastructure of the motile zoospore (Round, 1971). Among the several reasons for this restricted approach to phylogeny is, admittedly, my own inexperience as a phylogeneticist, but it is justifiable for another, more valid reason, in spite of the justice of Klein and Cronquist's (1967) comment that a "problem has been the tendency for specialists in any given field to assume that the item of their speciality is, by itself, almost sufficient to serve as a primary base for erecting a phylogeny." There is considerable value in assessing such evidence as will be produced purely on its own merits, without initially taking into account preexisting notions. This is more than a useful academic exercise, since it avoids the human tendency to interpret all evidence in such a way that it fits, and thereby supports, the current concensus. If this latter statement seems a little supercilious, I need only remind the reader of the confusion mentioned already concerning the centriole, deriving from attempts to fit experimental observations into prevailing dogma. Any tentative conclusions I draw on this restricted basis can be compared with other, more expert opinions mentioned in the second part of Chapter 8; the reader can then modify or discard my ideas as he sees fit. The concept of evolution applied to cell structure is, in my opinion, exciting and productive. Cell systems must have evolved just as organisms must have evolved. It may be just as reasonable to conclude that the existence of a centriole in some nonmotile cells represents an inherited character from a previously flagellate progenitor as it is to conclude that the human coccyx represents an appendage derived from our own ancestors who possessed tails. I believe this approach may aid considerably our understanding and appreciation of the intricacies of cytoplasmic structure and function, as indeed the concept of evolution has added immeasurably to all aspects of biology.

Nevertheless, in spite of such convictions, I have approached the writing of this book with some misgivings and a feeling of trepidation. I am not a botanist, nor to my regret have I ever even attended a lecture course in botany. Hope-fully, the consequences of this neglect will not be too obvious in the following pages, and I trust that more erudite practitioners of that discipline will treat my lapses and naiveté with some indulgence. This book is not in any sense a botany textbook, and it arose, like so many others of its kind, from my employer's insistence that I offer a coherent set of lectures to students not overwhelmed by active involvement in the pursuit of knowledge. The book has become a personal record of a biologist's fascination with the beauty and variety of algal cells, and his sense of wonder at how they grow, reproduce, and order their lives. In view of my lack of qualifications, the reader will naturally wonder why someone so ill-equipped should presume to attempt such a book. By default, it seems, as I know of no one else who wants to try! Further impetus has been provided by a sense of frustration at the ever-increasing mass of uncollated data on cell structure being published. In the last few years, biologists have had the chance to repeat the superb light microscopical work carried out in the nineteenth and early twentieth centuries, but now using such techniques as scanning and transmission electron microscopy, time-lapse and interference light microscopy, all coupled with a vast range of other tools available to the cell biologist. The response of biologists—and particularly botanists—has been exuberant, indiscriminate, and often disappointing (see Brown's, 1960, comments). It seems logical to begin using these new techniques after their novelty has worn off in attempts to clarify basic aspects of cell structure and function—how cells divide, reproduce, differentiate, and adapt to their environment. Instead, electron microscopy on plant cells is usually concerned with limited and often desultory observations on a vast range of disparate subjects, and it often fails to reveal the potential of many of these cells in studying various basic phenomena of life. Another irritant is the criticism leveled at electron microscopists, often deserved, for their overly optimistic and bountiful discussion on cell structure and function based upon insufficiently documented observations. So often the mere proximity of one organelle to another is sufficient to provoke a burst of speculation that may be fanciful, ingenious, and usually unverifiable. Indeed, it has become an occupational hazard of the careful microscopist that, when after laborious effort he is able to achieve some correlation of behavior of the cell with its internal organization, his achievement may then be taken as mere confirmation of a hypothesis (usually one of many) suggested earlier by others whose own speculations were not circumscribed by the necessity for rigorous documentation and analysis. Having made these critical comments, I seem to be placing my own head upon the block because I am about to interpret the results of many such ultrastructural observations. Yet by its very nature, electron microscopy is concerned with the images of dead cells—dehydrated, embedded, and then sectioned. To make sense of the two-dimensional images of cellular ultrastructure thus arising, we have to speculate and use our imagination. Otherwise, our efforts are sterile. In fairness to the reader, therefore, I shall try to present interpretations as opinions distinct from reasonably established facts; the reader should *never* assume that these interpretations are necessarily valid, either wholly or partly, and must treat them with a considerable degree of suspicion.

I have always believed that one of the most profitable ways of utilizing electron microscopy to investigate the relation of structure and function involves the detailed study of sequences of intracellular events combined, where possible, with observations on living cells. Any subsequent speculation on the nature and function of cell organelles is much more justifiable, since their appearance, disappearance, disposition, and alteration can sometimes be clearly associated with specific cellular phenomena. For this reason, I shall usually confine my discussion to various aspects of the structure and life cycles of algae which have been carefully documented in some detail, emphasizing light microscopy where it seems appropriate. Regrettably, few algae have been tackled this way, perhaps because the work is tedious and difficult, and usually specialized techniques have to be developed to undertake it. I shall not attempt to catalogue or review the numerous papers concerned purely with ultrastructural morphology. Catalogues of bare facts rarely make stimulating reading. I make no apology for the heavy emphasis upon cell division. Not only is this subject fascinating in its own right, but also the processes of mitosis and cytokinesis have clearly defined objectives that involve considerable morphological changes within the cell. Thus cell division offers an invaluable insight into the relation of structure and

function, and comparison of mitotic and cytokinetic systems has turned out to be unexpectedly revealing in the study of the phylogeny of intracellular systems. Further topics that have proved profitable for ultrastructural analysis include asexual (e.g., formation and fate of zoospores) and sexual reproduction.

Since at this stage, the number of Green Algae studied in depth is very limited, the reader will undoubtedly discover with vexation that I have omitted reference to some particularly interesting species, and all I can suggest is that he go out and do something about it! For example, the fascinating work on morphogenesis in *Acetabularia* has not been mentioned, mainly because electron microscopy of this refractory organism is restricted and sporadic, and also because it has been reviewed elsewhere. Nevertheless, I hope even the present limited results will prove interesting, particularly to those not familiar with these organisms. The lack of coherent information on algae generally can be no better illustrated than by the fact that (at the time of writing) *not one* detailed ultrastructural account has appeared concerning cell division in the Brown Algae and only one on a Red Alga (McDonald, 1972). Let us hope this situation will soon change, for the study of all these cells will inevitably be exciting and rewarding to those willing to undertake the work involved.

This general lack of information raises another problem which may be serious. In almost all cases, the work I draw upon here has been obtained from observations on a single—or at best a few—species in any family or order. Hopefully, they will be reasonably "representative" of related organisms as well. "Representative" is a usefully vague word, however, and so the reader should always allow for the considerable variability that can exist in behavior and structure between closely related species or genera. Thus, for example, when conjugation in one (unidentified) species of *Spirogyra* is described, the reader should not assume that other members of the Zygnemoideae are identical or even greatly similar in all respects to the example given. Several examples come to mind of the differences that may exist in the same process undergone by closely related species: (1) fertilization in *Chlamydomonas* seems to proceed in at least two distinct ways in different species (Section 2.1b); (2) there are minor but significant differences during mitosis in the two species of *Oedogonium* studied, particularly with respect to the behavior and appearance of the persistent nucleolar material (Section 5.1b); and (3) *Spirogyra* and *Mougeotia* show appreciable differences in their mitotic and cytokinetic apparatus (Chapter 6). So generalizations, as always, are rather risky, but may also be valuable if an appropriate spirit of caution prevails when they are used.

In introducing the various algae described, I shall try to keep my general comments as brief as possible without being cryptic. Readers not familiar with the organisms, and especially those wishing to find out more about them, will need to consult more detailed textbooks. An absolutely invaluable compendium of information for advanced readers is Fritsch's (1935) classic: *Structure and Reproduction of the Algae* (Volume 1). The excellent book, *Morphology of Plants,* by Bold (1967) summarizes a great deal of data on the Green Algae and puts them in the context of the whole of the plant kingdom, including the other great divisions of algae, the fungi, and higher plants. Many other valuable reference books could be cited (e.g., Smith, 1950) which are available on most library shelves. Some of these will be mentioned in the text, particularly in Chapter 8.

2

THE VOLVOCALES

The Volvocales include a wide variety of genera, many of which have obvious close affinities with fairly simple flagellate cell types exemplified by the familiar *Chlamydomonas*. Indeed, organisms such as *Chlamydomonas* are thought by many to represent modern forms of the progenitors of all the Green Algae (Chapter 8). The reader is referred to Scagel et al. (1965, Chapter 15) and Klein and Cronquist (1967) for a discussion of this view. As may be expected with such an ancient family of organisms, considerable diversification has taken place within it; Fritsch (1935) mentions that there are, for example, almost 150

7

species of *Chlamydomonas* alone (more recently Dixon, 1970, says that over 500 species have been accepted), and this genus displays considerable morphological variation (Ettl, 1971). Other related organisms are colorless and appear to have lost their chloroplasts, becoming saprophytic.

In the Volvocales, Chlamydomonadineae (Fritsch, 1935) is an especially interesting suborder, since within its many genera it clearly reveals how a simple cell type has given rise to increasingly complex colonies of unicells, an advance in evolution matched by an increasing division of labor between cells of the colony and a tendency toward more complex specializations in reproductive behavior. Yet, throughout this "volvocine line of evolution" (Chapter 8; see Scagel et al., 1965), the vegetative cell unit in the colonial forms remains very similar to some flagellate, unicellular members of the suborder (e.g., *Chlamydomonas*).

We can follow this line of evolution (Fig. 8.18) by referring to common extant species. Thus, *Chlamydomonas* could represent the basic cell type. All cells in a population undergo vegetative reproduction whereby 2, 4, 8, or more (i.e., 2^n, where n is the number of division cycles) individuals are normally released following repeated division of a mother cell. Sexual reproduction is (usually) isogamous (the gametes are the same size; see below), and all individuals participate in it under appropriate conditions (e.g., the mixing of appropriate strains). Furthermore, some species show what may be a primitive colonial tendency: under certain conditions, they form "palmella" stages, whereby vegetative cells become nonflagellate and proliferate while randomly embedded within a gelatinous matrix (which may serve to protect them from desiccation). These palmella stages closely resemble some genera of the suborder Tetrasporineae, and the individual cells rapidly revert to the flagellate form following suitable stimulation (e.g., flooding the colony with water), as indeed do some genera of the Tetrasporineae.

Increasing evolutionary advancement in the volvocine line is marked by an increasing size of the colonies or coenobia, in which the cells are specifically oriented with respect to one another. The number of cells in each coenobium is determined during its early development by the sequential cycles of division in the reproductive

Figures 2.1–2.6
VOLVOCINE EVOLUTION

2.1, 2.2. *Gonium sacculiferium,* face and side view. × 2,650.

2.3. *Gonium pectorale.* × 1,300.

2.4. *Pandorina morum.* × 1,900.

2.5. *Platydorina caudata.* × 1,000.

2.6. *Eudorina unicocca.* × 950.

These colonial organisms, photographed in an opaque medium to reveal more clearly their mucilaginous investment, illustrate evolution in the Volvocales. *Gonium* forms simple planar colonies, whereas *Pandorina* consists of a ball-like cluster of cells. *Platydorina* is a slightly twisted plate-like array (with characteristic projections of mucilage), and the species of *Eudorina* illustrated here is a spheroid of identical cells (cf. Figs. 2.7, 2.8).

SOURCE. Dr. R.C. Starr, unpublished micrographs.

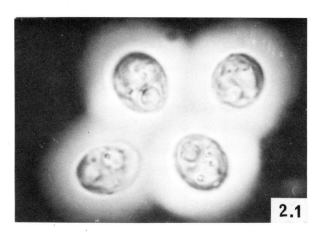

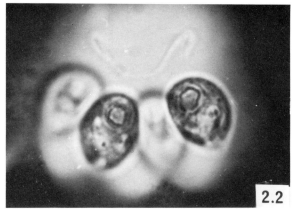

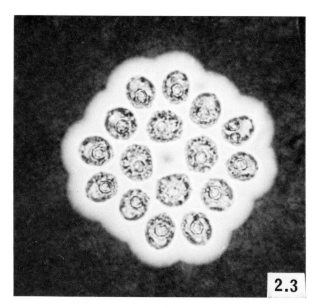

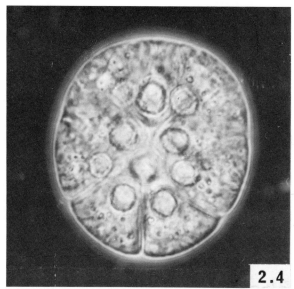

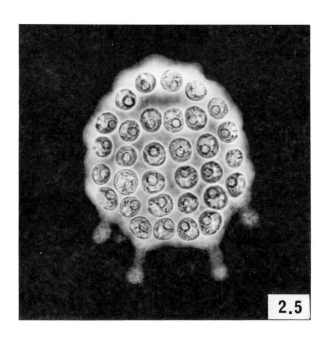

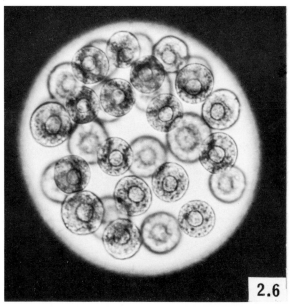

cells of the mother coenobium, and it is exactly (or close to) a power of 2 (i.e., 2^n again), as would be expected if daughter colonies each arise from an individual cell undergoing a succession of synchronized divisions. While the number of cells in a colony is usually fairly specific for each species, it is nevertheless sometimes variable. For example, colonies of *Gonium pectorale*, normally 16-celled (Fig. 2.3), may produce instead 4- or 8-celled colonies under certain cultural conditions. Furthermore, if smaller colonial species are disrupted artificially into unicells, the latter often survive and reproduce, whereas individual somatic cells of larger colonial forms (e.g., *Volvox*), usually appear incapable of survival (e.g., Kochert, 1968; p. 439). Thus, with increasing size of the colonies, individual cells appear to become increasingly subordinate to and dependent upon the whole organism.

As the number of cells per colony increases, the colonies assume highly characteristic morphologies. Simple forms (such as *Gonium* and *Platydorina;* Figs. 2.1–2.3, 2.5) are often planar or have the cells clustered together in a ball-like mass (e.g., *Pandorina;* Fig. 2.4). The well-known larger forms such as *Eudorina* (Figs. 2.6–2.8), *Pleodorina* (Fig. 2.9), and *Volvox* (Fig. 2.10) are spheroids and have their constituent cells arranged as a single layer.

Another important specialization is encountered as the colonial forms become larger. In simpler forms (e.g., *Gonium* and *Pandorina*), all the cells of a colony undergo vegetative reproduction via cell division, and thus all give rise to daughter colonies. However, as the colonies get bigger the vegetative cells show an increasing specialization whereby only some, larger than the others, undergo the cell divisions that lead to vegetative reproduction, while others in the same colony are incapable of further division and eventually die. Allied with this specialization is the appearance of polarity, an obvious example of which is the tendency of reproductive cells to be grouped at one end of the colony. Polarity is also manifested by the direction the entire organism swims and by the orientation of the individual cells (see Fritsch, 1935); cell components such as eyespots (Section 2.1) may show a gradation in development along the colony (Goldstein, 1967). Thus, in the small spherical colony (usually containing 32 cells) of certain strains of *Eudorina elegans*, most of the cells (28) reproduce

Figures 2.7–2.10
VOLVOCINE EVOLUTION

2.7. *Eudorina cylindrica,* vegetative colony. × 380.

2.8. *Eudorina elegans* var. *carteri,* homothallic, sexually induced colony. × 430.

2.9. *Eudorina (Pleodorina) californica* var. *californica.* vegetative colony. × 200.

2.10. *Volvox carteri* f. *nagariensis,* vegetative colony. × 72.

Differentiation of large, reproductive versus smaller, usually nonreproductive cells is apparent in the small colonial species, *E. cylindrica* (cf. Fig. 2.6). During sexual reproduction, the smaller cells can form sperm packets, as in the variety of *E. elegans* shown here. The remainder of the large cells are oogonia in this homothallic strain. *E. elegans* var. *elegans,* however, is heterothallic: colonies are either male or female, and all cells form gametes. The vegetative colony of *E. californica* shows an increasing proportion of smaller somatic cells (cf. Figs. 2.6, 2.7), typical of the larger colonial Volvocales. Vegetative colonies of *Volvox* (Fig. 2.10) contain mostly tiny somatic cells. In the two specimens shown, vegetative reproduction via gonidia is at slightly different stages. The daughter colonies in the left-hand spheroid were about to be released, and already their own gonidia had probably started cleaving. There are only about 10–16 gonidia per colony.

SOURCE. Figures 2.7–2.9: Dr. M. Goldstein, published in *J. Protozool. 11,* 317 (1964). Figure 2.10: Dr. R. C. Starr, published in *Devel. Biol.* (Suppl.) *4,* 59 (1970).

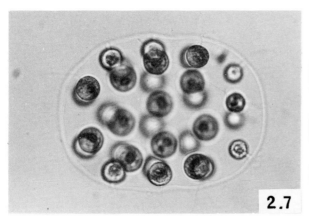

2.7

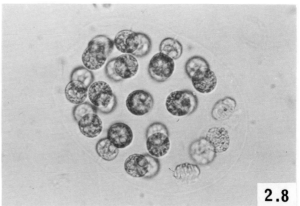

2.8

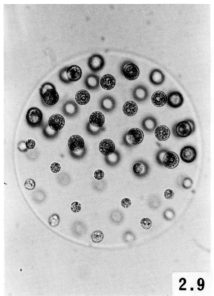

2.9

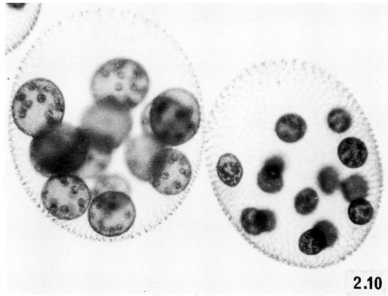

2.10

vegetatively, and the remainder (4), all grouped at one end, do not (Fritsch, 1935). In other strains of *E. elegans,* vegetative cells are almost all equivalent and all undergo reproduction, although this may be delayed in four smaller cells at one end of the colony (Goldstein, 1964, 1967; see also Fig. 2.7). Even in the simpler *Pandorina* (Fig. 2.4), a similar vegetative specialization may be encountered when a small anterior cell fails to give rise to a colony (Fritsch, 1935; p. 100). Coenobia of larger species of *Eudorina (Pleodorina)* contain around 128 cells as a rule, and in *E. californica* (Fig. 2.9), approximately half the cells in the anterior portion of the coenobium are smaller and nonreproductive. This specialization of vegetative reproduction reaches its peak in the various species of *Volvox,* whose giant and beautiful colonies may contain up to approximately 20,000 cells. However, of these cells, only a very few (16 or less) much larger individuals, the "gonidia," give rise to daughter colonies (Fig. 2.10). The gonidia themselves arise from an unequal cleavage of most or all cells in the growing colony at a particular stage in its development around the fourth division (i.e., that which leads to the 16-cell stage). From then on, the gonidia do not divide (or else cleave off only a few somatic cells; Starr, 1970a; p. 68) but merely enlarge, while the remainder of the cells in the colony continue to divide many times until the full cell complement of the vegetative coenobium has been attained. Eventually these fully formed daughter colonies undergo inversion (a fascinating process described later) and are released from the parental coenobium, which falls apart and dies. At some stage during or after these events, the few gonidia begin their own division cycles to form new coenobia.

It would be of great significance to understand how the determination of somatic versus reproductive potential is brought about, for example, in *Volvox* at the fourth division cycle. Some work has been done on this subject with *Pleodorina californica* by Gerisch (1959) and with several species of *Eudorina* by Goldstein (1967). These fascinating experiments cannot be reviewed here in detail, but they seem to point to the existence in reproductive cells of some unknown cytoplasmic constituent (an organelle or substance) which is unequally divided between daughter cells during cleavage. This unequal partitioning is apparently initiated at the first cleav-

2.11

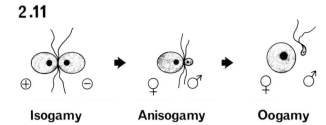

Isogamy　　**Anisogamy**　　**Oogamy**

Figure 2.11
EVOLUTION OF SEXUAL REPRODUCTION

In *isogamy,* the gametes are indistinguishable in size and general morphology; however, ultrastructural and physiological differences can often be demonstrated between the conjugants, by convention designated the plus (+) and minus (−) strains. In *anisogamy,* the conjugants are clearly of different sizes, and by convention, the larger is called the female. In *oogamy,* the female cell is far larger than the tiny flagellate male gamete, and it is almost always nonflagellate.

Figure 2.12
INVERSION IN VOLVOX

The large reproductive cells, the gonidia, are normally nonflagellate and have their basal bodies (small, paired black dots) facing the outside of the colony, in the same position as that of the smaller, flagellate somatic cells. After the first gonidial cell division (top line), the basal bodies move toward the center of the daughter cells prior to the second and subsequent divisions. (Analogous movement also takes place in unicellular Volvocales.) At a certain stage (second line; see text), unequal divisions give rise to the next generation of gonidia, which project outward from the forming colony. The basal bodies in all cells of this developing daughter colony lie on the inside of the spheroid, the plane of divisions being strictly controlled spatially. After cell division has ceased in somatic cells, the colony turns itself inside out (Fig. 2.41), so that the newly emerging flagella (bottom line) will be on the outside of the daughter colony; the gonidia are thereby concurrently moved to the inside (see top line).

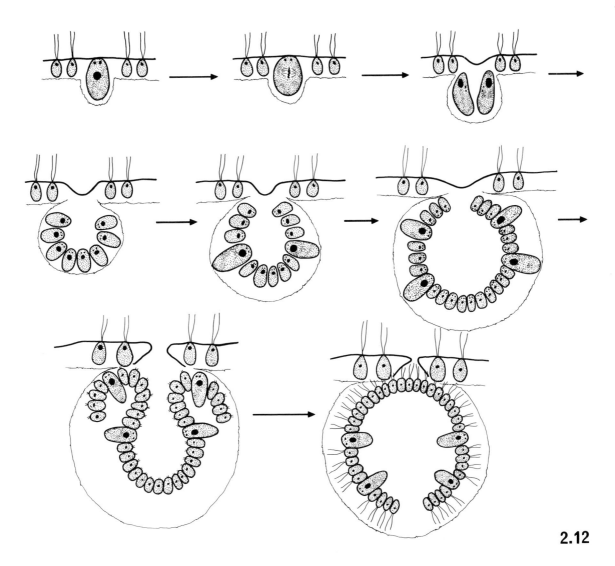

2.12

age. It should be mentioned here also that in many colonial forms the individual cells of the colony are known to remain interconnected by cytoplasmic bridges for a large part of the colony's life span.

More advanced forms of sexual reproduction commonly coincide with the enlargement of colonies and development of the specialized variations in vegetative reproduction discussed above. Both homothallic and heterothallic species are found in most genera. (In homothallic species, gametes of both mating types or sexes can be produced from one cell or plant body; in heterothallic species, plants or cells are of only one mating type or sex and need the other complementary organism to undergo sexual reproduction.) Two features are especially notable in the development of advanced forms of sexual reproduction.

First, there is a progression from isogamy through anisogamy to oogamy (Fig. 2.11). In isogamy, the gametes are of the same size and morphology. This concept is based upon light microscopy, but it will soon be clear (e.g., Section 2.1b and later) that there are often important physiological and ultrastructural differences between the two gametes of each conjugating pair. In anisogamy, the gametes are of unequal size, and by convention the larger gamete is the female. In oogamy, the gametes are of vastly different size and morphology; the female cells are few in number, very large, and almost invariably nonmotile, whereas the numerous male gametes are tiny and actively motile. Most species of *Chlamydomonas* are isogamous, and all cells in a culture partake in sexual reproduction; complementary "plus" and "minus" strains are well known in some heterothallic species. Anisogamy appears to be an increasing tendency in larger colonial forms such as *Pandorina,* but the situation may be variable (Fritsch, 1935, p. 113; see also Coleman, 1959). For example, while conjugating gametes of some species may be of appreciably different sizes; at other times in the same species, conjugation may occur between pairs of both smaller and larger gametes. *Eudorina, Pleodorina,* and *Volvox* show marked oogamy. (Oogonia in *Volvox carteri,* incidentally, are reported to be flagellate; Kochert, 1968).

Second, just as in vegetative reproduction, the more advanced the organism, the greater the tendency toward specialization of sexual repro-

Figures 2.13, 2.14
STRUCTURE OF CHLAMYDOMONAS

2.13 Diagrammatic representation. Some organelles labeled in Figure 2.14 are recognizable here. Others out of the plane of sectioning in Figure 2.14 include the eyespot (*es;* see Fig. 2.16), pyrenoid *(py)* surrounded by starch plates, and contractile vacuole *(cv),* part of which is visible in Figure 2.14.

2.14 *C. reinhardii.* Interphase cell, showing typical distribution of some organelles including the nucleus *(n),* chloroplast *(c),* nucleolus *(nc),* Golgi bodies *(g),* flagella *(f)* and flagellar apparatus near components of the contractile vacuole *(cv),* cell wall *(w),* mitochondria *(m),* endoplasmic reticulum *(e),* and assorted vacuoles *(v).* × 13,000.

SOURCE. Figure 2.14: Drs. U.W. Goodenough and K.R. Porter, published in *J. Cell Biol.* 38, 403 (1968).

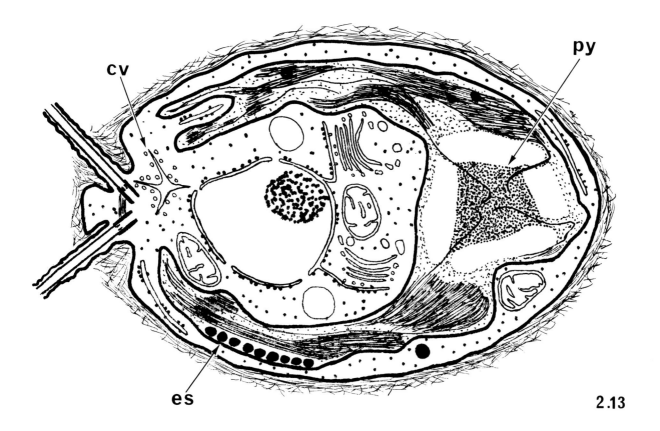

2.13

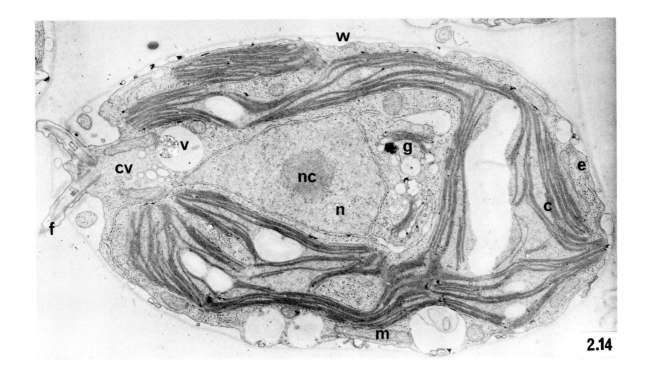

2.14

ductive cells. In monecious species of *Eudorina,* for example (Fig. 2.8), most of the cells in a colony give rise to oogonia directly, and the rest, by a repeated series of cell divisions, form "sperm packets"—flat bundles of tiny male gametes (e.g., see Goldstein, 1964). (Some species of *Eudorina* are reportedly anisogamous; Smith, 1950.) In *Pleodorina,* gametes are formed only from the posterior half of the coenobium (cf. its vegetative reproduction), and in some species of *Volvox,* only a few cells of the sexual colony, analogous to the gonidia, give rise to sperm packets and/or oogonia, depending upon whether the species is monecious or diecious (Section 2.2). In other species of *Volvox,* special male or female colonies may be formed which in male strains, for example, produce an equal number of sperm packets and nonsexual somatic cells. However, it needs to be pointed out that evolution of advanced forms of sexual reproduction in Green Algae is not always clearly correlated with increasing complexity of vegetative organization. Some of the unicellular Volvocales are markedly anisogamous; in particular, the female gamete of the unicellular *Chlorogonium oogamum* is a large, nonflagellate ameboid cell, while the male gamete is tiny, elongated, and flagellate.

It is perhaps worthwhile stating at the outset something about sex in Green Algae which will become more obvious during this book. Many algae use the sexual process for two different purposes: (1) it allows the organisms to utilize the obvious benefits of sex—the creation of tremendous variability in the genetic pool of the species, from which suitable variants can be selected for the successful exploitation of varied environmental conditions; and (2) it allows the formation of resistant spores that can survive unfavorable conditions in an unstable environment (e.g., a pond subject to drying up) which would wipe out vegetative populations. Thus, we can see the advantages of both homothallism and heterothallism. Heterothallic organisms, being self-sterile, are obliged to find suitable mates when they want to form spores, and hence they automatically gain the benefits of genetic variability conferred by sex. However, homothallic organisms in some situations are better equipped to face environmental stress, since they have retained the option of self-fertility—i.e., they are not prevented from forming spores by the lack of a suitable mating strain in their immediate vicinity. This option is

Figures 2.15, 2.16
CHLOROPLASTS OF CHLAMYDOMONAS

2.15 *C. reinhardii;* freeze-etched preparation of isolated chloroplasts. Four distinct surfaces can be recognized by the type and distribution of particles upon them. The *u* and *s* portions of each face represent the unstacked and stacked regions of the faces. *Bu* and *Bs* faces are continuous with one another as are *Cu* and *Cs* faces. The *B* and *C* faces are complementary to one another following the cleaving process. The terminology is that used by Branton and Park (1967). Note how the distribution of particles, characteristic of each face, changes when the membranes become stacked (i.e., note the transition between *Bu-Bs* and *Cu-Cs,* where the membranes come together at the small arrows). Large arrow indicates the direction of shadowing. × 77,000.

2.16 *C. eugametos.* Eyespot composed of a single layer of densely stained granules inside the chloroplast, which is appressed to the plasmalemma at this region. Note also the stacking of the thylakoid membranes in the chloroplast (cf. Fig. 2.15). × 74,100.

SOURCE. Figure 2.15: Drs. U.W. Goodenough and L.A. Staehelin, published in *J. Cell Biol. 48,* 594 (1971). Figure 2.16: Drs. P.L. Walne and H.J. Arnott, published in *Planta 77,* 325 (1967).

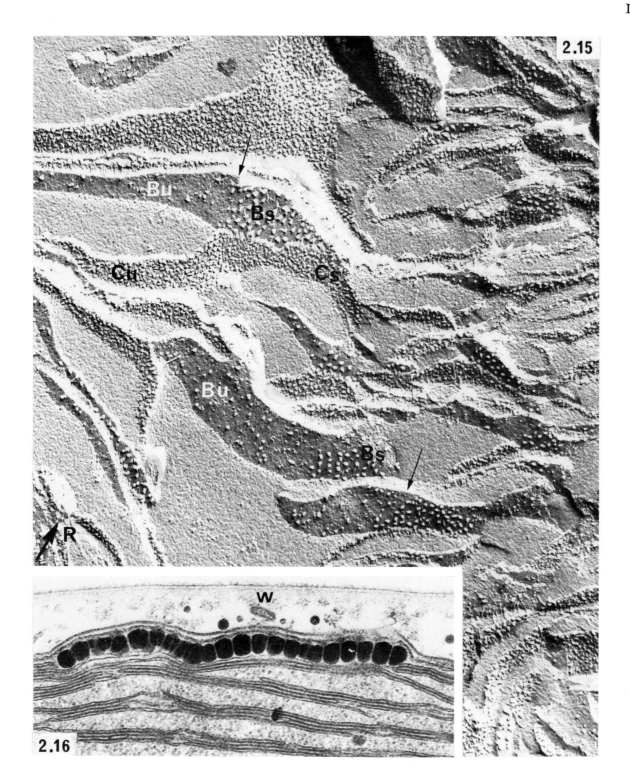

particularly useful for colonial forms, when male and female gametes are produced on the one vegetative structure. Another way of surmounting the difficulty of finding compatible mating strains in heterothallic species is by developing the ability to form "parthenospores" or asexual spores (also called "akinetes") directly from vegetative cells. However, it should not be forgotten that obligatory heterothallism may be a small but distinct disadvantage under certain circumstances, particularly for organisms with limited powers of motility (e.g., desmids). Many species of algae have either homothallic or heterothallic strains. For example, *Cosmarium botrytis* (Section 6.4) and *Eudorina elegans* (Fig. 2.8) are obtainable from the Indiana Culture Collection (Starr, 1964) in either form.

Another complicating factor arises as the size of the colony increases. When the number of constitutive cells is small, the coenobium is usually flat or perhaps slightly curved, but as the number increases, colonies become more spherical. Such colonies when first formed have their flagella inside the sphere, a consequence of the location of the basal bodies in the original cells allied with a precise spatial control of the planes of the sequential cell divisions (Fig. 2.12). These spherical colonies, therefore, have to undergo the unique and fascinating process of "inversion" whereby, at a certain stage in development, the young spherical colonies turn themselves completely inside out! This intriguing process will be discussed in some detail in Section 2.2.

We can get some idea of how the process of inversion evolved by reference to fairly small colonial species such as *Eudorina elegans* (Goldstein, 1967). Four to seven successive divisions of a vegetative cell form a "plakea," a curved plate of cells which is cup- or saucer-shaped. This plakea then undergoes inversion, giving rise to a motile spherical colony. Thus, it appears that as the colonial tendency becomes more developed in the Volvocales, the colonies become increasingly curved as they get larger. For maximum efficiency of their flagellar apparatus, the colonies developed the complex process of inversion, probably from a small flexing movement necessary to bring the flagella from the concave side of the plakea to the convex. Such a rudimentary form of inversion seems to occur during the development of sperm packets in some species of *Volvox* (see later). Perhaps the sperm packet in

Figure 2.17
STRUCTURAL COMPONENTS OF BASAL BODIES AND FLAGELLA

The arrangement of microtubules and ancillary structures seen in cross sections at levels 1–10 are illustrated on the right. Note that the two single central microtubules terminate in the transitional region and are not templated by the basal body. In the transitional region, two of the fused triplet microtubules of the basal body become continuous with the fused doublets of the flagellum, and here is situated the stellate structure characteristic of certain plant flagella (cf. Figs. 2.18–2.20).

SOURCE. Redrawn after Dr. D.L. Ringo, *J. Cell Biol. 33*, 543 (1967).

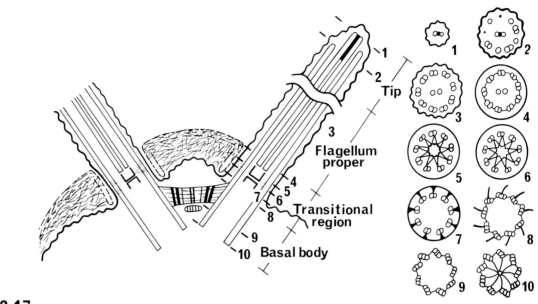

Tip

Flagellum
proper

Transitional
region

Basal body

2.17

Volvox is representative of the more primitive colonial form of this species, and its behavior is in accord with a general principle, mentioned elsewhere in this book—that differentiation of motile cells, both sexual and asexual, is often accompanied by the appearance of various primitive characteristics.

The flat colonial organism *Platydorina* (Fig. 2.5) shows an interesting behavioral variation concerned with inversion. Apparently, the colony does undergo inversion from a plakeal stage, as does *Eudorina;* however, it soon flattens out again, perhaps because the interior cells of the colony are alternately oriented in opposite directions (see Smith, 1950, p. 101). I have not been able to discover how these interior cells become oriented in such a manner, but obviously, once such a reorientation has been achieved, a flat colonial form is the most efficient for utilization of all the flagella of the colony in movement.

In this chapter I shall be concerned mainly with *Chlamydomonas* and *Volvox,* two organisms from the suborder Chlamydomonadineae which have been studied in some detail ultrastructurally. I shall also briefly mention *Tetraspora* from the suborder Tetrasporineae, whose close affinity to *Chlamydomonas* will be seen to extend to the ultrastructural level. Unfortunately, little ultrastructural work has been done on many of the intermediate members of the Volvocales of interest. Nevertheless, by comparing the structure and function of cells in both *Chlamydomonas* and *Volvox,* we can perhaps get some idea of what to expect in these intermediate organisms and see the morphological differentiation that has accompanied the evolution of colonial forms in this order.

2.1.Chlamydomonas

Chlamydomonas is familiar to botanists and indeed to most biologists. Some of the numerous species are commonly encountered in nature and may be extremely prolific under appropriate environmental conditions. They are often easily maintained in laboratory culture and consequently are widely used in research and for teaching purposes. A fair amount of ultrastructural work has been done on these cells and some of their various subcellular components. Since many of these features are common to other algae

Figures 2.18–2.23
FLAGELLAR STRUCTURE IN GREEN ALGAE

2.18. Flagellar apparatus, vegetative cell of *Chlamydomonas reinhardii* (cf. Fig. 2.17). Note also the striated fiber interconnecting the basal bodies. × 39,000.

2.19. Cross section of a basal body, equivalent to level 10 in Figure 2.17. × 110,000.

2.20. Cross section of a flagellum, equivalent to level 5 in Figure 2.17, obtained from a young colony of *Volvox tertius.* × 110,000.

2.21. Longitudinal section through the flagellum/basal body of a cell of *V. tertius* (cf. Fig. 2.17). × 80,000.

2.22. Cross section of the unusual flagellum containing only a single central microtubule from the spermatozoid of the chlorococcalean alga *Golenkinia minutissima.* × 90,000.

2.23. Longitudinal section of the flagellar structure illustrated in Figure 2.22. The central tubule can be seen to undulate. × 60,000.

SOURCE. Figure 2.20: Author's micrograph, published in *Planta 90,* 174 (1970). Figures 2.22, 2.23: Dr. Ø. Moestrup, published in *Br. Phycol. J. 7,* 169 (1972).

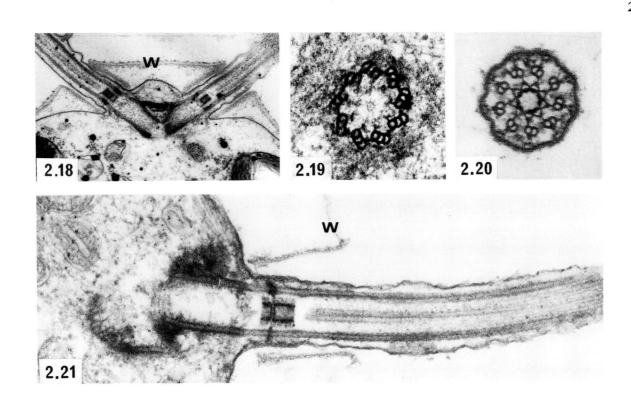

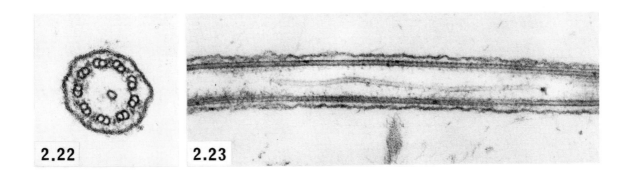

(albeit often with modification), and since *Chlamydomonas* probably exemplifies a relatively primitive type of cell organization, similar to that from which many other members of the Chlorophycophyta were originally derived (Chapter 8), I shall spend a little time on discussion of the structure of these cells. Papers dealing with different aspects of cell ultrastructure in the Volvocales include those of Belcher (1968), Lang (1963a), Underbrink and Sparrow (1968), Lembi and Lang (1965), Berkaloff (1966), Walne (1967), and Levine and Ebersold (1960). Recently, Schötz et al. (1972) have completely reconstructed the organization of a gamete of *C. reinhardii* from serial sections. Their results will be mentioned further below.

Most species of *Chlamydomonas* have quite small vegetative cells (ca. 20μ in length) which are usually more or less oval in shape with two flagella arising from the anterior end (Figs. 2.13, 2.14). Some closely related genera (e.g., *Carteria*) are quadriflagellate. Adjacent to the flagella are two (or more) contractile vacuoles which pulsate alternately in living cells, slowly enlarging and then suddenly discharging their contents into the external medium. The wall is thin, but in *C. reinhardii* it has been shown to possess a surprisingly complex structure containing seven layers (Roberts, Gurney-Smith, and Hills, 1972); glycoproteins containing hydroxyproline have been detected within it. The presence of cellulose in this wall, reported in the review of Levine and Ebersold (1960), has been disputed by several other authorities (see Roberts, Gurney-Smith, and Hills, 1972). The wall is presumably not very rigid, and consequently the cells take up water continuously from the external medium through the semipermeable plasmalemma. These contractile vacuoles, therefore, probably serve to expel the accumulating water and thereby prevent the cells from ballooning up and eventually rupturing. Algal cells contained within rigid cell walls do not seem to possess contractile vacuoles; however, on those occasions when these cells become naked (e.g., during transformation into a zoospore; as an example, see *Oedogonium*, Section 5.1c), contractile vacuoles immediately appear all over their surface. In cells viewed with the electron microscope, these vacuoles are surrounded by numerous tiny vesicles whose origin is not obvious; presumably, they enlarge before coalescing with and emptying their contents into

2.24

Figure 2.24
MICROTUBULAR SYSTEM IN CHLAMYDOMONAS

Diagrammatic representation of the peripheral system of microtubules that radiate out from the basal body complex and enclose the cytoplasm. The full extent of the system is not yet determined, but one end of the microtubules terminates close to the large striated fiber that interconnects the basal bodies.

SOURCE. Redrawn after Dr. J.L. Ringo, *J. Cell Biol. 33*, 543 (1967).

Figure 2.25
FLAGELLAR STRUCTURE IN CHLAMYDOMONAS REINHARDII

Inset shows a whole cell on a supporting film, stained with uranyl acetate. The flagellar membranes have been stripped off and the tubular components have frayed out while remaining attached to the cell. The larger micrograph shows an intact flagellum (*f*) with some of its fine hairs or mastigonemes (*ma*) and the tubular components of a frayed flagellum, all negatively stained. The central pair of microtubules is labeled $t_{1,2}$. Note the fine spokes (small arrow) and additional material (large arrow) attached to the doublet tubules. × 69,000; *inset* × 3,740.

SOURCE. Dr. J.M. Hopkins, published in *J. Cell Sci. 7*, 823 (1970).

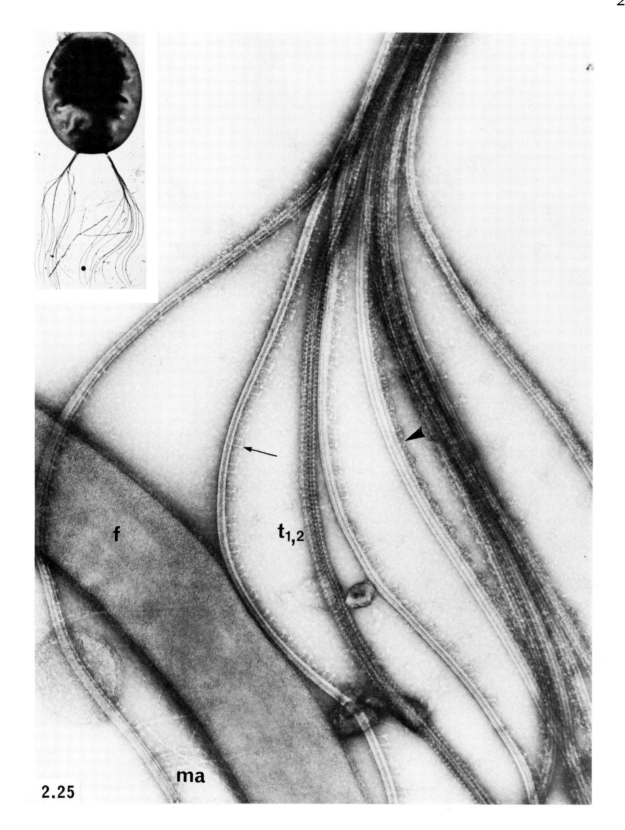

f

t$_{1,2}$

ma

2.25

the contractile vacuole. In *C. reinhardii,* vacuoles occupy about 8% of the cell's volume (Schötz et al., 1972). There are usually two Golgi bodies per cell, adjacent to the nucleus (*ca,* 10% of the cell's volume). The 10–15 mitochondria are elongated and branched, winding about the cell; some are always sited near the flagellar bases.

A large chloroplast, variously described as being cup-, urn- or H-shaped, fills much of the cell and contains one large, conspicuous pyrenoid or in some species several smaller ones. In *C. reinhardii,* the highly lobed chloroplast occupies about 40% of the cell (Schötz et al., 1972). Pyrenoids are ubiquitous and usually conspicuous inclusions in the chloroplast of most algae, including the Green Algae, and the bryophyte *Anthoceros.* Some genera of algae and almost all higher plants lack them (for example, they are absent from the chloroplasts of *Microspora* and the Charales; see Sections 4.3 and 7.1). They usually consist of a dense, granular matrix, occasionally crystalline in some region, which may be traversed by varying arrangements of membranous lamellae continuous with the photosynthetic lamellae of the chloroplast. Pyrenoids vary considerably in size, and in Green Algae they usually are surrounded by platelets of starch. They appear to contain considerable quantities of protein, but their functions are rather obscure (Brown and Arnott, 1970). Recently Rosowski and Hoshaw (1971) described a method for the isolation of these organelles from *Zygnema,* and Holdsworth (1971) has investigated the nature of pyrenoid protein from *Eremosphaera.* It is possible that details of pyrenoid ultrastructure may prove important taxonomically (e.g., Brown and McLean, 1969; Stewart, Mattox, and Floyd, 1973). For a general review on pyrenoids, see Griffiths (1970). On the outer surface at one side of the chloroplast is usually an orange-red pigmented body—the eyespot or stigma—which in *C. eugametos* (Fig. 2.16) is composed of a layer of about a 100 osmiophilic, polygonal granules (Walne and Arnott, 1967) within the limiting chloroplast membrane.

Because of the ease of growing these organisms and isolating mutants, some species of *Chlamydomonas* have been extensively used for research into chloroplast structure and biogenesis. One of the first convincing demonstrations confirming the existence of chloroplast DNA was the work of Ris and Plaut (1962), who used *C.*

Figures 2.26, 2.27
MITOSIS AND CYTOKINESIS IN CHLAMYDOMONAS REINHARDII

2.26. Late metaphase/early anaphase spindle. The nuclear envelope *(ne)* is essentially intact except for polar fenestrae *(pf),* toward which spindle microtubules *(t)* converge. Note how the cytoplasm is subtly differentiated at the polar fenestra. Microtubules of the metaphase band are indicated by the arrow. × 37,000.

2.27. Early cleavage. The cleavage furrow *(cf)* is growing inward between daughter nuclei *(n).* Two sets of microtubules are coplanar with the cleavage furrow, the cleavage microtubules *(ct),* and the internuclear microtubules, all cut transversely (some arrowed). × 29,000.

SOURCE. Drs. U.W. Goodenough and K.R. Porter, published in *J. Cell Biol. 38,* 403 (1968).

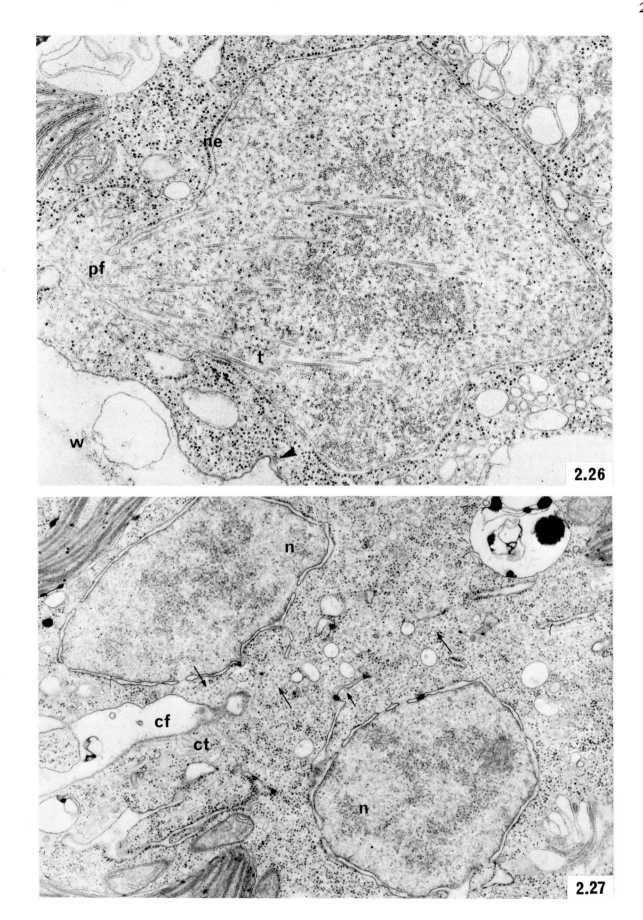

2.26

2.27

moewusii. Prolamellar bodies, analogous to those in etiolated plastids of higher plants, have also been reported in *C. reinhardii* (Friedberg, Goldberg, and Ohad, 1971). This latter organism has been used by Goodenough and Staehelin (1971) in freeze-etch studies to show that the distribution of particles unique to chloroplast membranes could be related to whether these membranes were unstacked or stacked in the thylakoid configuration (Fig. 2.15). Stacking was not found to be a prerequisite for carrying out photosynthesis, since an "ac-5" mutant could be grown under conditions whereby stacking did not occur but photosynthesis was still carried out. In another mutant, it was demonstrated that chloroplast fractions could be isolated in either the stacked or the unstacked configuration depending upon the salt concentration used in the isolating medium. Such studies are obviously relevant to the problem of accounting for the complexity and development of organization in chloroplast membranes. Wild-type and mutant strains of *Chlamydomonas* are also widely used in biochemical and physiological studies of chloroplast development and function (for example, Volume 44 of the *Journal of Cell Biology* contains six such papers).

The flagellar apparatus on this organism has also received much attention. Ringo's (1967) excellent paper on *C. reinhardii* is a source of much of the following material, and he lists many earlier relevant references. Two forms of flagellar motion propel the cells through the medium. The commonest is normally a planar, synchronous beat that is best compared to the human breast stroke and that moves the organism jerkily in the direction of the flagellar bases. Sometimes, however, synchronous waves are propagated along the flagella, moving the organism smoothly but more slowly in the opposite direction.

The basic structure of the flagella, basal bodies, and ancillary components (Figs. 2.18–2.21) is shown diagrammatically in Figure 2.17 (redrawn from Ringo, 1967; see also Hobbs, 1971). From here on in this book, I shall consider the centriole and basal body as homologous structures and use the former term when the organelle is not or has not recently been attached to a flagellum (e.g., when it is associated with the nucleus or in the spindle) and the latter term when it is associated with a flagellum. Many of the principal structural features indicated here for *C. reinhardii*

Figure 2.28
CYTOKINESIS IN CHLAMYDOMONAS REINHARDII

Later cleavage. The cleavage furrow *(cf)* has grown inward all around the cell, but fastest between the basal bodies *(bb)* and daughter nuclei *(n)*. It is still lined by cleavage microtubules (some arrowed). The chloroplast is also furrowing in the plane of cell cleavage, partitioning the pyrenoid *(py)* into daughter cells. × 24,000.

SOURCE. Dr. U.W. Goodenough, published in *J. Phycol.* 6, 1 (1970).

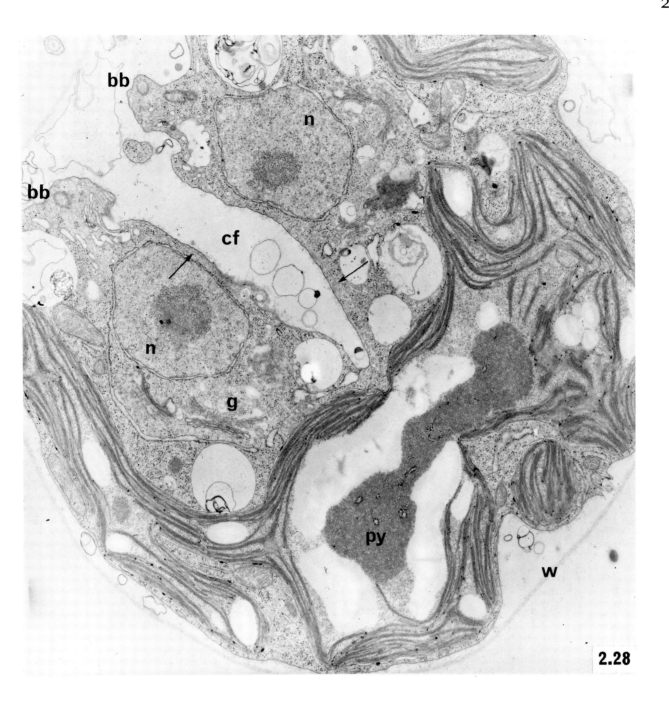

2.28

are probably very common if not universal among Green Algae, although there exist minor differences in certain areas. The flagellar membrane may be covered with fine hairs (e.g., Ringo, 1967; Hopkins, 1970; Fig. 2.25) often called "mastigonemes"—a term that has been rather confusingly applied to a variety of hair-like flagellar appendages. The striking stellate structure at the junction of the basal body and flagellum (Figs. 2.17, 2.20; Lang, 1963b) has been widely reported in Green Algae and also occurs in some other algae and lower plants (Manton, 1965; Ringo, 1967). A few Green Algae may lack it (Moestrup, 1972). In contrast, it has never been encountered in basal bodies of protozoa or animal cells (cf. Gibbons and Grimstone, 1960), and Manton (1965) discusses the phylogenetic implications of its existence. The basal "cartwheel" (Fig. 2.19) is a consistent feature of centrioles and basal bodies generally, and during centriole formation this structure may be involved in assembling the triplet tubules into their proper spatial configuration: tubules appear sequentially (i.e., inner, middle, and then outer tubules) on the tips of the nine spokes of the cartwheel during basal body replication in *C. reinhardii* (Johnson and Porter, 1968; Fig. 2.73). In other organisms, the cartwheel structure appears before the basal body is formed (Perkins, 1970), and release and dispersion of large numbers of similar cartwheels may precede the generation of thousands of basal bodies (Mizukami and Gall, 1966). Further, in some cells the cartwheel structure may be lost once the functional flagellum has been formed (see Section 7.5). Dippell (1968), however, reports that in *Paramecium* newly forming basal bodies initially lack the basal cartwheel structure and acquire it later, after the first tubules have appeared. In *Chlamydomonas* the two basal bodies may be accompanied by a third, adjacent and parallel to one of them; Friedmann, Colwin, and Colwin (1968) report two such extra basal bodies in gametes of *C. reinhardii*. These extra (apparently nonfunctional) basal bodies are not uncommon in various algae, having been reported in *Chromulina* (Manton, 1959), *Prymnesium* (Manton, 1964a), and *Volvox* (Pickett-Heaps, 1970a; see Section 2.2), and they may represent newly formed organelles.

The orientation of the central pair of tubules in different flagella of many multiflagellate (ciliated) organisms is often strikingly parallel, and

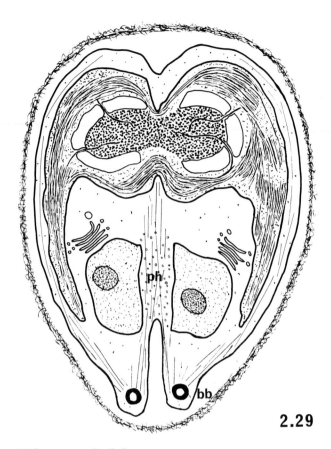

2.29

Figure 2.29
CYTOKINESIS IN CHLAMYDOMONAS

Diagrammatic representation. The cleavage furrow, circumferential around the cell, grows fastest through two sets of microtubules which together constitute the phycoplast *(ph)*. One set—the cleavage microtubules as well as peripheral microtubules—is oriented toward the basal body complexes *(bb)* at the edge of the cell. The other set—the internuclear microtubules—is shown as small circles, oriented perpendicular to the cleavage microtubules.

Figures 2.30, 2.31
SEXUAL REPRODUCTION IN CHLAMYDOMONAS MOEWUSII

2.30. Preliminary clumping of gametes mediated by agglutination of flagella. × 1,300.

2.31. A later stage than that in Figure 2.30. Gametes are now paired and move unidirectionally, with the nonfunctional set of flagella adhering to the surface of the other gamete. × 1,300.

SOURCE. Drs. R.M. Brown, Jr., C. Johnson, and H.C. Bold, published in *J. Phycol. 4,* 100 (1968).

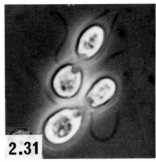

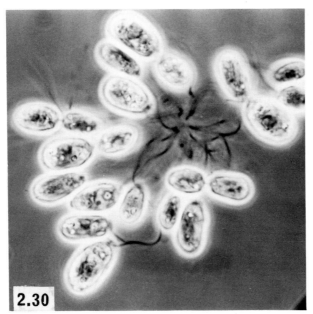

it usually appears perpendicular to the plane of beating (Fawcett and Porter, 1954; see discussion in Ringo, 1967), but this may not be the case in *C. reinhardii.* It is worth recalling that not *all* functional flagella have this central pair of tubules. The flagella of gametes of the diatoms *Lithodesmium* (Manton and von Stosch, 1966) and *Biddulphia* (Heath and Darley, 1972) have no tubules, and the flagella of the spermatozoid of *Golenkinia,* which functions quite normally, has only a single, rather sinuous tubule (Figs. 2.22, 2.23; Moestrup, 1972). Randall and co-workers (Randall et al., 1964; Warr et al., 1966; Randall, 1969) have isolated nonmotile mutants of *Chlamydomonas* in whose flagella these two tubules are disorganized or absent (some other nonmotile mutants had flagella normal in appearance). Just near the tips of the flagella, the outer doublets irregularly become single and are then terminated; usually the central pair projects a short distance beyond the outer tubules (Fig. 2.17).

It has now been well established that in flagella generally the component microtubules are not all equivalent, since clear-cut chemical differences are easily detectable between them (Behnke and Forer, 1967; Burton, 1968; Tilney and Gibbons, 1968). The central two tubules have distinctly different properties from the fused outer doublet tubules, which themselves are composed of different A and B tubules. Isolated flagella are frequently used as a source of microtubular protein, and ever finer fractionations of the tubules into separable proteins continue to be reported (Witman et al., 1972; Witman, Carlson, and Rosenbaum, 1972). The central pair of tubules may be more closely related to other cytoplasmic microtubules (for example, they appear to bind colchicine; see Shelanski and Taylor, 1968) than the outer tubules, which are appreciably different (they don't bind colchicine). The basal body (i.e., the centriole when it becomes involved in flagellar formation or function) clearly determines the exact spatial disposition of the nine outer doublet tubules of the flagellum, which are continuous with two of the three tubules of the basal body (Fig. 2.21); in contrast, there is no similar equivalent template for the central pair in the basal body. Even the two central tubules appear to be subtly dissimilar (Witman et al., 1972; Hopkins, 1970), and the further one investigates flagellar structure, the

Figures 2.32–2.34
SEXUAL REPRODUCTION IN CHLAMYDOMONAS MOEWUSII

2.32. Fertilization has been achieved by fusion of the apical papilla (at the flagellar base) of each conjugant. $\times$ 17,300.

2.33. Later stage in plasmogamy, with further merging of the protoplasts initiated at the apical papillae. The primary wall of the zygote has just begun to be secreted. $\times$ 10,800.

2.34. Part of a mature zygote, typically very dense. The resistant, ornamented wall is clearly visible. $\times$ 17,600.

SOURCE. Drs. R.M. Brown, Jr., C. Johnson, and H.C. Bold, published in *J. Phycol. 4,* 100 (1968).

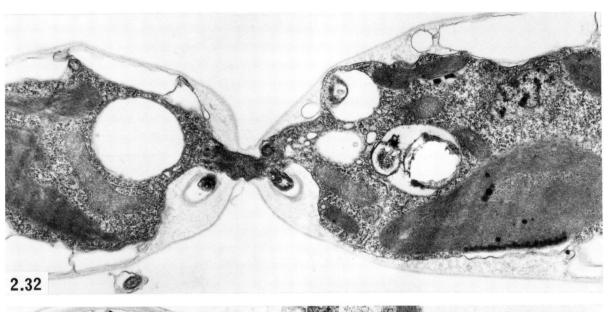

2.32

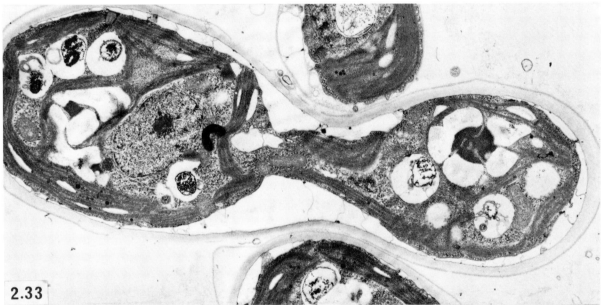

2.33

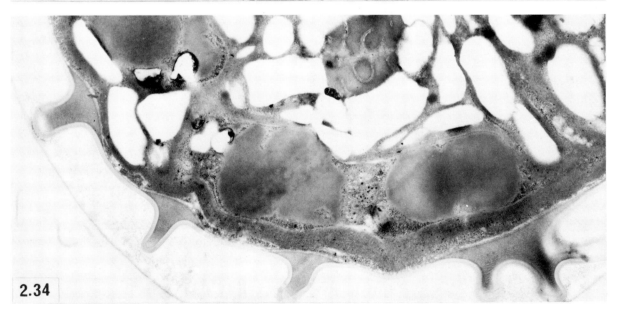

2.34

more complicated it seems to become (see Fig. 2.25, from Hopkins's 1970 study of the flagella of *C. reinhardii*). To what extent the components of the flagellar and cytoplasmic tubules are interconvertible can only be guessed.

Systems of striated fibers interconnecting basal bodies or emanating from basal bodies into the cytoplasm are common in flagellar systems of most organisms. Considerable variation is encountered in their dimensions, pattern of striations, and disposition with respect to the basal bodies. *C. reinhardii* possesses three such fibers (Ringo, 1967). Two smaller fibers interconnect the sides of the "proximal" ends (i.e., those away from the flagella) of the basal bodies, while a larger one interconnects the middle portion of the basal bodies (Figs. 2.17, 2.18). Ringo lists examples of such fibrous systems in other Green Algae. Some idea of the range in structure of these fibrous systems can be gained by comparing those of *Chlamydomonas* with the simple, spindle-shaped fiber between the two basal bodies in sperm of *Nitella* or *Chara* (Section 7.5) or with the massive and incredibly complex fibrous systems interconnecting and also emanating from the ring of basal bodies in zoospores of *Oedogonium* (Section 5.1c). The function of these fibrous rootlets is still a subject for conjecture; they obviously appear to have important mechanical properties and could serve in various ways, such as in coordination of the flagellar beat.

The flagellar apparatus of *C. reinhardii* contains another set of important components encountered, with numerous structural variations, in most motile cells of other Green Algae. Between the two basal bodies arise four sets of four microtubules which extend outward around the cell periphery, separating as they go. Their disposition is shown diagrammatically in Figure 2.24, and they obviously enclose the cytoplasm in a cage-like framework. Manton and her coworkers first described these "root" or "rootlet" microtubules in a number of algae (Manton, Clarke, and Greenwood, 1955; Manton, 1964b; Ettl and Manton, 1964), finding that the number of sets of rootlet microtubules, the number of constituent microtubules, and their arrangement are all fairly consistent and characteristic for each species of alga (see also Manton, 1965). These rootlet microtubules likely serve as a structural cytoskeleton that helps determine the shape of the cell and transmits the forces engendered by flagellar

Figures 2.35–2.38
FERTILIZATION IN CHLAMYDOMONAS REINHARDII

2.35. Fertilization tubule *(ft)* just forming in a gamete which still possesses its cell wall *(w)*. Choanoid body *(cb)* is immediately under a highly differentiated region of the plasmalemma. × 35,000.

2.36. Grazing section of the fertilization tubule, showing the numerous small projections of the plasmalemma. Note the choanoid body (arrow) at the base of the fertilization tubule. × 42,000.

2.37. Fertilization tubule connecting two gametes just after gametic union. × 30,000. *Inset* shows, at a higher magnification, the choanoid body at one end of the fertilization tubule, and the much smaller, somewhat analogous structure (arrow) in a similar position in the other gamete.

2.38. Gametic fusion well under way. Basal bodies *(bb)* of the two flagellar apparatuses are still quite separate, although this stage is described as an early stage of flagellar coordination. The choanoid body is still visible (arrow). × 17,000.

SOURCE. Drs. I. Friedmann A. L. Colwin, and L.H. Colwin, published in *J. Cell Sci. 3,* 115 (1968).

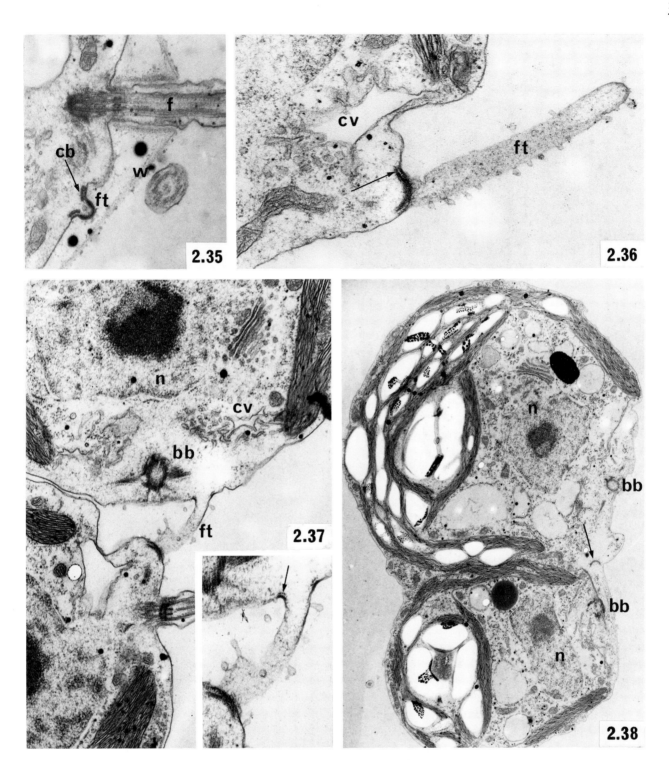

motion throughout the entire cytoplasm (Hoffman, 1970). An excellent example of the complex elaboration of such a simple rootlet system that has occurred is again provided by the zoospore and spermatozoid of *Oedogonium.* Its flagellar apparatus contains a ring of numerous, regularly spaced basal bodies, and a set of three such rootlet microtubules emanates around the cell from each region between basal bodies; these rootlet microtubules are probably firmly attached to the basal body complex. For example, when Hoffman and Manton (1962, 1963) and Hoffman (1970) isolated the flagellar apparatus of *Oedogonium,* rootlet microtubules always remained attached to the ring of basal bodies (Fig. 5.62). In my opinion, the cytoskeletal manchette tubules of sperm cells which grow from and remain attached at one end to the basal body complex (e.g., in *Chara;* Section 7.5) could also be considered specialization derived from the basic rootlet microtubular system.

The ends of the rootlet tubules are usually thickened and dense, and I suspect that these ends in fact are discrete structures—the "rootlet templates" (Pickett-Heaps, 1971a,b)—analogous to basal bodies in that they both act as some form of template for the extrusion of a specific spatial array of microtubules. When centrioles arise de novo during zoosporogenesis in *Oedogonium,* rootlet templates invariably are formed concurrently between them. Later, both these template structures are moved to the edge of the cell, and there they extrude their tubule systems concurrently (Section 5.1c). So I suspect that rootlet templates may often replicate along with centrioles, for example, during cell division, and I have argued (Pickett-Heaps, 1971a) that both centrioles and these rootlet templates can be considered as two highly structured and conspicuous forms of "microtubule-organizing centers" (MTOCs) which in the case of *Chlamydomonas* are probably associated with a much more ill-defined morphogenetic center situated around the flagellar bases. During cell division in *C. reinhardii* (Section 2.1a), it may be useful to imagine this whole region duplicating itself and then later becoming involved in the formation of specific arrays of microtubules associated with cell cleavage, which may be functional in ensuring that the two resultant centers (including basal bodies, rootlet templates, and even perhaps the contractile vacuoles) are segregated into the two daughter cells.

Figure 2.39
ZYGOTE OF CHLAMYDOMONAS REINHARDII

Zygote, 5.5 hours old. The nuclei *(n)* have fused, and the chloroplast *(c)* now contains two pyrenoids *(py),* indicating that the two gametic chloroplasts have also fused. The wall of the zygote is beginning to be secreted. × 14,000.

SOURCE. Dr. T. Cavalier-Smith, published in *Nature (Lond.)* *228,* 333 (1970).

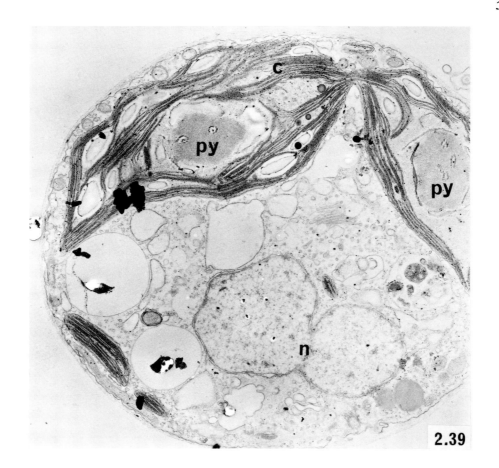

From now on, I shall frequently use the terms "centriole complex" and "basal body complex," which are meant to imply that the whole ill-defined region around the centriole or basal body (e.g., at the poles of the spindle) probably has important morphogenetic potentials (such as control over microtubule systems) which do not reside just in these conspicuous organelles themselves.

Another elegant set of experiments on *C. reinhardii* will be mentioned here, whose implications are highly significant and almost certainly can be applied to flagellar systems in general and perhaps to other cytoplasmic microtubular systems elsewhere in the cell. Rosenbaum, Moulder, and Ringo (1969) studied regeneration of flagella by cells whose flagella had been amputated by careful use of a homogenizer. They first established that the kinetics of regeneration were reproducible and that elongation started with virtually no lag after scission. They then concentrated upon cells that had been deprived of only one of the two flagella. They showed that almost immediately after amputation, the remaining flagellum shortened while the other reappeared and elongated. Sometimes, this remaining flagellum shrank to near zero length and immediately began to lengthen again; more often, it ceased shortening when it was about half normal length, roughly the same length as the regenerating flagellum. Then they both continued elongation concurrently but at a similar decelerating rate. This experiment clearly indicates that the components (microtubules?) of the two flagella are in quite labile equilibrium with one another, so that when one is removed, the other shortens and contributes to the extension of the new flagellum. The requirement for protein synthesis during regeneration was examined by treating deflagellate cells with cycloheximide (a potent inhibitor of protein synthesis). If both flagella had been removed, regeneration proceeded to a small and variable extent, a result which indicated the presence of a pool of flagellar precursors even in cells that previously had fully grown flagella. If only one flagellum had been removed in the presence of cycloheximide, the remaining one shortened, and regeneration of the missing one followed as before, but the two flagella did not reach their normal length. Presumably the pool of precursors within the cell

Figures 2.40–2.42
ASEXUAL REPRODUCTION IN VOLVOX CARTERI

2.40. Colony containing 16 large vacuolate gonidia (see also Fig. 2.10). × 200.

2.41. Daughter colony halfway through inversion. Cells are elongated. × 525.

2.42. Daughter colony just before undergoing inversion. Note the gonidia are on the outside (cf. Fig. 2.12). Lobes are forming in the somatic cell layer around the phialopore. × 390.

SOURCE. Figures 2.40, 2.41 Dr. R.C. Starr, published in *Devel. Biol.* (Suppl.) *4,* 59 (1970). Figure 2.42: Dr. G. Kochert, published in *J. Protozool. 15,* 438 (1968).

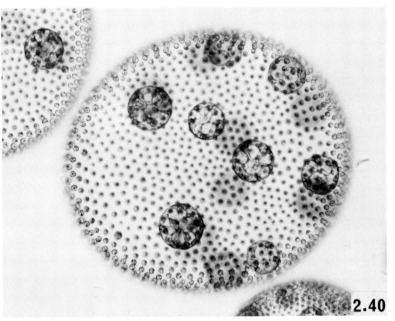

2.40

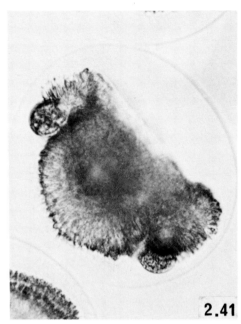

2.41

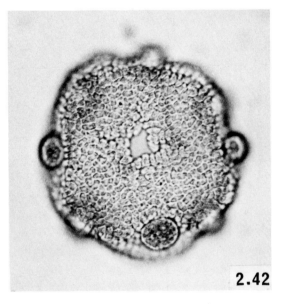

2.42

combines with the components of the shortening flagellum to produce two flagella of subnormal length; attainment of normal length and restitution of the precursor pool in the cell require protein synthesis. It seems likely that cycloheximide and the absence of protein synthesis did not affect the cell's ability to break down and reassemble the flagellar components. These workers then showed that in the presence of an appropriate concentration of colchicine, flagellar regeneration could be completely and reversibly inhibited in all experiments without any appreciable effect on protein synthesis. Furthermore, in cells minus one flagellum, the other shortened as normal, either partially or completely. Thus, shortening and reassembly of the flagella are separate processes. An arginine-requiring mutant of *C. reinhardii* was then exposed to tritiated arginine at different times after flagellar amputation. The results indicated that the maximum rate of incorporation of this amino acid (indicating the maximum rate of protein synthesis) occurred when the regenerating flagella were about half formed (i.e., when the pool of precursors in the cell was perhaps at its lowest size). Furthermore, radioautography of cells treated with pulses of labeled arginine showed that protein containing the label was added predominantly to the tip of the flagella (i.e., at the opposite end to the basal body).

These elegant experiments confirm in a quantitative fashion what biologists have begun to suspect about the nature of microtubular systems in the cell generally. The observations clearly indicate the presence of an intracellular pool of tubule subunits or precursors (in this case, flagellar components) which is in labile equilibrium with assembled polymers of the subunits. This situation seems pertinent to understanding the behavior of other microtubular systems (and in particular, the mitotic apparatus) where highly ordered arrays of microtubules are formed and broken down in relatively rapid and strictly controlled succession (see Inoué and Sato, 1967).

Recently, Ettl (1971) has compared the morphology of a considerable number of species of *Chlamydomonas*. While the structure of the flagella, nucleus, and cell wall was consistent throughout, differences were noted in the position and morphology of the contractile vacuoles, papillae, and mitochondria; the chloroplast and pyrenoid showed considerable variation in form.

Figures 2.43–2.48
SEXUAL REPRODUCTION IN VOLVOX CARTERI

2.43. Female colony containing about 40 oogonia. Note that they are smaller and more numerous than gonidia (cf. Fig. 2.40). × 200.

2.44. Male colony showing the typical 1:1 ratio of large androgonidia to small somatic cells. The androgonidia later form sperm packets (Fig. 2.45). × 500.

2.45. Male colony. Sperm packets have formed from the androgonidia (note their numerous flagella), but the somatic cells remain unchanged. × 500.

2.46. Intact sperm packet becoming attached to the surface of a female colony. × 490.

2.47. Sperm packet beginning to disintegrate into individual cells. × 490.

2.48. Individual sperm actively penetrating the matrix of the female colony. Note the fertilization pore at the site of attachment of the sperm packet to the female. × 490.

SOURCE. Figures 2.43–2.45: Dr. R.C. Starr, published in *Devel. Biol.* (Suppl.) 4, 59 (1970). Figures 2.46–2.48: Drs. W. Hutt and G. Kochert, published in *J. Phycol.* 7, 316 (1971).

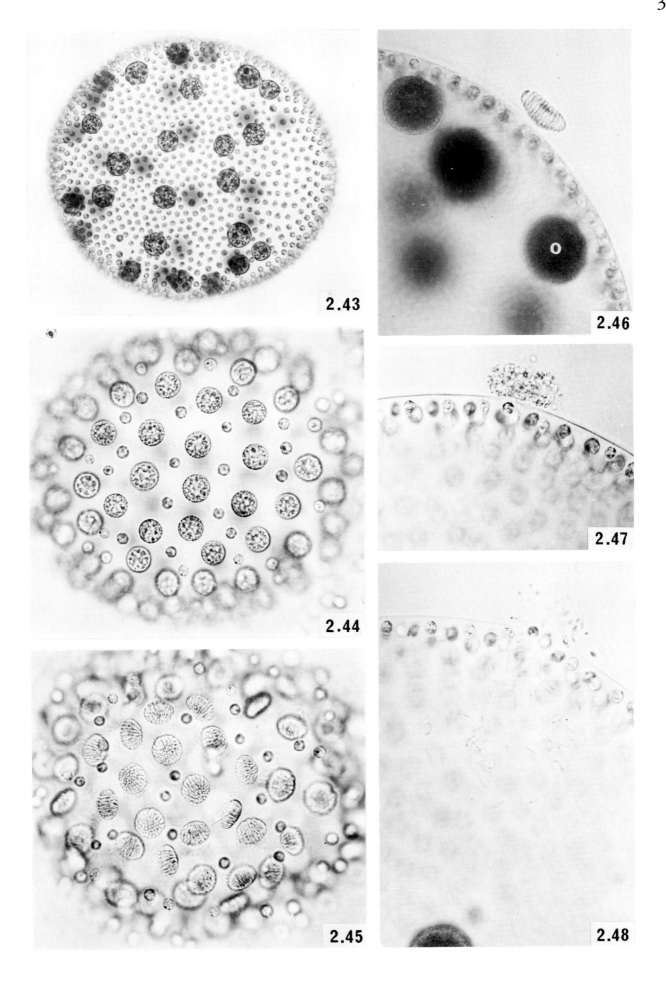

2.43

2.46

2.44

2.47

2.45

2.48

2.1a. Cell Division

Some species of *Chlamydomonas* and related organisms remain motile during cell division, but the majority lose or withdraw their flagella. Cleavage is always longitudinal, and usually the circumferential furrow grows through the cell unevenly, fastest from that end occupied by the basal bodies. In some species which have cell walls (e.g., *C. reinhardii,* described below in detail), the protoplast rotates during division; cleavage is still longitudinal through the protoplast, but obviously in these cases is also more transverse with respect to the original cell wall (Kater, 1929; Fritsch, 1935; Akins, 1941; Buffaloe, 1958). This rotation is clearly evident in time-lapse sequences of cell division (Schlösser, 1966). The contractile vacuoles also replicate and separate before mitosis, and the cleavage furrow later passes between them. The pyrenoid elongates, and both pyrenoid and chloroplast divide usually during later stages of cell division; the eyespot in the chloroplast passes on to one daughter cell, and a new one is formed in the other (Buffaloe, 1958). After the first mitosis and complete cell cleavage, synchronous mitoses and cleavages usually follow in the daughter cells, and still further divisions may ensue later. Thus, at the completion of the division cycle, between 2 and 16 individual daughter cells are usually formed, whereupon the parental cell wall is dissolved away (presumably enzymatically) to release motile daughter cells. These events have been well photographed in *C. reinhardii* using time-lapse cinematography (Schlösser, 1966).

One enigmatical problem can be raised here. In many genera of the Volvocales, reproduction is brought about by a rapid series of consecutive cell divisions. Mitosis itself commonly lasts about 10 minutes (Cave and Pocock, 1951; Rayns and Godward, 1965), and the interphase period between successive divisions may be very brief. The nucleus in reproductive cells of the complex colonial Volvocales (e.g., in the gonidia of *Volvox;* see later) is very large indeed before the onset of nuclear division, but the nucleus in the numerous daughter cells is far smaller (Fig. 2.63). Even the chromosomes themselves appear to become smaller during successive divisions. A typical case quoted by Godward (1966, p. 13) involves a reduction in size of one chromosome

Figures 2.49–2.51
SEXUAL REPRODUCTION IN VOLVOX CARTERI

2.49.   Mature zygote. × 320.

2.50.   Zygote germinating. × 320.

2.51.   Colony formed from the zoospore released from a zygote. It is smaller than normal vegetative colonies and contains a reduced number of gonidia—in this case, one tier of four. × 320.

SOURCE. Dr. G. Kochert, published in *J. Protozool.* 15, 438 (1968).

Figure 2.52
MUTANT SEXUAL REPRODUCTION IN VOLVOX CARTERI

An interesting mutant of *V. carteri* (see text), in which "somatic" cells have become fertile. In the male colony, for example, seen here, the androgonidia have formed normal sperm packets. The somatic cells have differentiated into smaller packets too (cf. Fig. 2.45). × 386.

SOURCE. Dr. R.C. Starr, published in *Devel. Biol.* (Suppl.) 4, 59 (1970).

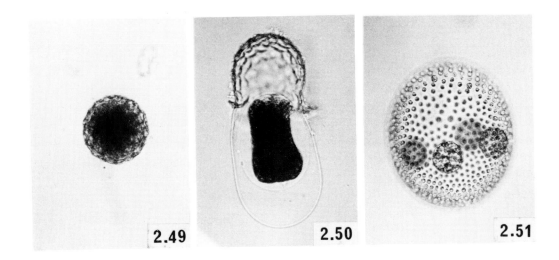

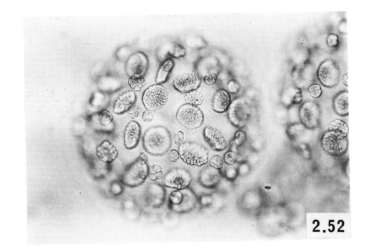

from 1.2μ to 0.6μ during five successive mitoses. However, the number of chromosomes per nucleus does not decrease, to the best of my knowledge. These observations suggest the possibility that reproductive cells in the Volvocales undergo a prolonged phase of DNA synthesis before mitosis begins, and that DNA synthesis does not necessarily intervene between later divisions; perhaps therefore, the chromosomes are initially polytene. Unfortunately, I do not know of any definitive reports concerning such a possibility, which incidentally may also be the case during reproduction in some members of the Chlorococcales (e.g., *Chlorella,* Section 3.2). In the case of *Chlamydomonas,* Buffaloe (1958) reports a specific and temporary inhibition of mitosis by intense light, and in this case, the cells become polyploid; upon return to normal conditions, somatic reductive divisions of an uncertain mechanism bring back the cells to normal ploidy.

Fritsch (1935, p. 93) mentions that after a series of divisions the daughter cells of *Chlamydomonas,* if elongated, usually become arranged parallel to one another before release from the parental wall. This phenomenon may be significant when considering the formation of coenobia in organisms such as the distantly related chlorococcalean alga *Scenedesmus* (Section 3.6a). In this case, the daughter cells, similarly arranged inside the mother cell wall, become linked together before release, thus giving rise to characteristic 2- to 16-cell colonies. One further point can be made here regarding the relation of the Chlorococcales to the Volvocales. In broadest terms, many of the unicellular Chlorococcales could be considered similar to nonmotile (premitotic?) forms of the Volvocales. A characteristic of many Chlorococcales is the formation of nonmotile "autospores" released from parental cells after cycles of sequential cell divisions (Sections 3.2–3.4); these autospores are exact replicas of the parental cell. *Brachiomonas,* which is closely related to *Chlamydomonas,* is a motile cell of distinctive form (Fritsch, 1935, p. 92). When *Brachiomonas* divides, the motile daughter cells acquire the distinctive form of the adult cell before they are released from the parental wall. Thus we have in the Volvocales more or less distinct examples of *motile* "autospores" arising after vegetative reproduction in a fashion exactly analogous to that displayed by the Chlorococcales. Indeed, autospores can probably be re-

Figures 2.53–2.56
VEGETATIVE CELLS AND GONIDIA OF VOLVOX TERTIUS

2.53. Section through colony. Parts of three gonidia can be seen. The young colony is filled with a fibrous matrix which becomes a more distinct envelope around cells. Note the difference in size between somatic cells and the gonidia (see also Fig. 2.40). × 280 approx.

2.54. Somatic cell. The eyespot *(es)* is more complex than in many unicellular Volvocales, and the wall *(w)* enclosing each cell is quite distinct from that which constitutes the wall of the coenobium *(cw)*. × 7,100.

2.55. Young colony, prior to inversion. Cell division may have ceased in this colony (it is difficult to determine for sure). The cells at this stage are appreciably smaller than the somatic cells of the parental colony (arrow). The chloroplasts are on the outside of the colony, the nuclei on the inside. × 1,000.

2.56. Young colony just after inversion. Note now that the chloroplasts are on the inside and the nuclei (with nearby short flagella just being extruded; small arrows) are on the outside. The cells are also conspicuously elongated (cf. Figs. 2.41, 2.57) but later round up to form the typical somatic cells (large arrow; Fig. 2.53). × 1,000.

SOURCE. Author's micrographs. Figures 2.54, & 2.56 published in *Planta* 90, 174 (1970).

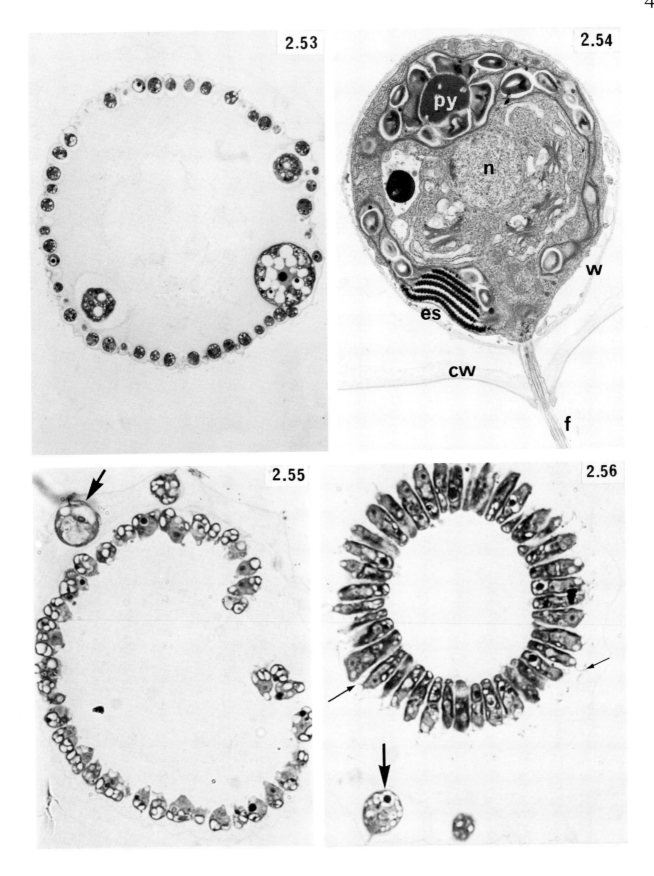

garded as nonmotile cells phylogenetically related to flagellate zoospores (Smith, 1950, p. 221).

The ultrastructure of cell division in *C. reinhardii* has been described by Johnson and Porter (1968) and Goodenough (1970). Basal body replication is completed before prophase, and the presence of four basal bodies in a cell is a sign of imminent division. (Other species may contain extra basal bodies during interphase; see earlier comments.) The assembly of new basal bodies was studied specifically in daughter cells about to divide again after the first cell division. These new basal bodies are very short, and the earliest stage observed showed nine single tubules attached to the tips of the basal cartwheel structure; a second and then a third tubule are added somewhat irregularly. A new striated fiber interconnecting the basal bodies is also formed in the daughter cells, and these events all occur in a small, ribosome-free region (the basal body complex) of cytoplasm around the preexisting basal bodies. Before the first division, the four basal bodies become paired, and the flagella become detached from their basal bodies; scission occurs at the "transitional" region (i.e., about level 6 in Fig. 2.17), and sometime afterward the protoplast rotates through 90°; Johnson and Porter were able to illustrate such a detached flagellum still projecting from the protoplast but some distance removed from the basal bodies. The tubules of the flagella break down as they are resorbed into the cell, but they may still be present after telophase of the first division (Johnson and Porter, 1968, Fig. 5). In the time-lapse cinematographic record of cell division in *C. reinhardii* by Schlösser (1966), the flagella are seen to be completely retracted into the cell before mitosis. After the basal bodies have become separated from their flagellum, they are indistinguishable from centrioles. They are reportedly not thereafter associated with the spindle in any way, remaining instead near the cell periphery, where they later become involved with the cytokinetic apparatus.

Mitosis itself appears quite normal in most respects; unfortunately, we have no information on the ultrastructure of prophase, anaphase, or telophase spindles. The nucleolus disperses during division, and the nuclear envelope remains essentially intact around a typical diamond-shaped metaphase spindle (Fig. 2.26). A promi-

Figures 2.57–2.60
INVERSION IN VOLVOX TERTIUS

2.57. Sectioned colony, halfway through inversion. Note elongation of the cells (cf. Figs. 2.41, 2.56). Forming flagella can be seen (arrow). × 1,100 approx.

2.58. Section from colony shown in Figure 2.57; the region shown here is at the folding edge. Note how the cells are interconnected at their chloroplast ends, which are narrowed at the fold. Further away from the fold, the cells are not so pointed. Flagella (*f*) are just being extruded. × 4,300 approx.

2.59, 2.60. Pattern of circular or spiral striations on the plasmalemma around the cytoplasmic interconnections (large arrow) between cells of colonies before, during, and after inversion. The plane of sectioning is approximately in the plane of the plasmalemma in Figure 2.59 (note other striations nearby; small arrows), whereas the striations have been sectioned transversely in Figure 2.60 (arrow). Figure 2.59, × 34,000; Figure 2.60, × 25,000.

SOURCE. Author's micrographs. Figures 2.59, 2.60 published in *Planta 90,* 174 (1970).

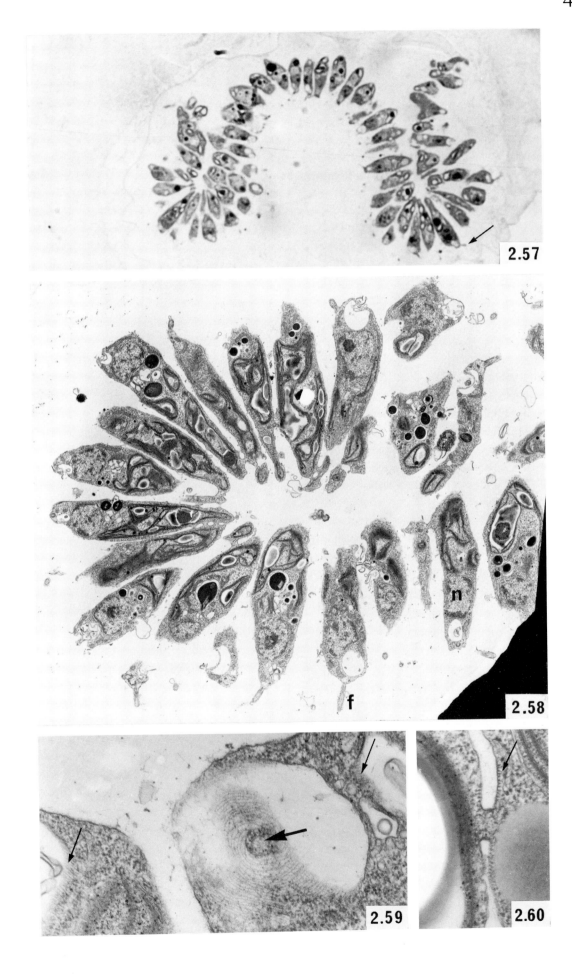

2.57

2.58

2.59

2.60

nent hole or "fenestra" is formed at each pole toward which spindle microtubules are oriented, and nuclear pores appear modified during mitosis, losing some of their internal structure. The cytoplasm surrounding these polar fenestrae is somewhat differentiated, appearing free of ribosomes and larger cytoplasmic organelles (cf. the basal body complex). Adjacent to the nucleus, which always moves over to one side of the cell during mitosis, Johnson and Porter also noted a characteristic invagination of limited extent in the cell membrane. This invagination is roughly in the plane of the metaphase plate of chromosomes and arches over the mitotic nucleus. A "metaphase band" of about four microtubules arranged in a row is always situated parallel to the invagination, and thus perpendicular to the axis of the spindle; these microtubules persist throughout mitosis. Their origin, function, and fate remain obscure.

The cleavage apparatus in *C. reinhardii* has turned out to be unexpectedly complicated, but it is analogous in many ways to that found in other Green Algae (Figs. 2.27, 2.29). At least two distinct sets of microtubules are involved. After telophase, the re-formed daughter nuclei come back fairly close together, and a network of endoplasmic reticulum interconnects them. Among these membranes appears an array of parallel "internuclear" microtubules, oriented perpendicular to the spindle axis and confined to the cytoplasm immediately between the nuclei. Another set of microtubules becomes prominent before cleavage proper commences. This second array of "cleavage" microtubules is coplanar with the internuclear microtubules but at right angles to them, and the two systems interpenetrate one another between the nuclei. Many cleavage microtubules line the growing cleavage furrow and appear to radiate across the cell from the basal body complexes (see also Fig. 2.28). The cleavage furrow, apparently initiated from the membrane invagination next to the metaphase band, soon extends circumferentially around the cell and grows inward, fastest from the end of the cell occupied by the two pairs of basal bodies, which are invariably arranged on either side of the furrow. As cleavage progresses, the nuclei (formerly situated near the cell surface) move into the center of the cells, and the whole cytoplasm rotates through 90° (see earlier). Thus, degenerating flagella (i.e., dissociated from basal bodies and

Figures 2.61–2.64
REPRODUCTIVE CELLS OF VOLVOX TERTIUS

2.61.   Oogonium. These cells are similar to gonidia (Fig. 2.63) except that they usually contain more reserve material (oil droplets, starch, etc.). × 920.

2.62.   Sperm packets, male colony. Some somatic cells (arrow) can also be seen. × 600 approx.

2.63.   Gonidium, typically highly vacuolate, with a peripheral chloroplast and a large central nucleus (compare with the nuclei of somatic cells; small arrows). The gonidium possesses contractile vacuoles as normal; one is just visible (large arrow). × 1,400 approx.

2.64.   Section grazing edge of a sperm cell (see Fig. 2.66), clearly showing the cytoskeleton of numerous longitudinally oriented microtubules (t). × 42,000 approx.

SOURCE. Author's micrographs. Figure 2.61 published in *Planta 90*, 174 (1970).

47

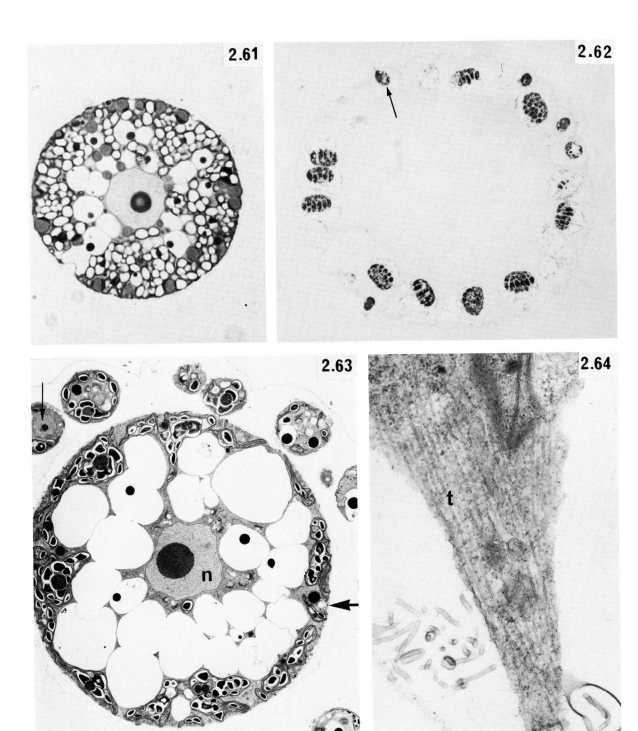

marking the original situation of the cell's anterior pole) are seen some distance from the new flagellar bases, often immediately adjacent to the chloroplast—an arrangement never encountered in interphase (Johnson and Porter, 1968, Fig. 5). The cleavage microtubules, initially somewhat randomly arranged, also become more closely aligned with the ingrowing furrow. Other microtubules, oriented toward the basal body complexes, extend around the cell's periphery.

Later stages of cytokinesis (Fig. 2.28) have been described in detail by Goodenough (1970). Once the cleavage furrow has grown inward an appreciable distance from the anterior end of the cell, the chloroplast develops a deepening circumferential constriction precisely located in the plane of cell cleavage. The dense granular matrix of the pyrenoid becomes elongated across the cleavage furrow in the chloroplast. Finally, the chloroplast divides in two, with the pyrenoid being partitioned fairly equally between the halves. Then the cleavage furrow, still clearly associated with the cleavage microtubule system, extends between daughter chloroplasts and completes cytokinesis.

When the nuclei undergo a second mitosis, they move to the cell surface near the older mother cell wall. Then nuclear and cell division proceeds as described for the first mitosis. Cleavage can take place in either of two orientations, both perpendicular to the plane of the first cell division. These are termed "longitudinal" and "equatorial" cleavages. In both cases, prior to cytokinesis the pair of basal bodies moves to the appropriate new position in the cell where the next cleavage furrow will form. It is not clear whether or not basal body replication precedes the second mitosis; each new daughter cell of the tetrad may receive only one basal body and then form another later on. On some occasions, the cells divide yet again to form eight daughter cells. After cell division is complete, the two basal bodies in each cell form a new flagellum, and the actively motile daughter cells are later released from the mother cell wall.

The complexity of the cytokinetic apparatus, utilizing at least two sets of microtubules, in such a relatively simple organism as *Chlamydomonas,* was originally a little puzzling. As Johnson and Porter (1968) conclude, these microtubules must have some special significance, and they suggest that the microtubules may somehow facilitate

Figure 2.65
LATE ANAPHASE IN FORMING SPERM PACKET OF VOLVOX AUREUS

The closed spindle has polar fenestrae *(pf)* through which chromosomal tubules penetrate (cf. Fig. 2.26). Note, however, that the basal body complexes *(bb)* lie adjacent to the chromosomes *(ch)* (not at the poles), and extranuclear microtubules (arrows) run between them. × 23,000.

SOURCE. Drs. T.R. Deason and W.H. Darden, Jr., published in *Contributions in Phycology,* Lawrence, Kansas, Allen Press, 1970, p. 67.

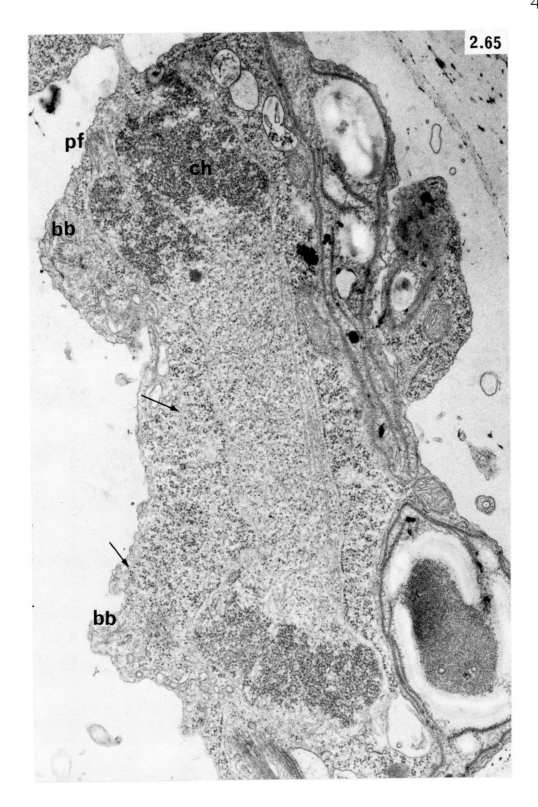

cleavage by providing a rigid cytoskeletal frame-work against which furrowing could occur. I find this hypothesis unattractive because many organisms are able to cleave in somewhat similar fashion without such microtubular arrays. Johnson and Porter also suggest that these microtubules may exert a directive force during cleavage. I am much more inclined to favor this view; it is obvious that the cleavage furrow would have great difficulty in growing in any other plane except that defined by the two sets of microtubules in the cell. Thus, these tubules could ensure that after such vital cytoplasmic components as the nucleus, chloroplast, and basal body complex (the latter perhaps marking the site of the more diffuse morphogenetic center postulated to exist earlier) have replicated, the cleavage apparatus will unfailingly partition one complement of these major organelles into each of the daughter cells. How could such a cytokinetic system have arisen in the first place, particularly since it is not recorded in many other cells that cleave by in-growing furrows? As a guess, I should say that the cleavage microtubular system obviously could have evolved from a modification of the microtubule-organizing properties of the basal body (morphogenetic center) complex. (In this regard, Johnson and Porter noted that both the poles of the spindle and the region around the basal bodies are subtly differentiated from the surrounding cytoplasm; e.g., they are ribosome-free; Fig. 2.26). This hypothesis receives some support from the sequence of events recorded during cell division in sperm packets of *Volvox,* where the basal body complex demonstrates its ability to form microtubular systems in the cell by creating its own rudimentary "spindle" alongside and parallel to the mitotic spindle (Fig. 2.65; Section 2.2a). However, the origin of the inter-nuclear system of microtubules is much more difficult to account for, and as yet no satisfying hypothesis can be put forward.

The existence of this complex cytokinetic system in *Chlamydomonas* has other important phylogenetic implications which are still somewhat difficult to specify in detail owing to a lack of relevant information. It is fairly widely accepted that *Chlamydomonas*-like organisms long ago gave rise to the other Green Algae, probably through three main lines of evolution: the volvocine, the ulotrichalean, and the siphonous lines (Chapter 8). Therefore, we could reasonably ex-

Figures 2.66, 2.67
SPERM PACKETS OF VOLVOX TERTIUS

2.66. These cells are highly differentiated in comparison with somatic cells. Apart from their elongated form, associated with a highly developed cytoskeleton of microtubules (Fig. 2.64), notice how the organelles have become stratified in each cell: at the top is the pyrenoid *(py)*, then the Golgi bodies *(g)*, vacuoles with dense inclusions *(v)*, the nucleus *(n)*, contractile vacuoles *(cv)*, mitochondria *(m)*, eyespot *(es)*, and finally the flagellar apparatus *(f)*. The chloroplast and peripheral microtubules run most of the length of the cell. × 11,000 approx.

2.67. Sperm packet sectioned transversely, but slightly skew to show the disposition of some of the stratified organelles visible also in Figure 2.66. Each cell is separated from the others by a wall *(w)*. As far as I can tell, in mature sperm packets the cells are not interconnected. The peripheral microtubules are clearly visible in each cell. × 32,000 approx.

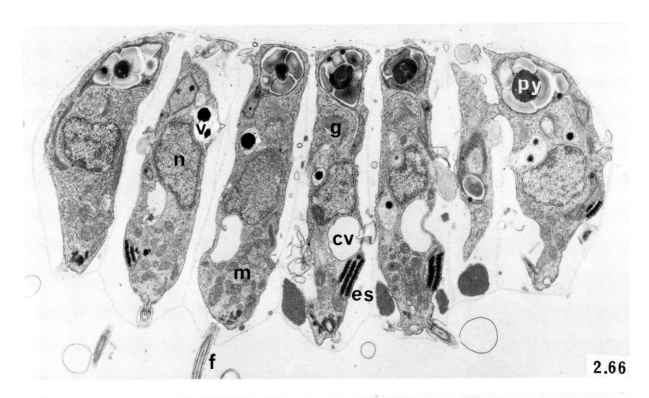

2.66

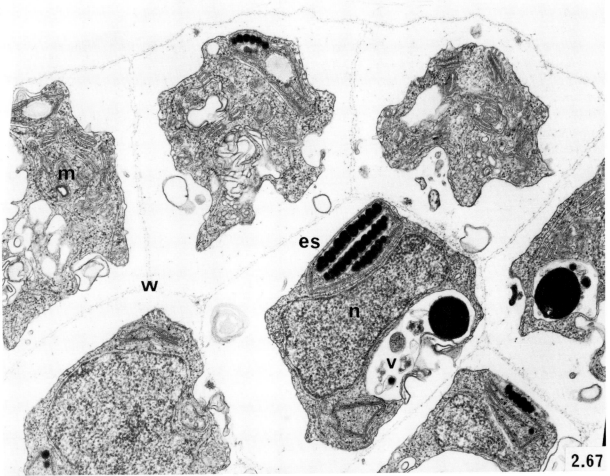

2.67

pect to encounter this type of cytokinetic apparatus elsewhere, and if we do, we might conclude that it is probably a primitive cytokinetic system. I have repeatedly stressed (Pickett-Heaps, 1969a, 1972a) the importance of "phycoplast" microtubules in cytokinetic systems of Green Algae; the distinguishing feature of these cytokinetic tubules is that they are always oriented *in the plane of cytokinesis*. This is in striking contrast to the "phragmoplast" system of a few Green Algae and all higher plant cells, where microtubules are oriented *perpendicular* to the plane of cytokinesis (Chapter 8). I should interpret the two systems of cytokinetic microtubules in *Chlamydomonas* as representing either the primordial phycoplast system or a more specialized development of it. Guessing on the basis of the small amount of data we have at our disposal, I should predict that the phycoplast will be present (often in modified form) in most (if not all) members of the Volvocales and Chlorococcales (Chapter 3), many of the Ulotrichales (Chapter 4), and all members of the small group of the Oedogoniales (Chapter 6). However, certain other quite complex species in these orders have been shown to cleave without needing phycoplasts (e.g., *Ulva, Klebsormidium;* see later). In many members of the Chlorococcales and Ulotrichales that have phycoplasts, the centriole complexes are associated with the mitotic spindle, forming the polar organizing centers; after nuclear division, these complexes often migrate around the daughter nuclei, away from where the poles have been, as the interzonal spindle collapses. The daughter nuclei concurrently come back together again, often flattening against one another (a quite characteristic feature of phycoplast-forming cells), and the centriole complexes then become located close to one another, on either side of the phycoplast system; this is fairly closely analogous to the situation during cleavage in *Chlamydomonas.* Pursuing the analogy further, ultrathin sections of the phycoplast in some members of the Chlorococcales could be interpreted as containing two sets of coplanar tubules, one radiating out in fan-like array from the centriole complexes, and the other essentially parallel and confined roughly between the nuclei—again corresponding to the systems present in *Chlamydomonas.* There may even be a third group of tubules which encircles the cell in the plane of cleavage near the cell envelope (cf. the metaphase band).

Figures 2.68, 2.69
SCANNING ELECTRON MICROGRAPHS OF TETRASPORA SP.

2.68. Both flagellate and nonmotile cells are shown. × 2,600.

2.69. A typical biflagellate, chlamydomonad-type individual. × 8,500.

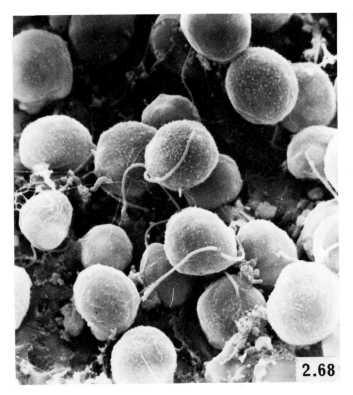

2.68

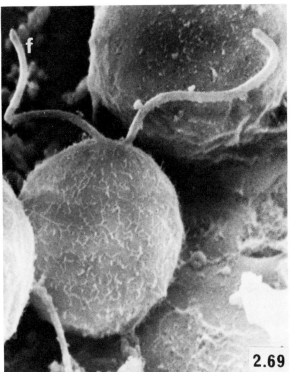

2.69

However, I must stress that although I consider this a reasonable interpretation of the phycoplast's tubular system, it is perhaps overly optimistic, and three-dimensional reconstructions or high-voltage microscopy of thick sections will be necessary to confirm it.

There is another apparently useful property of these cytokinetic microtubular systems, both phragmoplast and phycoplast. They can be used to collect vesicles together in a planar array in the cell, so that after mitosis the vesicles can fuse together to form a new cross wall. This development has given rise to the "cell plate" found, for example, in the Oedogoniales, some members of the Ulotrichales, the Charales, and all higher plants where presumably cytokinesis has evolved from cleavage to formation of the cross wall via the cell plate. The advantages of using a cell plate rather than a cytoplasmic furrow are not yet obvious, but the method is fairly consistently observed in more advanced organisms. However, so far as I know, it is confined to the Green Algae and higher plants.

It should be apparent from the foregoing discussion that the study of mitosis and cytokinesis in Green Algae may be of considerable importance in understanding the relations of families with one another and is beginning to provide us with an excellent example of the possible phylogeny of ultrastructural systems (Chapter 8). Thus, as will be seen in the following chapters, by studying selected examples of Green Algae, we can begin to get an idea of how evolution may have taken place at the ultrastructural level, altering cytoplasmic systems within the cell. Since many Green Algae may have evolved from a *Chlamydomonas*-like cell, I have devoted some time to discussing this particular organism for reasons which I hope have now become obvious.

2.1b. Sexual Reproduction

While many species of *Chlamydomonas* undergo isogamous sexual reproduction, some are anisogamous, and related unicellular genera (e.g., *Chlorogonium*) may even be oogamous. However, even isogametes in a conjugating pair may possess subtle but important differences, often clearly discernible at the ultrastructural level. That the isogametes are not equivalent is, of course, obvious from other evidence. In many species of *Chlamydomonas,* for example, sexual

Figures 2.70–2.72
INTERPHASE AND MITOTIC CELLS OF TETRASPORA SP.

2.70.   Vegetative cell with its own wall, but still enclosed by parental cell wall after division. Morphology is typically chlamydomonad, with the large chloroplast *(c)* enclosing much of the cell. Plane of sectioning is approximately right angles to that in Figure 2.14. $\times$ 14,000.

2.71.   Prometaphase spindle. The basal body complex *(bb)* has moved along the cell's surface; see site of original flagellar insertion *(fi)* in the wall. The spindle appears curved at this stage, with both intranuclear and extranuclear microtubules being oriented toward the basal body complex situated at the polar fenestra. $\times$ 23,000 approx.

2.72.   Late anaphase spindle, considerably elongated. Chromosomal microtubules penetrate through the polar fenestrae, one of which is always situated near the edge of the cell. $\times$ 16,000.

SOURCE. Author's micrographs, published in *Ann. Bot. 37,* 1017 (1973).

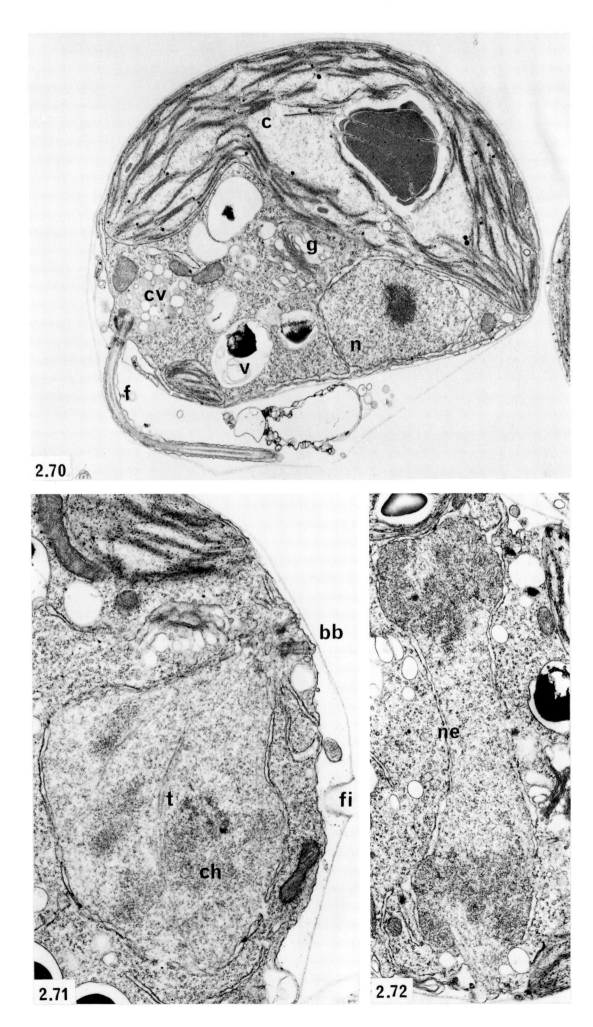

reproduction can take place only after complementary mating types (usually designated "plus" and "minus") have been mixed under appropriate conditions. Close investigation (e.g., by "labeling" one of these strains with large starch grains induced by optimal culture conditions) reveals that conjugating pairs contain one gamete from each strain. Another example of the differences that may exist between complementary mating strains is the ability of the extract of one mating type to induce preliminary manifestations of sexual behavior (e.g., clumping) in the other strain (Bold, 1967, p. 30). This clearly indicates that chemical differences involved in sexual response must exist between such strains. Thus the term "isogamy" can be misleading and should be taken to indicate that isogametes are only superficially similar—for example, in size and gross morphology. This statement can be equally true in homothallic clones (i.e., when the progeny arising from a single cell, often haploid, contain the two complementary mating types). For example, during sexual reproduction in *Hydrodictyon* (Section 3.7b), the multitude of isogametes formed in one coenocytic cell is derived ultimately from a single uninucleate zoospore; however, one gamete in each pair of conjugants always possesses certain structures not present in the other. For an excellent review of sexual reproduction in *Chlamydomonas* (and many other algae), see Wiese (1969); this review covers many interesting aspects, such as chemotaxis and agglutination of gametes, which are not discussed here.

Brown, Johnson, and Bold (1968) have carefully investigated sexual reproduction in *C. moewusii* with both light and electron microscopy. Soon after compatible mating types are mixed together, cells pair up; then others aggregate around these pairs, forming characteristic clumps (Fig. 2.30; see also Wiese, 1969). Adhesion initially occurs at the tips of the flagella, where strands of amorphous material can be detected in negatively stained preparations of sexually active cells. This material probably is a glycoprotein involved in attracting and/or holding the gametes together. Clumps of gametes then begin to sort themselves out again into pairs (Fig. 2.31), and the membranes of the flagella fuse progressively along their length from the tips. Thus, several complements of flagellar tubules may be encountered within the one membrane, but sometimes such fusion is transitory (e.g.,

Figures 2.73, 2.74
CELL DIVISION IN TETRASPORA SP.

2.73.   Cytokinesis. The phycoplast *(ph)* and cleavage furrow *(cf)* pass between daughter nuclei. The forming basal body (arrow) is shown in the inset. × 11,000. *Inset.* Forming basal body. The nine peripheral tubules are only single at this stage and are attached to the arms of the basal cartwheel (cf. Figs. 2.17, 2.19). × 170,000.

2.74.   Mitosis in recently formed daughter cells. The lower spindle is clearly at metaphase, with a basal body at one pole (single arrow). The metaphase invagination of the cell membrane is in the plane of the metaphase plate of chromosomes *(ch)*. The stage of division of the other nucleus is unclear, and basal bodies are at the outer edge of the cell (large arrow). Note also the eyespot (double arrow). × 11,000.

SOURCE. Author's micrographs, published in *Ann. Bot. 37,* 1017 (1973).

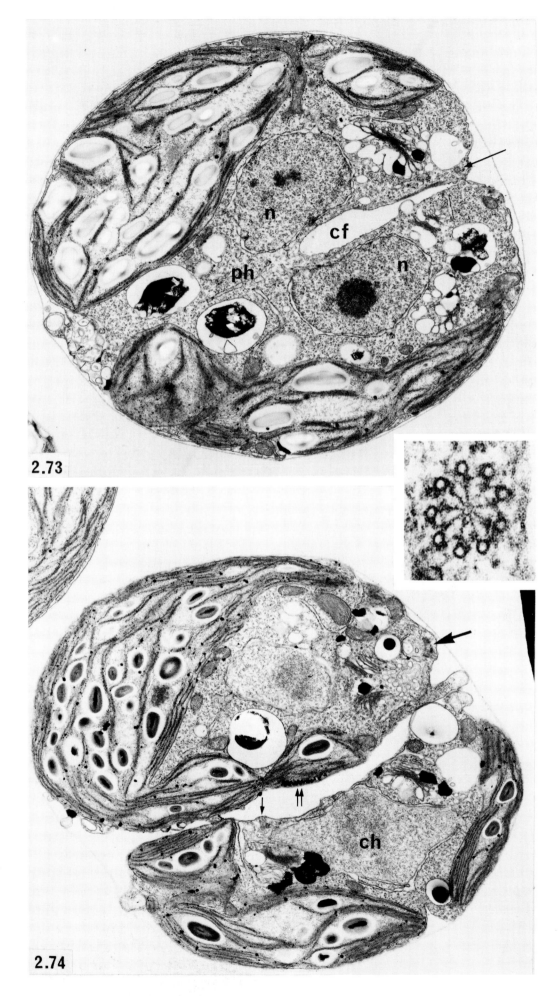

2.73

2.74

when two conjugants are attached temporarily to other gametes). The gametes approach each other so that their apical papillae (situated between their flagellar bases) become apposed. The papillae then elongate, and the cell wall around them dissolves. This elongation is characteristic of mature gametes and is never observed in vegetative individuals; it has to be precisely coordinated and concurrent in each gamete to ensure successful conjugation. The papillae during extension become increasingly dense and filled with fibrous material. Apparently, jostling of the cells effected by movement of the paired flagella brings the elongating papillae close enough together to allow cytoplasmic fusion to be initiated (Fig. 2.32).

After "syngamy" (fusion of the protoplasts, or fertilization), the flagella slowly separate; then one of the two pairs starts beating again, the other pair apparently remaining immobile and attracted in some way to the surface of the fused gametes which now, strictly speaking, should be called a "zygote." Motion of the pair of conjugants is unidirectional and persists for several hours, after which the paired gametes cease moving and clump together a second time. The protoplasts of cells now merge fully with one another, and in the process the joined papillae are discarded. Soon, the primary wall of the zygote is secreted (Fig. 2.33), and the flagella disappear owing to scission at the transitional zone (as occurs during cell division; Section 2.1a). Brown et al. have evidence that the chloroplasts unite (see also below) and "caryogamy" (nuclear fusion) soon follows, being initiated by coalescence of the outer unit of the nuclear membrane. The zygote is now diploid.

As the zygote matures, it secretes a secondary wall which develops a spiky ornamentation characteristic of the species (Fig. 2.34) and becomes increasingly dense so that many organelles become difficult or impossible to discern. Lipid and starch are accumulated, while the chloroplasts containing the two original pyrenoids condense markedly.

Friedmann, Colwin, and Colwin (1968) have investigated some ultrastructural aspects of fertilization in *C. reinhardii* and demonstrated significant differences between this organism and *C. moewusii,* described above. Gametes of *C. reinhardii* do not possess a cell wall, and a distinctive structure, the "fertilization tube" (about 2μ

long and 0.2μ in diameter; Figs. 2.35, 2.36) is extended from the apical papilla, probably only in one of the two mating types. At the base of this tubule, situated slightly to one side of the plane of the basal bodies, is the "choanoid body," a dense, ring-like structure just under the plasmalemma. The authors describe it as "an obliquely truncated, hollow cone," and it persists even after fusion of the gametes is completed (Fig. 2.38). The fertilization tubule appears devoid of obvious contents, but its bounding membrane, the plasmalemma, possesses numerous small projections (up to 0.5 mμ in length) radiating outward from the axis of the tubule (Fig. 2.36). In appearance it is quite unlike the apical papilla of *C. moewusii* described by Brown, Johnson, and Bold (1968). Fertilization is initiated by the typical clumping phenomenon in vivo, although this was not described by Friedmann, Colwin, and Colwin (1968). Cytoplasmic fusion apparently commences when the tip of the fertilization tubule in one gamete ("plus") contacts an equivalent region near the flagellar base of an individual of the complementary strain ("minus"). At the region of contact (aptly described by one of my students as an "erogenous zone"), the plasmalemmas fuse, and there is also a small, rather inconspicuous ring-like structure, possibly the homologue of the choanoid body in the "minus" strain (Fig. 2.37). The fertilization tubule then shortens and expands, losing its projections. Soon the apical ends of the gametes are directly opposed to one another; then the cells merge laterally (Fig. 2.38) and completely by a slow jackknife movement as their membranes fuse along their length. The two original protoplasts remain distinguishable for some time thereafter.

Initially after fusion of the gametes, the flagella beat haphazardly and the zygote does not move. However, the flagellar bases soon approach one another, and then the flagella all become coordinated and the quadriflagellate zygote can begin to move about. How this coordination between two different sets of flagellar apparatuses can be brought about is not yet clear, although the basal fibers apparently intermingle at this stage. Friedmann, Colwin, and Colwin confirmed in a number of conjugating pairs of gametes that only one strain (labeled with large starch grains) possesses the choanoid body and fertilization tubule.

Since the chloroplasts of *C. reinhardii* contain only one pyrenoid and eyespot, Cavalier-Smith (1970) has been able to demonstrate clearly that the chloroplasts of gametes do fuse after conjugation, since the large, single chloroplast in the zygote contains two pyrenoids and eyespots (Fig. 2.39). This observation strengthens the possibility that recombination of genetic material in the chloroplast could occur, as required by earlier observations that certain mutations affecting the chloroplast are transmitted during sexual reproduction by non-Mendelian genes within the chloroplast (Kirk and Tilney-Bassett, 1967).

Germination of zygotes of *C. moewusii* has been briefly described by Brown, Johnson, and Bold (1968), who were severely hampered in ultrastructural work by the cytoplasmic density of the zygotes and their refractory nature towards electron microscopical processing techniques. (This problem incidentally is invariably encountered with zygotes of all other Green Algae examined.) The authors were not able to investigate meiosis, which precedes germination and thereby restores the haploid vegetative condition. The cells cleave and grow flagella before release as dense motile organisms filled with lipid and starch.

2.2. Volvox

The genus *Volvox* represents the ultimate expression of colonial development in the volvocine line of evolution and one which is probably a dead end (Scagel et al., 1965). The colonies (coenobia), in some species easily large enough to be seen with the naked eye, are beautiful microscopical objects, actively motile and containing up to about 20,000 cells (Fritsch, 1935); the largest species are correspondingly rather fragile. The spherical or ellipsoidal colony (Fig. 2.40) consists of a single layer of cells embedded in mucilage, which is probably proteinaceous since various proteolytic enzymes can dissociate colonies completely into single cells (Kochert, 1968; Starr, 1969). Most drawings of colonies show that the chlamydomonad cells are interconnected by fine cytoplasmic threads, but these intercellular connections are not present during the whole life span of the colony in some species (see Fritsch, 1935, pp. 104 et seq.). The colonies contain only two types of cells: somatic and reproductive. The somatic cells are small, regularly spaced, and flagellate and usually far outnumber the reproductive cells. These latter cells vary appreciably in size, degree of differentiation, and disposition in the colony. Reproduction in *Volvox* has been the subject of a number of investigations (see Fritsch, 1935) and has recently been described in considerable detail for *V. carteri* and *V. aureus* by Starr (1969), Kochert (1968), and Darden (1966). The reader is referred to these papers for more detailed information. Starr and Flaten (1970) have also made a superb time-lapse cinematographic record of the life cycle of *V. carteri*.

Vegetative (i.e., asexual) reproduction is brought about by single-celled gonidia, which are considerably larger than somatic cells. These gonidia usually constitute a very small proportion of the total number of cells in the colony (numbering around 16 or less in colonies containing hundreds or thousands of cells, depending upon environmental conditions), and they project inward in mature colonies. After cell division has ceased in somatic cells of a forming colony, and usually after the colony has inverted (see below), the gonidia initiate vegetative reproduction of new daughter colonies by undergoing a rather prolonged phase of cell division. Following each set of mitoses, the orientation of cleavages is always strictly controlled, successive cleavages following a precisely predetermined pattern. The gonidia are probably not flagellate, at least in *V. carteri* (Kochert, 1968) and *V. tertius* (personal observations), and their contents are oriented as in the surrounding somatic cells; in particular, the flagellar bases are directed outward (Fig. 2.12). The succession of gonidial cleavages then forms a hollow sphere of cells whose flagellar bases and nuclei become oriented inward. If these colonies are not becoming sexually differentiated, the fifth set of cleavages is asymmetrical (in the species studied in detail), so that, in principle, 16 larger and 16 smaller cells are formed (in practice, fewer than 16 larger ones are normally thus created). In *V. carteri*, two distinct sets of divisions one after the other, in anterior and posterior sets of cells, give rise to these larger gonidia (Kochert, 1968), but such subtle complications need not concern us here. Kochert and Yates (1970) have evidence for the existence of an ultraviolet-labile morphogenetic material that is involved in development of the gonidia. Irradiation with ultraviolet light causes a decrease in the number of gonidia formed, proportional to the dose given. This defect is never transmitted to offspring, and therefore no genetic mutation is involved. The larger cells arising from these asymmetrical divisions normally do not undergo any further division at this stage, although Starr reports that they may cleave off a couple of somatic cells; instead, they form the new gonidia now situated on the outside of the developing colony. In contrast, the smaller cells continue to divide, thereby giving rise to the multitude of tiny somatic cells; the colony meanwhile grows appreciably in size. Also as a result allied with a subtle rearrangement of cells early in colony formation (Starr and Flaten, 1970), a hole, the "phialopore," appears in the young colony at its outer surface (Fig. 2.42).

Next follows the fascinating and poorly understood phenomenon of inversion, whereby the whole colony turns itself inside out (see Fritsch, 1935, for early references). Although *Volvox* itself was first observed by Leeuwenhoek about 1700, inversion was apparently not discovered until 1908 by Powers (Kelland, 1964). This process (Fig. 2.41; Section 2.2b) is especially dramatic when visualized by time-lapse cinematography (Starr and Flaten, 1970). Inversion in some species starts with the formation of four (or more) lobes around the phialopore (Kochert, 1968; Darden, 1966) which split apart and then curl back over the edge of the colony. They continue to move over the colony's surface until inversion is complete; the gonidia are now on the inside of the colony. Soon afterward the colonies become actively motile. Some species of *Volvox* undergo a slight variation of this behavior. Inversion commences with the appearance of an equatorial constriction around the colony; then the lips of the phialopore flex back over the colony to achieve inversion (Kelland, 1964; Bonner, 1971).

The daughter colonies are by now fully formed, but continue to expand, losing their protoplasmic connections in some species. They secrete their own mucilage envelope and acquire eyespots, and soon they are released from the parental colony as the latter dies and falls apart. By now, the gonidia are usually arranged in a definite pattern within the new colony, for example, in two tiers of four gonidia in some strains of *V. carteri* (Kochert, 1968), and they soon commence another round of asexual reproduction.

Sexual reproduction in *Volvox* is oogamous and has been most intensively studied in *V. carteri* and *V. aureus* by Starr (1969), Darden (1966), and Kochert (1968). As usual, monecious (homothallic organisms) and diecious (heterothallic organisms) species exist, and usually gonidia are not present alongside sexually differentiating cells.

Female colonies produce oogonia instead of gonidia (Fig. 2.43) and in slightly higher numbers (e.g., normally ca. 40 in *V. carteri*); however, Darden (1966) says that "newly released vegetative colonies with undivided gonidia function as such" (i.e., oogonia) for *V. aureus*. In *V. carteri*, the oogonia are dense and flagellate and arise in much the same way as do gonidia, by an unequal cell division of some of the cells of the colony at around the 32-cell stage. The disposition of the oogonia within the mother colony, however, is much more irregular than that of gonidia (Fig. 2.43; Starr, 1969).

Male gametes are small, elongated, and comparatively highly differentiated cells that arise in plakeas or sperm packets. In *V. carteri*, males differentiate quite differently in many ways from female colonies. The growing male colony undergoes a usual sequence of divisions, but the male initial cells, the "androgonidia," are differentiated at the *last* cell cleavages undergone by the forming coenobium. The male spheroid then contains anywhere between 64 and 512 cells (depending upon cultural conditions), of which half, appreciably larger than the somatic cells with which they are paired, are androgonidia (Fig. 2.44) and will form sperm packets. Then the colony inverts as normal, and next the androgonidia go through another cycle of cell divisions, this time to form the slightly curved, plate-like plakeas of sperm (Fig. 2.45). The sperm packets remain intact after release from the parental colony while moving around in the medium and are markedly attracted to female colonies in vivo, undoubtedly under the influence of a chemotactic agent emanating from the female. The packets appear to hover over the surface of the female colony (Fig. 2.46) and then disaggregate (Fig. 2.47), a process beautifully recorded by Hutt and Kochert (1971). At this point, they form a fertilization pore in the female colony by secreting a proteolytic enzyme, a response apparently triggered by the presence of the female. The individual sperm actively penetrates the mucilaginous matrix of the female colony to achieve fertilization (Fig. 2.48). If sperm packets contact vegetative colonies, they soon swim off and apparently never penetrate the colony (Kochert, 1968).

Although the creation of sperm packets is characteristic of *Volvox,* the sequence of events outlined above is not precisely followed by all species. For example, in *V. aureus* (Darden, 1966), male colonies have no gonidia, but a proportion of gonidia-like cells in the posterior two thirds of the colonies divides to give a plate of 32 cells; these latter colonies then undergo a rudimentary form of inversion and then differentiate into 32 sperm. Development of these colonies is not entirely synchronous in a given parental coenobium.

In some species of *Volvox,* oogonia will become vegetative gonidia if they are not fertilized (Starr, 1969). Fertilized oogonia enlarge and turn orange, secreting a thick ornamented cell wall. These cells are eventually released following the death of the maternal coenobium. Zygospores (Fig. 2.49) are extremely resistant structures. For example, a common method of obtaining *Volvox* involves the collection of earth from the bottom of dried-up ponds. Upon addition of the earth to water, colonies often appear within a few weeks (Starr, 1969). During germination (Starr, 1969; Kochert, 1968; Darden, 1966), the zygote becomes paler and elongated, and then its wall splits to release a large, weakly motile zoospore (Fig. 2.50). This cell soon begins a succession of divisions (the first two presumably being meiotic unless meiosis has already occurred in the zygote) and follows essentially the developmental course outlined earlier. The colonies thus arising are smaller than usual and possess only a few (e.g., one tier of four) gonidia (Fig. 2.51).

2.2a. Physiology and Genetics of Reproduction in *V. carteri* and *V. aureus*

Owing to the work of Starr and his collaborators, we have some intriguing information concerning the physiology of reproduction in *V. carteri,* and these workers have succeeded in isolating a number of interesting developmental mutants. This work has been excellently reviewed recently (Starr, 1970a), and I shall do no more than briefly summarize it here. Many of the Volvocales are probably haploid through most of their life cycle, and since they undergo typical meiosis and are easy to grow rapidly, they lend themselves to genetic studies, as has been demonstrated in *Eudorina elegans,* for example, by Mishra and Threlkeld (1968).

Male colonies appear spontaneously in *V. carteri,* often in older cultures or after transference to fresh medium. The few male sperm packets, upon release, appear to induce the formation of still more males in previously vegetative colonies of the male strain. That male sperm packets release some inducing substance is clearly revealed by induction of oogonia in vegetative female strains exposed to sexually active male colonies or to filtrate from cultures containing sperm packets. This induction is never spontaneous. The filtrate also induces sexuality in vegetative male colonies. Bioassay methods have been developed using this reaction of the female to the inducing substance, which has thereby been shown to be nondialyzable, heat labile, and destroyed by pronase while it is unaffected by trypsin, chymotrypsin, and RNAase. Single-celled gonidia in the female strain do not immediately differentiate into oogonia, but instead have to go through one complete cycle of colony formation; thus these cells need about 30 hours' exposure to the inducer (cf. with *V. aureus* mentioned below). The inducer, which appears to be a secondary metabolite or a breakdown product of disintegrating sperm packets, is amazingly potent. Under optimal conditions, it gives 100% induction of females at a concentration of 10^{-10} gm/liter, or to put it differently, one male colony can potentially induce about 55 million gonidia, or one sperm cell can induce about 200 gonidia. There are also different varieties of inducer from different strains of *V. carteri;* thus, three subtly different inducers can be obtained from Nebraskan (U.S.A.), Indian, and Australian strains of the species. They naturally induce the females in their own strains; the Indian inducer works on both others, while both others work on the Indian strain. However, the Australian strain cannot induce females in the Nebraskan strain, and neither can the Nebraskan strain induce Australian females.

Starr (1970a) also describes several mutants in *V. carteri* isolated in his laboratory. One female strain is highly insensitive to the inducer, needing 10^4 times the normal concentrations to achieve at best partial induction of oogonia.

However, at the higher, ineffective dilutions, the inducer does have some effect as shown by its tendency to increase the number of daughter gonidia arising in vegetative colonies. Starr has also isolated a "spontaneous female" which appears without the need of the male inducer; these mutants were difficult to grow but soon revealed that they were permanently female. Following sexual reproduction when the mutants were crossed with normal males, the expected ratio of 1:1 male to female strains was obtained from germinating zygotes, but all the females were spontaneous (the males were unaffected). This phenomenon therefore results from a definite sex-linked mutation at the gene level (rather than from some transient biochemical "mistake"). Mutant males have also been isolated in two strains in which the normal 1:1 ratio of androgonidia to somatic cell is not maintained. One of the mutants, the R1 strain, apparently undergoes the important unequal differentiation division earlier than normal and thus produces fewer androgonidia. A similar result is obtained differently by the R2 mutant, in which not all cells participate in this final division. Thus the R2 strain contains a number of large "somatic" cells, in addition to small somatic cells and androgonidia in the ratio of 1:1. The inheritance of these characteristics after sexual reproduction is complex and difficult to understand, but again there is little doubt that they represent true genetic mutations. Starr describes the behavior of another stable female mutant that produces many more gonidia than normal (up to 60 per colony). This trait is passed on through the sexual cycle and, moreover, male colonies formed from the cross of this female mutant with normal males form many more sperm packets than normal; indeed some differentiated male colonies that acquired the trait consist almost entirely of androgonidia. Starr has found yet another interesting variant, derived originally from a colony containing 50–60 small spheroids, which when isolated and cultured was again revealed as a stable mutant. In these colonies, the gonidia develop as normal. However, many of the somatic cells also enlarge and then undergo cleavage, eventually giving rise to smaller spheroids (16–200 cells), and these, as well as the true gonidia, can propagate the species vegetatively. When sexually induced, both gonidia and the small spheroids can form oogonia. After being mated with normal males, the resultant male progeny are most interesting. These, when next undergoing differentiation into male colonies, first form the usual 1:1 ratio of "somatic" cells to androgonidia. But, both these types of cells then differentiate, the smaller, apparently somatic cells, giving rise to small sperm packets (Fig. 2.52) alongside the normal sperm packets derived from the androgonidia! In other words, the distinction between reproductive and nonreproductive cells has been partially lost.

This work by Starr and his associates opens up all sorts of fascinating possibilities for the investigation, both genetically and physiologically, of reproduction in *Volvox*. Although the implications of the results described above cannot yet be evaluated in detail, we can look forward to a change in this situation in the near future.

Darden (1966) has investigated sexual differentiation in *V. aureus*, which shows some interesting differences in detail from *V. carteri*. As mentioned earlier, in the former species undivided vegetative gonidia apparently function as females, to which male sperm packets are attracted. It became apparent that the spontaneous appearance of a few male colonies was soon followed by the induction of male sexuality in around 50% of the colonies. As before, the existence of a chemical inducer was established—relatively heat-stable but destroyed by trypsin as well as pronase (cf. the inducer in *V. carteri*)—and a bioassay system for it was developed. Males were induced to the full capacity of the culture (ca. 50%) by as little as 30 minutes' exposure to the inducer, and even 5 minutes' exposure had some effect. Ely and Darden (1972) have devised methods for purifying the inducer considerably. Vegetative colonies were most susceptible to the inducer at around the middle of their developmental period (i.e., at about the 48-hour stage in colonies that are released around 90 hours after the gonidia commence division). Starr (1970a) says that this maximum of 50% induction is attributable to the method by which the assay was conducted.

I should emphasize again that these detailed and elegant studies have been confined to very few species. Numerous variations in structure and behavior are being documented as more species of *Volvox* are investigated. For example, the species *V. pocockiae* resembles *V. aureus* in many ways, but male colonies are dwarf and contain no sterile somatic cells (Starr, 1970b); in *V. africanus* (Starr, 1971), gonidia give rise to daughter colonies at different times, so that one tier of four daughter colonies is considerably further developed than the next tier.

2.2b. Cellular and Colonial Structure

Some aspects of the ultrastructure of *V. carteri* and *V. tertius* have been described by Kochert and Olson (1970a), Olson and Kochert (1970), and Pickett-Heaps (1970a). Somatic cells of mature colonies closely resemble those of *Chlamydomonas* (Fig. 2.54), and only few differences need be recorded. Each cell has its own wall (Kochert and Olson say this is not a true wall), but it is also embedded in a layered matrix that holds the whole colony together. The flagellar apparatus appears very similar to that of *Chlamydomonas*. Extra basal bodies, probably incompletely formed, are situated near the flagellar bases of most types of cells of *V. tertius* and in gonidia and sperm packets of *V. carteri*. The basal cartwheel structure, however, is reportedly missing in the latter organism from functional basal bodies (Olson and Kochert, 1970). Basal bodies (centrioles) are conspicuous features of nonflagellate gonidia in both species. Eyespots consist of several layers of densely stained granules.

All cells go through morphologically differentiated stages during their life cycle and make an interesting comparison with *Chlamydomonas*, for example. The gonidia are large, highly vacuolate cells with a large nucleus and thick cell wall (Figs. 2.53, 2.63). Their chloroplast is extensively lobed and contains numerous pyrenoids. These larger cells are, therefore, morphologically quite different from the smaller colonial cells they later give rise to (Kochert and Olson, 1970a). Endosymbiotic bacteria have been reported in the cytoplasm of various cells of *V. carteri* (Kochert and Olson, 1970b). Cell division in gonidia has not yet been described ultrastructurally. A few observations (unpublished) I have made on cell division in forming colonies of *V. tertius* indicate that the spindle is centric (i.e., has centrioles associated with it). Cleavage utilizes a phycoplast, but most importantly, it is not complete (Deason, Darden, and Ely, 1969). Cells remain joined by bridges of cytoplasm (Figs. 2.58, 2.60), which are probably most important in enabling accomplishment of the complex pro-

cess of inversion. It is unclear from published reports whether these interconnections always persist in mature colonies; they do not in *V. tertius* (Pickett-Heaps, 1970a). Thus, while cleavage occurs essentially as has been described for *Chlamydomonas*, it has been subtly modified during evolution of the colonial habit so as to maintain intercellular continuity. The connections have been described by Dolzmann and Dolzmann (1964), Bisalputra and Stein (1966), Ikushima and Maruyama (1968), and Pickett-Heaps (1970a); they are not equivalent to the plasmodesmata of higher plants, being much larger and usually containing a strand of endoplasmic reticulum and sometimes other small organelles.

INVERSION

While the fascinating process of inversion was probably first described by Powers (1908) and confirmed by several other investigators some time ago (see earlier), only a few attempts have been made to explain this phenomenon, most notably those of Kelland (1964). Before inversion, somatic cells are rounded and somewhat irregular in outline (Fig. 2.55). Kelland reports that many cells develop slow-moving, elongated projections from the initially outer (i.e., nonflagellar) end before and during inversion, an observation confirmed in our laboratory. (These projections are just visible in Fig. 2.41.) Inversion in species such as *V. tertius* and *V. aureus* commences as splits develop at the phialopore and then the lips thus formed curl back over the colony. The cells at the phialopore elongate radially as the lip curls over; these cells round out slightly before becoming very narrow and elongated again, along with the rest of the cells of the colony, after inversion (e.g., Fig. 2.56). The other cells in the colony become more elongated as the bend reaches and then passes them. In other species, such as *V. globator*, in which inversion commences with the appearance of an equatorial constriction, the cells at the phialopore do *not* elongate until the postinversion stage; rather, they become flattened circumferentially, and their protoplasmic connections stretch considerably, and apparently passively, as the colony passes through the pore. Kelland also observed individual cells and fragments from ruptured colonies. Single cells also change shape, becoming elongated. Fragments of colonies tend as far as possible to form clumps, with the elongated cells oriented radially and the flagellar bases directed outward. These and other experiments indicate clearly that changes in shape of the individual cells are intimately associated with, and probably involved in causing, the movements of inversion.

Inversion normally lasts about 40 minutes (Darden, 1966; Kochert, 1968; Starr, 1969; Starr and Flaten, 1970). Cells at this stage have short flagella (Fig. 2.58; Pickett-Heaps, 1970a), one apparently always longer than the other (Kochert, 1968) as is also the case in some other members of the colonial Volvocales (Goldstein, 1964). Starr (1969) says that in his strain of *V. carteri* the second flagellum appears much later, after inversion is finished.

I find it puzzling that no one to my knowledge except Kelland (1964) and myself (Pickett-Heaps, 1970a) emphasizes the marked elongation which cells undergo before and during inversion and which lasts until well after it is over (Fig. 2.56). Kochert (1968) vaguely mentions a "structural differentiation" of cells prior to inversion, and Bonner (1971) also refers to changes in cell shape which pass through the colony in waves. However, elongation of cells appears obvious in all light micrographs I have seen of inverting colonies in vivo (e.g., Fig. 2.41) and is striking in sectioned material (Figs. 2.57, 2.58; Pickett-Heaps, 1970a). I have suggested that this elongation could be associated with a proliferation of longitudinally oriented microtubules, similar to the arrays present in the elongated cells of sperm packets (Fig. 2.64), and that elongation is most likely important in achieving inversion. This postulate has been supported experimentally by treating *V. globator, V. carteri,* and *V. tertius* with the microtubule-disrupting drug colchicine, which prevents inversion from continuing (Renner and Pickett-Heaps, unpublished data). If colchicine is then removed after about a 1-hour treatment, the colonies cannot undergo or continue inversion; however, they later form daughter colonies from their gonidia as usual, showing conclusively that inversion has not been stopped merely by killing the colonies with the drug. Colchicine treatment does not cause the elongated cells to round up; they remain quite columnar in shape, and so I assume (see later) that the cells must also be fairly rigid to achieve inversion.

The numerous cytoplasmic connections are confined to the chloroplast end (i.e., before inversion, the outer facing end) of the interconnected cells of *V. tertius* (Fig. 2.58; Pickett-Heaps, 1970a). Thus they could obviously provide the means by which the intercellular communication necessary for the coordinated cell movements of inversion can be achieved. They also must serve to hold the colony together; the elongated cells, "hinged" at one end, are able to split apart at the other end, occupied by the nucleus, as the colony flexes. There is another striking feature, so far unique, associated with all these cytoplasmic connections before, during, and just after inversion in *V. tertius*. This consists of a system of regularly spaced circular or spiral striations or microfilaments attached to the plasmalemma and concentric around each connection (Figs. 2.59, 2.60). Their nature is unknown, but I suspect they are contractile in some way, since the ends of the cells at the cytoplasmic connections become narrowed and indeed rather pointed during inversion at the region of flexing (Fig. 2.58). Many micrographs (e.g., Figs. 2.57, 2.58) indicate that on either side of the folding portion of the colony the cells are not so pointed in this fashion. We (Renner and Pickett-Heaps, unpublished data) have attempted to show a correlation of these microfilaments with those of the contractile ring that lines the cleavage furrow of various animal cells by using the drug cytochalasin B. This drug, which in many animal cells causes the cleavage furrow to relax and disappear and the microfilaments to disperse (Schroeder, 1970), does indeed stop inversion; however, concentrations that are thus effective also appear toxic to the colonies, and therefore the experiments remain inconclusive (since inversion can be always stopped by any agent that kills the colonies). Unfortunately, I have not had the chance to confirm the existence of the striations in other species of *Volvox,* a necessary step before we attribute much significance to them.

We should note some interesting similarities of inversion to certain morphogenetic processes in animal embryos (see also Kelland, 1964). For example, the neural tube forms during urodele neurulation when a flat plate of presumptive neural cells rolls into the tube form (Burnside, 1971). Two main events are involved in this movement: the cells elongate, apparently owing to changes in their complement of longitudinally oriented microtubules, and they constrict apically where bundles of microfilaments interconnect localized regions of the cell's periphery. Cellular elongation, mediated by microtubules, also seems necessary in achieving avian gastrulation (Granholm and Baker, 1970). Spooner and Wessells (1972) studied salivary gland morphogenesis, which involves the formation of clefts in previously undifferentiated tissue. Microfilaments are apparently involved in cleft formation, since treatment with cytochalasin B causes (1) disruption of the microfilamentous systems and (2) loss of the clefts. Colchicine treatment at moderate concentrations halts morphogenesis, with the tissue remaining unchanged after prolonged exposure to the drug (just as is the case with inversion), although the cells remain columnar in shape. However, I remain somewhat confused by these authors' results. They investigated the recovery in colchicine of tissue pretreated with cytochalasin B; cleft formation proceeded normally. They conclude that microtubules are not important in this morphogenesis, but leave unexplained the fact that they used in the recovery experiments a concentration of colchicine (0.1μ gm/m1) which they just previously say has no effect on morphogenesis, although this same concentration is reported to disrupt cytoplasmic microtubules. This leaves unexplained why the higher concentration of colchicine was effective in the first place.

In summary, it should be possible to explain inversion at least partly in terms of these three phenomena: (1) longitudinal rigidity acquired before inversion through microtubules (causing cell elongation); (2) interconnections being confined to one end of the cells, the inside end during flexing; and (3) a wave of coordinated contraction of the cells' membrane, possibly associated with the microfilaments around the interconnections. We should not overlook the possibility that the cytoskeleton of microtubules could also provide a transport system along the cell (cf. Granholm and Baker, 1970); I am intrigued by the usual movement of the pyrenoid along each cell at the point of flexion (Figs. 2.57, 2.58). Gross movement of cytoplasmic organelles could cause (or be caused by) such contraction. It is possible to make models along these lines that exhibit some of the necessary properties, but I have not yet been clever enough to construct a spherical model that works. Doubtless some readers will prove more skillful.

2.2c. Structure of Reproductive Cells

The structure of oogonia and sperm packets has not been investigated in detail, but some points are worth noting. Oogonia are large, vacuolate, and quite like gonidia (cf. Figs. 2.61, 2.63); in *V. tertius,* however, they contain much reserve material such as starch and lipid droplets.

Sperm packets, in contrast, consist of tiny elongated cells (Figs. 2.62, 2.66; Pickett-Heaps, 1970a). Cell division in male colonies that gives rise to sperm packets (Fig. 2.65) has been described by Deason and Darden (1971). In many respects, mitosis resembles what has already been described for *Chlamydomonas* (Section 2.1a). The flagella of the male initially are withdrawn before the first of the series of synchronous divisions. During mitosis, the basal bodies apparently separate alongside the daughter chromosomes, but outside the closed spindle, alongside the normal mitotic apparatus. The disposition of the basal bodies during mitosis appears variable, however; in one of their micrographs (Deason and Darden, 1971, Fig. 14), one pair is closely associated with the fenestra at one pole, whereas the other basal body in the micrograph is some distance from the second pole. These results can also be compared with spindle structure in *Tetraspora* (Section 2.3). Cytoplasmic connections are common in forming sperm packets (Deason, Darden, and Ely, 1969), but disappear during maturation of the sperm (in *V. tertius;* Fig. 2.66).

Mature sperm are highly differentiated cells. They are not only small, but elongated and narrow (Figs. 2.62, 2.66, 2.67), a form which is probably related to a highly developed cytoskeleton of numerous longitudinally oriented microtubules (Fig. 2.64). Individual cells are separated from each other by a thin wall (Figs. 2.66, 2.67). Most conspicuous is the way the organelles in each cell are stratified (see Fig. 2.66 and its caption).

2.3. Tetraspora

Some unicellular members of the Volvocales, particularly species of *Chlamydomonas,* can grow as palmella stages, during which vegetative cells become immobilized and loosely held together in a gelatinous matrix; motility can be quickly reacquired under the influence of a suitable stimulus (see page 8). The palmella stages closely resemble some species of *Tetraspora*. Indeed, *Tetraspora* can probably be considered a form similar to *Chlamydomonas* in which the premitotic loss of motility has become increasingly prolonged. In many genera, vegetative reproduction proceeds by cleavage of the parental colonial cells without any intervening motile stages, but in others, the two to eight daughter cells are flagellate and actively motile upon release for a variable period. Fritsch (1935) emphasizes the "marked plasticity" of vegetative reproduction which confirms the affinity of *Tetraspora* with unicellular Volvocales. Such a loss of motility appears to be a common evolutionary trend which has given rise to other groups of Green Algae (e.g., Chlorococcales; Fritsch, 1935). Many authorities (e.g., Fritsch, 1935) put *Tetraspora* and its relatives in a family, the Tetrasporaceae, included in the Volvocales; others (e.g., Scagel et al., 1965) consider them distinctive enough to constitute a separate order (Tetrasporales). No one, however, seems to doubt their close affinity with and derivation from the unicellular Volvocales. I feel it appropriate to include a brief description of cell division in *Tetraspora* in this chapter on the Volvocales.

Characteristic morphological features of many species of *Tetraspora* are the "pseudocilia," pairs of fine processes (often very difficult to detect) arising from each cell and projecting from the colony. Fritsch (1935; p. 125) considers the name "pseudocilia" an unfortunate one, claiming these structures are not functionally equivalent to flagella or cilia. To me, they sound much like channels or some remnant left in the cell wall or colonial mucilage once the (functional?) flagella, appearing briefly after cell division, have been withdrawn. Cultures of a species of *Tetraspora* I have examined (Figs. 2.68, 2.69; Pickett-Heaps, 1973c) contain numerous actively motile cells whose flagellar apparatus is indistinguishable morphologically from that of *Chlamydomonas*. The quiescent cells (Fig. 2.68) are not noticeably colonial and they contain no flagella or any other structure that could be equated with pseudocilia. However, the pseudocilia of *T. lubrica* have been described in several papers (Lembi and Herndon, 1966; Lembi and Walne, 1969, 1971) and have differences in their substructure that could be related to their lack of motility. These differences include a lack of the central pair of tubules

(functional flagella lacking these tubules are known in diatoms; see page 30) and a general disorganization of the outer pairs which irregularly become single (cf. level 2 in Fig. 2.17). Several points remain unclear—for example, whether these flagella are completely immobile from the time of formation or more briefly, perhaps prior to retraction. (The cells of this species are normally immobilized by their inclusion within the colony.) Neither is the length of these pseudocilia mentioned, and one interpretation of those authors' micrographs might relate them to disintegrating flagella in the process of being retracted. To establish unequivocally a difference between pseudocilia and true flagella will (in my opinion) necessitate (1) examination of them immediately after extrusion, and (2) a definitive comment about whether they are invariably motionless even at this stage. In other respects, cells of *Tetraspora* (Fig. 2.70) are often quite similar to those of *Chlamydomonas*, and some species at least contain typical eyespots (Pickett-Heaps, 1973c).

2.3a. Cell Division

Cell division in one unidentified species of *Tetraspora* (Cat. No. LB 234, from the Indiana Culture Collection; Starr, 1964) has been described by Pickett-Heaps (1973c). An early description of division in *T. lubrica* is that of McAllister (1913). As might be expected, mitosis and cytokinesis closely resembled those described already for *Chlamydomonas* (Figs. 2.71–2.74; cf. Section 2.1a; Johnson and Porter, 1968).

There are, however, a couple of notable differences. First, in contrast to *Chlamydomonas*, no evidence was found that basal bodies always replicate before division. Second, the basal body complex is always close to at least one of the polar fenestrae of the spindle (Fig. 2.71). The possibility exists that *Tetraspora* may be "unicentric," with the other pole devoid of basal bodies, which could be formed late in mitosis. The clearest examples of centrioles forming (e.g., Fig. 2.73) were always discovered in cells at later stages of mitosis and particularly during cytokinesis. However, this possibility remains unproved. The flagella have always been withdrawn by prophase, and during prophase the nucleus moves toward the basal body complex from which increasing numbers of microtubules radiate. By prometaphase (Fig. 2.71), the basal bodies have moved along the cell surface away from the site of the original flagella insertion, which is approximately coplanar with the metaphase plate of chromosomes and the metaphase invagination in the cell surface (e.g., Fig. 2.74; cf. Johnson and Porter, 1968). Microtubules (focused at one pole upon the basal body complex) penetrate through the polar fenestrae into the spindle, which is often curved (Fig. 2.71), since both poles at this stage lie near to the cell surface. Metaphase (Fig. 2.74), anaphase (Fig. 2.72), and telophase follow as normal, with the basal body complex remaining associated with the spindle until cytokinesis. Cleavage resembles that in *Chlamydomonas* and utilizes a typical phycoplast (Fig. 2.73). At some stage during later stages of division, the protoplast rotates within the cell wall (Pickett-Heaps, 1973c). Division in daughter nuclei follows in much the same fashion, again with basal bodies associated with one (or possibly both) polar fenestrae (Fig. 2.74).

In one highly important respect these events differ from what has been reported for *Chlamydomonas* by Johnson and Porter (1968), who state quite definitely that basal bodies are not associated with that spindle. However, I have pointed out (Pickett-Heaps, 1973c) that their assertion is not unequivocally supported by the few micrographs of spindles they published, and consequently, I personally would like to see more direct evidence before accepting this statement. Such a difference between these two genera is surprising and might have considerable phylogenetic implications (Chapter 8).

Note added in proof. Coss (1974) has now shown with high voltage electron-microscopy of thick sections, that *C. reinhardi* does indeed have basal bodies very near the polar fenestrae (see also Chapter 8).

3

THE CHLOROCOCCALES

The Chlorococcales comprise a rather diverse group of families. As Fritsch (1935) points out, most attempts to classify them seem unnatural, and they are likely to be polyphyletic. Many unicellular forms closely resemble various species of *Chlamydomonas* morphologically. However, most genera show considerable or total suppression of motility during their life cycles. This trait can be simply understood in terms of the nonmotile stage undergone by organisms such as *Chlamydomonas* immediately preceding cell division (Section 2.1a), becoming greatly protracted, while motility may appear only briefly after cell division (e.g., *Chlorococcum*) or not at all (e.g., *Chlorella*) in normal circumstances. In unicellular forms (as in *Chlamydomonas*), repeated divisions in a parental cell may give rise to two, four, eight, etc. daughter cells, but in the Chlorococcales the behavior and nature of these latter are highly varied. In many genera, the daughter cells are actively motile (i.e., they are true zoospores) for a varying period of time after release. The number of zoospores formed may be relatively small (up to 128) or extremely numerous, and the zoospores may link together to form daughter colonies possessing a highly characteristic morphology (e.g., *Pediastrum, Hydrodictyon*). The released daughter cells in other genera are nonmotile miniatures of the parental cell (e.g., *Kirchneriella, Tetraedron*), and the number of these "autospores" formed is fairly constant in some genera and highly variable in others. The autospores themselves may be colonial (i.e., linked together before release from the parental cell, as in *Scenedesmus*). However, the mechanisms by which colonies are formed may be quite different, since in organisms such as *Pediastrum* and *Hydrodictyon* free-swimming zoospores actively unite to form a colony and retract their flagella, while in *Scenedesmus,* for example, the nonmotile autospores emerge from the parental cell already linked together. Sexual reproduction, reported in a number of genera, is usually rather simple, isogamy being the general rule; oogamy is rare (Moestrup, 1972). Gametes, too, may develop independently without undergoing conjugation (e.g., forming azygospores in *Hydrodictyon;* Section 3.7b), and this also indicates a primitive stage in the development of sexual reproduction (Fritsch, 1935).

Although the ultrastructure of cells undergoing vegetative and sexual reproduction has been investigated in only a few genera, the observations have revealed several intriguing phenomena. In particular, the phycoplast is the common cytokinetic apparatus in all species so far described, and this system has undergone various modifications during evolution in keeping with the increasing complexity of life styles adopted by advanced members of the group (e.g., *Hydrodictyon*).

The order Chlorococcales can be subdivided as follows (Fritsch, 1935); the genera named in parentheses will be described in some detail in the following pages:

Figures 3.1, 3.2
CHLOROCOCCUM ELLIPSOIDEUM

3.1. Large vegetative cell before autospore or zoospore formation. The nucleus is big, and the cells are apparently acentric. × 17,000.

3.2. Cleavage of parental cell into flagellate (*f*), naked zoospores. × 15,000.

SOURCE. Dr. H.J. Marchant, unpublished micrographs.

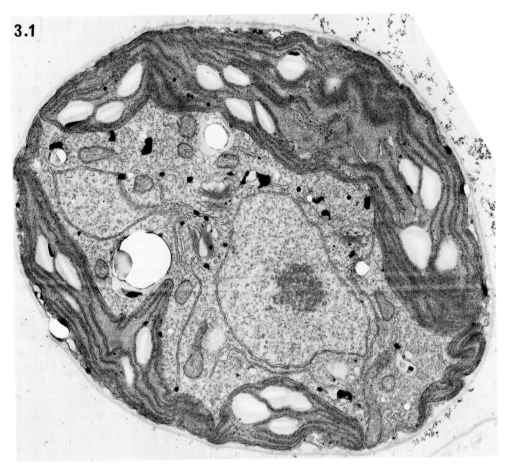

3.1

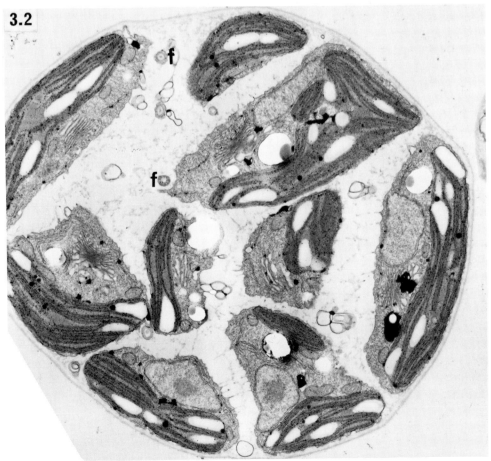

3.2

Family 1: Chlorococcaceae *(Chlorococcum)*
Family 2: Eremosphaeraceae
Family 3: Chlorellaceae *(Chlorella, Tetraedron)*
Family 4: Oocystaceae
Family 5: Selenastraceae *(Kirchneriella, Ankistrodesmus)*
Family 6: Dictyosphaeriaceae
Family 7: Hydrodictyaceae *(Pediastrum, Hydrodictyon, Sorastrum)*
Family 8: Coelastraceae *(Scenedesmus)*

3.1. Chlorococcum

Chlorella and *Chlorococcum* are superficially very similar genera; the key feature distinguishing them is the presence or absence of a motile stage in the life cycle. *Chlorella* has never been observed to produce motile cells, whereas vegetative multiplication in *Chlorococcum* can result in either autospore or zoospore production. These two organisms together illustrate the origin of the chlorococcalean genera in the siphonous lines of evolution (see Chapter 8). A tendency of chlamydomonad types of progenitors to lose motility (cf. the Tetrasporales), coupled with a tendency for their nuclei to divide without cytokinesis necessarily following immediately afterward, can easily account for the origin of the diverse genera now lumped together in the Chlorococcales. The mention of *Chlorococcum* in this book, necessarily brief owing to lack of relevant information, is useful in this regard.

Vegetative cells of *Chlorococcum* may become aggregated into gelatinous masses (cf. the Tetrasporales and the palmella form of *Chlamydomonas*). As in certain other genera, the fine structure of the pyrenoid may be a useful taxonomic feature (Brown and McLean, 1969). The cells (Fig. 3.1) often become quite large before zoospore (Fig. 3.2) or autospore formation. The vegetative cells of *C. ellipsoideum* examined in our laboratory are acentric, i.e., they have never been found to contain centrioles (Marchant, unpublished information). It therefore appears as if the basal bodies necessary for flagellar formation (Fig. 3.2) must arise de novo during zoosporogenesis. It will be most interesting to discover whether mitosis is different preceding zoospore or autospore formation (e.g., whether spindles are acentric or centric, and whether the cytoplasm of a mitosing cell rotates prior to zoosporo-

Figures 3.3, 3.4
FORM OF THE CHLOROPLAST AND MITOCHONDRION IN CHLORELLA FUSCA

3.3.   Reconstruction from serial sections of the chloroplast *(c)* and the complex single mitochondrion. The centrioles are indicated by the arrow. In front of them is the large nucleus (removed from the model for this picture).

3.4.   Reconstruction of part of the mitochondrion (all other organelles removed) from another cell. Individual thin sections suggest that the cell contains numerous mitochondria. Only by reconstructions such as this is the complex reticulated form revealed.

SOURCE. Drs. A.W. Atkinson, Jr., P.C.L. John, and B.E.S. Gunning, published in *Protoplasma 81*, 77 (1974).

3.3

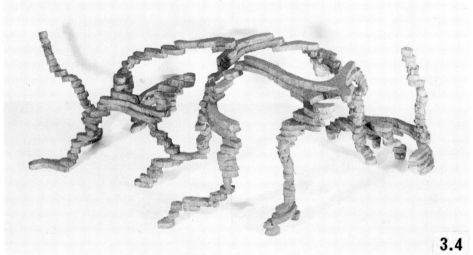

3.4

genesis, as it does in *Chlamydomonas* and *Tetraspora*). Presumably, zoospores are naked (Fig. 3.2), whereas autospores may be enclosed by a wall before release.

3.2. Chlorella

Chlorella has never been observed to produce motile cells (zoospores) during vegetative reproduction. Instead, the daughter cells are small replicas of their nonmotile parental cells (i.e., they are true autospores—cf. *Kirchneriella* and *Tetraedron;* Sections 3.3, 3.5). This genus exemplifies the suppression of motility typical of the Chlorococcales in formerly motile progenitors perhaps resembling *Chlorococcum* or *Chlamydomonas.*

Chlorella is a common and well-known organism that has been widely used for biochemical and physiological studies (e.g., on photosynthesis) because of its ease of culture and ready availability. Some species grow symbiotically in various organisms (e.g., *Hydra, Paramecium,* and some other protozoa and lichens). The cells are very small and structurally quite simple, containing a relatively large cup-shaped chloroplast and a tiny nucleus. The Golgi body is often quite conspicuous. Gunning and his collaborators (personal communication) have done some reconstruction of whole cells of *Chlorella* from serial sections. One interesting discovery they have made—which reveals how misleading thin sections can be—is that each cell studied contains only one mitochondrion (Figs. 3.3, 3.4), which is highly branched and fenestrated, permeating the cell. It lies close to and mostly inside the large chloroplast.

3.2a. Cell Division

As is typical in the Chlorococcales, vegetative reproduction is effected by cleavage of enlarged cells into a number of smaller autospores, usually numbering between 2 and 16. Because of the small size of the nuclei and the rapidity of nuclear division, mitotic stages are difficult to visualize, leading Wanka (1968) to conclude that nuclei elongate and split without utilizing a spindle apparatus (see also Wanka and Mulders, 1967). This rather risky deduction is incorrect, even allowing for the variability in spindle structure mentioned below. Wanka (1965) earlier

Figures 3.5–3.8
CHLORELLA PYRENOIDOSA (ACENTRIC STRAIN)

3.5. Premitotic cell. The peripheral chloroplast has cleaved. On the opposite side of the cell are some microtubles (arrow, perhaps like the metaphase band of *Chlamydomonas;* Fig. 2.26), toward which the nucleus is drawn out. × 30,000.

3.6. Metaphase. Spindle is closed, and in this particular strain (see text), the interphase and mitotic cell is acentric. Note the invagination in the cell opposite the cleavage in the chloroplast (as in Fig. 3.5), very like that observed in *Chlamydomonas* (Fig. 2.26). × 27,000.

3.7. The end of the first cleavage inside the parental wall. Whether this cell would have divided further before autospore release is unknown. × 27,000.

3.8. Four autospores, enclosed by their own wall, still within the parental wall. This strain does not seem to possess the trilaminar (sporopollenin) layer illustrated in Figures 3.11–3.15. × 30,000.

SOURCE. Dr. H.J. Marchant, unpublished micrographs.

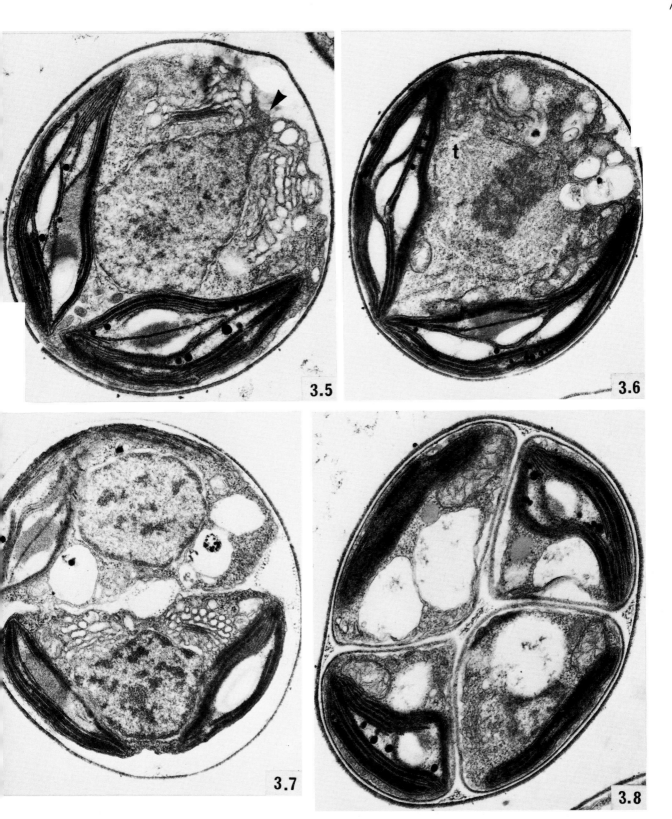

showed that, prior to the successive series of mitoses which eventually gives rise to these autospores, the nucleus may contain all the DNA required to carry out the divisions. Thus, in cells about to form 16 nuclei, the premitotic nucleus apparently contains 16 times the DNA of the normal interphase nucleus, and so the successive rapid mitoses appeared to be reductive in nature. However, Wanka and Mulders (1967) also have evidence that nuclear divisions follow each phase of DNA replication. John et al. (1973) say that the period of DNA synthesis (the S period of the cell cycle) is single and continues during the first mitosis, up until the second and final mitosis in *C. fusca.* This matter has also been mentioned in Section 2.2a.

Before mitosis in *C. pyrenoidosa,* the pyrenoid diminishes in size and divides, while a cleavage furrow appears in the chloroplast, predicting the eventual plane of cytokinesis (Fig. 3.5; Marchant, unpublished data). Opposite this cleavage, a ridge on the nucleus approaches the cell surface, near which microtubules appear (Fig. 3.5). The spindle itself, although small, appears quite typical (Fig. 3.6). The cell may possess a metaphase band of microtubules and the invagination in its surface (Figs. 3.5, 3.6) analogous to those of mitotic cells of *Chlamydomonas* and *Tetraspora* (Sections 2.2a, 2.3a). The nucleolus disperses and the nuclear envelope remains intact throughout mitosis except for the development of typical polar fenestrae. We have not encountered a "perinuclear envelope" (cf. Section 3.3a) in our strain (211-8b) of *C. pyrenoidosa* from the Cambridge Culture Collection, but Wanka and Mulders (1967), using the same strain obtained from the University of Gottingen, illustrate an unmistakable perinuclear envelope (their Figs. 4A,B) and later confirm its existence (Wilson, Wanka, and Linskens, 1973). Furthermore, we have never found centrioles in this strain (Figs. 3.5–3.8). In contrast, Atkinson et al. (1971), working with strain 211-8p of the same species, have illustrated typical persistent centrioles in vegetative cells which during cell division are associated with both the spindle and the cleavage (see below) apparatus. For example, these centrioles duplicate and separate at prophase, establishing the poles of the spindle. This interesting variation between the two strains points up again the unimportance of centrioles in both mitosis and the organization of cleavage and other cyto-

Figures 3.9, 3.10
CYTOKINESIS IN CHLORELLA FUSCA (CENTRIC STRAIN)

3.9. Daughter nuclei coming together in the center of the cell. Centriole complexes *(cn)*—absent in some strains of *Chlorella* (cf. Figs. 3.5–3.8)—are moving together (cf. *Kirchneriella, Tetraedron*). A typical phycoplast *(ph)* system of microtubules is growing across the cell from between nuclei. × 38,000 approx.

3.10. As for Figure 3.9, but now a cleavage furrow is growing through the phycoplast. × 48,000 approx.

SOURCE Drs. A.W. Atkinson, B.E.S. Gunning, P.C.L. John, and W. McCullough, unpublished micrographs.

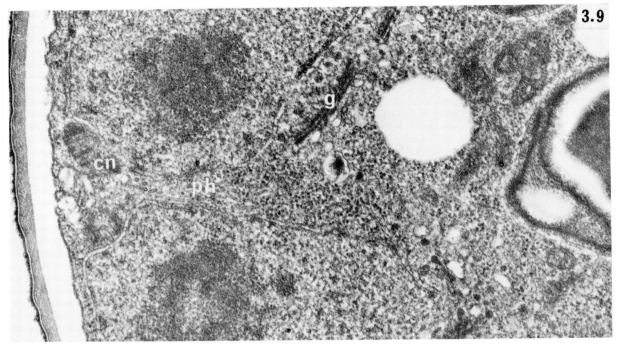

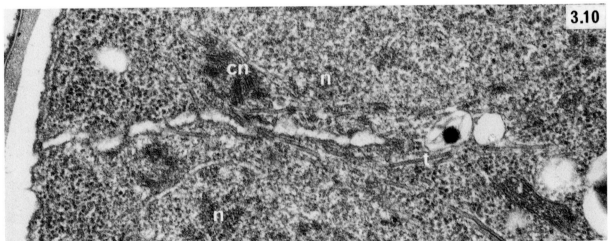

plasmic microtubules. Furthermore, it confirms the tendency of members of the Chlorococcales to suppress motility, a trend that could easily be accompanied by a tendency for the organisms to either lose their centrioles entirely during the cell cycle or not express structurally the information necessary to form them (as appears to be the case with *Chlorococcum;* Section 3.1).

After mitosis, the cell cleaves (Figs. 3.7, 3.9, 3.10). Later, the two daughter cells may go on to divide again (Fig. 3.8), in which case the axes of the two spindles of the daughter nuclei are parallel to the first cleavage. Phycoplasts (Figs. 3.9, 3.10) are present during cleavage (Atkinson et al., 1971; Wilson, Wanka, and Linskens, 1973), but it is not clear whether cytokinesis is effected by the outer cell membrane furrowing inward (Soeder, 1965) or whether fusion of vesicles derived from the prominent Golgi bodies gives rise to the cleavage furrow. Bisalputra, Ashton, and Weier (1966) and Wanka (1968) suggest the latter to be likely. As might be expected from the foregoing description, cell division in *Chlorella* is sensitive to disruption by colchicine (Wanka, 1965). Wilson, Wanka, and Linskens (1973) say that cytokinesis does not commence until the tetranucleate stage in cells that will ultimately form 16 daughter cells. When only four autospores are to be formed, however, partial cleavage may precede the second mitosis (Gunning, personal communication), as is the case with *Kirchneriella* (Section 3.3).

There are two important reasons why one might expect *Chlorella* to show morphological variations described above: (1) the organism probably has a haploid life cycle like many other unicells and (2) it shows no evidence of undergoing sexual reproduction. These two facts mean that the organisms have no diploid buffering of genetic changes and no chance of accomplishing genetic recombination. Genetic buffering allows the accumulation of neutral or even deleterious genetic alterations in diploid organisms, thus tremendously increasing their potential variability (Raper and Flexer, 1970). Asexual haploid organisms such as *Chlorella* circumvent this disadvantage to some extent by undergoing rapid vegetative reproduction, so that any advantageous mutations can quickly be exploited. It follows that strains and even clones derived from unicells will accumulate a number of nonlethal genetic alterations over a period of time. As anyone who has

Figures 3.11–3.15
TRILAMINAR (SPOROPOLLENIN) LAYER IN THE WALL OF CHLORELLA FUSCA

3.11. Whole mount of a cell wall after acetolysis (treatment with concentrated sulfuric acid and acetic anhydride at 90°). This residue is extremely inert chemically and represents the outer layer of the wall, identified as sporopollenin. × 5,000.

3.12. Trilaminar plaques on the surface of autospores. In surface view (large arrow), these appear to grow and fuse with one another; see adjacent cell on the right. Trilaminar layer of the mother cell wall is indicated by the small arrow. × 30,000.

3.13. As for Figure 3.12, showing plaques (arrow) apparently growing and fusing outside the plasmalemma of the autospores. × 67,000.

3.14. Later stage of wall formation. The inner, fibrillar layer of the autospore wall *(w)* is being secreted inside the trilaminar layer. The inner layer of parental wall (arrow) is being broken down. × 67,000.

3.15. As for Figure 3.14, but the inner layer of the parental wall has been completely broken down leaving only its outer layer, which ruptures to release the autospores, now with their own wall system complete. × 53,000.

SOURCE. Drs. A.W. Atkinson, Jr., B.E.S. Gunning, and P.C.L. John, published in *Planta 107.* 1 (1972).

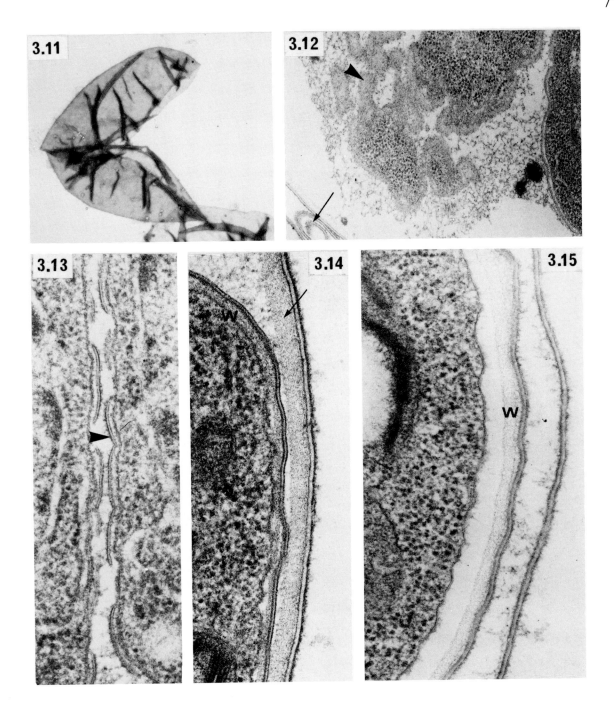

worked with *Chlorella* knows, it is highly variable physiologically. Fritsch (1935, pp. 182 et seq.), mentions that "the stability of [these] cultures is only illusory, since new micromutations are constantly arising, although masked in the pure line derived from a single cell." Some variability at the ultrastructural level is hardly surprising, and the presence or absence of centrioles, the perinuclear envelope, etc. could be associated with the appearance of such micromutations.

Atkinson, Gunning, and John (1972), investigating further the difficulty they experienced in trying to make naked protoplasts from *C. fusca*, made an interesting and important discovery. These cells were enclosed by a trilaminar layer on their outermost surface, inside of which was the cell wall proper, the latter being digestible with cellulose and other polysaccharidase preparations. The outer sheath was extremely inert chemically, and since it was the only known organic component of plant walls that could withstand acetolysis (treatment with concentrated sulfuric acid and acetic anhydride at about 95°; see Fig. 3.11), they identified it as sporopollenin. This complex material is typically found in the walls of spores and pollen of higher plants, contributing greatly to their viability and their preservation (e.g., in peat bogs) over long periods of time.

The way this sporopollenin sheath arises is of interest. Before and during mitosis, granular material accumulates in the lumen of endoplasmic reticulum. Similar material accumulates between autospores after cleavage of the cytoplasm. Wall secretion proper starts with the appearance on the cell surface of trilaminar plaques (Figs. 3.12, 3.13), which appear to grow, perhaps by some self-assembly process, and fuse with one another until they enclose each autospore (Fig. 3.14). Next, the fibrillar wall is secreted under this layer (Figs. 3.14, 3.15). During these events, the fibrillar layer of the mother wall is broken down until it has almost entirely disappeared, a common phenomenon in autospore-forming cells, leaving intact the original trilaminar layer of the mother cell (Figs. 3.12, 3.15), which later ruptures to release the autospores. Atkinson et al. (1972) emphasize that the autospores are never exposed directly to the external environment without being enclosed by either (or both) trilaminar sheaths. They speculate that the broken-down components of the fibrillar mother wall

Figures 3.16–3.18
MITOSIS AND CLEAVAGE IN KIRCHNERIELLA LUNARIS

3.16. Metaphase. Daughter nuclei after the primary mitosis are lying on either side of the partially formed cleavage furrow (*cf;* see Fig. 3.19). Note the typical chlorococcalean spindles, enclosed by the perinuclear envelope *(pe)* with centrioles *(cn)* situated at the polar fenestrae. In this organism, the perinuclear envelope becomes interposed between the nucleus and the Golgi bodies *(g)* during mitosis (cf. *Tetraedron;* Fig. 3.27). × 44,000 approx.

3.17. Telophase in one of two daughter nuclei. The interzonal spindle is dispersing along with the perinuclear envelope. Note particularly the segment of nuclear envelope (small arrow) becoming interposed between the re-forming nucleus and the centriole complex (large arrow), thereby sealing the polar fenestra. Some microtubules are already back near the primary cleavage furrow. × 33,000 approx.

3.18. Higher magnification micrograph of the phycoplast and cleavage furrow in Figure 3.19. There seem to be two sets of tubules in the phycoplast; one lines the furrow across the cell, and the other encircles the cell in the plane of cytokinesis. Both centriole complexes *(cn)* are characteristically sited at the phycoplast. × 42,000.

Inset. Cross section through the very short centriole, in this case one at the fenestra of an anaphase spindle. Many of the outer tubules appear only as doublets instead of the usual triplets. × 150,000.

SOURCE. Author's micrographs. Figures 3.17, 3.18 published in *Protoplasma 70*, 325 (1970).

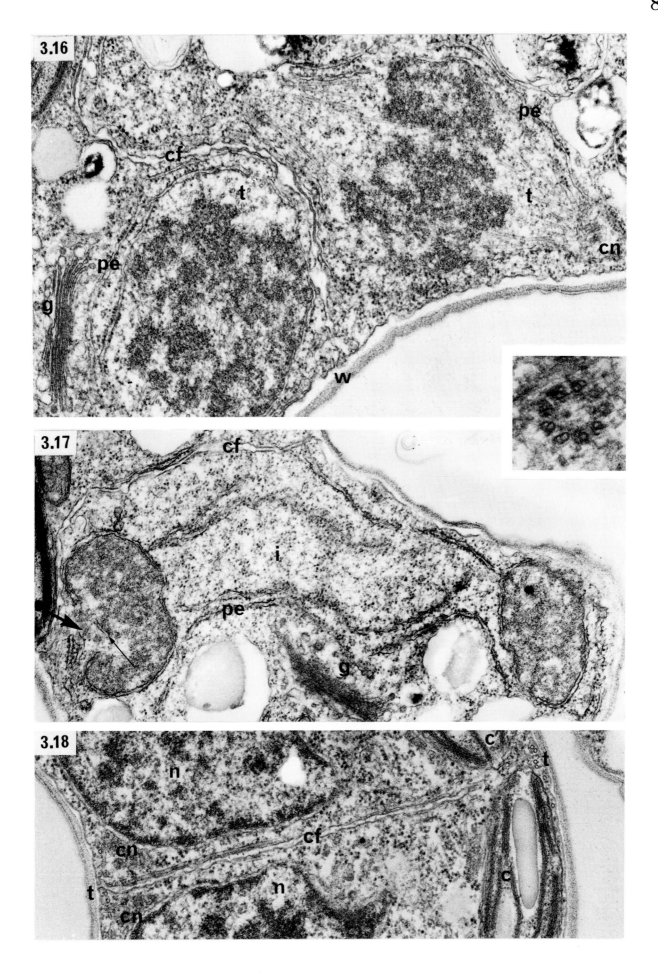

may be reutilized by the autospores. These authors found a similar component in the wall of another strain of *Chlorella,* and also in *Scenedesmus* and *Prototheca* (a pathogen in man resistant to many drugs and lysosomes), but it was absent in several other species of *Chlorella* (e.g., see Figs. 3.5–3.8). Such a sporopollenin layer might well prove common in Green Algae. As will be seen (Section 3.6a), the outer sheath of this species of *Scenedesmus* is assembled in a fashion similar to that described above, and *Pediastrum* also possesses a similar resistant sheath (Section 3.8a). The synthesis of sporopollenin is probably related to the ability of such cells to produce secondary carotenoids, and *Chlorella* may be a useful organism to study the biosynthesis of these compounds.

3.3. Kirchneriella

Kirchneriella lunaris is a small, crescent-shaped member of the family Selenastraceae. The simple, uninucleate vegetative cells have a parietal chloroplast lying around their outer edge which contains a pyrenoid at one tip of the cell (Fig. 3.19). This organism provides a classic example of vegetative reproduction by autospore formation. The nucleus undergoes two successive mitoses, and then the cell cleaves to give rise to four autospores (Figs. 3.21–3.23) which are miniatures of their parental cell. (In some conditions, more may be formed by another mitotic and cytokinetic cycle.) Sexual reproduction or motile stages have not been reported to my knowledge (Pickett-Heaps, 1970b).

3.3a. Cell Division

The major events leading up to the liberation of the four autospores can be summarized as follows (Pickett-Heaps, 1970b):
1. Primary mitosis
2. Primary cleavage (Fig. 3.19), which bisects the cell, but which is variable in extent and probably rarely complete
3. Synchronous secondary mitoses (Fig. 3.16) in the daughter nuclei arising from the primary mitosis
4. Secondary cleavages (Fig. 3.20), during which the partly formed primary cleavage is completed and two new cleavages appear between the daughter nuclei arising from the

Figures 3.19, 3.20
CLEAVAGE IN KIRCHNERIELLA LUNARIS

3.19.   Binucleate cell undergoing the primary cleavage which is rarely completed until after the daughters have divided again (Fig. 3.16). Note the centriole complexes close to each other on either side of the phycoplast (Fig. 3.18). × 21,000.

3.20.   Tetranucleate cell undergoing the cleavages that would have given rise to four autospores (Fig. 3.23). The centriole complexes are again close to the phycoplasts. The middle, primary cleavage has obviously not completely cut the chloroplast after the first mitosis (cf. Fig. 3.19). Note that there is apparently a second small pyrenoid forming in the peripheral chloroplast (arrow). × 19,000.

SOURCE. Author's micrographs, published in *Protoplasma 10,* 325 (1970).

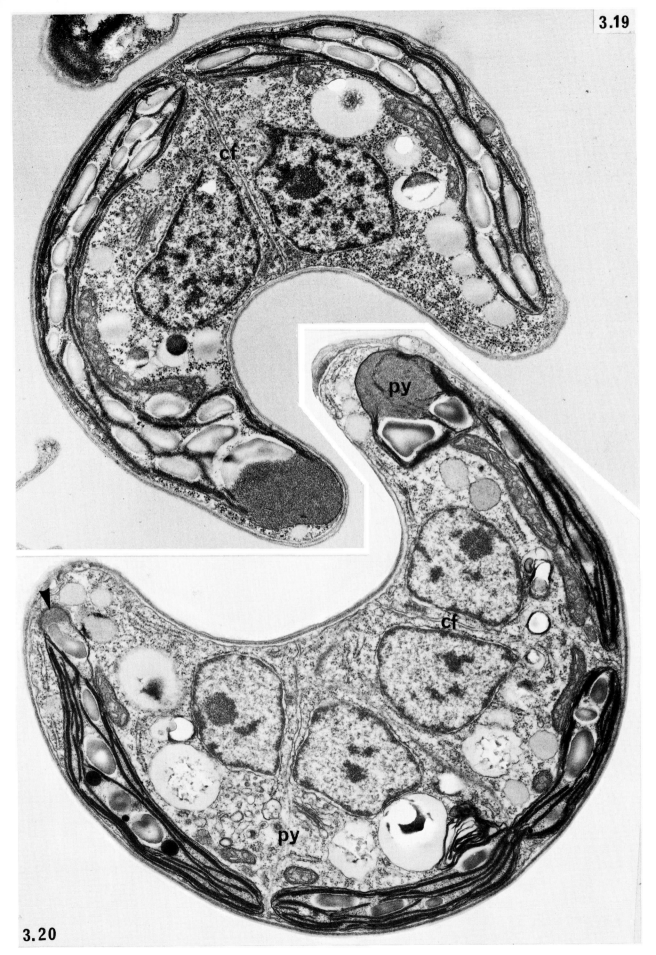

secondary mitoses; these partition the original cell into four

5. Secretion of the wall around the newly formed autospores (Fig. 3.24), followed by their release from the parental wall (Fig. 3.23).

Mitosis and cleavage, whether primary or secondary, involve analogous cytoplasmic systems and can be described under two headings: mitosis and cytokinesis.

MITOSIS

Centrioles have not been observed in interphase cells (of *K. lunaris*), which are, therefore, very likely acentric (Pickett-Heaps, 1970b). However, centrioles appear near the nuclei as they go into prophase. These organelles seem to be degenerate or rudimentary; they are very short and always display some doublet tubules among their triplets (*inset* Figs. 3.16, 3.17). I do not believe this observation can be ascribed merely to an artifact of thin sectioning, since I have never heard of centrioles elsewhere displaying such a consistent variation in structure. These centrioles are surrounded by ill-defined amorphous material, and the centriole complex forms the polar centers during mitosis (Fig. 3.16).

As the cells go into prophase the nucleolus disperses, the chromosomes condense, and the two centriole complexes separate to opposite sides of the nucleus. A spindle-shaped sheath of extranuclear microtubules now develops, focused upon the centriole complexes. A layer of endoplasmic reticulum begins to envelope these extranuclear microtubules, and it concurrently becomes interposed between the nuclear envelope and the Golgi bodies (Fig. 3.16), which always lie adjacent to one another during interphase. Thus, the extranuclear spindle is soon almost entirely enclosed by this perinuclear envelope (as in Fig. 3.31), which then persists, as in some other members of the Chlorococcales, throughout the rest of mitosis.

During prometaphase, as the extranuclear sheath of microtubules disappears, microtubules penetrate the nucleus through fenestrae which appear in the nuclear envelope adjacent to the centriole complexes. These complexes also enter the nucleus, coming to lie just inside the polar fenestrae. The nuclear envelope persists intact,

Figures 3.21–3.24
CLEAVAGE AND AUTOSPORES IN KIRCHNERIELLA LUNARIS

3.21–3.23. Live cells. In Figure 3.21, the first mitosis and partial cleavage has finished; the clear area in the center of the cell contains two nuclei. In Figure 3.22, the cytoplasm has cleaved into four lunate autospores, still contained within the parental wall. In Figure 3.23, the parental wall has almost completely broken up to release the autospores. All × 1,750.

3.24. Autospores prior to release. Each has secreted its own wall, and the parental wall is thin and probably undergoing dissolution. × 14,000.

SOURCE. Author's micrographs, published in *Protoplasma 10*, 325 (1970).

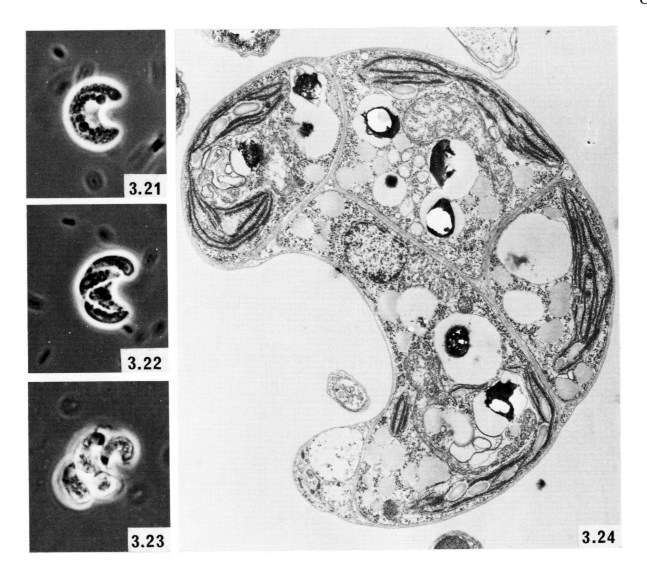

3.21

3.22

3.23

3.24

except for the fenestrae, throughout mitosis. Metaphase (Fig. 3.16) and anaphase are quite typical. By telophase, the spindle has undergone considerable elongation, and the nuclear envelope contracts tightly around the chromosomes, separating them from the centriole complexes (Fig. 3.17) and the remnants of the spindle apparatus, which then disappear.

CYTOKINESIS

After the primary mitosis, the widely separated daughter nuclei move to the middle of the cell at its inside edge and become appressed to one another (Fig. 3.19). Their attendant centriole complexes always now move adjacent to one another, close to the wall. A typical phycoplast is then formed (Fig. 3.18), with microtubules proliferating between the nuclei in the plane of future cleavage. Two rather distinct arrays of tubules are probably involved, one (composed of usually between 2 and 10 individual tubules) encircling the cell near the wall, and the other more randomly extended across the plane of the incipient division thus defined. The peripheral chloroplast often develops a pronounced constriction where the plane of division passes through it (Fig. 3.18), which is probably not completed at this stage. Meanwhile, a furrow grows through these phycoplast microtubules; its origin and mode of formation are not obvious from electron microscopy, but I believe that it is an invagination of the plasmalemma (Fig. 3.18). This primary cleavage is probably always incomplete. The phycoplast microtubules disappear from around it during the secondary mitoses (Fig. 3.16), suggesting that they may be incorporated into the newly forming spindles.

Much the same sequence of events follows the secondary mitoses. The two pairs of nuclei, usually separated by the partly formed primary cleavage, move to the center of the cell, and phycoplast microtubules appear as before between them all (i.e., adjacent to the incomplete primary cleavage as well; Fig. 3.20). Cleavages now proceed to completion, after which walls are secreted around the four autospores (Fig. 3.24). The original pyrenoid seems to persist throughout these events, and new pyrenoids can sometimes be detected within the chloroplasts of the other three autospores at this stage. Finally, the parental wall disintegrates to release the four crescent-shaped daughter cells (Fig. 3.23).

Figures 3.25–3.28
REPRODUCTION BY AUTOSPORES IN ANKISTRODESMUS FALCATUS

3.25. Longitudinal section through cleaving cell. Two of the cell's four nuclei (n) are visible in this section, and the site of the three cleavage furrows is indicated by three sets of arrows. × 9,600 approx.

3.26. Group of individual cells showing range of sizes. Two cells are clearly binucleate (arrows). × 1,300 approx.

3.27. Group of autospores still enclosed within their parental wall. × 1,300 approx.

3.28. Cleavage furrow (cf) growing among phycoplast microtubules (t) in a cell undergoing cleavage (as in Fig. 3.25). × 51,000 approx.

SOURCE. Author's micrographs. Figures 3.25, 3.28 published in *Cytobios* 5, 59 (1972).

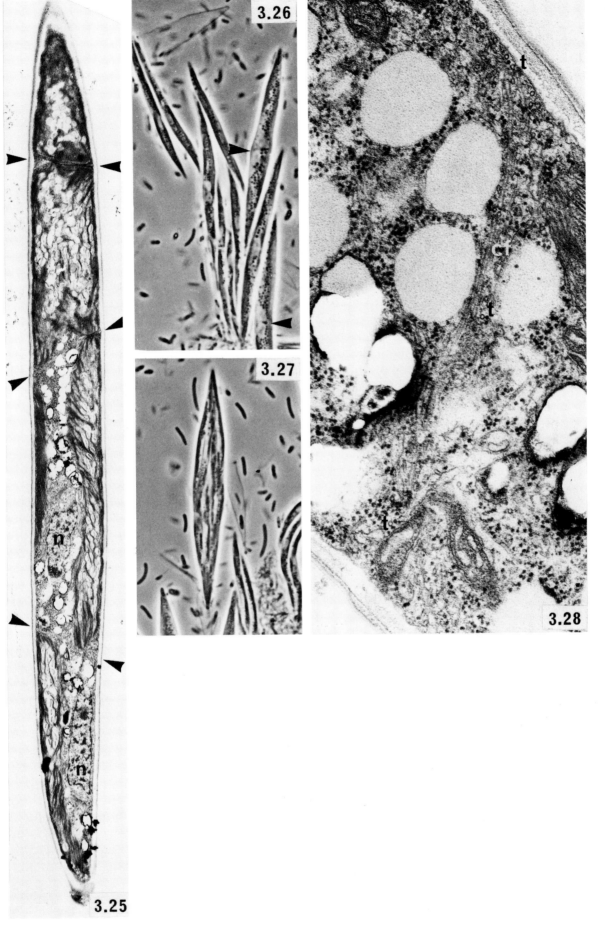

The significance of the enigmatic centrioles in these mitoses and cleavages remains unclear. For reasons argued elsewhere (Pickett-Heaps, 1969a, 1970b), I cannot believe that centrioles per se play any important role in the elaboration of the spindle or phycoplast microtubular systems. Since they apparently are formed in *Kirchneriella* only during mitosis and cell cleavage, and since they also appear to be somewhat degenerate in structure, I conclude that the appearance of these centrioles represents a secondary manifestation of the activities of the spindle-organizing centers. I have suggested previously that in animal cells, for example, spindles are formed by microtubule-organizing centers (MTOCs) intimately associated with inactive centrioles, and that when these MTOCs duplicate and separate during prophase, the centrioles follow suit. Such a hypothesis follows from the necessity to explain why the centriole is not needed for spindle formation in so many plant and other cells (Pickett-Heaps, 1970b). In *Kirchneriella,* then, we may have a variation in this behavior, perhaps an inheritance from motile ancestors in which flagellate zoospores (i.e., requiring basal bodies) used to be formed after cell cleavage. The brief appearance of ancillary structures near persistent centrioles only during mitosis and cell cleavage will be shown in *Tetraedron* (Section 3.5a) and *Microspora* (Section 4.5a). I interpret these structures also to represent elements of a flagellar apparatus formed after cleavage but not usually utilized by the cells concerned. It will be remembered that in many extant genera of the Chlorococcales, motile zoospores are released upon cell cleavage. If *Kirchneriella* does in fact possess the capability of forming motile cells, further elaboration of its rudimentary centrioles would appear necessary. Examples of centrioles forming de novo within diffuse centrosomal structures involved in spindle formation have been reported in the diatom *Lithodesmium* (Manton, Kowallik, and von Stosch, 1970), the protozoan *Labyrinthula* (Perkins, 1970; Porter, 1972), and during spermatogenesis in some bryophytes (Moser and Kreitner, 1970).

3.4. Ankistrodesmus

The cells of *Ankistrodesmus* are long and needle-shaped (Fig. 3.26), although many species

Figures 3.29, 3.30
INTERPHASE AND EARLY MITOSIS IN TETRAEDRON BITRIDENS

3.29.   Scanning electron micrograph showing morphology and size range of vegetative cells. The smallest individuals are recently released autospores. × 2,900.

3.30.   Early prophase. The numerous cytoplasmic microtubules are focused upon the polar centrioles. The perinuclear envelope *(pe)* is forming around the extranuclear sheath of tubules *(t).* × 33,000.

SOURCE. Author's micrographs, published in *Ann. Bot. 36,* 693 (1972).

3.29

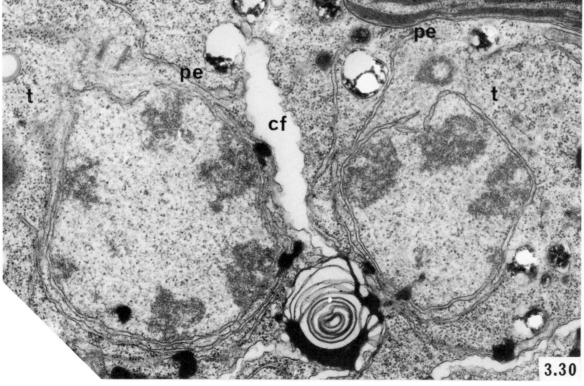

3.30

are pleomorphic, their shape being dependent upon growth conditions. As with *Kirchneriella*, motile stages and sexual reproduction have not been observed in this genera to my knowledge.

3.4a. Cell Division

I have briefly looked at cell division in *A. falcatus* for a comparison with its close relative, *Kirchneriella*. My results are regrettably incomplete (for example, I have been able to examine only two mitotic nuclei), but suggest that centrioles are transitory structures in this organism too and that cytokinesis is undoubtedly similar. However, after the uninucleate vegetative cells have undergone their first mitosis, partial cleavage, as in *Kirchneriella*, does not seem to occur. Cleavages in quadrinucleate cells were common in my samples (Fig. 3.25). Typical phycoplasts were unquestionably involved (Fig. 3.28). These were often quite steeply inclined to the longitudinal cell axis, and subsequent cytoplasmic cleavage was also longitudinally oriented. Thus, the long, slender autospores released (usually four per parental cell) are miniatures of their parent (Fig. 3.27).

3.5. Tetraedron

Cells of *Tetraedron* are usually polyhedral (Fig. 3.29) in shape, with their corners bearing simple pegs or more complex spines; some species are flattened and angular. Vegetative cells are relatively simple and contain a parietal chloroplast which may be segmented prior to cell cleavage, and which in turn contains the single large pyrenoid. The pyrenoid disappears during or after the final stages of cytoplasmic cleavage, as is the case with many other chlorococcalean genera(e.g., *Hydrodictyon, Scenedesmus, Pediastrum*). In cultures, the polyhedral cells vary considerably in size (Fig. 3.29), the larger ones containing many (16–32) nuclei (Fig. 3.34). Reproduction normally proceeds by autospore formation whereby vegetative cells cleave into (usually) uninucleate individuals which are, or soon become, polyhedra like the parental cells. The number of autospores released varies considerably, increasing to around 32 when growth conditions are especially favorable (Smith, 1918). In some species investigated in detail, reproduction by akinetes (Davis, 1966a) or motile zoospores (Starr, 1954b) has been observed.

Figures 3.31–3.33
MITOSIS IN TETRAEDRON BITRIDENS

3.31. Prophase nuclei. The perinuclear envelope now completely encloses the extranuclear sheath of microtubules (small arrows), except at the Golgi bodies *(g)*. Few microtubules are left in the cytoplasm. The nuclei both contain small bundles of tubular components (large arrows) which are not related to cytoplasmic microtubules. See text. × 36,000.

3.32. Synchronous metaphases. The cell is partially divided by an incomplete cleavage furrow (arrow). × 8,200.

3.33. Typical metaphase spindle with microtubules focused toward the centrioles located at the polar fenestrae. The perinuclear envelope *(pe)* is two layers thick at one point. × 28,000.

SOURCE. Author's micrographs, published in *Ann. Bot. 36*, 693 (1972).

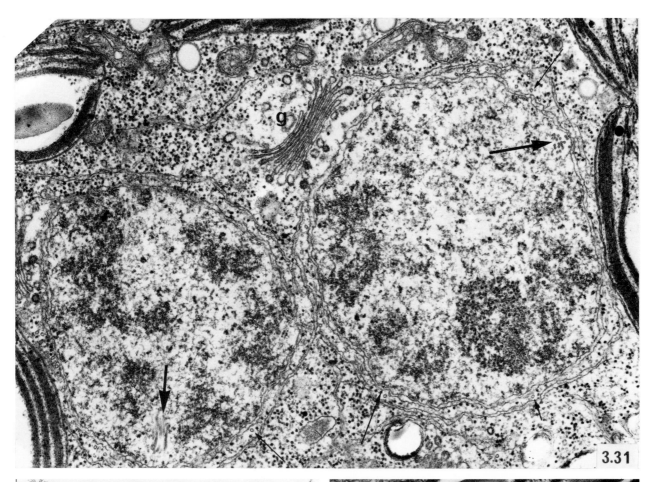

3.31

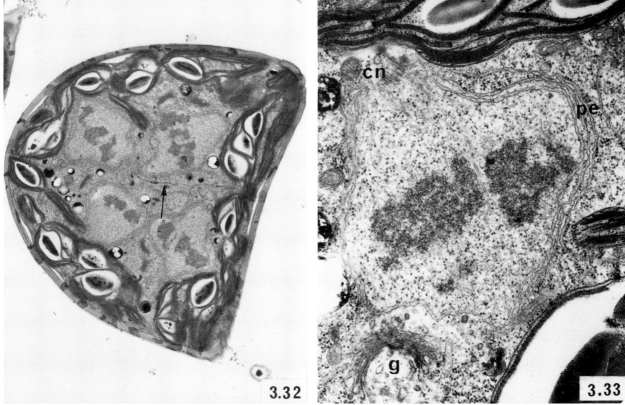

3.32

3.33

In some members of the Hydrodictyaceae, polyhedral cells arise from the zoospores released by germinating zygospores or azygospores (Sections 3.7, 3.8). These polyhedra are initially uninucleate, but repeated nuclear divisions accompanied by considerable cell enlargement soon render them coenocytic; subsequent cytoplasmic cleavage in the polyhedra gives rise to numerous motile zoospores which later link up characteristically to form colonies. These polyhedral stages often resemble the vegetative cells of *Tetraedron,* and indeed, several early descriptions of "species" of *Tetraedron* can probably be more correctly ascribed to other organisms (Fritsch, 1935).

Ultrastructurally, vegetative cells (of *T. bitridens;* Pickett-Heaps, 1972b) are usually multinucleate, and pairs of centrioles, persistent throughout the cell cycle, are associated with the nuclear envelope (Fig. 3.34). In multinucleate cells, one or more partially formed cleavage furrows associated with arrays of phycoplast microtubules (see below) usually run inward from the outer cell membrane (Fig. 3.32). The chloroplast is usually parietal, but quite often elements of it penetrate among the nuclei in association with adjacent cleavage furrows (Fig. 3.32). One distinctive and very common cytoplasmic inclusion consists of loosely arranged bundles of microtubule-like components (Fig. 3.34, large arrow). The constituent tubules themselves are appreciably smaller than cytoplasmic microtubules (ca. 180 Å versus 250 Å) and seem to consist of a helical array of subunits. These tubule bundles are found inside nuclei (Fig. 3.31), in mitotic spindles, and in the cytoplasm. Their significance is not known, and there is no reason to associate them in any way with the ubiquitous and typical cytoplasmic microtubules.

3.5a. Cell Division

Cell division and autospore formation have been studied under the light microscope in *T. minimum* by Smith (1918) and using electron microscopy in *T. bitridens* by Pickett-Heaps (1972b).

MITOSIS

Mitosis is highly synchronous within cells (Fig. 3.32) and commences as the centrioles replicate and become associated with increasing numbers of microtubules (Fig. 3.30). The pairs

Figure 3.34
CYTOPLASMIC CLEAVAGE IN TETRAEDRON BITRIDENS

Cleavage furrows *(cf),* growing through arrays of phycoplast microtubules, have rather irregularly begun to cleave the cell. Elsewhere, phycoplasts *(ph)* are forming between nuclei near centrioles (small arrows). There is a large bundle of tubular structures (cf. Fig. 3.31 and see text) at the top of the micrograph (large arrow). × 14,000.

SOURCE. Author's micrograph, published in *Ann. Bot. 36,* 693 (1972).

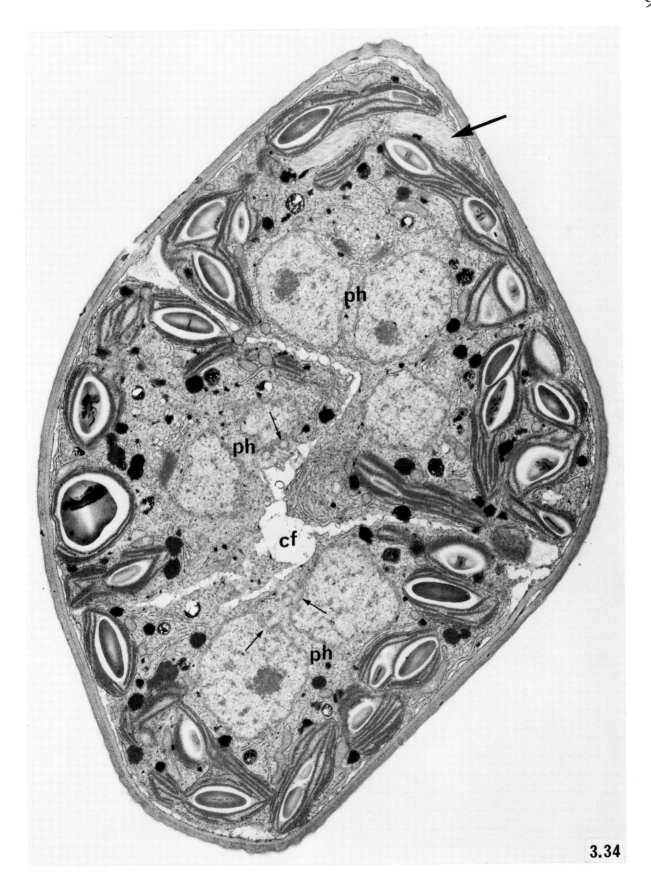

3.34

of centrioles, surrounded by amorphous material, separate and move around the nucleus. By prophase, the centriole complexes mark the two poles or foci of radiating arrays of extranuclear microtubules, many of which approach the nucleus tangentially. Also during prophase, while the chromosomes condense and the nucleolus disperses, the forming spindle becomes increasingly ensheathed by a layer of endoplasmic reticulum (Figs. 3.30, 3.31), so that soon the spindle microtubules are sandwiched between the nucleus and this perinuclear envelope (Fig. 3.31). At the poles, the perinuclear envelope is distended by the accumulation of microtubules focused upon the centriole complexes. During prometaphase, the nuclear envelope adjacent to the centriole complexes is pushed inward (Fig. 3.30), presumably by microtubules, until it ruptures, forming polar fenestrae. Thereupon the nucleus is invaded by microtubules as the extranuclear spindle disappears. The diamond-shaped metaphase spindles, enclosed by one or sometimes two (Fig. 3.33) perinuclear envelopes, are quite typical in structure, with the centriole complexes lying near or just inside the polar fenestrae. During anaphase, the daughter chromatids separate as usual, and the spindle elongates considerably. The nuclear envelope remains essentially intact throughout mitosis, and at telophase it contracts tightly around the chromosomes, excluding from them the centriole complexes and the interzonal spindle apparatus. The centriole complexes remain associated with the nucleus as it expands and acquires a nucleolus, reverting to the interphase condition, while the interzonal spindle apparatus disperses completely.

CLEAVAGE

Cytoplasmic cleavage is basically similar to that described in Section 3.3 for *Kirchneriella.* Microtubules proliferate and extend across the cell in the plane of cleavage (Fig. 3.35), *i.e.,* oriented transversely between daughter nuclei which are flattened against one another. Centriole complexes also always move adjacent to these phycoplasts, along which cleavage furrows, apparently derived initially from invaginations of the plasmalemma, appear to extend to transect the cell (Fig. 3.34). Cleavage usually appears to proceed in a rather haphazard fashion. According to Smith (1918), it does not start until just before

Figures 3.35, 3.36
CYTOPLASMIC CLEAVAGE AND AUTOSPORES IN TETRAEDRON BITRIDENS

3.35.   Numerous phycoplast *(ph)* microtubules are extending throughout the cytoplasm from between nuclei, thereby defining the planes of cell cleavage. A partially formed furrow *(cf)* has already begun cleaving the cytoplasm. Centrioles *(cn)* and often Golgi bodies *(g)* are closely associated with nuclei. Note particularly a "rootlet template"—three adjacent dense tubular structures (arrow)—close to one of the centrioles. I suspect this is a transitory structure whose appearance is related to potential differentiation of zoospores, a usual product of cytoplasmic cleavage in more primitive algae and some other members of the Chlorococcales (see also Fig. 4.55). × 23,000. *Inset* shows a group of live autospores still contained within their parental wall. × 1,800 approx.

3.36.   Group of autospores each enclosed by its own wall, inside the parental wall *(pw),* which is disintegrating. Pyrenoids *(py)* have re-formed. × 5,500.

SOURCE. Author's micrographs, published in *Ann. Bot. 36,* 693 (1972).

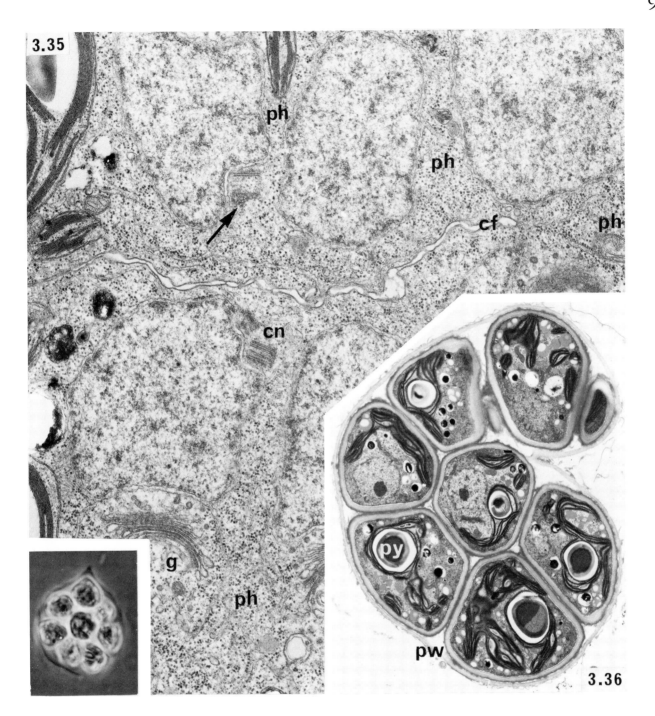

3.35

ph

ph

cf

ph

cn

g

ph

py

pw

3.36

the penultimate mitoses. Thus, in cells that will release eight autospores, the primary cleavage which divides the nuclei into two equal groups is observed at the four-nuclei stage. Smith says that the orientation of this primary cleavage is across the shortest diameter of the cells. Secondary cleavages later grow at right angles to the primary cleavage, and the final cleavages are inclined at about 45° to the secondary cleavages. Pyrenoids disappear after the first cleavage and reappear in the autospores (Fig. 3.36). My own ultrastructural work indicates that partially formed cleavage furrows are common in larger cells; because I was working with very thin sections, I was unable to confirm directly Smith's description of the spatial progression of cleavages, which become even more complicated in cells that contain 16 or 32 nuclei. My own feeling that cleavage is not tightly controlled, either temporally or spatially, is strengthened by the discovery of several binucleate autospores (sometimes even containing their own miniature phycoplasts) being released from parental cells. There can be no doubt that the cytokinetic mechanism used in *Tetraedron* is quite typical of that used in all other members of the Chlorococcales, namely, the formation of a phycoplast followed by growth of a cleavage furrow through this phycoplast. Once the cells have been cleaved, the uni- or binucleate fragments secrete a wall around themselves to form the nonmotile (Fig. 3.36) autospores. As in *Kirchneriella,* dissolution of the parental wall releases these miniatures of the parental cell (Fig. 3.35, *inset*). This behavior can be contrasted with that of some other members of the Chlorococcales, which release naked, motile zoospores formed by analogous cytoplasmic cleavage.

3.6. Scenedesmus

Scenedesmus is a common and usually conspicuous member of the family Coelastraceae (Fritsch, 1935) in the Chlorococcales. The genus is readily identifiable because the elongated vegetative cells are rarely single (Fig. 3.38) and more usually are stuck to one another side by side, forming a planar or slightly curved colony (Figs. 3.37–3.44 et seq.). The number of cells in such coenobia varies from 2 to usually 8 or 16 (rarely more), 4 being perhaps the most common number. One or more curved spines often arise

Figures 3.37–3.43
LIVE CELLS OF SCENEDESMUS

3.37. *S. quadricauda* No. 276. Two-celled coenobium. Each cell has a single conspicuous pyrenoid next to the small nucleus. × 600.

3.38. As for Figure 3.37, this single cell has two nuclei (arrows). × 600.

3.39. *S. quadricauda.* Four-celled coenobia of different sizes. The central two cells of the largest coenobium are undergoing cleavage. × 450 approx.

3.40. *S. quadricauda.* A few seconds before this micrograph was taken, the small daughter colony (large arrow) had been forcibly ejected from the now empty cell of the parental colony (small arrow). The daughter colony is slowly expanding. It flattens out, and the spines, previously folded back when inside the parental wall (Fig. 3.73), are becoming more erect. × 400 approx.

3.41. *Scenedesmus* sp. *b.* Four-celled coenobium. Note the pattern of spines in this species. × 500.

3.42. As for Figure 3.41. One of the inner cells is empty, having released a daughter colony. One outer cell has just undergone the first transverse cleavage (arrow; see Fig. 3.63). The other two cells are considerably shrunken inside the mother cell wall, having cleaved into four daughter cells, which are forming new coenobia. × 760.

3.43. As for Figure 3.41, showing later stage of daughter colony formation. × 460.

SOURCE. Author's micrographs, paper in preparation.

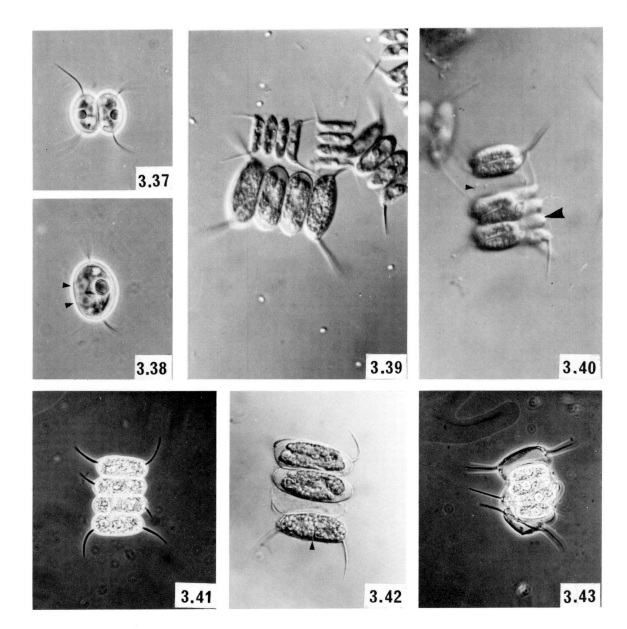

from each end of the outer cells of a colony (Figs. 3.39, 3.40, 3.44*A*, 3.52), and more rarely, from the inner cells as well (Fig. 3.41). However, considerable variation in the size, number, and disposition of spines can be found in a given species grown under different conditions. Further surface ornamentation can sometimes be detected with the light microscope (e.g., by drying colonies on a microscope slide) in the form of tufts of fine bristles of varying sizes attached to the wall (Trainor and Burg, 1965b). The general morphology of colonies, even in a given species, is highly variable (Fritsch, 1935, p. 176) and has been much studied recently (e.g., Trainor, 1964, 1969, 1971; Trainor and Roskosky, 1967; Trainor and Rowland, 1968). Sometimes when the cells are solitary, they resemble the genus *Chodatella* (Fott, 1968). The degree of development of bristles and spines has a considerable effect upon the buoyancy of the colony; the more ornamentation there is, the more buoyant the colonies become (Conway and Trainor, 1972). Trainor (1966) reports a phototactic response of species of *Scenedesmus* which possess bristles, and so these morphological features may aid in keeping the organisms efficiently illuminated in natural conditions. Each cell contains a conspicuous, usually single pyrenoid in its chloroplast (Figs. 3.37, 3.38). Vegetative reproduction takes place by a modified form of autospore formation. The nucleus divides (Fig. 3.38), and the cell sometimes undergoes cleavage (Smith, 1914), which may be only partial if further mitoses follow. Other accounts (and my own ultrastructural results) indicate that cleavage is delayed until at least four nuclei are present. As in many other chlorococcalean algae, several cycles of nuclear division may take place before the final cleavages give rise to 2^n daughter cells inside the parental wall. When these cells form a colony (the usual situation), the daughter cells are firmly stuck together by the time they are released from the parental wall (Fig. 3.40), in marked contrast to what occurs in some other colonial chlorococcalean algae such as *Hydrodictyon* and *Pediastrum* (Sections 3.7a, 3.8a). Asexual reproduction via motile zoospores and isogamous sexual reproduction have only recently been discovered in *Scenedesmus*. Trainor (1963) reported that nitrogen deprivation could induce the formation of biflagellate zoospores in cultures of *S. obliquus* and *S. dimorphus*, and Trainor and Burg (1965a) de-

Figures 3.44*A–C.* SCANNING ELECTRON MICROGRAPHS OF SCENEDESMUS QUADRICAUDA NO. 276

3.44*A.*   Vegetative colony showing spines, combs, and the large circular props (see Figs. 3.47, 3.-56). × 4,000.

3.44*B.*   Surface morphology of the warty layer in this species (cf. Fig. 3.50). × 13,000.

3.44*C.*   The tubules comprising the terminal spines (cf. Figs. 3.49, 3.58). × 10,000.

SOURCE. Author's micrographs, paper in preparation.

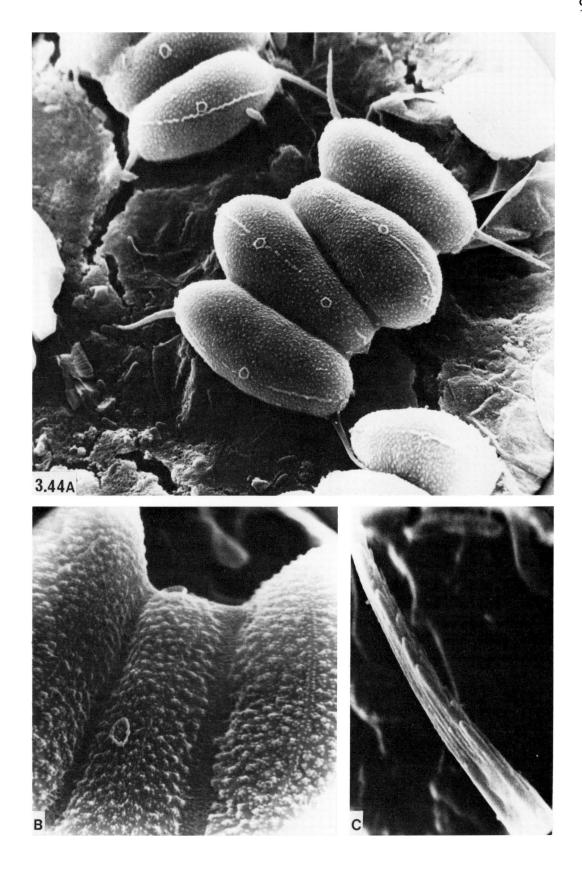

3.44A

B

C

scribed how similar biflagellate gametes in the former species clumped and underwent sexual reproduction; the resultant zygotes could be induced to germinate by adding a nitrogen source to the medium.

STRUCTURE OF THE WALL

Several recent papers have described aspects of the wall structure in this genus, revealing that it is complex and has various unusual structural features. It is also apparently highly variable in morphology, both within and between species. Bisalputra and Weier (1963) showed that the wall of *S. quadricauda* contains three main layers, an inner cellulosic layer around each cell of the coenobium, an outer pectic layer which surrounds the whole colony, and a thin middle layer of membrane-like structure around each cell. The outer pectic layer was shown to contain a hexagonal meshwork of electron-dense material, and this net-like or reticulate structure is supported by a system of tubular structures or "props" which holds the net off the cellulosic layer of the wall. There are two types of props: small ones, which are common, and larger ones more complex in morphology. The long curved spines, characteristic of the genus (Figs. 3.37–3.43), were shown to consist of clusters of close-packed tubules which decreased in number along to the tip. These workers summarized their results in a diagrammatic representation of the wall, and these results were confirmed in a later paper (Bisalputra et al., 1964). Scanning electron micrographs of *Scenedesmus* were later published by Higham and Bisalputra (1970), showing the distribution of various ornamented features over the surface of the coenobium.

We (Staehelin and Pickett-Heaps, 1975; Pickett-Heaps and Staehelin, 1975) have confirmed these results, but question some of the interpretations derived from them. The reticulate layer is illustrated in Figures 3.53, 3.54, and 3.56; Figure 3.54, in particular, reveals that this layer is highly complex structurally. The smaller props, square in cross section, connect with the reticulate layer at intersections of the ridges which form the net-like patterning. The larger, more complex type of prop is seen in Figures 3.55 and 3.56 (also Figs. 3.44*A*, 3.44*B*, 3.47). This outermost layer clearly does not adhere to the surface of the coenobium (Figs. 3.55, 3.57). The connotation "pectic" layer for this structure

Figures 3.44D–3.48 ORNAMENTATION OF SCENEDESMUS SP. B AND S. QUADRICAUDA NO. 276

3.44*D*.   *S. quadricauda* No. 276. This scanning electron micrograph clearly shows the combs on the coenobium. × 7,500 approx.

3.45.   Morphology of a young coenobium, sp. *b*. Note evidence (arrows) of the combs (Figs. 3.44 D–3.48) which run around the cells. × 6,000.

3.46.   Transverse section of a coenobium (sp. *b*) showing typical disposition of the combs (small arrows), which in this orientation show up as single tubular components (see Fig. 3.47). Note also that the sporopollenin sheath (a single layer in this species) adheres closely to the cells except where they are joined (large arrows). The surface of this sheath is shown in Figure 3.50 (it differs from the double sheath found around cells of *S. quadricauda;* cf. Figs. 3.52–3.57). × 8,000.

3.47.   This section has approached a coenobium (sp. *b*) tangentially, cutting the comb transversely. The comb is obviously composed of a row of tubular components. The surface of the sporopollenin sheath (arrow), seen also in Figure 3.50, is covered with numerous tiny warts. A larger prop has also been cut in cross section (see arrows in Figs. 3.55, 3.56). × 26,000.

3.48.   A comb in longitudinal sectionl (sp. *b*). × 16,000.

SOURCE. Author's micrographs, papers in preparation.

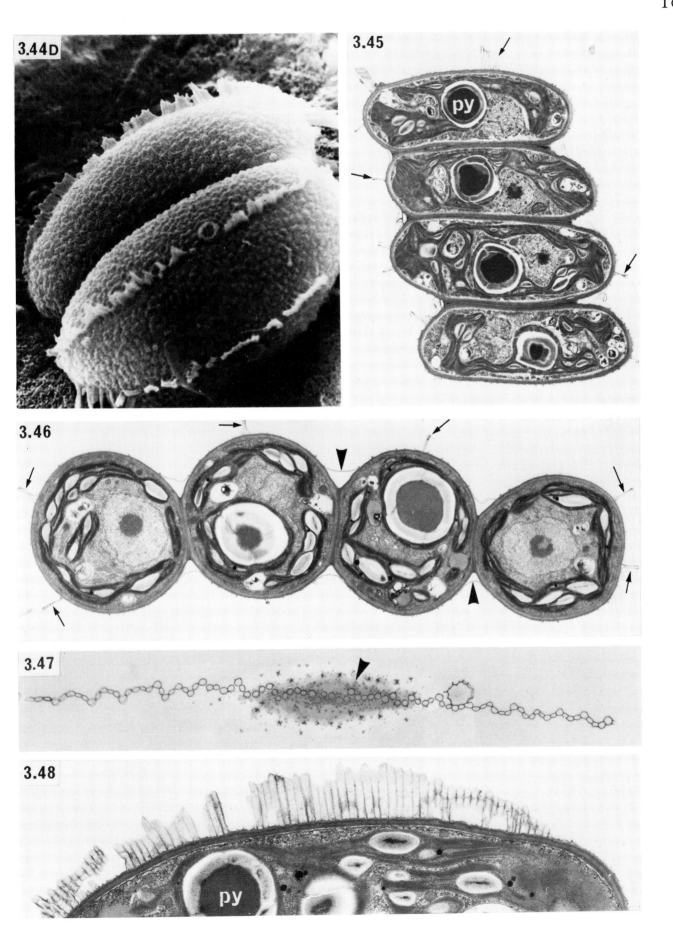

(Bisalputra and Weier, 1963), although frequently recurring in the literature on this and other species, may be misleading. Our preparations for transmission microscopy show the region between this layer and the coenobium to be empty (which means only that any material present has been extracted from it during processing). However, while some species of *Scenedesmus* may be enclosed in a gelatinous sheath of more diffuse mucilage (e.g., Figs. 1 and 2 in Bisalputra and Weier, 1963), the ultrastructural localization and appearance of the sheath seem to be undetermined at present. Freeze-etched preparations do show the existence of pectin-like materials in this region for *S. quadricauda,* and preparations for light microscopy using India ink demonstrate a zone of exclusion of the ink particles from around the coenobium. Bisalputra and Weier say this pectic layer is very stable; when daughter colonies are released from coenobia, the layer is discarded partly intact to form ghosts that accumulate in the culture. Atkinson, Gunning, and John (1972) note the similarity of the outer wall of *Scenedesmus* to that of *Chlorella* (Section 3.2). I too have subjected whole cells of *S. quadricauda* to acetolysis, and these ghosts (and nothing else) survive intact (Fig. 3.76); their components are clearly identifiable with the electron microscope (Staehelin and Pickett-Heaps, 1975). I do not know of any pectic material that could possibly withstand such treatment. Thus, the term "pectic" is misleading and almost certainly incorrect, and I believe it should be discarded; I shall use the term "reticulate" layer hereafter. It is now also clear that the "membrane-like" structure described by Bisalputra and Weier (1963) forms a trilaminar sheath that encloses each cell of the colony individually (Figs. 3.39, 3.71). This sheath is also acetolysis-resistant, and obviously it is analogous to the trilaminar layer around *Chlorella,* which Atkinson, Gunning, and John (1972) have found to contain sporopollenin (Section 3.2). The coenobia are firmly stuck together by a densely staining, acetolysis-resistant adhesive material (Figs. 3.55, 3.57) that appears to be integrated into the trilaminar layer (Fig. 3.68). The vegetative cell wall is laid down inside the trilaminar sheath. The tubular nature of the spines is illustrated in Figures 3.58 and 3.75. This information is summarized in Figure 3.59*A*.

For comparison with *S. quadricauda,* we have also investigated the structure of other spe-

Figures 3.49–3.51
ORNAMENTATION OF SCENEDESMUS SP. B IN FREEZE-ETCHED PREPARATIONS

3.49. The tubules of this spine (cf. Fig. 3.58) are composed of subunits arranged in semicrystalline form. × 69,000.

3.50. The surface of the outer sheath. The site of a sheared-off spine, indicated by the characteristic ridge on the surface, is shown by the double arrow. Part of the comb is seen end-on (large arrow). Note how the sheath is stretched flat (small arrows) between the cells (cf. Fig. 3.46, large arrows). × 12,000.

3.51. The tubules of the comb are also composed of regular arrays of subunits. × 73,000.

SOURCE. Dr. A. Staehelin, paper in preparation.

cies of *Scenedesmus*. One remains unidentified and will be referred to hereafter as species *b;* the other is a strain of *S. quadricauda* supplied by the Cambridge Culture Collection (No. 276/4A) and will be referred to with the annotation "No. 276."* Their ornamentation is similar, if not identical, and is illustrated in Figures 3.44–3.51 and summarized in Figure 3.59*B.* Terminal spines (Fig. 3.44*A*) were less common in these species; upon freeze-etching, the walls of the spines' tubules (Fig. 3.44*C*) were seen to contain regularly arrayed subunits (Fig. 3.49). More commonly, the cells had "combs" running around the coenobia (Fig. 3.44*D*). These combs were usually symmetrically arrayed, two to each end cell and one on each inner cell, as in Figure 3.46. The combs consisted of a single, somewhat irregular row of tubules (Figs. 3.44*D,* 3.46–3.48) which, upon freeze-etching, appeared to contain subunits similar to those in the spines (Fig. 3.51). The whole coenobium was enclosed in a single resistant sheath, closely appressed to the cells except where they were joined (Fig. 3.44*B,* 3.46). No props were present, although some more complex projections from the cell surface (e.g., in Figs. 3.44*A,* 3.44*B,* 3.47) bore a resemblance to the larger props of *S. quadricauda.* This surface layer was covered by a pattern of fine warts (Figs. 3.44*B,* 3.47, 3.50). The adhesion site between cells appeared identical in *S. quadricauda* and species *b.* A comparison of these species, therefore, indicates how surface morphology may be altered by perhaps quite small alterations in the disposition and arrangements of basically similar components. Further research will almost certainly reveal further different combinations of these features. Many species of *Scenedesmus* may possess tufts of bristles instead of such spines, and these too have been shown by negative staining to consist of a regular array of subunits (Trainor and Massalski, 1971; Marcenko, 1973). Marcenko also has evidence that these bristles are proteinaceous. Massalski

*My attention has recently been drawn to a paper by Komarek and Ludvik (1972) who have used the ornamentative features of *Scenedesmus* to classify and distinguish between some thirteen species. Their paper is an excellent example of the use of fine structure in taxonomy. They reclassify both the *Scenedesmus* described above, one as *S. pannonicus* (C.C.C. No. 276/4a), and the other as *S. longus* (I.U.C.C. No. 614). Unfortunately, I cannot do justice to their observations at this stage, but readers interested in *Scenedesmus* are recommended this paper.

Figures 3.52–3.54
SCENEDESMUS QUADRICAUDA

3.52.   Two young coenobia. The area indicated by the arrow is seen in more detail in Figure 3.53. × 3,500.

3.53.   The reticulate nature of the surface of these cells can just be seen. × 7,000.

3.54.   This freeze-etched preparation shows clearly the complex substructure of the reticulate surface layer (cf. Figs. 3.53, 3.56). The arrows indicate regions where the larger props probably connect with this layer (cf. Fig. 3.55). × 40,000.

SOURCE. Figure 3.54: Dr. A. Staehelin, paper in preparation. Figures 3.52, 3.53: Author's micrographs, paper in preparation.

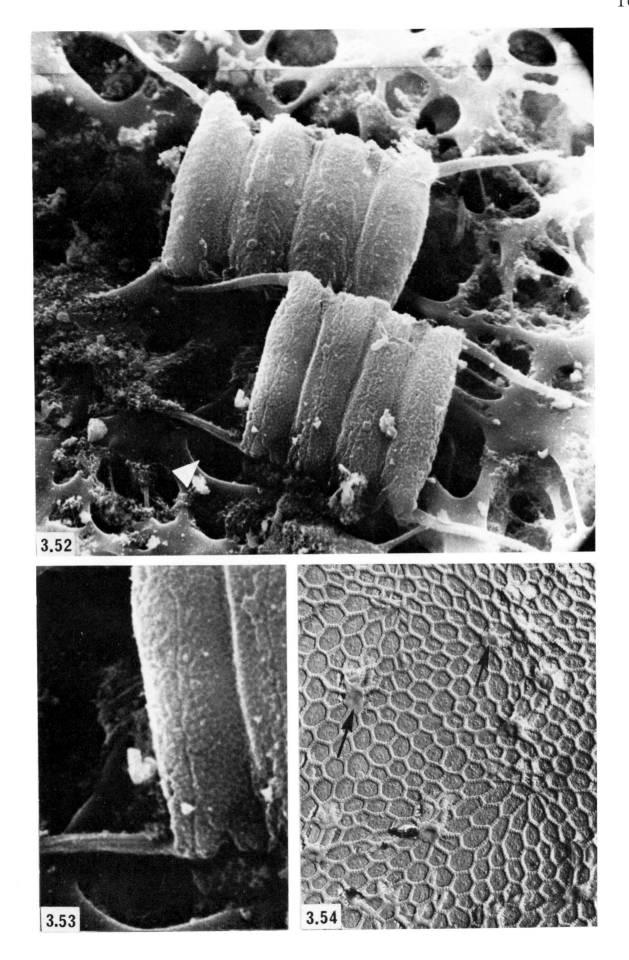

3.52

3.53

3.54

and Trainor (1971) used negative staining to illustrate yet another feature on the surface of a particular strain (culture 16) of *Scenedesmus.* These "capitate" appendages consist of an elongated stalk of varying length surmounted by a circular cap, distributed rather irregularly over the surface of the cells. We should finally note that a species of *Chlorella* examined by Clémençon and Fott (1968) has an outer layer that is warty and bears some resemblance to those described for *Scenedesmus* sp. *b,* above (this layer was not shown in enough detail to determine whether it was trilaminar, as indeed seems likely).

3.6a. Cell Division and Colony Formation

Interphase cells of *Scenedesmus* are quite normal in ultrastructural morphology. Centrioles are persistent, usually being found close to the nucleus, near the wall. The numerous Golgi bodies are mostly arrayed around the nuclear envelope. During interphase, the cells grow considerably (e.g., Fig. 3.39), and the nucleus also enlarges; as the cells prepare for mitosis, a most characteristic change is observed which lasts until cleavage is over. Localized dilations of the endoplasmic reticulum accumulate densely staining granular material (Figs. 3.60, 3.61, 3.64) in quite large amounts. This phenomenon has been observed in all three species I have examined. At the light microscope level, the cells become "granular" before division (Smith, 1914). The paired interphase centrioles replicate as usual and separate around the nucleus, and the prophase nucleus is enclosed by an extranuclear spindle. The nucleus breaks up during prophase. A distinct perinuclear envelope, characteristic of other chlorococcalean algae, is not always formed, although the spindle is enclosed by some irregularly arranged elements of endoplasmic reticulum. Polar fenestrae appear at prometaphase, and the metaphase spindle is typical (Fig. 3.60). The microtubules appear to focus upon an ill-defined mass just inside the fenestrae (an MTOC?) and not upon the centrioles themselves (Fig. 3.62). The spindle elongates at anaphase (Fig. 3.61), and telophase is normal.

Daughter nuclei come back close together after telophase, and the phycoplast is formed between them (Pickett-Heaps, 1972a; Fig. 3.63). However, I do not think cleavage is initiated at

Figures 3.55–3.58
MORPHOLOGY OF SCENEDESMUS QUADRICAUDA

3.55. This longitudinal section shows the spines at each corner and demonstrates how the outer, reticulate layer of the sheath (Figs. 3.53, 3.54) is loosely held above the surface of the coenobium, supported by numerous small and some larger and more complex props (arrow). The morphology of this surface ornamentation is summarized in Figure 3.59a, and is quite different from that of species *b* and *S. quadricauda* No. 276 (see Figs. 3.44–3.48, 3.50). × 8,700.

3.56. Section tangential to the surface of a coenobium, showing the smaller and larger (large arrow) props in cross section. Note how the smaller props connect with the reticulate surface layer (small arrows). × 30,000.

3.57. Transverse section of a coenobium. Note the differences in surface morphology between this species and that in Figure 3.46, in particular, how the surface layer is held off the coenobium. × 6,900.

3.58. Spine in longitudinal section. × 16,000.

SOURCE. Author's micrographs, paper in preparation.

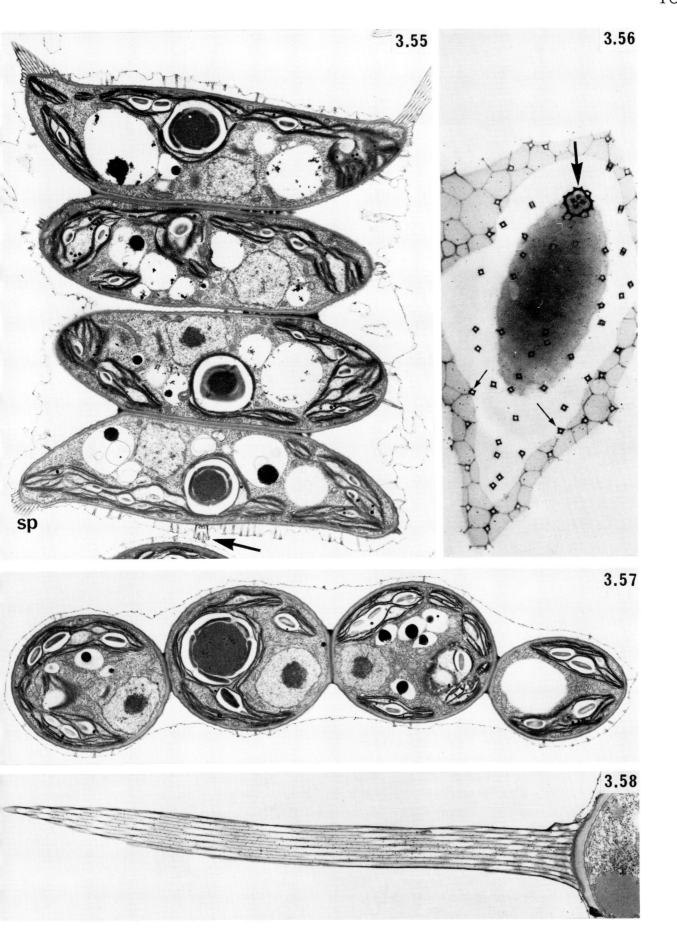

3.55

3.56

sp

3.57

3.58

this stage (Fig. 3.38) unless the future coenobium is going to be two-celled. All quadrinucleate cells discovered had no trace of cleavage immediately after the second mitoses whose spindles are apparently oriented across the cell (Fig. 3.61; Smith, 1914). Cleavage is illustrated in Figure 3.63; as usual, a furrow, whose origin is indeterminate, grows across the cell through the phycoplast. The Golgi bodies at this and later stages are invariably seen to be forming vesicles with dense contents (Figs. 3.63, 3.64, 3.66). The pyrenoid is still present at this stage, being partitioned into one of the daughter cells (Fig. 3.63; see also Smith, 1914).

The cleavages that follow the final set of mitoses have so far proved difficult to study ultrastructurally. Cells fixed at this stage and when the autospores were forming were always shrunken and very dense. The pictures presented here (Figs. 3.65–3.70) are the best I have been able to obtain from numerous fixations. Cytoplasmic detail is difficult to discern, and some organelles suffer obvious damage (e.g., the mitochondria; Fig. 3.66). Shrinkage of the cell is not entirely artifactual, being always clearly visible in live cells at this stage (Figs. 3.39, 3.42). Neither can the density of the autospore be ascribed to fixation conditions alone, since adjacent, nondividing cells (e.g., in the same coenobium) were excellently preserved.

These final cleavages are important because their orientation and the subsequent movement of the daughter cell protoplasts inside the wall profoundly affect the way the cells link together. Smith (1914) says that after the first transverse cleavage (Figs. 3.42, 3.63) the protoplasts may rotate inside the wall (perhaps like *Chlamydomonas*) and that the next two cleavages are diagonal. (In *S. acutus* and *S. quadricauda* Smith reports that the second mitoses take place after this first cleavage.) Smith then describes a period of "maturation" during which the autospores elongate somewhat. The net result of these cleavages and movement is that the four autospores become roughly parallel, with their axis aligned approximately with that of the mother cell. The pyrenoid later disappears, and each daughter cell reconstitutes a new one (Smith, 1914).

As mentioned above, the ultrastructure of cells during these important stages has largely resisted analysis. Shrinkage of the cytoplasm starts soon after the first cleavage furrow is com-

Figure 3.59
STRUCTURE OF COENOBIUM OF SCENEDESMUS QUADRICAUDA

3.59A. Diagrammatic representation of the structure of a coenobium of *S. quadricauda* (culture No. 614 from the Indiana Culture Collection; Starr, 1964). Two of the cells are empty following release of daughter coenobia. The cells are enclosed by two distinct layers. The outermost is the reticulate layer, loosely held off the trilaminar sheath by the props. The trilaminar sheath adheres closely to the wall of the cells. By the time daughter colonies are released, the cell wall proper has been broken down completely; the outer layers, props, and spines remain intact, forming ghosts in the culture. The *inset* shows how the cells are interconnected. The cell wall *(w)* is bounded by the plasmalemma *(pm)* and the trilaminar sheath *(ts)*. The coenobial adhesive *(ca)* binds adjacent trilaminar sheaths together.

3.59B.. Diagrammatic representation of the structure of a coenobium of *S. quadricauda* (culture No. 276/4A from the Cambridge Culture Collection, England). The cells in this strain possess combs as well as spines. The warty layer *(wl)* is applied directly on the trilaminar sheath, except where the cells are joined to one another at the coenobial adhesive.

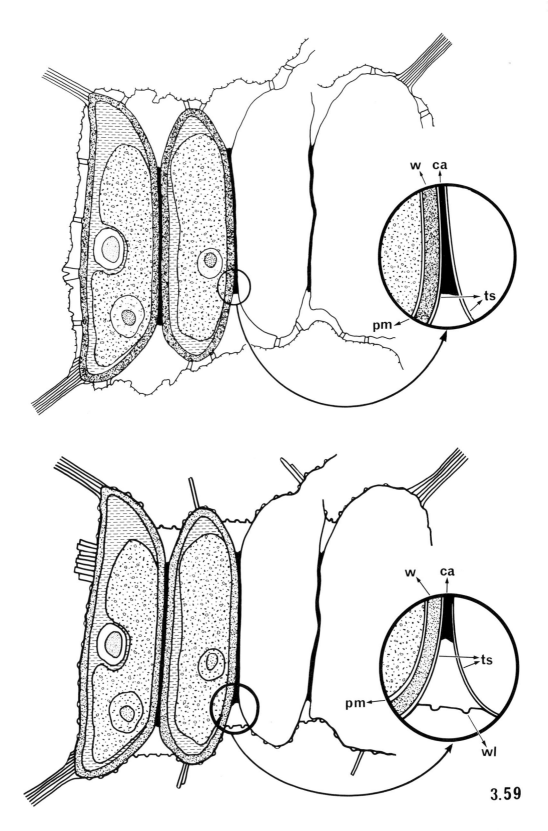

3.59

pleted, and soon much granular material collects in the furrow and outside the daughter cells, but inside the parental wall (Figs. 3.65–3.70). Microtubules can be found, often in groups, near the plasmalemma of cells apparently undergoing "maturation" (Fig. 3.66); their possible role in the ensuing morphological changes is unclear and is suggested more by analogy with other members of the Chlorococcales such as *Hydrodictyon*, *Pediastrum*, and *Sorastrum*, whose cells undergo shape changes mediated by microtubules as they form colonies (Sections 3.7–3.9). As the autospores shrink, the parental wall becomes more diffuse (Figs. 3.65, 3.71), a change which culminates in its complete disappearance by the time the daughter coenobia are fully formed (Figs. 3.73–3.75).

Formation of the daughter coenobia has been investigated in *S. quadricauda* and species *b*; only the former is described here, since both are essentially similar allowing for the differences in eventual ornamentation. Smith (1914) also gives a detailed account of the events in *S. quadricauda*. In some subtle way, the autospores are dissimilar; the two or more cells forming the central cells of the new colony have two adhesive sites along their length, where they stick to each other and the outer cells. The latter, however, have only one such site, and from regions on their ends arise the spines of the new colony. I have been unable to decide whether the autospores are all initially equivalent, or whether the central cells are indeed morphologically differentiated from the outer cells. A little reflection on the part of the reader will soon reveal serious problems with either possibility.

It seems most likely that the cells adhere to one another at precisely situated regions of the cell membrane, presumably differentiated in some way. Figure 3.66 shows a very early stage in the secretion of the coenobial adhesive. Later, the trilaminar layer first appears at these sites, and the layer is incorporated into the dense adhesive (Fig. 3.68). The spines are apparently fully formed early in colony formation (Fig. 3.67). Once the cells are stuck together, trilaminar plaques appear irregularly over their surface (cf. *Chlorella*) and grow (Figs. 3.67, 3.70, 3.71) until they merge with one another and the first-formed part of the trilaminar sheath at the adhesive sites (Fig. 3.69). Concurrently, the outer (i.e., future reticulate) layer also begins to condense some-

Figures 3.60–3.62
MITOSIS IN SCENEDESMUS QUADRICAUDA NO. 276

3.60. Metaphase. Typical diamond-shaped spindle, with centrioles (large arrows) at polar fenestrae. Note dense material accumulating in the lumen of the endoplasmic reticulum (small arrows). × 24,000.

3.61. Late anaphase. The orientation of this spindle is at right angles to that in Figure 3.60, and the cell may have been undergoing the second synchronous mitoses. As in Figure 3.60, the endoplasmic reticulum contains granular material (arrows). × 13,000.

3.62. As for Figure 3.60, showing the spindle microtubules converging toward dense amorphous material (an MTOC?) at the polar fenestra, with a pair of centrioles nearby. × 40,000.

SOURCE. Author's micrographs, paper in preparation.

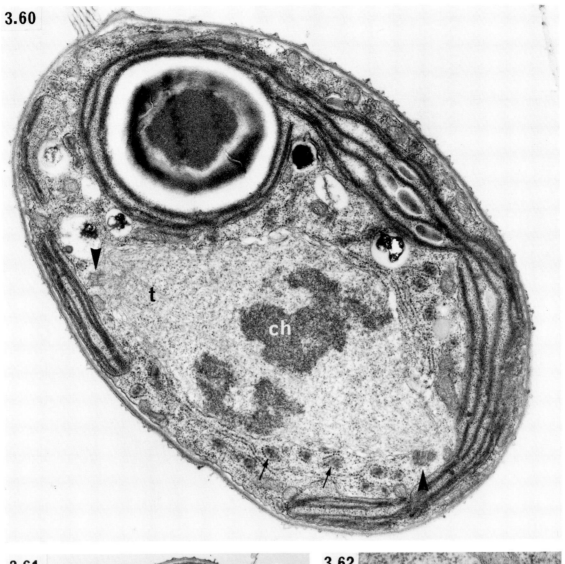

3.60

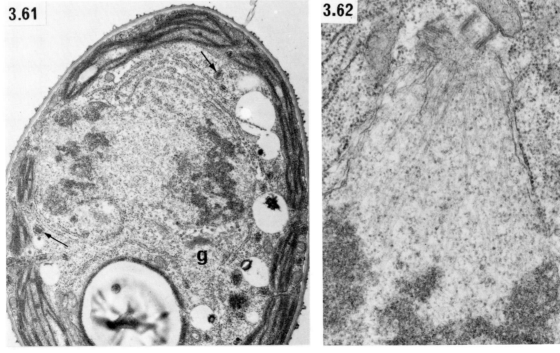

3.61

3.62

what irregularly (Figs. 3.67, 3.70, 3.71). This layer, of course, does not become attached to the trilaminar layers or adhesive sites (Fig. 3.69). Once it is complete, props appear and elongate (Figs. 3.69, 3.71, 3.72).

The previously shrunken autospores now expand to fill the trilaminar sheath of the mother cell, whose wall proper has disappeared (Figs. 3.74, 3.75). The spines are bent over and folded inward (Fig. 3.73). Release of the daughter coenobia (Fig. 3.40) is sometimes quite explosive. The trilaminar layer, which is apparently elastic (Smith, 1914), appears to tear quickly rather than be eroded away (see also Figs. 3.75, 3.77), and the daughter colony pops out and immediately begins unfolding and flattening (Figs. 3.40, 3.75). The spines instantly unfold and later slowly adopt their normal position. These events suggest that perhaps increase in turgor of the daughter cells may be instrumental in achieving their release.

From this account, many details of colony formation are obscure. It seems a reasonable interpretation that some products produced in the endoplasmic reticulum are secreted during and after cytokinesis, possibly via the Golgi bodies, and that this material gives rise to the granular extracellular aggregate. This material further seems likely to contribute somehow to the surface ornamentation of the daughter colonies (it has clearly all disappeared when the colonies are fully formed). However, more precise information regarding the constitution, origin, and arrangement of the various features that form this marvelously intricate wall is lacking. The spines, for example, resist acetolysis, but not 6N NaOH; the trilaminar layer resists both (Staehelin and Pickett-Heaps, 1975). The subunits on the spines and walls have no counterpart in the trilaminar sheath or reticulate layers. Whether the by-products of the breakdown of the parental wall are reutilized by the daughter colony (Atkinson, Gunning, and John, 1972) remains unclear. *Scenedesmus* apparently illustrates the potential of self-assembly phenomena controlled perhaps by subtle differentiations of the cell surface.

3.7. Hydrodictyon

The genus *Hydrodictyon* is most interesting for a number of reasons. The giant coenocytic vegetative cells of the two common species are either

Figures 3.63–3.65
CLEAVAGE AND AUTOSPORE FORMATION IN SCENEDESMUS QUADRICAUDA NO. 276

3.63. Cleavage furrow *(cf)* forming in the phycoplast *(ph)*, many of whose microtubules are focused upon the pair of centriole complexes (paired arrows) of the daughter nuclei *(n)*. Vesicles forming at the Golgi bodies *(g)* have characteristic dense contents. × 33,000.

3.64. As for Figure 3.63, showing the typical change in the Golgi bodies' secretory product at cytokinesis. The lumen of the endoplasmic reticulum *(e)* is still distended with granular material. × 40,000.

3.65. Cleavage is complete in the left cell, and the dense autospores have shrunk away from the disintegrating parental wall (compare the wall of the two cells; paired large arrows). Meanwhile, much granular material is collecting between the wall and the autospores (small arrow). × 22,000.

SOURCE. Author's micrographs, paper in preparation.

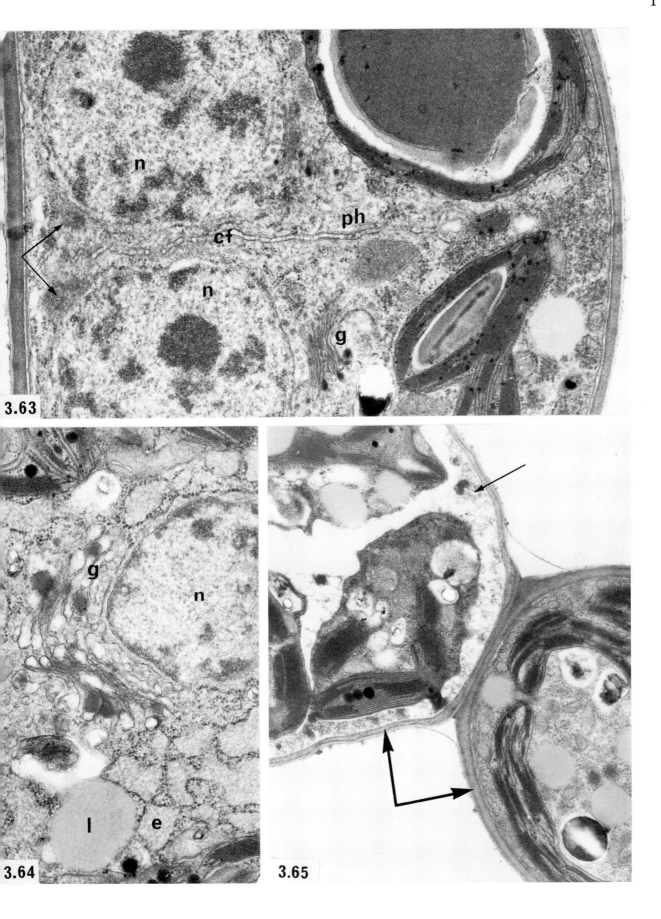

3.63

3.64

3.65

spherical *(H. africanum)* or cylindrical *(H. reticulatum),* and normally these cells are linked together to form net-like colonies or coenobia. These vegetative "water nets" are large and striking objects, sometimes appearing in vast quantities. They were recorded thousands of years ago by the Chinese in what must constitute one of the earliest records of any alga known (Tilden, 1968; p. 374). The vegetative cells may grow very large and provide a valuable experimental material, for example, in the study of ion transport into and out of the vacuoles of living cells (cf. *Chara* and *Nitella;* Chapter 7). A second, completely different morphological form of the cell arises following germination of zygospores or azygospores. These "polyhedra" (see below) are so unlike the vegetative coenobia that, not surprisingly, they caused much confusion in some early attempts to classify these algae (see Section 3.5). For an excellent and most comprehensive review of this genus and its reproductive behavior, the reader is referred to Pocock's (1960) classic paper. In this chapter, I shall be concerned with *H. reticulatum,* which has been intensively studied at the electron microscope level (Marchant and Pickett-Heaps, 1970, 1971, 1972a–d). No equivalent work has yet been attempted on *H. africanum.*

It is quite important at the outset to grasp the intricacies of the life cycle of *H. reticulatum,* which is not really as complicated as may first appear (Fig. 3.78). During the normal asexual reproductive cycle, each giant cell in a vegetative net itself gives rise to a whole new net (Figs. 3.79, 3.80). This rather remarkable feat follows a complex set of cytoplasmic cleavages which cuts up the coenocyte's cytoplasm into thousands of tiny uninucleate swarmers or zooids (another name for zoospores). These flagellate cells are confined for the short period of their motile existence within their parental cell wall. Just before they retract their flagella, they aggregate and adhere to one another in a regular pattern (Fig. 3.79). Later on, the parental cell wall gelatinizes, and during further growth each tiny vegetative cell enlarges enormously, becoming coenocytic as the original single nucleus undergoes successive divisions. The pattern of adhesion achieved after the motile phase (Figs. 3.79, 3.80) is maintained thereafter, and thus the enlarging vegetative cells remain linked together as a net-like colony (Fig. 3.81).

Figures 3.66, 3.67
FORMATION OF COENOBIA IN SCENEDESMUS QUADRICAUDA

3.66.   Cells just beginning to link up at their outer edges (large arrow). The Golgi bodies *(g)* are still apparently secreting dense material similar to that collecting between the plasmalemma and the thinning mother cell wall. The autospores are typically dense and shrunken. The mitochondria *(m)* show evidence of damage. A group of microtubules can be seen near the cell membrane (small arrow). × 30,000.

3.67.   Later stage in coenobial formation. The spines *(sp)* are formed, and the coenobial adhesive between cells (large arrow) is now conspicuous. Plaques of the trilaminar sheath (small arrows) are beginning to appear irregularly over the autospores. × 30,000.

SOURCE. Author's micrographs, paper in preparation.

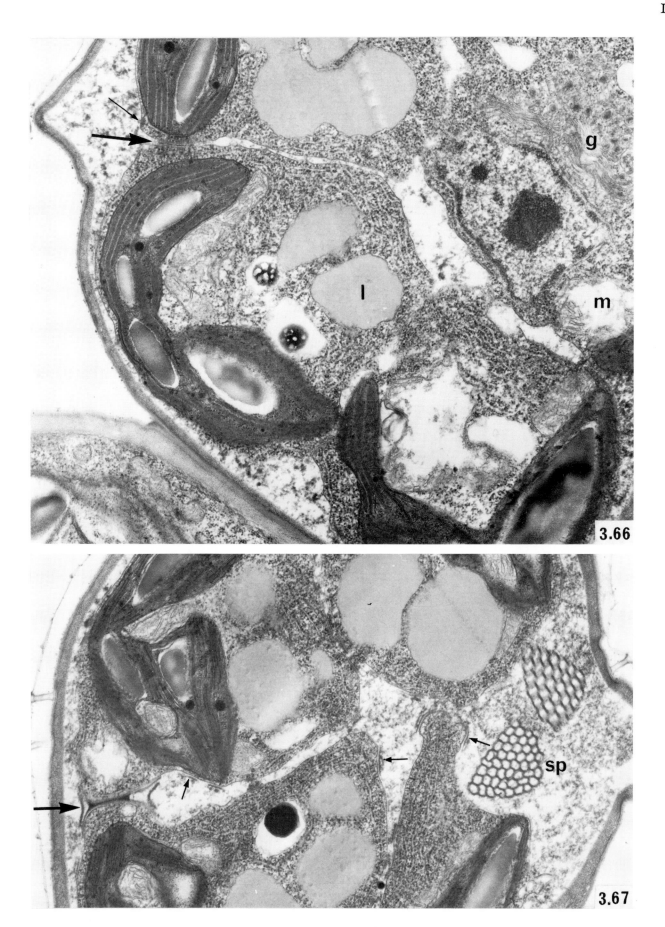

3.66

3.67

Cytoplasmic cleavage may, under the appropriate environmental conditions, give rise to small "gamete zooids." These have no tendency to form nets and are usually released from the parental wall. Most often (in our experience with *H. reticulatum*), these gametes either die or turn into resistant azygospores. Sometimes, however, they undergo sexual reproduction (often with gametes from the same coenobium), conjugating in pairs and then forming a zygospore. Both azygospores and zygospores may later germinate, releasing large motile but sluggish zoospores—a single one from azygospores, four to eight from zygospores—following meiosis. These zoospores soon retract their flagella and begin laying down a thick wall with several peg-like outgrowths. The cells turn into the characteristic polyhedra and grow steadily, becoming coenocytic as before. Cytoplasmic cleavage and formation of zooids follows in polyhedra much as in vegetative cells, and a net is also subsequently formed from zooids within a vesicle presumably derived from the polyhedral wall. However, this "germ" net contains far fewer cells than the net usually formed from vegetative cells, and furthermore it is usually flat—in striking contrast to the cylindrical form of nets derived from vegetative cells. Each cell of this flat germ net can subsequently give rise to typical cylindrical vegetative nets.

The structure of coenocytic vegetative cells is essentially very simple. Inside the wall is a thin layer of cytoplasm enclosing the enormous central vacuole (Fig. 3.82). The chloroplast is probably a single fenestrated cylinder containing numerous pyrenoids and starch grains. The small interphase nuclei, containing a prominent nucleolus, are scattered among other smaller organelles; each nucleus possesses a pair of persistent centrioles in a slight depression of its surface, and Golgi bodies are also closely associated with the nuclear envelope.

3.7a. Vegetative Reproduction

MITOSIS

Nuclear division in *H. reticulatum* (Marchant and Pickett-Heaps, 1970) is semisynchronous in vegetative cells, and although the mitosing nuclei are small, mitotic stages can be discerned in vivo (Fig. 3.83). At the ultrastructural level, mitosis is typical of other members of the Chlorococcales and is initiated by replication of the centrioles;

Figures 3.68–3.71
SECRETION OF THE SHEATH AROUND CELLS OF SCENEDESMUS QUADRICAUDA

3.68. Adhesion of the autospores (stage equivalent to that in Fig. 3.67). The trilaminar sheath is first formed at the adhesion site and does not extend far past it at first (paired arrow). This sheath can be clearly distinguished from the plasmalemma of each cell. × 58,000.

3.69. Later stage of coenobial formation. The sites of adhesion are clearly visible (large arrows) at the outer edges of the cells. There is, of course, no adhesive between the future outer cells of the coenobium, here seen with their four sets of spines *(sp)* folded up inside the mother cell wall. Both the inner trilaminar sheath and the outer reticulate layer are fully formed. Note how the outermost layer is not incorporated into the coenobial adhesive, whereas the trilaminar sheath is (Fig. 3.68). Props are also just starting to be formed (small arrow; see Fig. 3.72). × 16,000.

3.70. Earlier stage than that in Figure 3.69 in formation of the surface layers. Irregular plaques of the trilaminar sheath (cf. Figs. 3.12, 3.13) underlie the almost continuous reticulate layer. × 35,000.

3.71. As for Figure 3.70. Note the incipient prop (arrow). The reticulate layer could possibly be forming from condensation of the granular material. The mother wall characteristically appears granular as it breaks down. × 70,000.

SOURCE. Author's micrographs, paper in preparation.

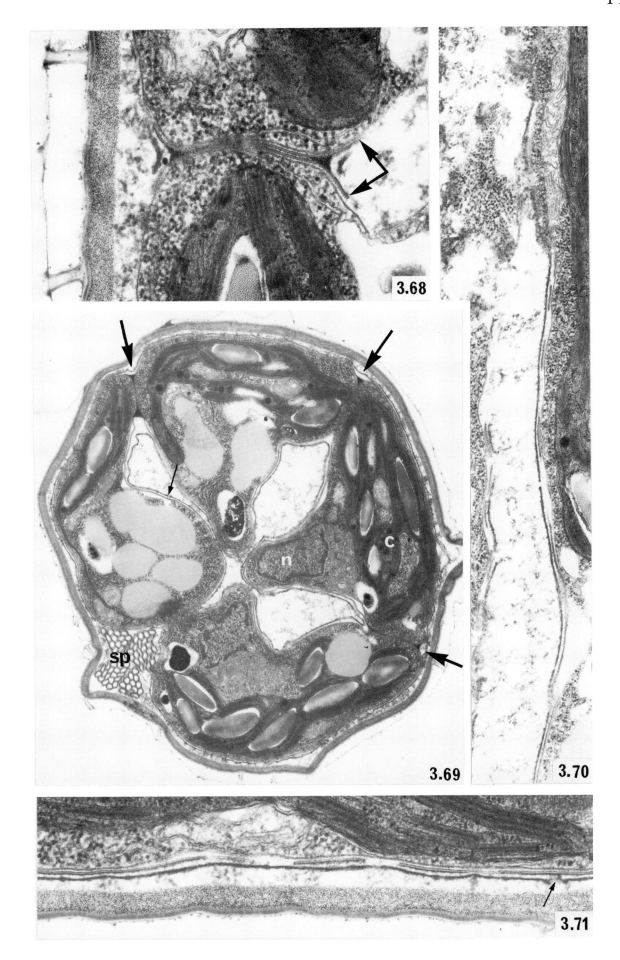

3.68

3.69

3.70

3.71

the pairs then separate as microtubules proliferate between them to establish the poles of the spindle (Fig. 3.84). At prophase, the nucleolus disperses; an extranuclear sheath of microtubules forms, which becomes enclosed by the perinuclear envelope. Polar fenestrae in the nuclear envelope allow the extranuclear microtubules to invade the nucleus and give rise to the intranuclear spindle (Fig. 3.85). After anaphase, when the spindle has elongated, the re-forming nuclear envelope isolates the interzonal apparatus from the daughter nuclei and is interposed between the nucleoplasm and the centrioles (Fig. 3.86). Cytokinesis as such does not follow mitosis; instead, the enlarging cells accumulate thousands of nuclei before giving rise to new daughter nets.

CYTOPLASMIC CLEAVAGE

Formation of the new cylindrical daughter nets is initiated by cytoplasmic cleavage which eventually gives rise to uninucleate zoids (Marchant and Pickett-Heaps, 1971). Some of these events can be clearly followed under the light microscope in vivo (Figs. 3.87–3.89). An early sign of incipient cleavage is the breakdown and disappearance of the pyrenoids, followed by an intensification in the green color of the chloroplast (Fig. 3.87). A thin membranous structure, the "vacuolar envelope" (see below), is then cleaved off the entire inside surface of the cytoplasm; this can sometimes be discerned in living cells by examining an optical section through the middle of the cell, particularly if Nomarski optics are used. Next, the cytoplasm progressively cleaves, leading to the characteristic "pavement stage" (Fig. 3.88). At this point, the cytoplasm has been cut up into thousands of closely fitting, angular segments, each of which contains a nucleus surrounded by elements of the chloroplast. Finally, often after an appreciable delay, a pair of flagella is extruded from each segment, which rounds up (Fig. 3.89) and becomes an actively motile zoid with paired contractile vacuoles (Fig. 3.93). The zoids then swarm about, confined between the parental cell wall and the vacuolar envelope, which maintains some turgor throughout the period of swarming. The significance of the vacuolar envelope will soon become apparent.

Sectioned material reveals much more about these processes. With the light microscope, the vacuolar envelope (Figs. 3.90, 3.92) and subse-

Figures 3.72–3.74
FORMING COENOBIA OF SCENEDESMUS QUADRICAUDA

3.72. Two partly formed props are clearly beginning to push apart the reticulate layer and the trilaminar sheath. × 28,000.

3.73. Spine on a fully formed autospore, before release of the coenobium. Note how it is bent so as to be folded inside the mother cell, whose wall now has completely disappeared. The large arrows indicate the reticulate layer of the *mother* cell; the smaller arrows, the same layer in the *daughter* cell. × 20,000.

3.74. Three fully formed daughter coenobia folded up inside the remains of the parental colony. Note that the mother cell walls have been completely broken down. One daughter colony has already been released. The spines *(sp)* of the mother colony remain intact, attached to the trilaminar sheath. × 6,600.

SOURCE. Author's micrographs, paper in preparation.

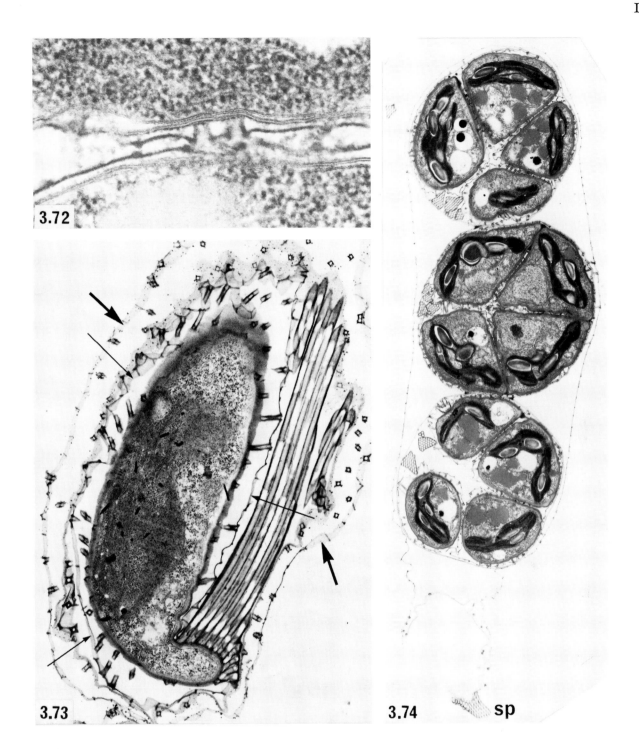

3.72

3.73

3.74 sp

quent radial cleavages of cytoplasm (Fig. 3.91) can be clearly detected, as can the disappearance of the pyrenoids. The wall, too, suffers partial dissolution (Fig. 3.92), which is complete once the daughter net has been formed. Ultrastructurally, the pyrenoids disintegrate as intrusions of cytoplasm penetrate into them; their dense, granular cores break up until only a small portion remains and their array of starch plates disperses. Concurrently, microtubules proliferate parallel to and just under the tonoplast; then a cleavage fissure or furrow grows through these tubules (Fig. 3.95). Its origin and mode of formation is unclear, but eventually it cleaves off the entire tonoplast to form the thin vacuolar envelope which usually contains little more than a few ribosomes and tubules in an extremely thin cytoplasmic layer sandwiched between the two membranes (Fig. 3.96). Soon after the vacuolar envelope has been formed, arrays of microtubules appear between all the nuclei (Fig. 3.96), and again a steadily ramifying complex of cleavage furrows grows through these tubules until the cytoplasm is eventually cleaved into the close-fitting uninucleate fragments (the pavement stage; Figs. 3.94, 3.96). Later, typical flagella and contractile vacuoles appear in these zooids, along with four sets of rootlet microtubules (analogous to those in *Chlamydomonas,* for example) extending around the cell's periphery. The events of cleavage are summarized in Figure 3.97.

Cytoplasmic cleavage in *H. reticulatum* can be considered as being mediated by the typical phycoplast system, whereby numerous proliferating microtubules define the path of the growing cleavage furrows. Obviously, however, utilization of this system is in no way related to the previous history of mitoses in the coenocyte. These cells have also managed to modify the phycoplast system so that it will also form the vacuolar envelope. We may note that these phycoplast microtubules are not associated with the centriole complexes, as is the case with other members of the Chlorococcales mentioned in this chapter. Thus we have a good example of how a cytokinetic system, common to other members of the Chlorococcales, may have become adapted during evolution of this particular genus. It seems clear that the tonoplast must be cleaved off the cytoplasm before the zoospores can be formed. Otherwise, if radial cleavage proceeded

Figures 3.75–3.77
RELEASE OF DAUGHTER COLONIES IN SCENEDESMUS QUADRICAUDA

3.75. Rupture of the parental trilaminar sheath and reticulate layer, and initiation of unfolding in the daughter colony. The spines *(sp)* are already beginning to extend (cf. Figs. 3.40, 3.69). The adhesive sites on the daughter colony are marked by large arrows; the coenobial adhesive of the parental colony by small arrows. × 14,000.

3.76. Whole cells after acetolysis for 1 hour at 95° C. The only resistant structures are the spines and two outer layers of the wall, which also form the ghosts of parental cells after release of daughter colonies (cf. Figs. 3.11, 3.147, 3.148). × 300 approx.

3.77. Scanning electron micrograph of mother coenobium which has released two daughter colonies. × 2,700.

SOURCE. Author's micrographs, paper in preparation.

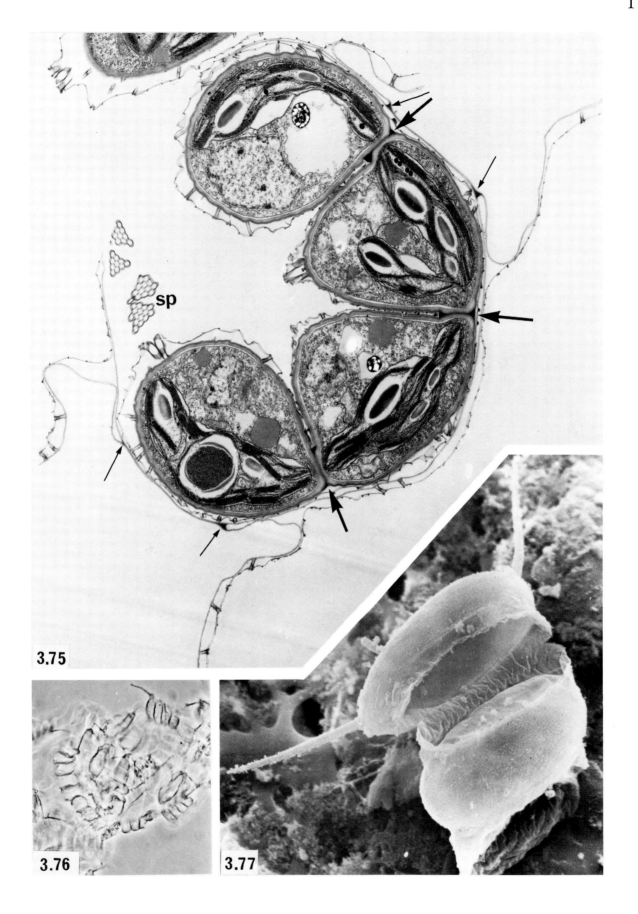

3.75

3.76

3.77

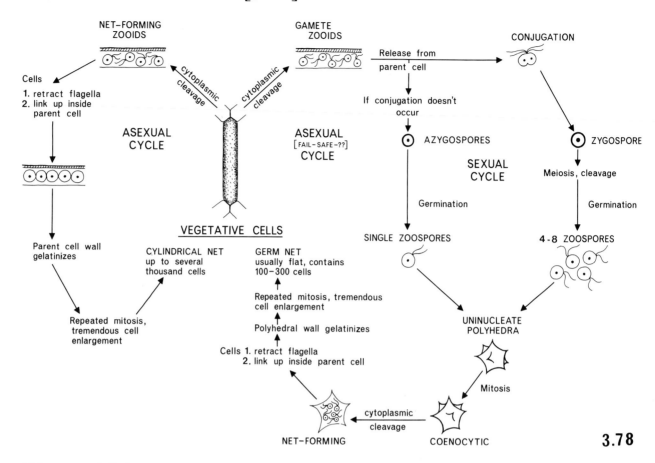

**Figure 3.78
LIFE CYCLE OF
HYDRODICTYON
RETICULATUM**

This life cycle is described in detail in the text, and various stages in it are illustrated in the following pages. Note particularly that cytoplasmic cleavage in vegetative cells can form either of two types of zooids (zoospore-like cells). Vegetative zoospores are retained within the parental wall and give rise to new *cylindrical* daughter nets (e.g., Figs. 3.79, 3.80). Gametes, however, are released from the parental cell. They re-form cylindrical nets only after having gone through a complex pattern of spore formation, germination, a polyhedral stage, and the *flat* germ net (Figs. 3.129–3.131).

SOURCE. Drs. H.J. Marchant and J.D. Pickett-Heaps, published in *Aust. J. Biol. Sci.* 24, 471 (1971).

**Figures 3.79–3.81
VEGETATIVE, CYLINDRICAL
NETS OF HYDRODICTYON
RETICULATUM**

3.79.   Newly formed net composed of hundreds of tiny cells. This print has been slightly overexposed to show the parental cell wall still enclosing the colony. The daughter net has formed in the shape of its parent cell. × 80.

3.80.   As for Figure 3.79, this slightly older net clearly reveals its constituent cells linked together. × 100.

3.81.   Vegetative net. The cells are still very small, each containing about four pyrenoids and several nuclei (invisible at this magnification). × 200.

SOURCE. Dr. H.J. Marchant, unpublished micrographs.

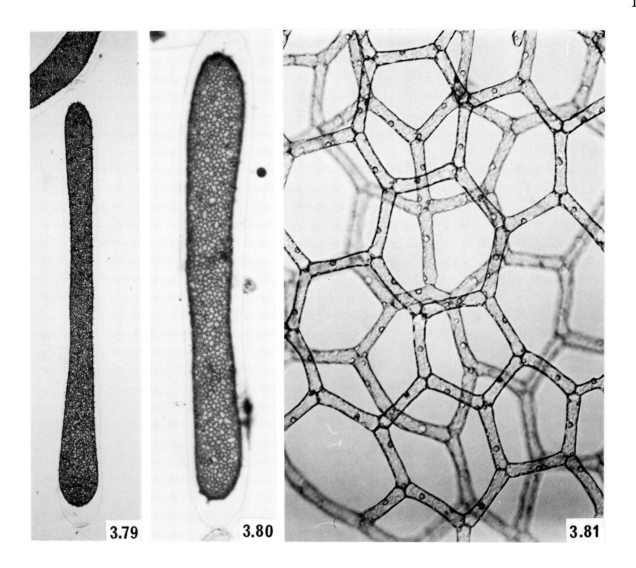

3.79

3.80

3.81

without this prior preparative step, the zoospores' surface would be a mosaic, partly covered by the tonoplast and partly by the plasmalemma, an arrangement that probably cannot be tolerated, since the properties of these two membranes are unlikely to be similar.

FORMATION OF CYLINDRICAL
DAUGHTER NETS

The way that the motile zooids link together to form beautifully regular daughter nets has long been a subject of interest among botanists. The events can easily be followed in live cells under the light microscope (Figs. 3.99–3.101) and are guaranteed to excite even the most lethargic student. However, it is not easy to time the induction of zoospores reliably enough for classroom demonstration unless the teacher is experienced in culturing these organisms (Marchant and Pickett-Heaps, 1972a).

After a period of intense activity during which they remain confined between the cell wall and the vacuolar envelope (Fig. 3.92), the live zooids more or less synchronously become quiescent. As their movements slow and finally cease, they shuffle about and eventually aggregate in a close-packed array upon the surface of the vacuolar envelope (Fig. 3.98). If carefully examined, their flagella can then be observed shrinking in length until each disappears. Most importantly, before settling, the zooids change their shape from oval to rhomboidal (Figs. 3.99, 3.102). Soon they have all adhered to one another, each usually to four neighbors (Fig. 3.101). The turgid vacuolar envelope acts as a mold during the aggregation of the masses of zooids. For example, if it is damaged or ruptured, irregular nets result. Zoospores released artificially from broken cells will also aggregate (Fig. 3.103), but never to form cylindrical nets.

McReynolds (1961) has performed experiments to investigate some aspects of net formation. In an interesting contrast to zoospores of *Pediastrum*, isolated zoospores of *Hydrodictyon* are apparently incapable of development into mature coenocytes. McReynolds interprets the results of several experiments to indicate that polarity of cell elongation is predetermined early in the zoospore's existence by its contact with other zoospores and the consequent resistance to growth engendered. We suspect (see later) that contact with other cells is predetermined by the

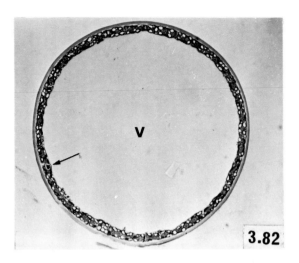

Figure 3.82
VEGETATIVE CELL OF HYDRODICTYON RETICULATUM

Cross section of a large, coenocytic vegetative cell. Inside the thick wall is the thin layer of cytoplasm containing the numerous nuclei, pyrenoids (arrow), and other organelles. This cytoplasm encloses the huge central vacuole *(v)*. × 640.

SOURCE. Drs. H.J. Marchant and J.D. Pickett-Heaps, published in *Aust. J. Biol. Sci. 24,* 471 (1971).

Figures 3.83–3.86
MITOSIS IN HYDRODICTYON RETICULATUM

3.83.   Live cell, flattened against the coverslip. Late anaphase and telophase nuclei (arrows) are visible through gaps in the extensive chloroplast. × 1,100.

3.84.   Early prophase. The centriole complexes *(cn)* are separating around the periphery of the nucleus as microtubules *(t)* proliferate between them. × 28,000.

3.85.   Metaphase. Paired centrioles lie in polar fenestrae. The typically chlorococcalean closed spindle is enclosed by the perinuclear envelope (arrows). × 21,000.

3.86.   Telophase. The nuclear envelope is reforming around the widely separated daughter chromosomes. × 22,000.

SOURCE. Drs. H.J. Marchant and J.D. Pickett-Heaps, published in *Aust. J. Biol. Sci. 24,* 471 (1971).

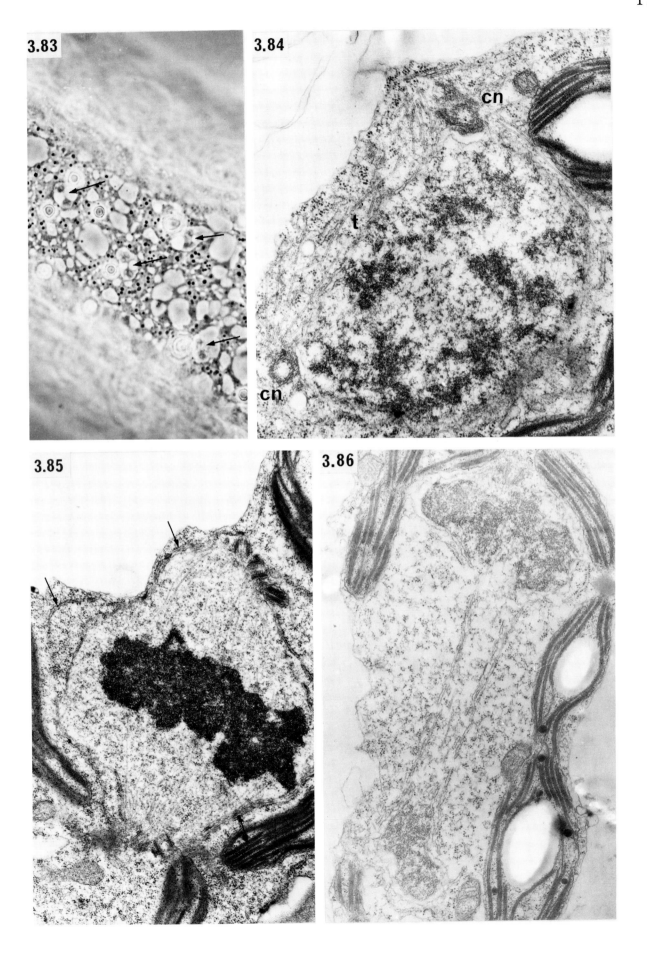

zoospore's internal asymmetry; i.e., that adhesion can occur only at sites on the cell surface predetermined by the cell's overall structure (see the cells in Fig. 3.130).

Electron microscopy suggests that after the flagella have shortened to about half their original length, the complement of flagellar tubules is assimilated into the cytoplasm intact (Fig. 3.107). Presumably, this is accomplished by fusion of the flagellar membrane with the plasmalemma. We believe we have actually seen this happening in living cells, but the details are difficult to discern. The flagellar tubules and basal bodies persist intact for a short time in the cytoplasm (Fig. 3.107). However, eventually their components dissociate and disappear. Net-forming zooids (in striking contrast to gamete zooids; Section 3.7b) always possess highly localized and characteristic arrays of parallel microtubules close to their plasmalemma (Figs. 3.105, 3.106). Precisely when these tubules appear is not clear; for example, phycoplast microtubules seem to disperse after zooid formation, but to distinguish clearly between these and the parallel microtubules present later is almost impossible. The cell membrane is usually distinctively flattened adjacent to the parallel microtubules, and thus we infer that the change from oval to rhomboidal shape undergone by the zooids in vivo (Figs. 3.99, 3.102) is related to the disposition of these tubules. Aggregation of the zooids seems to commence as these flattened regions of their membrane underlain by the tubules come in close contact (Figs. 3.105, 3.106). Amorphous material, difficult to detect, then appears in the narrow interstices between the flattened areas, and we suspect that this material acts as an adhesive. Soon after this, the cells are stuck together and start to enlarge (as in Fig. 3.130). As they begin to secrete their cell wall proper, the bands of peripheral microtubules all disappear, and the cell contents undergo considerable reorganization. The pyrenoid re-forms and very short centrioles now appear in small invaginations in the nuclear envelope (Fig. 3.109). To prove conclusively that these centrioles arise de novo is difficult, but this origin is considered likely, particularly since we have obtained clearcut examples of disintegrating basal bodies before these appear; furthermore, there is no trace of centrioles for the short intervening period of the cell cycle. The nucleus then commences its numerous series of mitoses, and the two vacuoles

Figures 3.87–3.93
ZOOSPORE FORMATION IN HYDRODICTYON RETICULATUM

3.87.   Live cell. At this early stage of zoosporogenesis, the pyrenoids have disappeared, and the chloroplast is fragmenting as cleavage progresses. × 1,000.

3.88.   The characteristic pavement stage in a live cell. Cytoplasmic cleavage into uninuclear fragments is complete. × 400.

3.89.   The zoospores in this live cell have rounded up and are just beginning to move about as their flagella become active. × 1,000.

3.90.   Radially oriented section of a cell equivalent to those in Figures 3.88–3.91. The vacuolar envelope (arrows) can just be discerned. × 900.

3.91.   This section of a cell is tangential and equivalent to the stage shown in Figure 3.88. A single nucleus can just be discerned in each fragment. × 900.

3.92.   Actively swimming zoospores were fixed and the parental cell sectioned radially for this micrograph. The zoospores, now rounded (cf. Fig. 3.89), are confined between the disintegrating parental wall (w) and the vacuolar envelope (arrow). × 1,200.

3.93.   Live, actively swimming zoospore released by artificial disruption of parental wall. Note the two flagella and contractile vacuoles (arrows). × 1,400.

SOURCE. Drs. H.J. Marchant and J.D. Pickett-Heaps, published in *Aust. J. Biol. Sci.* 24, 471 (1971).

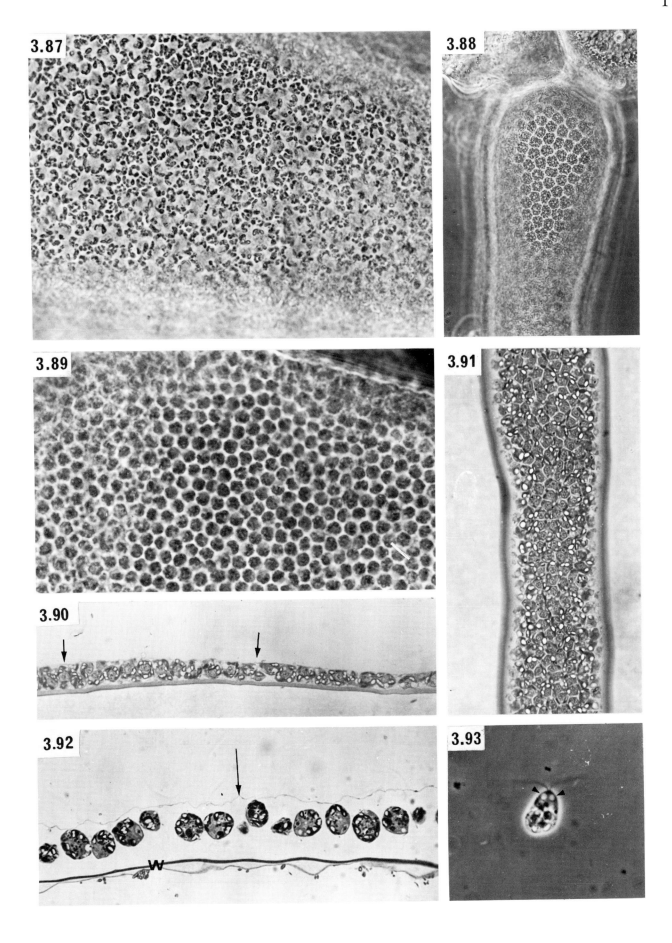

fuse and enlarge; thus the small elongated cells, firmly joined in the net-like pattern (Fig. 3.108), begin the cycle of growth that turns each into a giant coenocyte (Figs. 3.101, 3.104).

From these observations we can begin to guess at the mechanisms involved in net formation. The zooids may attract one another chemotactically (they can sometimes link together even after being released into the culture medium; Fig. 3.103), and they probably tend to aggregate in essentially planar arrays on the cylindrical vacuolar envelope which acts as a mold (see above and Pocock, 1960) whose curvature is insignificant compared with the size of the zoospores. We believe that the arrays of microtubules under the plasmalemma (similar to those of aggregating zooids in *Pediastrum*) are highly significant in the process of orderly aggregation (Section 3.8a). For example, colchicine treatment of aggregating zooids prevents their normal change of shape, and the treated cells, remaining oval, then aggregate in extremely irregular clumps, chains, etc. (Figs. 3.110, 3.111), which exhibit none of the order achieved in the absence of colchicine. Furthermore, these arrays of microtubules are not present in gamete zooids (Section 3.7b), which do not form nets. If we can presume that adhesive is secreted at specific sites at the ends of the cells, then flattening of these ends would obviously facilitate close packing and firm adhesion over a larger area of the cells than would be possible otherwise. The nature of the adhesive material is at present being investigated by enzyme treatments of aggregating zooids. Hawkins and Leedale (1971) indicate that they believe the cells are sticky all over their surface. We (Marchant and Pickett-Heaps, 1972a) find this difficult to accept, since one would then expect cells to clump together and subsequently adhere in any orientation. Instead, we feel the evidence indicates that the two ends of the cells are the regions that stick to one another.

In summary, we believe we understand some of the factors involved in net formation. But it is still not clear why some cells cannot project up from the planar array adopted by the others, unless the rhomboidal form acquired as the cells shuffle together is so rigid that the pattern of cells is unable to accommodate such a dislocation in a crystal-like planar lattice. Other aspects of net formation are discussed in Section 3.8a.

Figure 3.94
CYTOPLASMIC CLEAVAGE DURING ZOOSPOROGENESIS IN HYDRODICTYON RETICULATUM

The cleavage furrows are clearly dividing up the cytoplasm into uninucleate portions, having already cleaved off the vacuolar envelope (arrows), which appears irregular in this micrograph oriented tangential to the cell. These cleavages take place along phycoplast systems of microtubules (Figs. 3.95, 3.96). Note the absence of pyrenoids and the dense centriole complexes *(cn)* adjacent to nuclei. × 10,000.

SOURCE. Drs. H.J. Marchant and J.D. Pickett-Heaps, published in *Aust. J. Biol. Sci.* 24, 471 (1971).

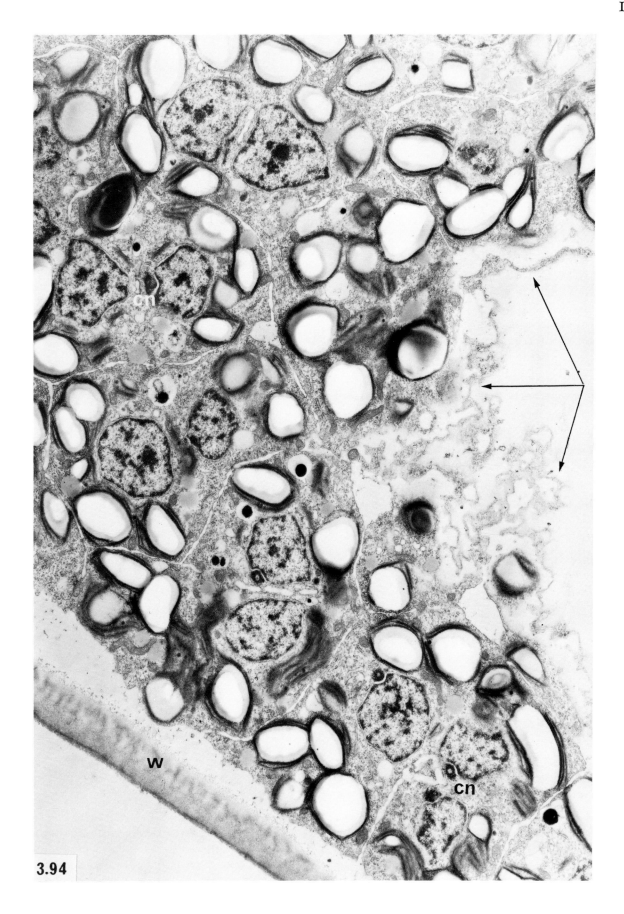

3.94

3.7b. Sexual (and Asexual) Reproduction
Involving Polyhedra

FORMATION OF GAMETE ZOOIDS

Vegetative cells produce gamete zooids in much the same way as they produce net-forming zooids, but much less frequently (except under certain environmental and experimental conditions), and the parental cell wall eventually ruptures to release the gametes. Conjugation in our specimens (Figs. 3.112, 3.113; Marchant and Pickett-Heaps, 1972b) occurred only among released gametes, never inside intact vegetative cell walls, and it was rare even when the released gametes clumped together for prolonged periods. The most obvious difference between gamete and net-forming zooids is the smaller size of the former; the gametes also usually contain more lipid and starch material.

At the ultrastructural level, these two types of zooids are similar in many respects, but important distinctions can be discerned between them. In particular, gametes *never* have the peripheral arrays of microtubules found in the net-forming zooids and important in net formation. Furthermore, a percentage of the gametes possesses an "apical cap," a small protrusion of cytoplasm located between the flagella, containing a densely stained collar of amorphous material (Fig. 3.114). As will be seen, this feature is possessed by only one of the two isogamous individuals in each conjugating pair, and it is never found in net-forming zooids.

CONJUGATION AND FORMATION OF SPORES

A gamete zooid failing to conjugate either dies or deposits a thick wall about itself to form an azygospore. Conjugation is difficult to follow in vivo as the tiny gametes are so active. However, initial contact of the cells is made near their flagellar bases (Fig. 3.112) and soon afterward the cells fuse laterally (Fig. 3.113). The quadriflagellate zygotes soon withdraw their flagella and secrete a cell wall to form zygospores.

At the ultrastructural level, a distinctive and probably important difference can be detected between gametes collected from conjugating and nonconjugating cultures. Among conjugating cells, the apical cap is extended to form a fertilization tubule (cf. *Chlamydomonas reinhardii*, Section

Figures 3.95, 3.96
CYTOPLASMIC CLEAVAGE DURING ZOOSPOROGENESIS IN HYDRODICTYON RETICULATUM

3.95. Formation of the vacuolar envelope *(ve)*. The cleavage furrow, growing through a phycoplast system of microtubules *(t)*, is giving rise to the very thin layer of cytoplasm bounded on its inside surface by the vacuolar membrane (arrows). × 25,000.

3.96. Radial cleavage of cytoplasm into uninucleate fragments (Figs. 3.91, 3.94) after the vacuolar envelope *(ve)* has been formed. As before, phycoplast microtubules *(t)* surround and predict the path of the ramifying cleavages. Note the mitochondrion "trapped" in one of the furrows (arrow). × 24,000.

SOURCE. Drs. H.J. Marchant and J.D. Pickett-Heaps, published in *Aust. J. Biol. Sci. 24,* 471 (1971).

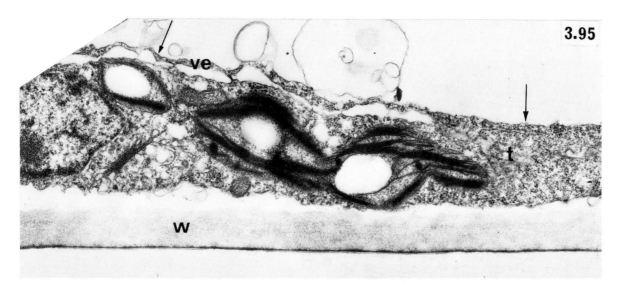

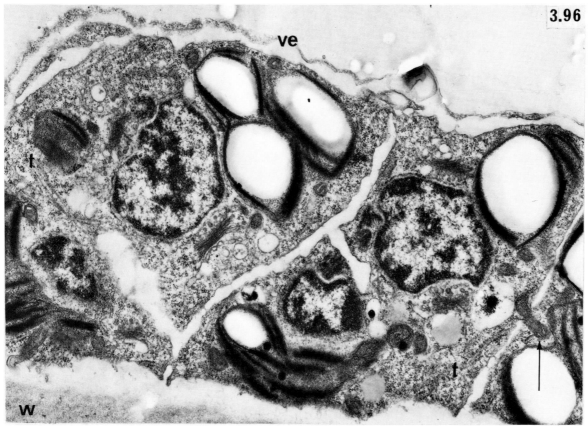

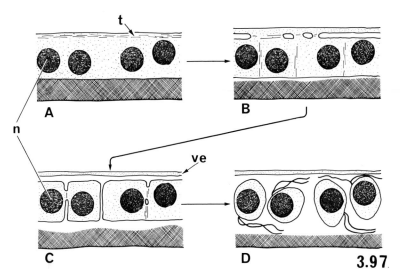

3.97

Figure 3.97
EVENTS LEADING TO ZOOSPORE FORMATION IN HYDRODICTYON RETICULATUM

After microtubules *(t)* appear just under the tonoplast *(A)*, the vacuolar envelope *(ve)* is cleaved off the inside of the cytoplasm *(B, C)*. Microtubules next appear between nuclei *(n)*, and further cleavage *(C)* gives rise to the uninucleate zoospores, which soon become flagellate *(D)*. Zoospores normally remain confined between the vacuolar envelope and the cell wall.

SOURCE. Drs. H.J. Marchant and J.D. Pickett-Heaps, published in *Aust. J. Biol. Sci. 24*, 471 (1971).

2.1b; Friedmann, Colwin, and Colwin, 1968), a thin protrusion of cytoplasm extending from the basal ring structure (Fig. 3.115). This extension has never been observed among nonconjugating cells (Fig. 3.114). Conjugation is apparently initiated when the fertilization tube contacts another cell at its apex (i.e., adjacent to the flagella), the latter cell, in our experience, invariably one without a basal ring or fertilization tubule (Fig. 3.116). Membrane fusion is shortly followed by the jackknifing and coalescing of the cells (Fig. 3.117). Caryogamy (nuclear fusion) probably occurs soon after this event, as many zygotes contain large, lobed nuclei with two nucleoli (Fig. 3.118), a condition never observed elsewhere in the life cycle; binucleate spores, however, are sometimes encountered. A wall is soon secreted around the cells (Fig. 3.119).

Figures 3.98–3.104
NET FORMATION IN HYDRODICTYON RETICULATUM

3.98. Very early stage in linkage of live zoospores, which are still rounded or somewhat angular (Fig. 3.99) as they become quiescent. × 560.

3.99. These live zoospores, just becoming quiescent (Fig. 3.98), are distinctly angular or rhomboidal in form as they begin to stick together. × 1,400.

3.100. Quiescent cells, now linked together, are beginning to elongate (see also Fig. 3.128). × 560.

3.101. A stage a little after that in Figure 3.100. Each cell has developed a small vacuole at either end of the central pyrenoid and nucleus (the latter invisible at this magnification). × 560.

3.102. Scanning electron micrograph of zoospores at the stage shown in Figure 3.99. The flagella are being retracted. Note the shape of the cells. × 4,000 approx.

3.103. The lateral adhesion often observed in zoospores artificially released from their parental cell and allowed to aggregate free in the culture medium. × 860.

3.104. Further elongation of cells (cf. Fig. 3.101) of a very young daughter net. × 1,400 approx.

SOURCE. Drs. H.J. Marchant and J.D. Pickett-Heaps, published in *Aust. J. Biol. Sci. 25*, 265 (1972).

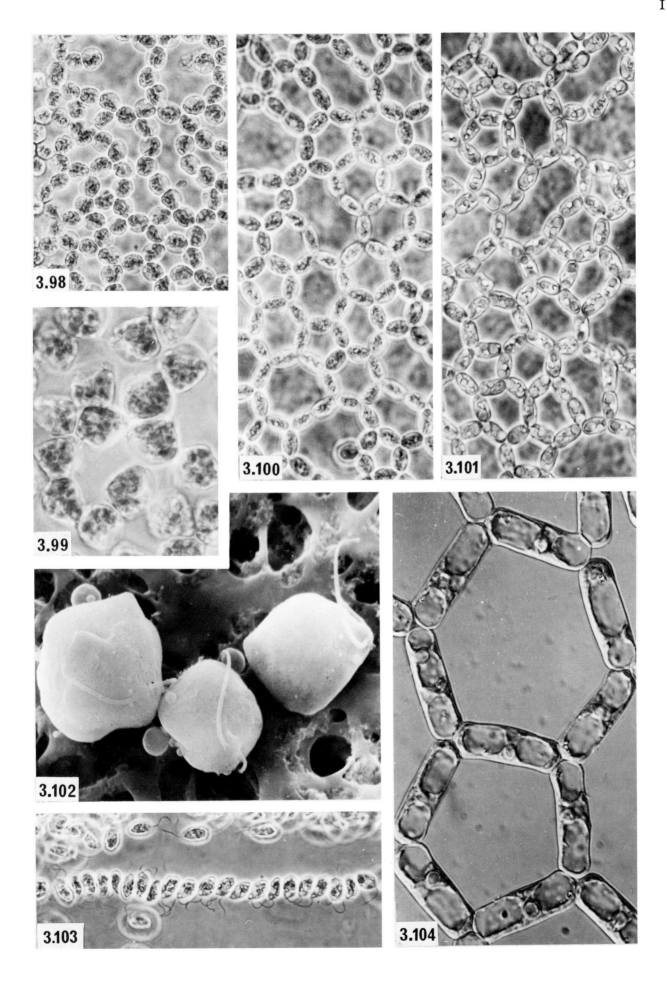

These observations suggest that, as in so many other gametes (and in particular, male sperm cells), at least one of the conjugants has to undergo some sort of structural modification or "capacitation" before conjugation can take place. This idea is briefly discussed in our paper (Marchant and Pickett-Heaps, 1972b).

DEVELOPMENT OF POLYHEDRA AND FLAT GERM NETS

While zygospores are generally larger than azygospores (Fig. 3.120), considerable variation in their size is often encountered. Both spores are resistant to desiccation and turn from green to orange as they age. These spores have proved difficult to examine ultrastructurally (Marchant and Pickett-Heaps, 1972c), mainly owing to the large amount of lipid they contain and the denseness of their cytoplasm. They are enclosed by a thick wall which weakens and balloons out upon germination to form a vesicle into which the zoospores emerge. We have not been able to observe meiosis, which according to Proskauer (1952) precedes germination of zygospores and thereby restores the haploid form of the vegetative cell cycle. Azygospores usually give rise to a single zoospore, whereas zygospores normally form four following meiosis and cytoplasmic cleavage (Fig. 3.121) mediated by the usual phycoplast. For release of the zoospore(s), the wall of the spore becomes eroded (Fig. 3.126) and then balloons outward (Figs. 3.121, 2.122). The biflagellate zoospores are large, full of lipid (Fig. 3.126), and very sluggish, seldom moving far when released from the vesicle of the spore. After a few minutes of motility, the zoospores withdraw their flagella and rapidly secrete a thickening wall around themselves. This wall immediately develops several outgrowths as well, and thus the cells soon acquire the characteristic, somewhat irregular polyhedral shape (Figs. 3.122, 3.123). The germinating zygospore in Figure 3.122 is unusual in that the four zoospores have precociously turned into polyhedra before being released from the vesicle.

The polyhedra (Fig. 3.123) now undergo a period of enlargement, during which usually around eight mitotic cycles take place, so that approximately 256 ($= 2^8$) nuclei are present. Cytoplasmic cleavage (Fig. 3.124) then forms unincleate zooids in a fashion similar to what has already been described in vegetative coenobia.

Figures 3.105–3.109
AGGREGATING ZOOSPORES OF HYDRODICTYON RETICULATUM

3.105. Zoospores just beginning to adhere to one another, with their system of peripheral microtubules (t) juxtaposed in adjacent cells. × 28,000.

3.106. Another section showing these peripheral microtubules. Note the trace of amorphous material (adhesive ?) between the cells. × 25,000.

3.107. Flagellar tubules retracted within the cytoplasm of a quiescent zoospore. × 16,000.

3.108. The complex adhesive site between two cells now firmly attached together and secreting their walls. × 24,000.

3.109. A pair of centrioles, apparently forming de novo on the nuclear envelope of a quiescent zoospore at the stage shown in Figure 3.130. The basal bodies of the flagellar apparatus earlier used by the zoospore seem to disintegrate at this stage. × 35,000.

SOURCE. Drs. H.J. Marchant and J.D. Pickett-Heaps, published in *Aust. J. Biol. Sci. 25*, 265 (1972).

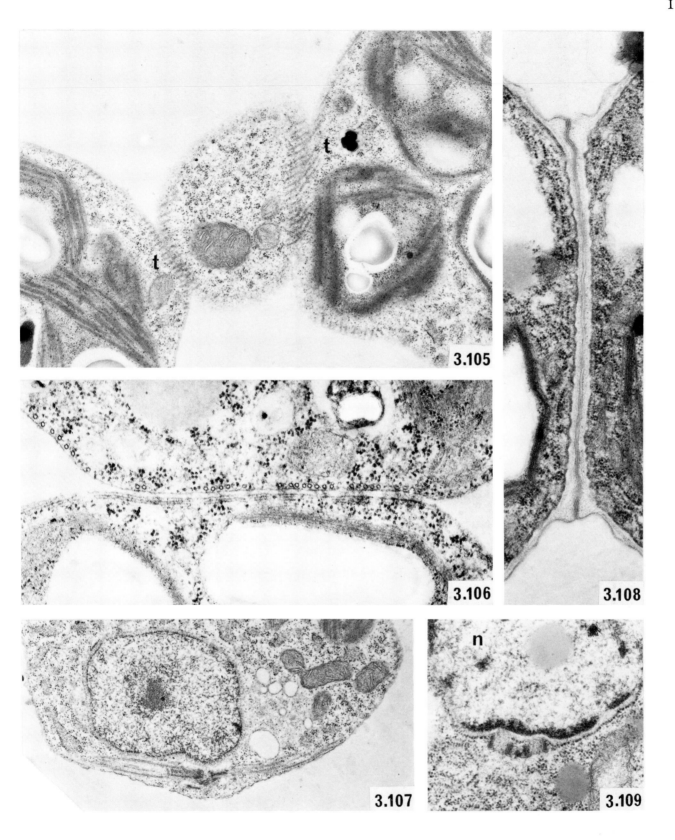

3.105

3.106

3.107

3.108

3.109

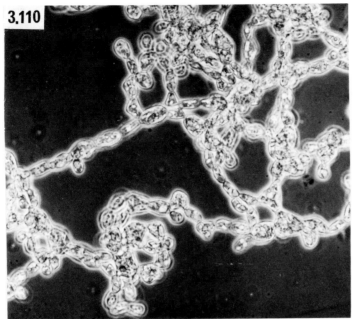

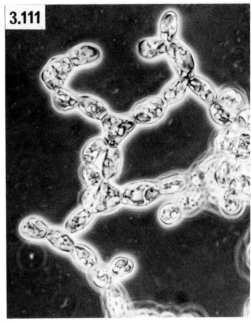

Figures 3.110, 3.111
COLCHICINE TREATMENT OF AGGREGATING ZOOSPORES OF HYDRODICTYON RETICULATUM

3.110.   Irregular linkage of zoospores following colchicine treatment has produced this tangled colony. × 1,000.

3.111.   As for Figure 3.110, showing more clearly the irregular form and adhesion of these cells. × 1,600.

SOURCE. Dr. H.J. Marchant, unpublished micrographs.

The vacuole in polyhedra often disappears before cytoplasmic cleavage; if it is still present, it is usually small, and the tonoplast is cleaved off as usual before zoospore formation, but the resultant vacuolar envelope never, in our experience, plays any significant role in subsequent net formation. Zoospores differentiate from the uninucleate fragments of the cytoplasm, during which time the polyhedral wall becomes much thinner (Fig. 3.127). The wall then ruptures, and a swelling vesicle balloons out of the rupture (cf. *Pediastrum;* Section 3.8a). The origin of the vesicle in unclear, but it is probably derived from the inner layer of the polyhedral wall. The active zoospores then swarm into the vesicle, leaving the old empty ghost cell wall attached to the vesicle (Fig.

Figures 3.112–3.117
SEXUAL REPRODUCTION IN HYDRODICTYON RETICULATUM

3.112.   Live gametes just becoming united at the base of their flagella. × 1,200.

3.113.   Live zygote; cytoplasmic fusion complete. Note the four flagella. × 1,200.

3.114.   Gamete, showing the apical cap (arrow) situated between the flagellar bases. × 43,000.

3.115.   Sexually active gamete from a conjugating culture. A fertilization tubule (arrow) is extended from the apical cap. × 50,000.

3.116.   Initiation of plasmogamy (cf. Fig. 3.112), which has occurred at the flagellar bases. Note the dense ring (arrow; see Fig. 3.115) derived from the apical cap. × 38,000.

3.117.   Jackknife fusion of the gametes after plasmogamy. The dense ring, formerly located at the base of the fertilization tubule, is still visible between the flagellar bases (arrow). × 18,000.

SOURCE. Drs. H.J. Marchant and J.D. Pickett-Heaps, published in *Aust. J. Biol. Sci. 25,* 279 (1972).

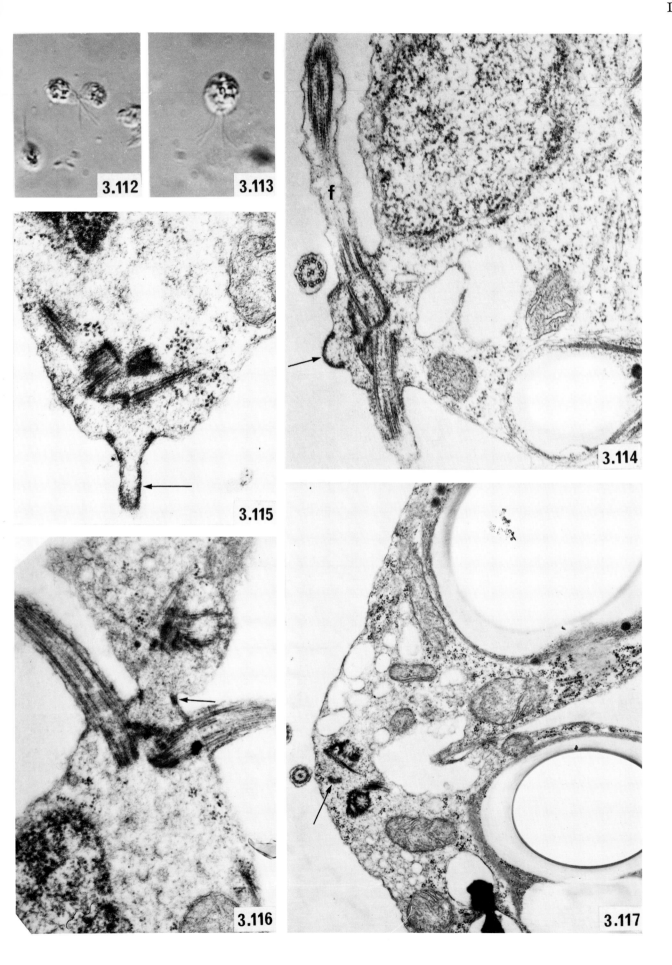

3.112

3.113

3.114

3.115

3.116

3.117

3.125). Sometimes, the vesicle is not formed and the zoospores instead swarm inside a greatly distended polyhedral wall.

After a period of activity inside the large and spherical vesicle, the zoospores begin to aggregate and jostle together (Fig. 3.125; Marchant and Pickett-Heaps, 1972c). Just as previously described for cylindrical net formation, their shape changes as they withdraw their flagella. In our experience, the resultant germ nets (Figs. 3.128, 3.129, 3.131) are predominantly flat, but irregularities or sack-like nets are not unusual, especially in old, crowded cultures or if the vesicle does not expand enough to allow the flat configuration to be attained. Pocock (1960) and Pringsheim (1861) say that germ nets are usually hollow, but we feel that given suitable conditions, their natural tendency is to be flat (Marchant and Pickett-Heaps, 1972c).

The cells' ultrastructure and differentiation before, during, and after (Fig. 3.130) net formation appear identical to what has already been described for their counterparts derived from vegetative coenocytes forming cylindrical nets (e.g., they contain the microtubule bands, adhere in the same fashion, and form centrioles de novo). We are confident, therefore, that essentially the same events and cytoplasmic mechanisms are involved in linkage whether the nets will be flat or cylindrical. The cells of the germ net soon elongate and grow as before; however, since the outer cells often grow faster than the inner ones, the nets tend to become saucer-shaped and increasingly tangled. There is often a conspicuous hole near the center of a germ net (Figs. 3.129, 3.131). Two prongs, quite reminiscent of those that characterize the outer cells of *Pediastrum* (Section 3.8a), develop in some cells that have not linked up in the normal fashion to four other individuals. These prongs are sometimes seen inside the colony (Fig. 3.131), but are most frequently encountered oriented outward in some of the peripheral cells (Figs. 3.129, 3.131, 3.132). Furthermore, the prongs are always on the side of the cell occupied by the chloroplast (cf. *Pediastrum*). We have discussed (Marchant and Pickett-Heaps, 1972d) the possibility, first suggested by Pocock (1960), that the prongs represent two unsatisfied linkage sites on the cell's surface. It appears more likely to me that adhesion to other zoospores mechanically suppresses the innate propensity of the cells to elon-

Figures 3.118, 3.119
ZYGOTES IN HYDRODICTYON RETICULATUM

3.118. Caryogamy (nuclear fusion) is indicated by the large size of this nucleus and the two nucleoli (*nc*) in a young zygote. × 20,000.

3.119. Zygote, or possibly an azygote, secreting its wall. × 11,000.

SOURCE. Drs. H.J. Marchant and J.D. Pickett-Heaps, published in *Aust. J. Biol. Sci. 25*, 279 (1972).

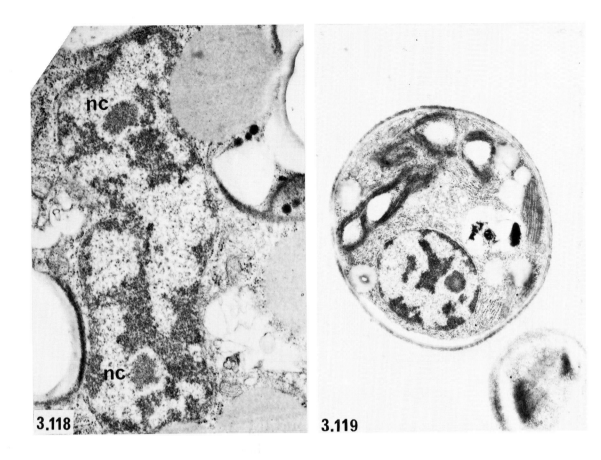

3.118

3.119

gate at these sites, as appears also to be the case with, for example, the inner cells of colonies of *Pediastrum.* These prongs are yet another indication of the close phylogenetic relation between *Pediastrum* and *Hydrodictyon.* Cells of germ nets usually reproduce asexually to give rise to new cylindrical daughter nets, as previously described. Further discussion of the mechanisms involved in the formation of planar nets is given in the section on *Pediastrum.*

3.8 Pediastrum

Pediastrum is a close relative of *Hydrodictyon* (both belong to the family Hydrodictyaceae), and therefore similarities between these two genera are not surprising. *Pediastrum* is a most characteristic and easily recognizable colonial organism in which 2^n cells (where n normally lies between 2 and 7) are joined together to form a flat, disc-shaped colony normally one cell thick. Quite often the individual cells are arranged in concentric circles (e.g., 5 and 10 cells surrounding a central one), and they are usually lobed (Figs. 3.133, 3.134). However, in the centrally situated cells, these two marginal lobes are poorly developed, whereas in the outermost cells, the two prominent pointed lobes extend radially outward from each cell. As a result, numerous rather irregular holes are often present in the central region of the colony, while an overall profile may resemble a deeply indented gear wheel. At high magnifications and under the scanning electron microscope, the surface of the cells may be seen to be ornamented, while fine bristles often emanate from the tips of the lobes of outer cells (see also Smith, 1916, p. 472).

Asexual reproduction in *Pediastrum* has many similarities to that in *Hydrodictyon.* Each cell in the colony grows and becomes multinucleate as a result of a series of synchronous mitoses of the original nucleus (Smith, 1916). Then the pyrenoid disappears as the whole cytoplasm cleaves to form biflagellate zoospores (Figs. 3.135*A,* 3.138) and the parental wall tears open in a characteristic fashion (Figs. 3.142, 3.148) to release these active cells, which still remain enclosed by a lenticular vesicle (Figs. 3.135*A–C,* 3.136). Smith (1916) describes in detail the spatial progression of the cleavages. After a brief period of motility, the zoospores aggregate in a planar

Figures 3.120–3.125
SEXUAL REPRODUCTION IN HYDRODICTYON RETICULATUM

3.120. Diploid zygotes *(z)* and a smaller haploid azygote *(az).* A central nucleus and a pyrenoid are visible in each cell. × 900.

3.121. Germination in a zygote. The cytoplasm has cleaved after the meiotic divisions. The wall is swelling (arrow) prior to release of the large, sluggish zoospores (cf. Fig. 3.126). × 900.

3.122. Later stage in germination. The zygote wall has ruptured, releasing the cell's contents. Normally four large, slow moving zoospores are freed, but in this case, the zoospores have already precociously turned into polyhedra. × 960.

3.123. Polyhedral coenocytic cell arising from the zoospore of a germinating zygote. × 3,000.

3.124. Zoospore formation in a live, mature polyhedron. The pavement stage shown here is equivalent to that in Figure 3.88. × 1,200.

3.125. Live zoospores swarming inside a thin vesicle, after release from a polyhedron. The vesicle is probably derived from the inner wall layer of the original polyhedron. Note its shape. × 170.

SOURCE. Drs. H.J. Marchant and J.D. Pickett-Heaps, published in *Aust. J. Biol. Sci. 25,* 1187 and 1199 (1972).

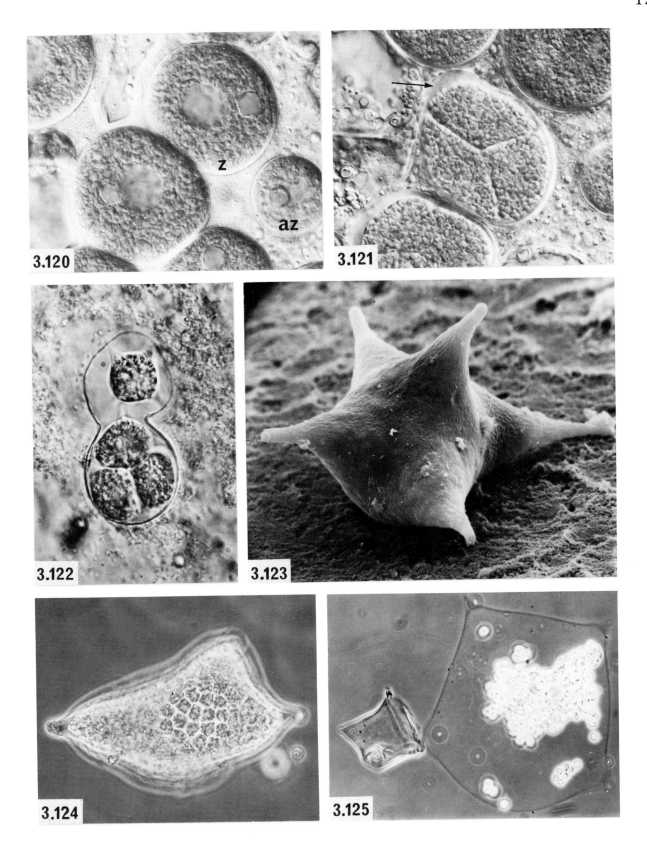

3.120

3.121

3.122

3.123

3.124

3.125

array and withdraw their flagella (Figs. 3.135B, 3.135C, 3.137). They soon adhere irreversibly to one another and enlarge as the cells develop the shape (Fig. 3.138) that gives the mature colony its characteristic form by the time it has been released from the vesicle. This reproductive cycle can often be induced (within 24 hours by transferring older—2 or 3 weeks—colonies to fresh medium) with sufficient reliability to make it a suitable phenomenon for demonstration to students. Even the most unresponsive class can be galvanized by the precision of the whole process, which is often essentially over within 30 minutes. Occasionally, two or more colonies appear in the one vesicle, each obviously containing a smaller number of cells than is usual. Similar behavior is normal during vegetative reproduction in the closely related *Euastropsis,* where the 2–32 sluggish zoospores released in one vesicle, as in *Pediastrum,* pair off and give rise to a number of two-celled colonies (Fritsch, 1935; p. 167).

Pediastrum simplex is known to have a sexual reproductive cycle, and in general terms, its life cycle resembles that of *Hydrodictyon* (Davis, 1967). Biflagellate isogametes are smaller than the colony-forming zoospores, but are formed and are then released in much the same manner as the latter. Conjugating pairs of cells turn into zygotes, which later germinate to produce numerous (Davis, 1967) zoospores, which in turn form polyhedra. After a period of growth and mitosis, the cytoplasm of the polyhedra cleaves to give another generation of zoospores, which this time aggregates into the typical plate-like colony (but see below for a variation of this behavior). Davis (1967) has demonstrated an asexual form of reproduction involving azygotes, and as might be expected from the life cycle of *Hydrodictyon,* these azygotes arise from gametes that have not undergone conjugation. Some azygotes, however, arise from larger motile cells (zoospores?). Davis's diagrams also indicate that conjugation commences by fusion of the cell membrane either laterally or at the end of the cells away from the flagella. These configurations are different from what has been observed in *Hydrodictyon* and suggest that perhaps gametes of *Pediastrum* do not have the copulatory apparatus (e.g., the basal ring and fertilization tubule between the flagella) already described for *Hydrodictyon.* Davis records another variation in behavior, where the polyhedra germinate to form a distinc-

Figures 3.126, 3.127
SEXUAL REPRODUCTION IN HYDRODICTYON RETICULATUM

3.126. Release of zoospores from a zygote. One flagellum can be seen (arrow). These zoospores are large in comparison with net-forming zoospores, and they contain much lipid *(l)* material. × 7,000.

3.127. Cytoplasmic cleavage in a polyhedron (cf. Fig. 3.124). Except for a lack of the vacuolar envelope, these zoospores are formed in similar fashion to those of cylindrical vegetative cells (Figs. 3.87–3.96). Note the flagella (arrows) being extruded and the inner wall layer (the future swarming vesicle) becoming diffuse. × 4,400.

SOURCE. Drs. H.J. Marchant and J.D. Pickett-Heaps, published in *Aust. J. Biol. Sci. 25,* 1187 and 1199 (1972).

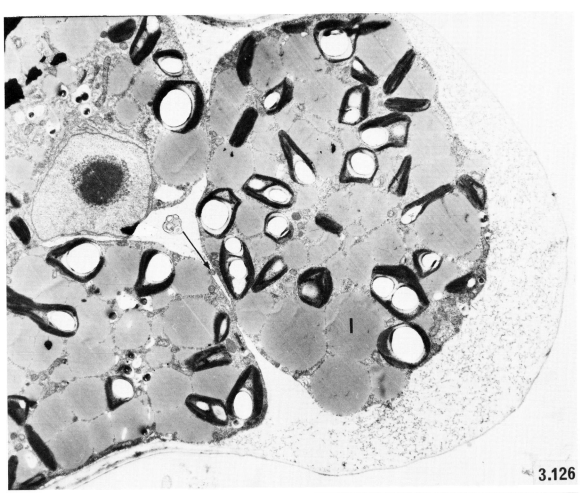

3.126

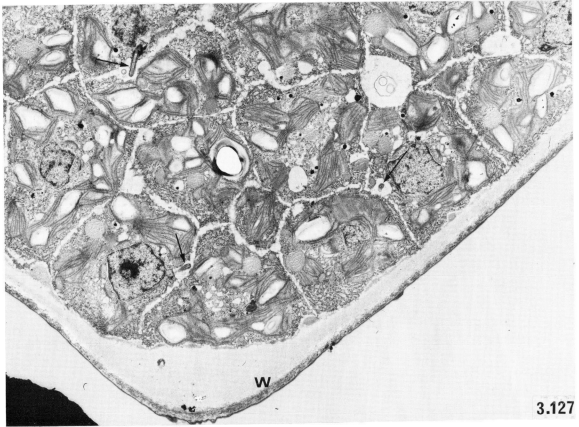

3.127

tive four-celled colony, each cell of which in turn gives rise to a spheroidal colony later on.

At the ultrastructural level, the cells (of *P. boryanum*) are quite simple in organization and little need be said here concerning them. Centrioles are persistent in the coenobia, and the cytoplasm contains numerous lipid droplets. Golgi bodies, as usual, are associated with the nuclear envelope.

3.8a. Cell Division and Colony Formation

Mitosis and cytoplasmic cleavage in *P. boryanum* closely resemble what has already been described for *Hydrodictyon* (Marchant, 1974a). Mitosis is synchronous in multinucleate cells (Fig. 3.139), and the mitotic spindles appear identical information and structure (including the possession of a perinuclear envelope) to those in *Hydrodictyon*. Likewise, cytoplasmic cleavage in multinucleate cells is mediated by typical arrays of phycoplast microtubules. However, in *Pediastrum* these microtubules are often associated spatially with centrioles; also, since there is no large vacuole, a vacuolar envelope is not formed during cleavage. After cytoplasmic cleavage is complete, the two centrioles elongate and then give rise to flagella. Release of the zoospores (Fig. 3.140, 3.141) is then effected when a rupture develops in the ornamented outer layer of the cell wall (Fig. 3.142), through which the inner layer of wall balloons out forming the vesicle inside of which the zoospores swarm.

Wall secretion and colony formation in *Pediastrum* (Fig. 3.143) have been recently investigated by a number of persons (Gawlik and Millington, 1969; Millington and Gawlik, 1970; Hawkins and Leedale, 1971; Marchant, 1974a). In essence, the process appears similar to what has been described earlier for *Hydrodictyon* (Section 3.7a), and the zoospores undergo analogous shape changes as they aggregate (Smith, 1916). Hawkins and Leedale fixed their material in osmium alone and so did not preserve microtubules; they consider that the chloroplast "is clearly a major factor in the orientation of the cells." Millington and Gawlik illustrate arrays of microtubules underlying the plasmalemma at the sites of contact of the zoospores (exactly as described for *Hydrodictyon*), and Marchant (1974a) has confirmed their results. However, Millington and Gawlik consider these microtubules more

Figures 3.128–3.130
FLAT GERM NET FORMATION IN HYDRODICTYON RETICULATUM

3.128.   Live, quiescent cells (just previously actively swarming; Fig. 3.125) aggregated in the typical planar array of the germ net, still contained within the distended swarmer vesicle (arrows). × 380.

3.129.   Germ net a little later in development, now free of the swarming vesicle. The meshes are somewhat irregular. The large central hole and the appearance of prongs (just visible; arrows) are characteristic of these nets (see also Figs. 3.131, 3.132). × 400.3.130.   Linked cells of very young germ net, equivalent to those in Figure 3.128 (and Fig. 3.98). The cells, not yet appreciably vacuolated, are enclosed within a thin cell wall. Note the subtle symmetry within the cells. The nuclei, although on either side of a mesh, are always centrally situated. × 8,100.

SOURCE. Drs. H.J. Marchant and J.D. Pickett-Heaps, published in *Aust. J. Biol. Sci. 25*, 1199 (1972).

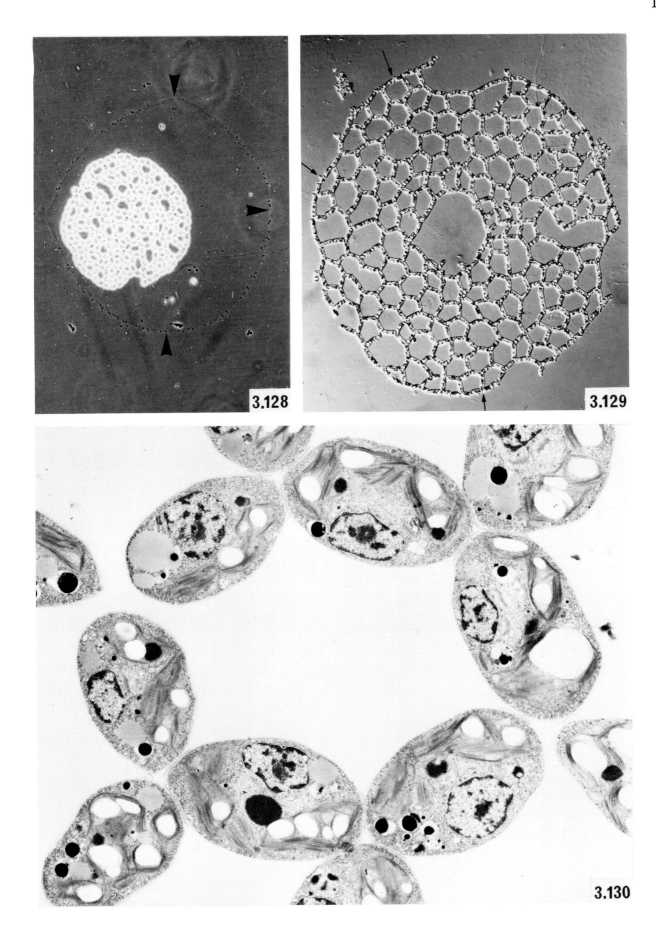

3.128

3.129

3.130

likely an expression of cell orientation, reflecting cell asymmetry, than a specific functional part of net formation. They also suggest that the microtubules "appeared to be more related to zoospore function than to wall formation," but what precise function these could have in the zoospores is not made clear (such microtubules are not present in zoospores of other Green Algae). Our own experiments, paralleling those conducted with aggregating zoospores of *Hydrodictyon* (Section 3.7a), throw some light on the zoospores' importance in colony formation. Colchicine treatment of zoospores prior to aggregation prevents their characteristic change in shape; this drug, as expected, prevents proper cleavage as well, and causes an undifferentiated mass of cytoplasm to be released from parental cells (Fig. 3.146). Subsequent to colchicine treatment of zoospores, regular linking up is never achieved; some cells aggregate irregularly and others remain unicells. We conclude that, as in *Hydrodictyon,* flattening of the zoospore's surface by microtubules and the change in the zoospore's shape are important in its orderly adhesion to its neighbors (Fig. 3.143). Mechanical disruption of colonies shortly after aggregation (when they are loosely held together) produces unicells, all of which develop the two prongs from one side (Fig. 3.145). (This experiment incidentally demonstrates that *all* cells of a colony possess prong-forming capability and not just those cells in the outer circle, as has been suggested by some earlier workers.)

Microtubules are not, however, involved only in flattening the sides of the zoospores. Colchicine treated unicells which are rounded up and have not linked together later enlarge, but they generally *remain* spherical even when the colchicine has been removed; they never develop their prongs. So development of cell asymmetry and general form (and perhaps differentiation of the plasma membrane) appears to be mediated by the peripheral microtubules as well, early in zoospore differentiation. This conclusion is strengthened by the observation that microtubules are seen running up into the prongs as they are formed (Fig. 3.144). In the genus *Sorastrum,* the relation of similar microtubules to prong formation is most pronounced (Section 3.9a; Figs. 3.159, 3.161). The disposition of these peripheral microtubules in the cell (e.g., the number of separate groups of them near the

Figures 3.131, 3.132
GERM NET OF HYDRODICTYON RETICULATUM

3.131. This scanning electron micrograph shows a double layer of cells near the central hole and prongs on some cells around the hole and at the periphery of the colony. × 770.

3.132. Some pronged cells at the outside of a germ net. × 2,700.

SOURCE. Drs. H.J. Marchant and J.D. Pickett-Heaps, published in *Aust. J. Biol. Sci. 25,* 1199 (1972).

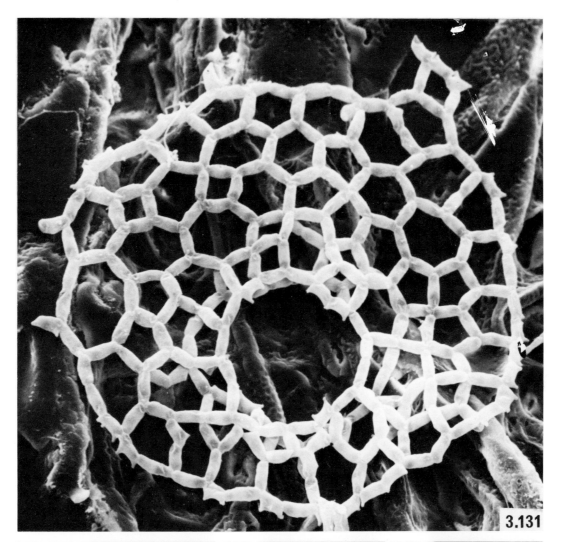

3.131

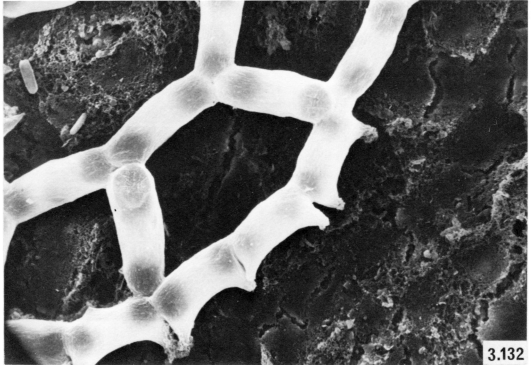

3.132

plasmalemma and how precisely their orientations match in adjacent aggregating zoospores) has not yet been precisely defined.

The surface of the cell adjacent to the chloroplast (which incidentally always later on produces the two prongs) always faces the outside in cells composing the concentric rings (Fig. 3.143). In our models of aggregating zoospores, we cannot explain yet why they adopt this configuration. These and other problems will probably have to be answered before we begin fully to understand the mechanics of colony formation.

Davis (1964) has ascribed the formation of a planar net to the restrictive influence of a flattened vesicle in which the zoospores are obliged to adhere to one another in a single plane. The research of Hawkins and Leedale (1971) and our own work indicate that this is an insufficient explanation, since the vesicle may be three or more times wider than the layer of zoospores contained within it; Ingold (1973) draws a similar conclusion. It will also be remembered that zoospores of *Hydrodictyon* released from polyhedra usually form planar nets even when they are contained within large spherical vesicles.

Millington and Gawlik note that wall deposition is initiated by the appearance of extracellular plaques of material which grow and eventually coalesce, the initial sites of these plaques being presumably templated by the plasmalemma. They suggest that there is a spatial correspondence between the sites of these plaques and clusters of ribosomes near the plasmalemma, but this latter correlation is not obvious in their published micrograph. Atkinson, Gunning, and John's (1972) results on the sporopollenin layer in *Chlorella* (Section 3.2) suggest that the outer warty and stable layer on the surface of *Pediastrum* (see Figs. 3.140, 3.142) also contains sporopollenin, as has been also described for *Scenedesmus* (Section 3.6). Ghost colonies always remain when the parental cells have undergone reproduction (Figs. 3.135–3.138, 3.142). I have reduced whole colonies to such ghosts by acetolysis (Figs. 3.147, 3.148), a result that strongly supports the existence of sporopollenin in this outer layer. Prescott (1970; p. 90) notes that the "walls" of *Pediastrum* are very resistant to decay and are found in semifossil form (as are pollen grains); he also says he has never observed this genus being parasitized by

Figure 3.133
PEDIASTRUM BORYANUM

This scanning electron micrograph shows beautifully how the cells are arranged in the 16-celled colony. Note that the cells are not exactly coplanar. Some inner cells have formed prongs projecting over other cells, suggesting that in these cells prong formation has been suppressed by cell adherence to one another (see also Fig. 3.145). × 4,000.

SOURCE. Dr. H.J. Marchant, paper in preparation.

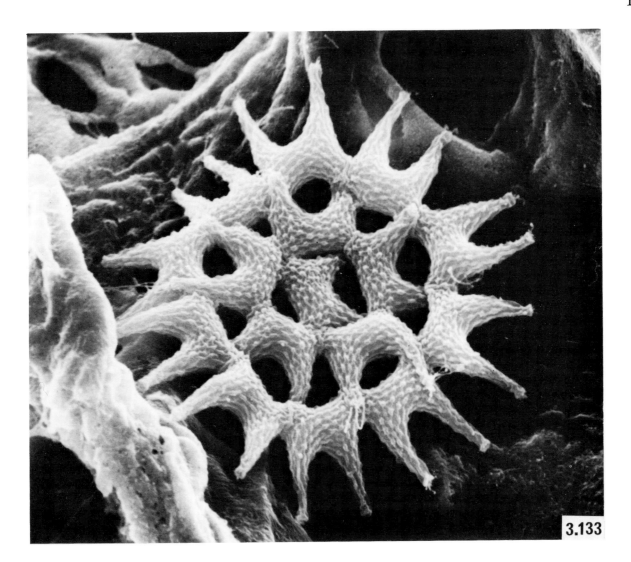

3.133

aquatic fungi or bacteria, a property obviously attributable to the outer wall layer.

3.9. Sorastrum

The cells of the genus *Sorastrum* are colonial, but unlike *Pediastrum* and *Hydrodictyon,* the colonies are spherical. The cells are interconnected to one another by their single stalk-like "stipe," and the axis of each cell is radially oriented. Each stipe is attached laterally at the end to several others, and the attachment zone forms a small sphere in the center of the colony. The cells are somewhat conical or pear-shaped, with their broadened outer portion bearing several large spines directed outward (two pairs in many species).

Vegetative reproduction is typically chlorococcalean (Davis, 1966b). After a period of enlargement, the cytoplasm of a parental cell cleaves into zoospores, which are released into a swelling vesicle through a rent in the wall. After a few (3–5) minutes of activity, they clump together, and by the time they have become quiescent, they have formed a new colony. During this process, the zoospores also undergo striking changes in shape as the horns and the stipe appear. The colonial organization appears to impose some restrictions upon the shape of maturing cells (this is true, for example, of *Pediastrum* too, in which the prongs of the central cells are stunted). Individual cells that have not formed part of a colony are appreciably more rotund than normal and are not unlike some cells of *Tetraedron.* Resistant orange akinetes have been described for this genus (Davis, 1966b), but I have not encountered any reports of sexual reproduction.

3.9a. Cell Division and Colony Formation

Ultrastructure and reproduction of *S. spinulosum* have been described by Marchant (1974b). Vegetative cells show no unusual cytoplasmic features. The four spines are situated in the corners of the somewhat pyramidal cells (Fig. 3.149) and in old cells contain only wall material (Fig. 3.160). The stipe is a larger cylinder of wall material (Fig. 3.160), thickened at its inner end where it is appressed to several others; each thereby terminates in a hexagonally faceted knob, and together these knobs constitute the

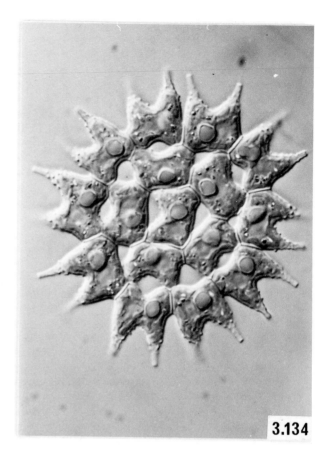

3.134

Figure 3.134
YOUNG COLONY OF PEDIASTRUM BORYANUM

This colony is similar to that in Figure 3.133. Each cell has a large pyrenoid, next to which is the less conspicuous nucleus, single at this stage. × 1,200.

SOURCE. Dr. H.J. Marchant, paper in preparation.

Figure 3.135
COLONY FORMATION, LIVE CELLS OF PEDIASTRUM BORYANUM

3.135*A.*   Numerous young daughter colonies, still within their swarming vesicle, clustered around the parental colony. One cell has just released a mass of active zoospores (large arrow). All cells remaining in the colony are undergoing cytoplasmic cleavage except one (small arrow), which still has its pyrenoid intact. × 1,000 approx.

3.135*B, C.*   Micrographs taken a few minutes after that above, showing two different stages in colony formation by the group of zoospores in Figure 3.135*A.* × 1,000 approx.

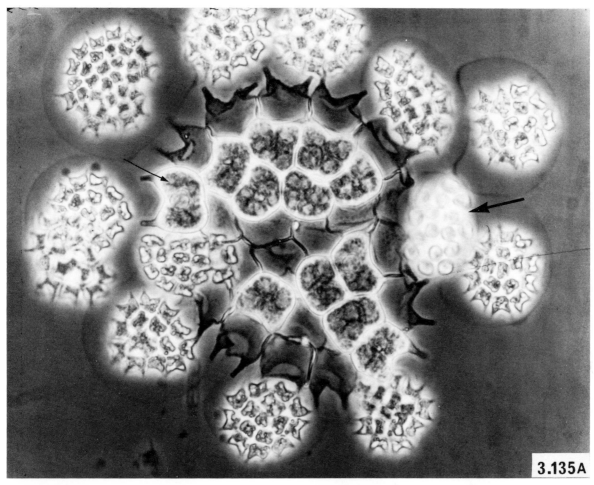

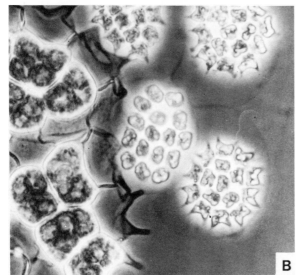

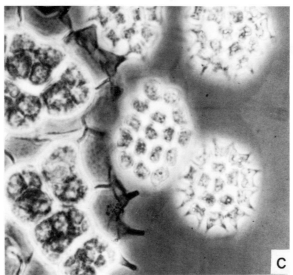

3.135A

B

C

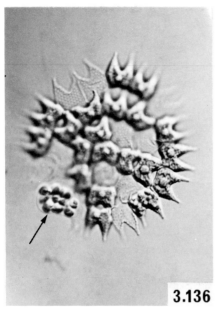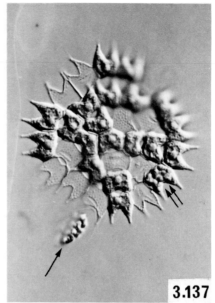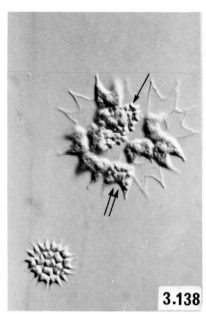

3.136 3.137 3.138

Figures 3.136–3.138
COLONY FORMATION IN PEDIASTRUM BORYANUM

3.136. This micrograph shows a mass of zoospores swarming inside the vesicle (arrow), as in Figure 3.135A, but photographed using Nomarski optics. × 600.

3.137. The same colony as that in Figure 3.136, a few minutes later. The zoospores have now become quiescent (cf. Fig. 3.135B, C). Particularly clear here is their planar configuration (single arrow). Zoospores are about to be released from another cell (double arrow). × 600.

3.138. A colony with a newly formed daughter colony nearby whose horns are fully formed. Another colony is being formed (single arrow), while zoospores await release elsewhere (double arrow). × 600.

SOURCE. Dr. H.J. Marchant, unpublished micrographs.

surface of the central hollow sphere (Figs. 3.153, 3.157).

Mitosis in the cells is typically chlorococcalean. The spindle is centric and enclosed by a perinuclear envelope. As usual, synchronous mitoses give rise to 2^n nuclei (the total number usually lies between 16 and 64). Cytoplasmic cleavage (Fig. 3.150) is again typical, involving phycoplasts; however, it is often extensive well before the final mitoses begin (as in *Tetraedron*

Figures 3.139, 3.140.
MITOSIS AND ZOOSPORE FORMATION IN PEDIASTRUM BORYANUM

3.139. Synchronous prophases. The typically chlorococcalean spindles are closed, centric, and enclosed within a perinuclear envelope. Note also how the cells of the colony are joined. × 13,000.

3.140. Sixteen flagellate zoospores still contained within the parental wall (w), which has become much thinner prior to zoospore release (cf. Fig. 3.139). Note the stable outer layer of wall material which is never eroded away. It probably contains sporopollenin (Figs. 3.142, 3.147, 3.148). × 8,000.

SOURCE. Dr. H.J. Marchant, paper in preparation.

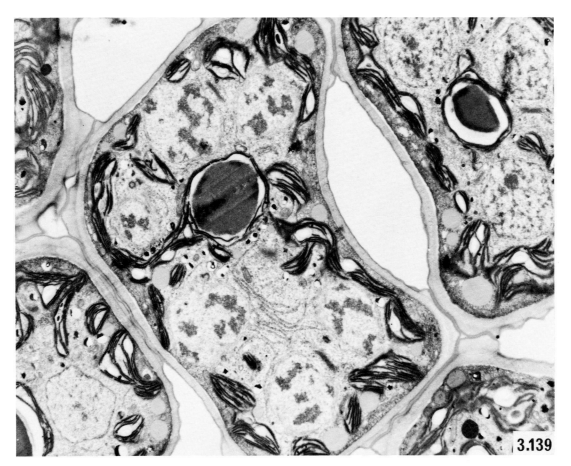

3.139

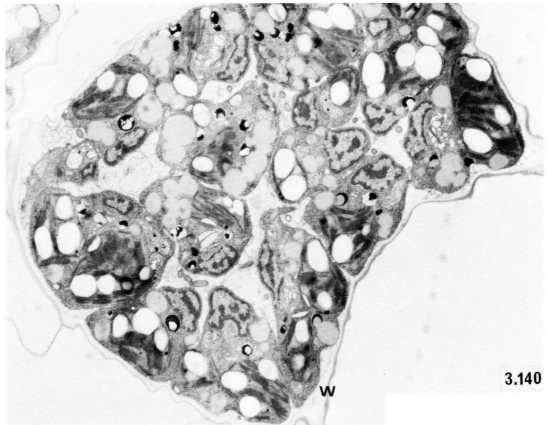

W

3.140

and *Kirchneriella*). After mitosis and cytokinesis are complete, uninucleate, flagellate zoospores differentiate, crowded inside the parental cell, whose wall undergoes the usual breakdown (Fig. 3.154). The outer layer of wall then ruptures to release this inner eroded layer as the extensible vesicle (Fig. 3.151) containing the zoospores, as in *Pediastrum* and the polyhedra of *Hydrodictyon*.

Colony formation is rapid and makes an interesting comparison with that of *Hydrodictyon* and *Pediastrum*. The cells seem to aggregate initially inside the swarming vesicle in much the same way in all three genera, but in *Sorastrum* apparently the different shape of the maturing zoospores and perhaps the different siting of adhesive areas on their surface cause the colony to adopt the spherical arrangement. The cells are pear-shaped as they begin to stick together, invariably with the flagellar (i.e., nuclear) end of each cell directed inward (Figs. 3.155–3.157). By the time the flagella have disappeared, the cells are cone-shaped (in distinct contrast to the colony-forming cells of *Pediastrum* and *Hydrodictyon*). First contact between these zoospores is lateral, and Marchant (1974b) is not sure whether this may indicate a weak lateral adhesion to one another at this stage. Soon the quiescent zoospores undergo profound and rapid changes in shape. The flagellar ends of the cells come in contact (Fig. 3.155), and then the cells constrict just behind these contact sites. The horns now appear (Fig. 3.152), always at the opposite end of the cell to the forming stipe, and the cells become slightly flattened laterally (cf. Fig. 3.149). Microtubules are undoubtedly involved in some of these changes; in particular, they run the length of the forming spines and along the edges of the cells (Figs. 3.159, 3.161), probably along most of the length of the cell. The polarity of these cells is similar to that of *Pediastrum*, whose horns also invariably arise from the "chloroplast" side of the cell, never the rounded "nuclear" side (Fig. 3.143). It is not difficult to imagine that the process of colony formation in *Hydrodictyon*, *Pediastrum*, and *Sorastrum* is basically similar, and as could be anticipated, colchicine treatment of *Sorastrum* affects colony formation much as it does in the other two genera (Figs. 3.110, 3.111). Once the cells are joined, the cytoplasm withdraws from both the stipe and the spines as more wall material is secreted at these sites (Figs. 3.155–3.158, 3.160).

Figures 3.141–3.144
COLONY FORMATION IN PEDIASTRUM BORYANUM

3.141. Structure of a motile zoospore. × 15,500.

3.142. Empty parental cell walls, showing the rents through which the zoospores escaped. × 3,000.

3.143. Colony formation. The quiescent zoospores have withdrawn their flagella and linked up in a planar array while still contained within the swarming vesicle. They are also beginning to develop the prongs on their periphery that are characteristic of the species. Note also how the nucleus is on the inner side of each cell. The colony has already acquired the 1–5–10 concentric arrangement of cells typical of 16-celled colonies (Fig. 3.133), although two of the outer cells are out of the plane of this section. × 7,000.

3.144. Development of prongs before the wall has been secreted. The prongs contain numerous microtubules *(t)*, which may be the cytoplasmic entities involved in the change of shape of the cells. × 25,000.

SOURCE. Dr. H.J. Marchant, paper in preparation.

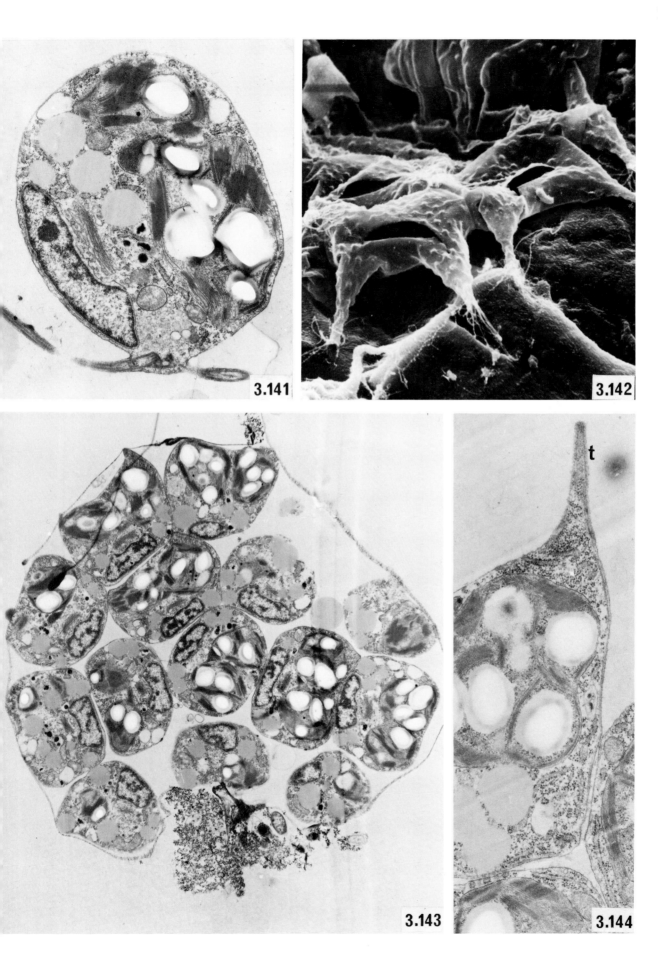

3.141

3.142

3.143

3.144

t

Davis (1966b) has proposed that the restricted size of the swarming vesicle imposes the shape on the new colony in *Sorastrum.* This attractive hypothesis, however, seems insufficient to explain some phenomena of colony formation (see also Section 3.8a). For example, it cannot explain the not infrequent appearance of two spherical colonies within the one vesicle, or the fact that when this vesicle grows much larger than the mass of aggregating zooids, the colony arising is still spherical.

Figures 3.145, 3.146
EXPERIMENTAL DISRUPTION OF COLONY FORMATION in PEDIASTRUM BORYANUM

3.145.   This colony, still contained within the swarming vesicle, had been dissociated soon after linkage by gentle pressure on the coverslip. Note that *all* the cells now have formed prongs. × 750.

3.146.   Inhibition of cytoplasmic cleavage by colchicine. This micrograph shows the release of a mass of undifferentiated cytoplasm from a parental cell following treatment of the colony with this drug. × 600.

SOURCE. Dr. H.J. Marchant, unpublished micrographs.

Figures 3.147, 3.148
ACETOLYSIS OF COLONIES OF PEDIASTRUM BORYANUM

3.147.   Ghosts remaining after all other components had been destroyed. These represent the outermost sheath, ornamented with warts, directly equivalent to the ghosts remaining after zoospores have been released from daughter colonies (cf. Figs. 3.136, 3.140, 3.142). × 1,000.

3.148.   Higher magnification of a ghost colony that had released zoospores before being acetolyzed. Note the characteristic tears in the wall (cf. Fig. 3.142). This ghost is no differently affected by the treatment than the adjacent younger colony. The wall component remaining probably contains sporopollenin (see also Figs. 3.11, 3.76). × 2,000.

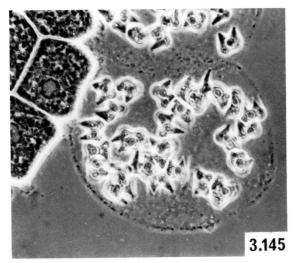

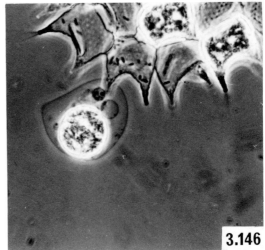

3.145

3.146

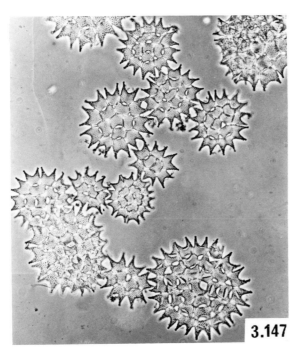

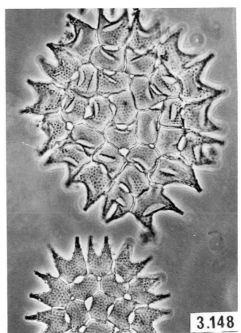

3.147

3.148

Figures 3.149–3.153
VEGETATIVE COLONIES OF SORASTRUM SPINULOSUM

3.149. Whole colony. Each cell is highly asymmetrical, with four spines directed outward, and an inwardly oriented stalk or stipe which is connected with the stipes from the other cells. × 3,000.

3.150. Live cells in which some cytoplasmic cleavage has already taken place. × 1,000.

3.151. Cells still enclosed within the parental swarming vesicle, beginning to change shape and clump together. × 1,200.

3.152. Young colonies in which the cells have acquired their vegetative morphology and stuck together. × 1,200.

3.153. Two slightly older colonies whose cells are equivalent to those in Figures 3.158 and 3.160. Note particularly how the cells are joined to each other at the center of the colony. × 1,200.

SOURCE. Dr. H.J. Marchant, paper in preparation.

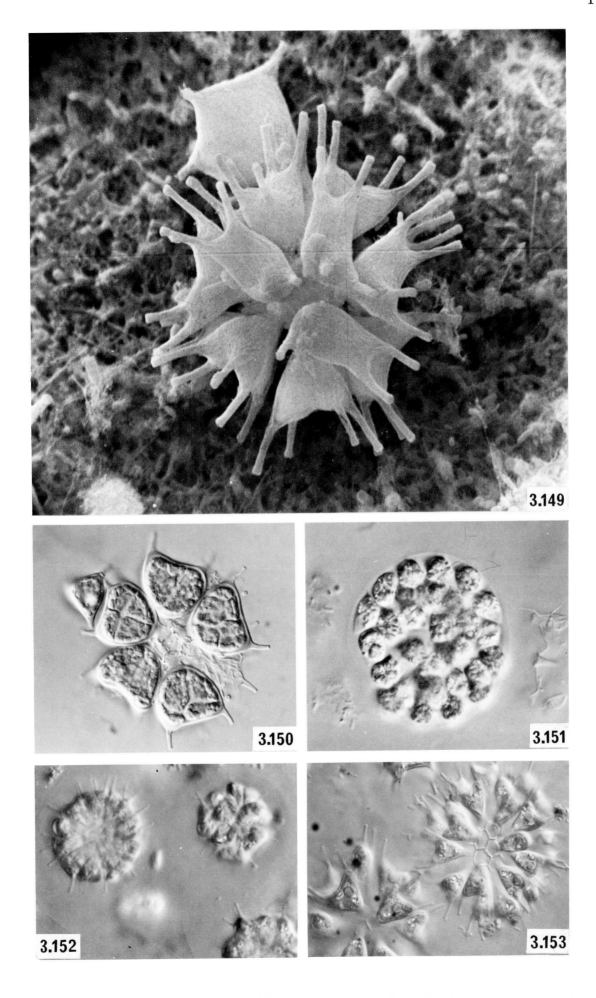

3.149

3.150

3.151

3.152

3.153

Figures 3.154–3.156
REPRODUCTION IN
SORASTRUM SPINULOSUM

3.154.   Fully formed zoospores (note flagella, *f*) inside parental cell before release. The wall (arrow) has begun to break down. × 8,000.

3.155.   Very early stage of colony formation, when the quiescent zoospores have just become interconnected at the tip of their stipe. The thin wall is continuous between cells at these points (arrows). × 14,000.

3.156.   A slightly later stage. As the cytoplasm withdraws from the stipe, the protrusion fills with newly deposited wall material. × 25,000.

SOURCE. Dr. H.J. Marchant, paper in preparation.

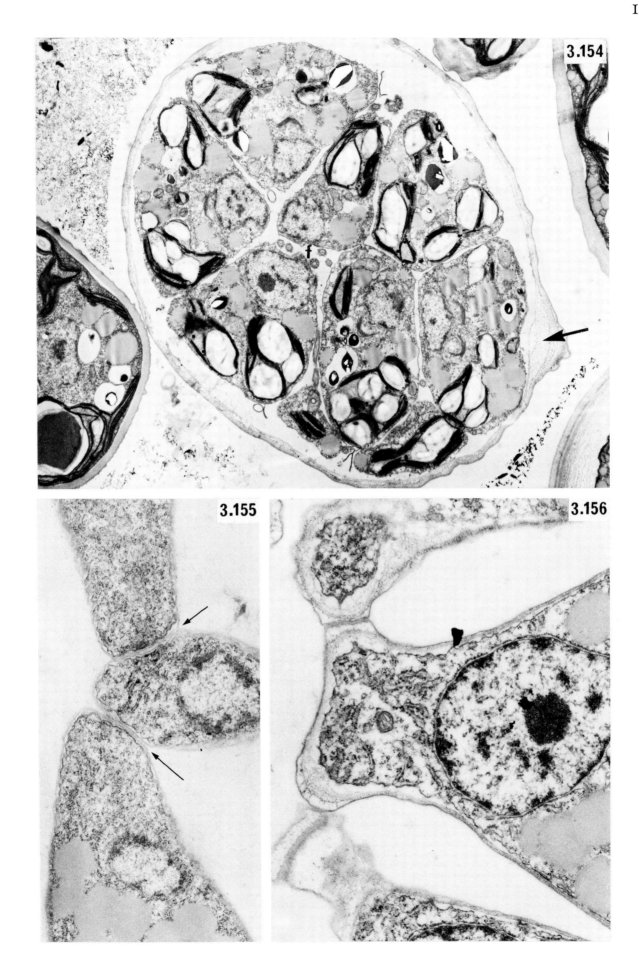

Figure 3.157
YOUNG COLONY OF
SORASTRUM SPINULOSUM

A slightly later stage in colony formation than that in Figure 3.156. The internal organization is the same in each cell. Their interconnection at the center of the colony (cf. Fig. 3.153) is obvious. $\times$ 17,000.

SOURCE. Dr. H.J. Marchant, paper in preparation.

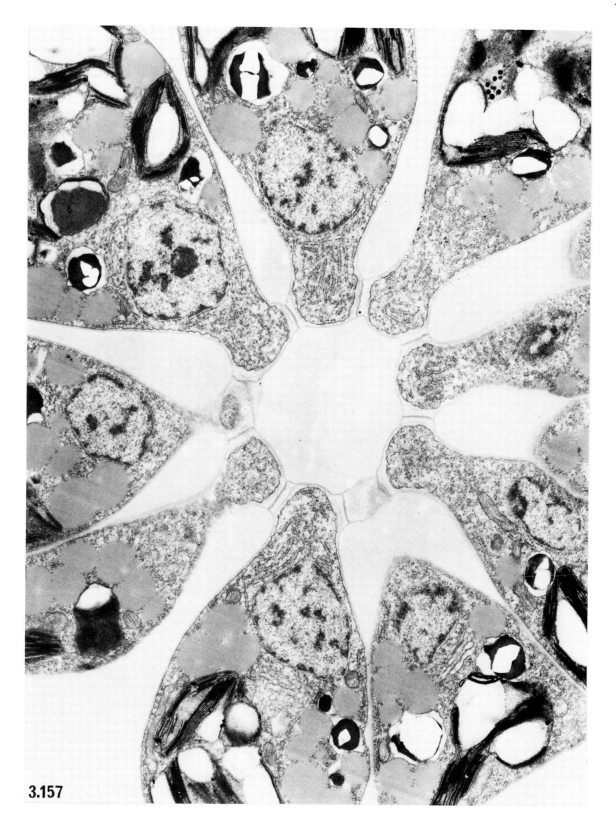

3.157

Figures 3.158–3.161
COLONY FORMATION IN
SORASTRUM SPINULOSUM

3.158.   A later stage (cf. Figs. 3.155–3.157) of cytoplasmic withdrawal from, and wall deposition in, the stipe. × 19,000.

3.159   Formation of prongs in very young colonies (similar to those in Fig. 3.152). Note the microtubules *(t)* inside the spine, which has no wall around it yet. × 40,000.

3.160.   Uninucleate cell in a young colony. The wall has thickened considerably, filling both spines and stipes. × 14,000.

3.161.   Cross section of a spine and the edge of cells equivalent to that in Figure 3.159. The microtubules are clearly involved in protrusion of the spine and probably the changes in shape of the cells. × 60,000.

SOURCE. Dr. H.J. Marchant, paper in preparation.

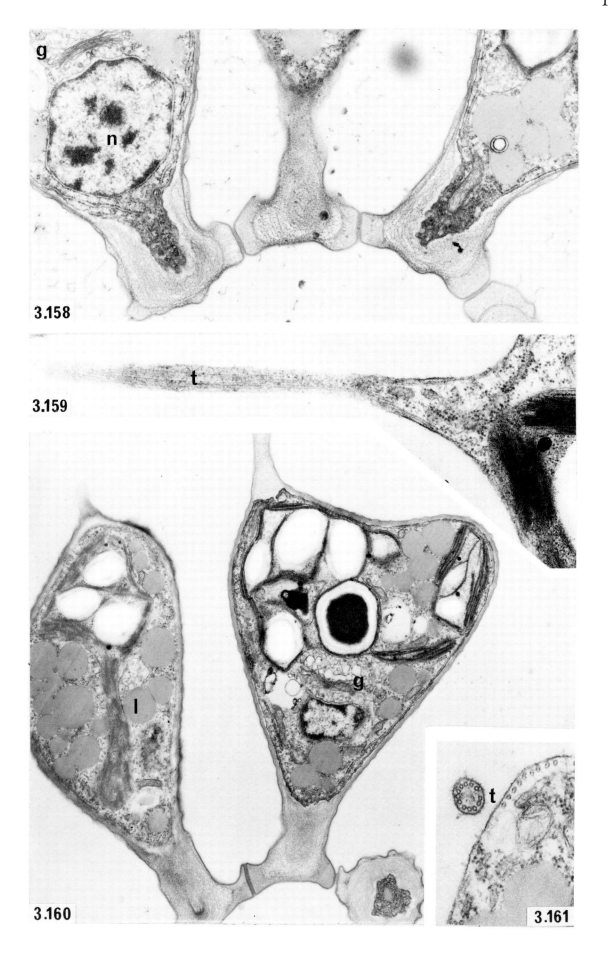

g

n

3.158

t

3.159

l

g

3.160

t

3.161

4

THE ULOTRICHALES

The Ulotrichales are a heterogeneous collection of algae which may not have close affinities with one another, a matter discussed in Chapter 8. A recent and much needed attempt at reclassifying the Ulotrichales has been proposed by Stewart, Mattox, and Floyd (1973). Throughout the rest of this book, I shall include in the Ulotrichales those more highly differentiated and advanced genera that Fritsch (1935) and others place in the separate order Chaetophorales. Many members of the Ulotrichales are polymorphic. For example, *Ulothrix,* normally considered a simple filamentous form, may be branched (Section 4.4). *Stigeoclonium,* normally branched and well known to be quite variable in its vegetative morphology, has now been shown (Campbell and Sarafis, 1972) in one species to vary widely in form, resembling five other genera, including *Schizomeris* (in this latter species, divisions in the filament take place in several planes so that the filaments may contain series of parallel cells; Fritsch, 1935). A distinction between the Ulotrichales and the Chaetophorales, based primarily on vegetative morphology (e.g., branching), seems artificial.

The Ulotrichales consist of, or else have evolved from, uninucleate, filamentous forms generally believed to have arisen long ago from *Chlamydomonas*-like predecessors (Chapter 8). Some genera contain multinucleate cells (e.g., the rhizoids of the Ulvaceae), and the genus *Sphaeroplea* is coenocytic, but the affinities of this latter organism are difficult to ascertain. The simpler Ulotrichales are unbranched filaments whose sole cell differentiation in many species consists of a holdfast basal cell; other species may not be even this differentiated. Fritsch (1935) considers these examples among the most primitive filamentous forms of Green Algae.

The more advanced members of the Ulotrichales (the Chaetophorales in many textbooks) have evolved considerable diversity of form, and many believe that these organisms gave rise to the progenitors of the higher land plants. An elaboration of the filamentous form arose from the acquired ability of the cells to divide in more than one plane; the latter restriction can give rise only to the simple filament. Branching, for example, takes place following an occasional division that is oriented off the axis of the filament, and extensive division in two planes gives rise to thalloid forms such as the Ulvaceae. The affinities of many such genera with the filamentous forms is clear. For example, *Ulva* first grows as a typical filamentous form upon germination of a zoospore, before becoming thalloid (Section 4.7), and I agree with Fritsch (1935, pp. 217 et seq.) that a relation to the Tetrasporaceae, suggested by some, is unlikely. Another important attribute acquired by advanced ulotrichalean algae (one considered "outstanding" by Fritsch, 1935), but not apparent in some species, is the acquisition of the "heterotrichous" habit, whereby a plant body can develop in two different perpendicular planes (Fritsch, 1942a). The plant differentiates

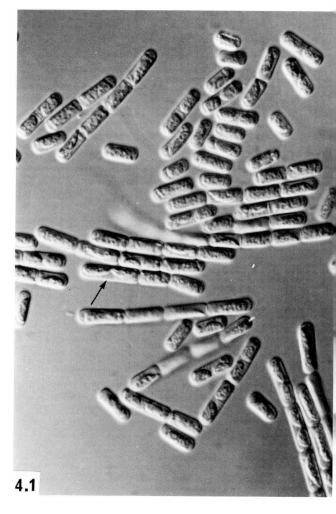

4.1

Figure 4.1
STICHOCOCCUS
CHLORANTHUS

The filaments easily break up into unicells. Careful examination of this micrograph will reveal many cells at various stages of division. One (arrow) may be at late anaphase, equivalent to that in Figure 4.6. × 1,200.

SOURCE. Author's micrograph, published in *Br. Phycol. J. 9,* 63 (1974).

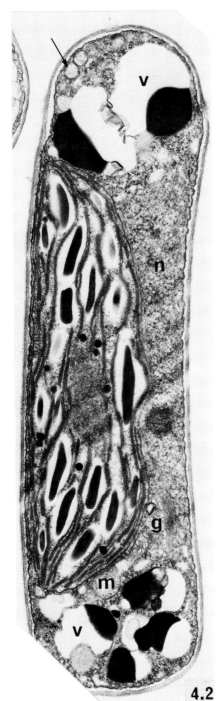

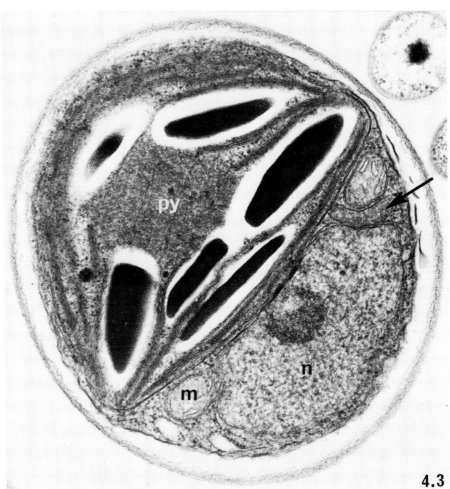

Figures 4.2, 4.3
INTERPHASE CELLS OF
STICHOCOCCUS
CHLORANTHUS

4.2. Longitudinal section. The nucleus (*n*) is flattened by the large chloroplast. Vacuoles (*v*) and small vesicles (arrow) are always found at the ends of the cells. The two mitochondria (*m*) run the length of the cell alongside the nucleus (see Fig. 4.3), and part of one can be seen here. × 22,000.

4.3. Transverse section. This shows the flattened nucleus flanked by the two mitochondria. The inclusion indicated by the arrow is tentatively identified as a peroxisome. ×40,000.

SOURCE. Author's micrographs, published in *Br. Phycol. J. 9,* 63 (1974).

into a "prostrate" system, growing over and usually attached to a substrate, and an "erect" system projecting from the prostrate portion of the organism. The development of heterotrichy is of considerable significance in understanding algal evolution, and it is characteristic, too, of Red and Brown Algae (Fritsch, 1949). Later regression of the erect, essentially filamentous system has given rise to several ulotrichalean genera that are now solely prostrate and often discoidal in form (e.g., *Coleochaete*; Section 4.3). Some genera have rhizoids differentiated from the prostrate portion of the thallus. As in the Volvocales, evolutionary advance in the Ulotrichales has been accompanied by a progression in sexual behavior from isogamy through anisogamy to oogamy. However, in my opinion, this statement is probably oversimplified and even incorrect as it stands. We (Pickett-Heaps and Marchant, 1972; see Chapter 8) have suggested that there are at least two distinct lines of evolution in the Ulotrichales: one containing the organisms that possess the phycoplast, and one containing the organisms that do not. The latter is the group in our opinion more closely related to the progenitors of the higher plants, and in this line, advanced oogamy has appeared in *Coleochaete*. So far, no oogamous forms of phycoplast-forming cells are known.

Unfortunately, little coherent ultrastructure work has been done on the Ulotrichales. This is surprising in view of their importance in attempts to establish the phylogeny of higher plants. Many members possess plasmodesmata traversing their end walls, while others do not; these plasmodesmata resemble those of higher plants but differ from those of the Oedogoniales (Chapter 5). This may have important phylogenetic implications, as has been discussed by Stewart, Mattox, and Floyd (1973). They point out that plasmodesmata occur only in algae that form cross walls using cell plates and that a cell plate unquestionably is a more advanced cytokinetic mechanism than furrowing (Pickett-Heaps, 1972a).

4.1. Stichococcus

Stichococcus is one of the simplest members of the Ulotrichales. If undisturbed during growth (e.g., when cultured on agar), the bacilliform cells are united into filaments which, however, are easily fragmented (Fig. 4.1)—the only method so far reported of vegetative reproduc-

Figures 4.4, 4.5
EARLY MITOTIC STAGES IN STICHOCOCCUS CHLORANTHUS

4.4.　Prophase. Chromosomes are appearing in the nucleus *(n)*. The chloroplast has a deep cleavage furrow into which the nucleus has become inserted (cf. Fig. 4.2); the latter has a Golgi body *(g)* at each future pole. A narrow band of microtubules (arrows) encircles the cell at the plane of division, and these tubules are embedded in dense material (spindle MTOCs?) near the nucleus. Part of one mitochondrion *(m)* can be seen. × 41,000.

4.5.　Metaphase. The nuclear envelope has dispersed, and the ill-defined chromosomes *(cb)* have formed a metaphase plate. Spindle microtubules *(t)* are oriented toward the poles, defined by Golgi bodies *(g)*, which may have replicated by this stage. × 32,500.

SOURCE. Author's micrographs, published in *Br. Phycol. J. 9*, 63 (1974).

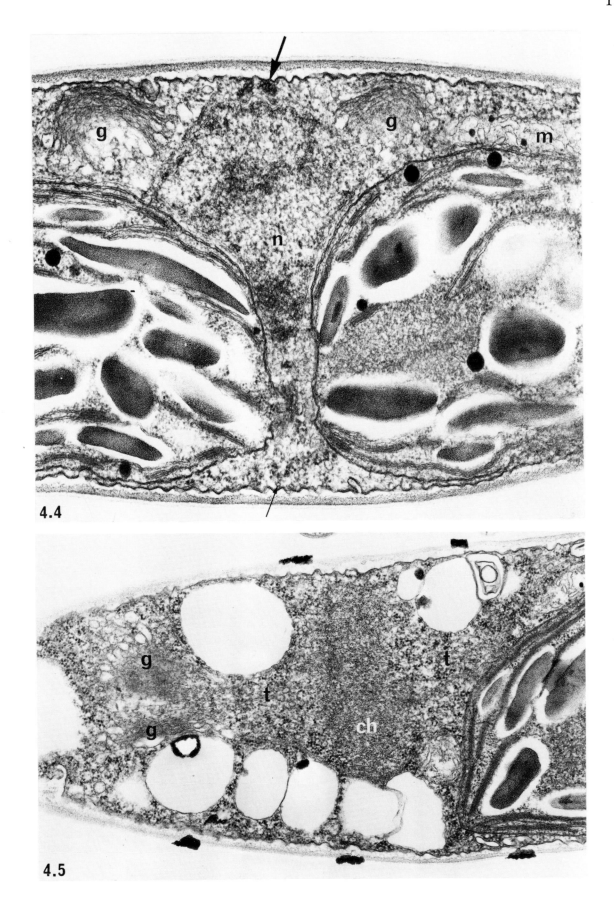

4.4

4.5

tion in this genus (Smith, 1950). Each cell contains a large chloroplast, which in turn possesses a pyrenoid (contrary to the statement by Smith, 1950; p. 145). The cells are so small, however, that little else can be discerned inside them using the light microscope (Fig. 4.1). In their ultrastructural organization (Figs. 4.2, 4.3; Pickett-Heaps, 1974b) the cells quite closely resemble those of *Klebsormidium* (Section 4.2), although they are much smaller. The chloroplast and pyrenoid fill much of the cell and the nucleus is rather flattened. Two mitochondria run the length of the cell on either side of the nucleus; at each end of the latter are one or more Golgi bodies. A single peroxisome, rather difficult to discern, lies near one mitochondrion. Several small vacuoles and vesicles fill the ends of the cell. Centrioles are absent.

4.1a. Cell Division

Premitotic cells are appreciably longer than the rest in the culture, and imminent cell division is indicated when a deep cleavage appears in the chloroplast (Fig. 4.1). (A cell possibly at `anaphase in live material is also indicated in Fig. 4.1.) At the ultrastructural level, a deep cleavage has appeared in the chloroplast (Pickett-Heaps, 1974b) by prophase, into which the nucleus becomes inserted (Fig. 4.4), rather as in *Klebsormidium* (Fig. 4.14). Chloroplast cleavage appears complete by metaphase (Fig. 4.5). Invariably, three to five microtubules encircle the cell in the plane of the chloroplast cleavage (Fig. 4.4); their significance is not known, and they always disappear by metaphase. Sometimes these microtubules are embedded in dense material (Fig. 4.4), which suggests the existence of microtubule-organizing centers (MTOCs) situated at prophase in the site occupied by centrioles in filamentous, centric cells (the latter, of course, replicate and move apart, marking the spindle poles).

By metaphase (Fig. 4.5), the nuclear envelope and nucleolus disperse and the chromosomes form a single mass across the center of the cell. Microtubules are few and difficult to see; thin sections rarely contain more than a few short segments of these organelles. Golgi bodies mark the poles of the spindle. The chromosomes separate during anaphase as normal, and Golgi bodies continue to mark the poles (Fig. 4.6). After telo-

Figures 4.6–4.8
ANAPHASE AND TELOPHASE IN STICHOCOCCUS CHLORANTHUS

4.6. Anaphase. Vacuoles *(v)* are beginning to collect between the widely separated daughter chromosomes *(ch)*. The positions of the polar Golgi bodies are marked by arrows. $\times$ 18,000.

4.7. Telophase. The nuclear envelope has re-formed around the daughter nuclei *(n)*. The cleavage furrow *(cf)* has begun to grow across the cell through interzonal vacuoles *(v)*. Part of the presumptive peroxisome is indicated by the arrow. $\times$ 34,000.

4.8. Late telophase. Cytokinesis is just complete. The nuclei are still far apart, and there is no phycoplast associated with the cleavage furrow. $\times$ 32,500.

SOURCE. Author's micrographs, published in *Br. Phycol. J. 9*, 63 (1974).

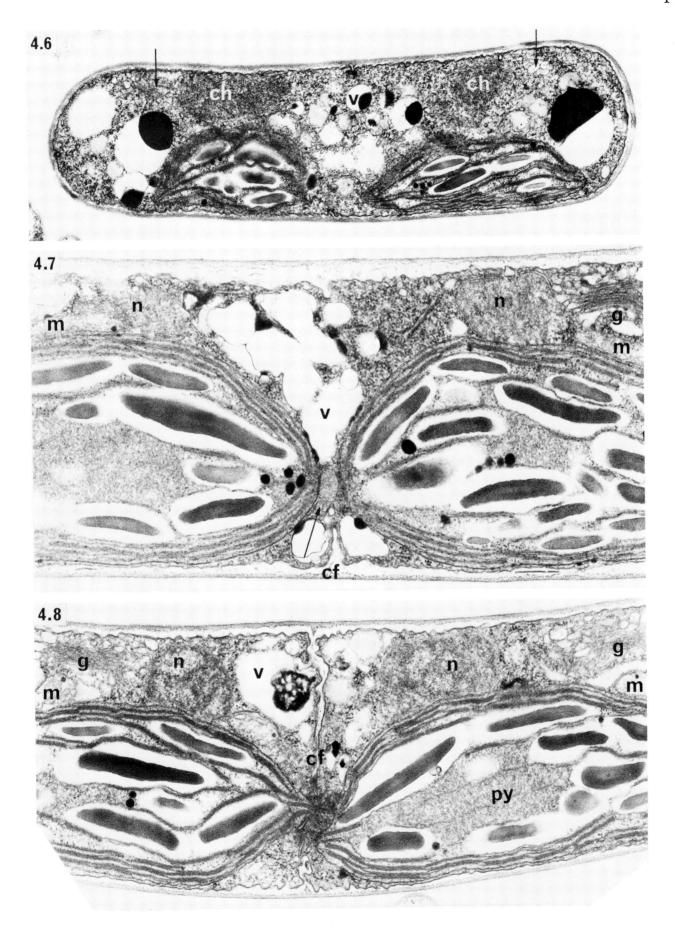

phase, the daughter nuclei remain far apart as their nuclear envelope is reconstituted, and no phycoplast is formed (Figs. 4.7, 4.8). Numerous small vacuoles collect in the interzone region between daughter nuclei, and cytokinesis is effected by an annular, ingrowing membrane that bisects the cell and passes among these vacuoles. Only after cytokinesis is complete do the nuclei often approach the new cross wall.

As will be seen, cell division as well as the organization of the interphase cell, are very similar in *Stichococcus* and in *Klebsormidium*. The only obvious difference is the lack of centrioles in *Stichococcus*. I have argued elsewhere (Pickett-Heaps, 1972h) that the presence or absence of centrioles may be of minor phylogenetic significance and is more likely related to whether the organisms form flagellate stages in their life cycle. *Stichococcus* is not known to produce zoospores or gametes. Recent work on *Raphidonema* (Pickett-Heaps, unpublished results) has shown that this genus is virtually identical to *Stichococcus* during division. The significance of these results should become more obvious later in this book. In particular, they tend to confirm our prediction (Pickett-Heaps and Marchant, 1972) that there are at least two natural groups in the Ulotrichales: those with and those without phycoplasts. The latter group includes *Klebsormidium, Stichococcus, Raphidonema,* and *Coleochaete,* and in our opinion, this group is the one more closely related to the line(s) of phyletic advance which originally gave rise to higher land plants (Chapter 8).

4.2. Klebsormidium

The taxonomy of *Ulothrix* and related organisms (e.g., *Uronema, Hormidium*) has been confusing, and in clarifying the situation, Mattox and Bold (1962) have made an important contribution. It has turned out rather unexpectedly that their suggested distinction between *Ulothrix* and its supposedly close relative *Klebsormidium,* based on morphological criteria, has been strongly vindicated, in my opinion, since these two genera have now also been found to have radically different mitotic and cytokinetic structures. It is important to emphasize both the obvious and the more subtle distinctions between *Klebsormidium* and *Ulothrix* (Mattox and Bold, 1962), since I suspect the implications of their differences,

Figures 4.9–4.13
INTERPHASE AND MITOTIC CELLS OF KLEBSORMIDIUM

4.9. *K. flaccidum.* Interphase, showing central nucleus flanked by terminal vacuoles and the single chloroplast. $\times$ 2,600.

4.10*A–G. K. flaccidum.* Metaphase to telophase in a live cell. The chloroplast is obviously deeply constricted by metaphase. The refractile dot at one end of the spindle (arrow) is the centriole complex. Note that during anaphase, the chromosome-to-pole distance does not decrease appreciably. The vacuole has appeared in the interzone by late anaphase *(D),* and later *(F)* the ingrowing furrow cuts the vacuole in two. Note that the daughter nuclei stay far apart during telophase, in contrast to phycoplast-forming algae (e.g., Figs. 4.46, 4.47). $\times$ 2,000.

4.11. *K. subtilissimum,* interphase cell. Note particularly the small single peroxisome (arrow) sandwiched between the nucleus and chloroplast. $\times$ 8,300.

4.12, 4.13. *K. flaccidum.* Both these cells were incubated in the DAB/H_2O_2 medium for the cytochemical demonstration of peroxidase (catalase) activity. Figure 4.12 was was poststained with lead and uranium; Figure 4.13 was not. In both examples, the dense reaction product is clearly confined to the peroxisome *(p).* Figure 4.12, $\times$ 40,000; Figure 4.13, $\times$ 45,000.

SOURCE. Figures 4.9, 4.10, 4.12, 4.13: Drs. K.D. Stewart, G.L. Floyd, K.R. Mattox, and M.E. Davis; Figures 4.9, 4.10 published in *J. Phycol. 8,* 176 (1972); Figures 4.12, 4.13 published in *J. Cell Biol. 54,* 431 (1972). Figure 4.11: Author's micrograph, published in *Cytobios 6,* 167 (1972).

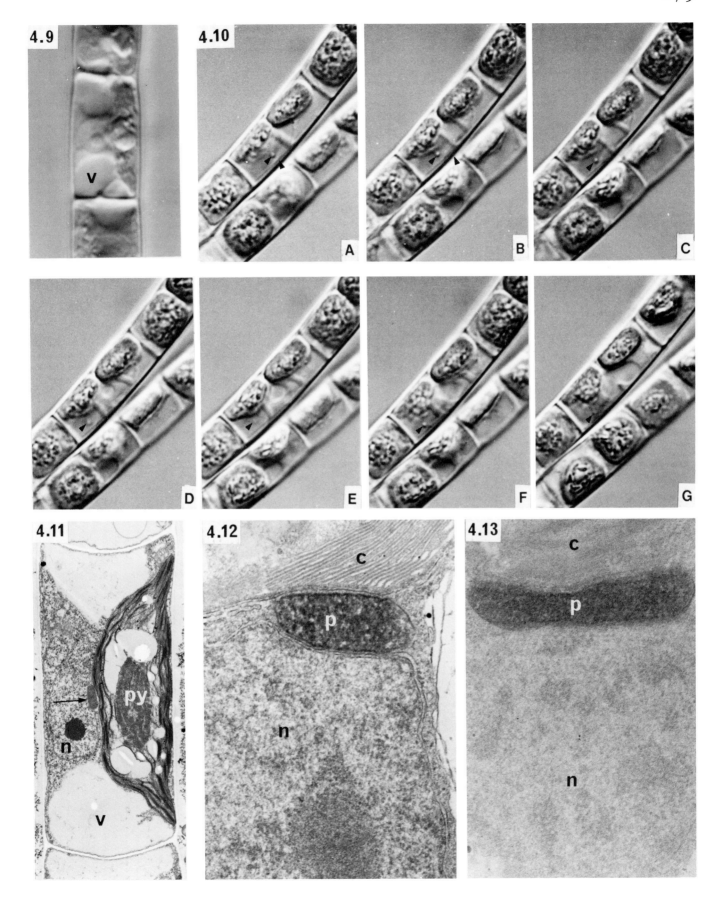

particularly during cell division, may be phylogenetically highly significant (Chapter 8). The name *Klebsormidium* is appropriate for the genus, previously called *Hormidium,* since this latter name had earlier been adopted for a genus of orchid (Silva, Mattox, and Blackwell, 1972).

Ulothrix characteristically contains a single parietal chloroplast which mostly or completely encircles the cell and usually contains more than one pyrenoid. In *Klebsormidium,* the single, plate-shaped chloroplast never encircles more than a small part of the cell lumen and contains at most one pyrenoid. Plasmodesmata, present in *Ulothrix,* are absent in *Klebsormidium. Ulothrix* readily forms a number of biflagellate, or more usually quadriflagellate, zoospores from each parental cell, with red stigmata, and the zoospores upon germination give rise to the holdfast of new filaments. *Klebsormidium* rarely produces zoospores, and never more than one per vegetative cell. These zoospores are biflagellate and lack stigmata; they give rise directly to new filaments without forming a holdfast. Other physiological differences between *Ulothrix* and *Klebsormidium* are summarized by Mattox and Bold (1962). Among other proposals, Mattox and Bold suggest that the species *Ulothrix subtilissima* was not *Ulothrix* at all and must be reclassified as *Klebsormidium subtlissimum.* My ultrastructural work on cell division in this organism (Pickett-Heaps, 1972h) unequivocally supports their decision and demonstrates how ultrastructural studies, particularly of cell division, may indeed be relevant to taxonomy. Mattox and Bold (1962) also list a reference regarding isogamous sexual reproduction in *Klebsormidium.*

The cell of *Klebsormidium* is simple in form (Floyd, Stewart, and Mattox, 1972b; Pickett-Heaps, 1972h) and contains two vacuoles occupying much of each end of the cell (Figs. 4.9, 4.11). The vacuoles are separated by a central bridge of cytoplasm containing the nucleus and Golgi bodies. The cell probably contains two elongated mitochondria (Floyd, Stewart, and Mattox, 1972b) similar to those in *Stichococcus.* Typical wall microtubules, oriented transversely to the cell axis, line the plasmalemma. The cell has one distinctive feature: sandwiched between the nucleus and the chloroplast is invariably found the cell's single, flattened peroxisome (Fig. 4.11), in which Stewart et al. (1972) have clearly demonstrated (Figs. 4.12, 4.13) a strong and spe-

Figures 4.14, 4.15
MITOSIS IN KLEBSORMIDIUM SUBTILISSIMUM

4.14.   Early prophase. The chloroplast is deeply indented, and between it and the nucleus is sandwiched the greatly elongated peroxisome *(p).* The future poles of the spindle contain centrioles *(cn),* and extranuclear microtubules *(t)* ensheathe the nucleus *(n).* ×26,000.

4.15.   Metaphase. The nuclear envelope has entirely dispersed, but some chromosomes are coated with membrane (arrow). × 14,000. approx.

SOURCE. Author's micrographs, published in *Cytobios* 6, 167 (1972).

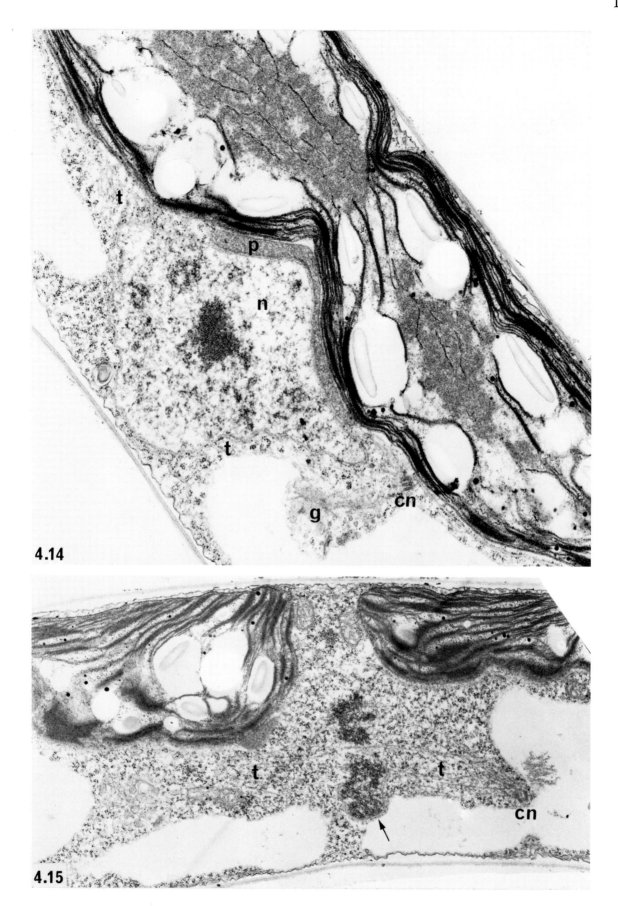

4.14

4.15

cific catalase reactivity using the DAB (diamino-benzidine) technique described by Frederick and Newcomb (1969). This type of peroxisome may turn out to be a distinctive generic feature and therefore valuable as a taxonomic criterion. The cell's persistent centrioles are located close to one edge of the peroxisome.

4.2a. Cell Division

Cell division in *K. flaccidum* and *K. subtilissimum* has been described by Floyd, Stewart, and Mattox (1972b) and by Pickett-Heaps (1972h). These accounts are essentially similar, although some minor differences may be noted.

Premitotic cells are considerably elongated, and imminent division is indicated by a cleavage appearing in the chloroplast and pyrenoid. As this cleavage deepens during prophase and metaphase, the nucleus and peroxisome remain appressed to the chloroplast and, therefore, become markedly deformed (Fig. 4.14). Wall microtubules disappear as the centrioles separate to establish the poles of the spindle at prophase, and numerous microtubules radiate from them, mostly ensheathing the nucleus, which becomes increasingly spindle-shaped and drawn out. The peroxisome also elongates considerably and is probably split in two by metaphase (Fig. 4.15). Thereafter, one end of each peroxisome, along with Golgi bodies, remains closely associated with a pair of centrioles that marks the poles of the spindle. (It will be remembered that in *Stichococcus*, too, the Golgi bodies mark the spindle poles.) The nucleolus disperses as the chromosomes appear, and the nuclear envelope breaks up during prometaphase and disperses completely. The open metaphase spindle is quite elongated and typical in most respects; however, membranous elements coat some chromosomes (Pickett-Heaps, 1972h).

Floyd, Stewart, and Mattox (1972b) beautifully illustrate anaphase in vivo (Fig. 4.10*A–G*), preceded by several rapid tilting movements of the spindle just before anaphase commences (cf. *Ulothrix*). Two unusual phenomena accompany chromosome separation during anaphase: (1) the chromosome-to-pole distance stays quite constant, and (2) two small vacuoles appear in the interzone between daughter chromosomes at early anaphase, and these immediately fuse and grow at the expense of the terminal vacuole as

Figures 4.16–4.19
LATE STAGES OF MITOSIS IN KLEBSORMIDIUM SUBTILISSIMUM

4.16. Late anaphase/early telophase. Cleavage has just started (small arrow). Note particularly the elongated spindle, whose poles are marked by large arrows. The chromosome-to-pole distance has not shortened appreciably (cf. Figs. 4.17–4.19). Portions of the elongated peroxisome (p) are visible near the chloroplast. × 8,500.

4.17. Detail of Figure 4.16, showing persistent chromosomal microtubules *(t)* lying between the daughter nucleus *(n)* and the centriole *(cn)*. Interzonal microtubules (arrow) are also persistent. × 31,000.

4.18. Very late telophase, cleavage under way. There is a vacuole *(v)* between daughter nuclei (cf. Fig. 4.10*D–G*), which have remained far apart, unlike the disposition of daughter nuclei in phycoplast-forming algae (e.g., Figs. 4.46, 4.47). Interzonal and chromosomal microtubules still persist (Fig. 4.19). × 8,700.

4.19. Detail of Figure 4.18, showing the persistence of chromosomal microtubules between the daughter nucleus and the centriole. Note also the portion of the peroxisome close to the centriole. × 26,000.

SOURCE. Author's micrographs, published in *Cytobios* 6, 167 (1972).

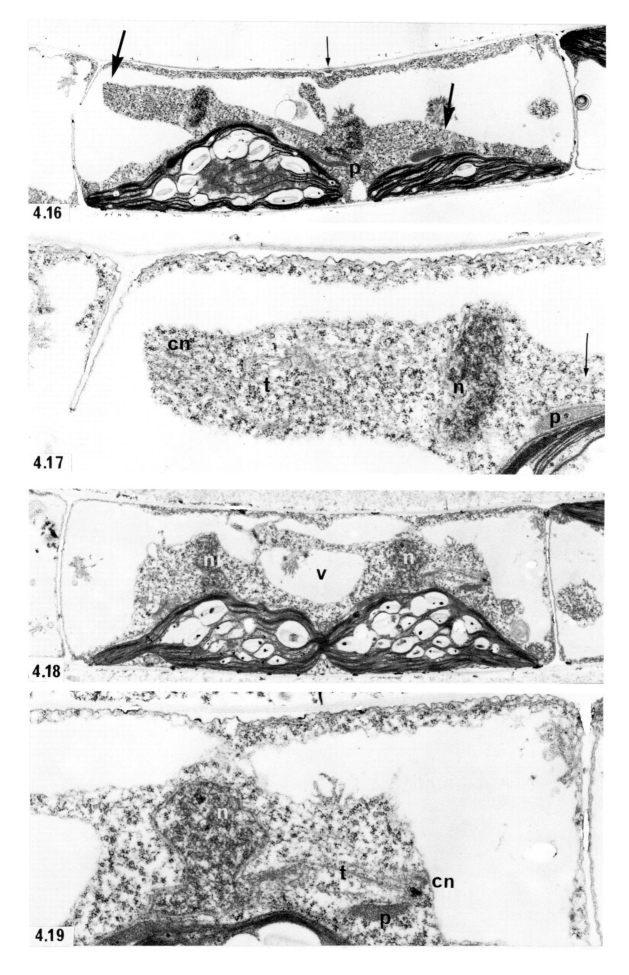

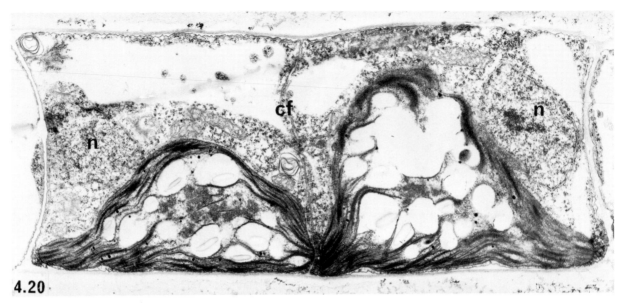

4.20

Figure 4.20
FINAL STAGE OF CELL DIVISION IN KLEBSORMIDIUM SUBTILISSIMUM

The daughter nuclei *(n)* are still widely separated as the cleavage furrow *(cf)* has just about completely divided the cell. × 14,000.

SOURCE. Author's micrograph, published in *Cytobios 6,* 167 (1972).

anaphase progresses (Figs. 4.10*A–G,* 4.16, 4.17). Figure 4.17 clearly shows how the anaphase spindle has elongated markedly while the chromosome-to-centriole distance has not appreciably shortened. My own results indicate that during anaphase the terminal vacuoles of the cell become joined into one essentially continuous system which includes the interzonal vacuole (Fig. 4.16).

Telophase is marked by more unusual features. The daughter nuclei stay widely separated throughout telophase (Figs. 4.18, 4.19) and cytokinesis (Fig. 4.20), and interzonal microtubules persist between them. The chromosomal microtubules do not all disperse once the nuclear envelope has re-formed (Fig. 4.19). Cytokinesis is brought about by the ingrowth of an annular furrow of the plasmalemma, visible by metaphase, into which vesicles may be discharged. No phycoplast or other microtubular system is involved. The ingrowing furrow finally cuts the

Figures 4.21–4.23
ZOOSPOROGENESIS IN KLEBSORMIDIUM FLACCIDUM

4.21*A–G.* Emergence of two zoospores. They move through a rupture in the wall, chloroplast first and flagella last *(f).* × 1,300.

4.22. Biflagellate zoospore after emergence. × 1,000.

4.23. Differentiating zoospore inside parental cell wall. The chloroplast is bulging into the wall papilla, an eroded region that ruptures to release the zoospore (Fig. 4.21). The nucleus *(n)* is still flanked by two vacuoles as in the interphase cell, and the two flagella *(f)* have been formed. × 16,000.

SOURCE. Drs. H.J. Marchant, J.D. Pickett-Heaps, and K. Jacobs. Figure 4.23 published in *Cytobios 8,* 95 (1973).

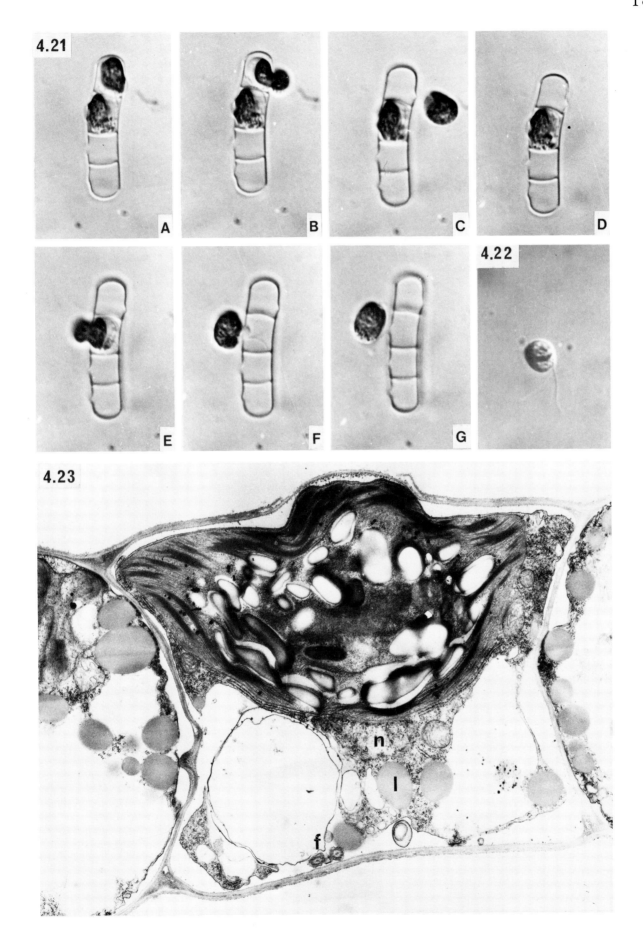

interzonal vacuole, mitochondria, and remnants of interzonal microtubules. After cytokinesis is completed, the daughter cells slowly restore their interphase morphology. The persistent microtubules near the centrioles eventually disperse, as the latter, along with the peroxisome, move back between the nucleus and chloroplast. Wall microtubules also return, and the terminal vacuoles become equal sizes.

The constancy of the chromosome-to-pole distance is unusual during mitosis. Chromosome movement in most cells appears to result from shortening of the chromosomal fibers allied with spindle elongation. In *Klebsormidium*, chromosomal separation apparently is brought about purely by spindle elongation, a phenomenon that must be taken into account in any general theory of spindle function. I have suggested that the persistence of the chromosomal tubules after telophase could be related to postmitotic cellular reorganization, as is the case with the desmids (Chapter 6). The light micrographs taken by Floyd, Stewart, and Mattox (1972b; Fig. 4.10 A–G) might appear to indicate that chromosomal separation is facilitated by enlargement of the interzonal vacuole during later anaphase, but since the whole vacuolar system probably becomes continuous at this stage, expansion of only one region of it is difficult to envisage. It is clear that cell division in *Klebsormidium* is similar to that in *Stichococcus*, but very different from that in its supposedly close relative *Ulothrix*.

4.2b. Zoosporogenesis

When we proposed our new phylogenetic scheme (Pickett-Heaps and Marchant, 1972) suggesting an affinity of *Klebsormidium* to the progenitors of higher plants, we also mentioned that the scheme was testable. We suggested, for example, that the zoospore of *Klebsormidium* might be different in cytomorphology from the zoospores of *Ulothrix* and of *Stigeoclonium*, and might rather resemble the zoospore of *Coleochaete* (Section 4.3c) and perhaps even the sperm of *Chara*. We have been agreeably surprised at the extent to which this prediction has been validated (Marchant, Pickett-Heaps, and Jacobs, 1973).

Experimental induction of zoosporogenesis in *Klebsormidium* has been investigated by Cain, Mattox, and Stewart (1973b; see also Mattox, 1971), who have published a detailed light mi-

Figures 4.24–4.28
STRUCTURE OF THE ZOOSPORE OF KLEBSORMIDIUM FLACCIDUM

4.24. Zoospore after release. Note particularly the microtubular band *(tb)* that runs from near the basal bodies *(bb)* and multilayered structure *(mls)* along the length of the cell (arrow). The usual contractile vacuole *(cv)* lies near the basal bodies. A collection of vesicles *(vs)*, whose significance is unknown, is always found at one end of the cell. × 19,000.

4.25. The microtubule band has been sectioned transversely near where it lies alongside the chloroplast (see arrow in Fig. 4.24). It contains about 21 tubules at this level (cf. Fig. 4.28). × 76,000.

4.26. The end of the microtubule band associated with a forming multilayered structure in a differentiating zoospore. Note that the microtubules are oriented almost at right angles to the nearby basal body. × 70,000.

4.27. This section contains one flagellum in longitudinal section, and consequently, the microtubule band is cut skew. × 35,000.

4.28. The end of the tubule band cut transversely (the flagellum now, of course, is cut skew; cf. Fig. 4.27). The multilayered structure is indicated by the arrow. There are approximately 34 tubules at this level in the band (cf. Fig. 4.25), some of which are obscured by contamination on the section. × 38,000.

SOURCE. Drs. H.J. Marchant, J.D. Pickett-Heaps, and K. Jacobs, published in *Cytobios 8*, 95 (1973).

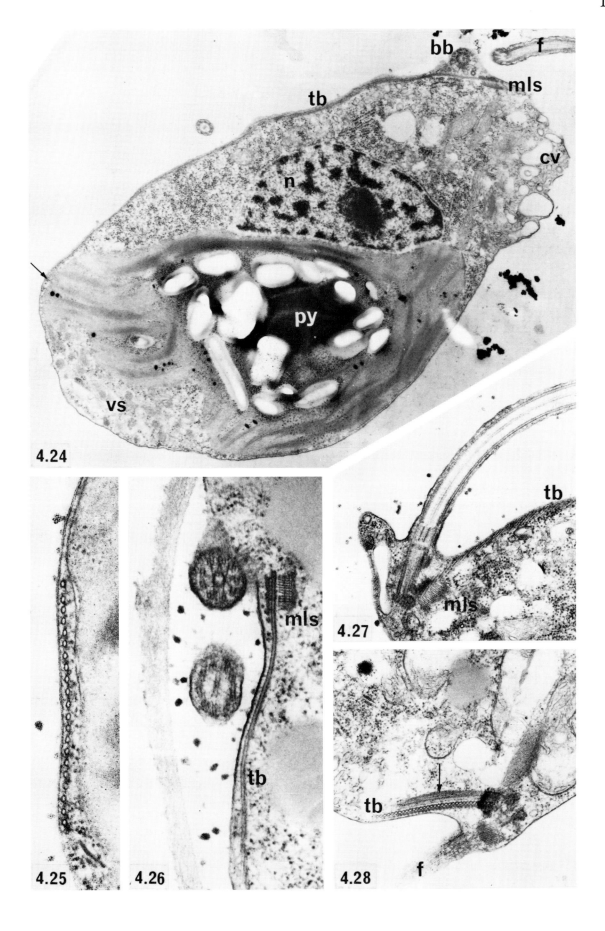

4.24

4.25

4.26

4.27

4.28

croscopic account of the differentiation and emergence of the zoospore (Cain, Mattox, and Stewart, 1973a). An early sign of incipient zoosporogenesis is the appearance of a papilla in the wall, at which site the wall is ruptured to release the mature zoospore (Fig. 4.21 *A–G*). The chloroplast becomes appressed to the papillar wall (Fig. 4.23), and later the zoospore emerges, usually chloroplast first (Fig. 4.21 *b, e, f*). The mature, biflagellate zoospore (Fig. 4.22) is smaller than the cell it occupies, and shrinkage of the terminal vacuoles (e.g., Figs. 4.9, 4.11) takes place early in zoosporogenesis.

At the ultrastructural level (Marchant, Pickett-Heaps, and Jacobs, 1973), the papilla can be seen to arise from a localized erosion of the wall (Fig. 4.23). The two centrioles migrate to the edge of the cell and form flagella, and the peroxisome moves away from the nucleus. Near the flagella bases appears a contractile vacuole (Fig. 4.24), and most importantly, a complex "multilayered structure" is also formed near the flagella bases. It is illustrated in Figures 4.26–4.28 and is described in more detail by Marchant, Pickett-Heaps, and Jacobs (1973). It contains an amorphous and lamellar layer, which surmounts a flat layer of about 34 (Fig. 4.28) dense microtubules; underneath the tubules is another amorphous layer (Fig. 4.26). The lamellae of the multilayered structure are oriented at approximately right angles to the tubules (Fig. 4.27). These dense microtubules are continuous with the flat band of tubules that runs along one side of the cell near the plasmalemma (Figs. 4.24, 4.26), apparently decreasing slightly in number along their length (Fig. 4.25). The two flagella are inserted skew to the band and at an angle to one another (Figs. 4.27–4.29). As a result, they emerge slightly from one side of the cell's apex (Figs. 4.24, 4.29). After a period of motility of up to 1 hour, the zoospores withdraw their flagella and secrete a wall without attaching to a substrate. The flagella appear to be incorporated into the cytoplasm directly after folding back along the cell's surface. It appears as if the flagella beat once or twice inside the cytoplasm immediately after retraction, as quite violent waves pass around the cell from the site of flagella insertion (Fig. 4.30*A–E*).

It is emphasized that the zoospore of *Klebsormidium* differs markedly in structure from those

Figures 4.29, 4.30
ZOOSPORES OF KLEBSORMIDIUM FLACCIDUM

4.29.   Scanning electron micrograph of a zoospore of *K. flaccidum,* showing the somewhat lateral insertion of the two flagella. × 32,000.

4.30*A–E*.   As the biflagellate zoospore *(A, B)* withdraws its flagella, a wave passes around the cell (arrows, *C, D*) before the cell becomes quiescent *(E)* and secretes a wall. × 1,200.

SOURCE. Drs. H.J. Marchant, J.D. Pickett-Heaps, and K. Jacobs, published in *Cytobios 8,* 95 (1973).

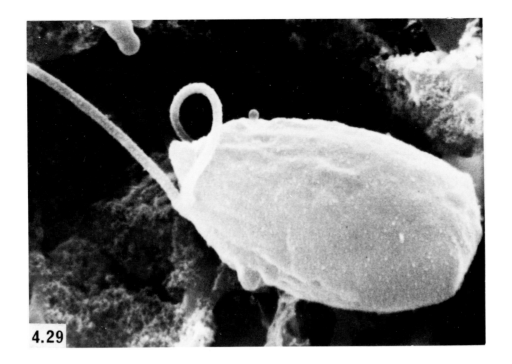

4.29

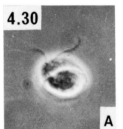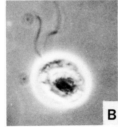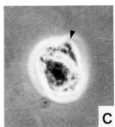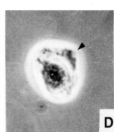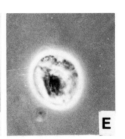

4.30

A B C D E

of the Chlorococcales, many of the Ulotrichales, and the vegetative cells of the Volvocales. The zoospores in these latter groups appear to be structurally related to one another, in particular having similar microtubular systems associated with two or four symmetrically inserted flagella. In contrast, the zoospore of *Klebsormidium* has a large band of microtubules and the multilayered structure; the flagellar insertion is asymmetrical. These features remind one of the zoospores of *Coleochaete* and the sperm of the Charales (which lack the multilayered structure) and the bryophytes.

4.3. Coleochaete

Coleochaete is the only known representative of the family Coleochaetaceae. As Fritsch (1935) emphasizes, like many other oogamous Green Algae (*Oedogonium, Chara,* etc.) its phylogenetic position is rather isolated, but he considers its inclusion in the Ulotrichales warranted for several reasons. It is a freshwater epiphyte, and some species consist of a mass of dichotomously branched filaments which may display heterotrichous differentiation; others take the form of discoidal colonies, usually only one or two cells thick. Characteristic of all species is the formation of a long, fine unbranched hair by certain cells in a colony. The uninucleate vegetative cells contain usually one large chloroplast and one (or more) pyrenoid. Cell division is confined to the apical cells of filamentous forms and the peripheral cells of discoidal species (Wesley, 1928; Fritsch, 1935).

Both asexual reproduction, via zoospores and aplanospores, and oogamous sexual reproduction are known in most species. Reproduction by zoospores is discussed in Section 4.3c. Sexual reproduction is most interesting and is of an advanced form. Spermatogenesis has been described in detail for several species and particularly for *C. scutata* (Wesley, 1930). In filamentous species, the antheridia arise in terminal cells. In *C. scutata* these appear following cell division in internal cells of the colony (which normally never divide except to regenerate damaged regions of the plant body). Such a cell commences sexual differentiation by undergoing first an equal division to form two antheridial mother cells. Each of these antheridial mother cells then divides twice asymmetrically, cutting off two an-

Figures 4.31–4.35
COLEOCHAETE SCUTATA

4.31.   Discoidal, slightly hemispherical vegetative colony. The two cells in the foreground were probably undergoing radial division. Note the hair *(h)* growing from the central cell. × 580.

4.32.   As for Figure 4.31, but this micrograph shows the arrangement of the cells in the colony. Some nuclei are visible (arrows). × 180 approx.

4.33.   Metaphase, radial division. The chloroplast already has a deep cleavage furrow in it (arrow). × 1,100.

4.34.   Two telophases (arrows), circumferential cell division. The daughter nuclei are far apart with a phragmoplast between them. × 1,100.

4.35.   Two peroxisomes *(p)* sandwiched between the nucleus *(n)* and the chloroplast *(c)* (cf. Figs. 4.11–4.13). Also visible is the interphase centriole *(cn)* close to the nuclear envelope. × 42,000.

SOURCE. Drs. H.J. Marchant and J.D. Pickett-Heaps, published in *J. Phycol. 9,* 461 (1973).

theridia, the second slightly smaller than the first. The contents of each antheridium differentiate into a sperm, which appears to be colorless (i.e., devoid of a chloroplast) in some species, but not *C. scutata*. The sperm then escapes in much the same fashion as does a zoospore. Wesley says that this method of spermatogenesis seems to be unique in the Green Algae and, in particular, that the complexity of antheridial development in *C. scutata* is surpassed only by the Charales. The oogonia are enlarged cells which in several species become enveloped by sterile vegetative cells. This latter feature is also an advanced characteristic of sexual reproduction which has the potential for forming a protected oogonium. We may finally note one more advanced characteristic of this genus. Fertilized eggs germinate to form a multicelled body which "is the nearest approach in the algae to the simplest sporophyte found in the liverworts" (Wesley, 1928; p. 1).

Ultrastructurally, the cells of *C. scutata* are not unusual in morphology (Marchant and Pickett-Heaps, 1973). The colony is discoidal in form and slightly hemispherical when young (Fig. 4.31), although older colonies tend to become more irregular. The central cell and often several others bear the characteristic hairs. The chloroplast is situated on the upper and outer walls (Figs. 4.32–4.34). Centrioles are persistent, and there are one or more peroxisomes flattened between the nucleus and chloroplast (Fig. 4.35), reminiscent of those in *Klebsormidium* (Figs. 4.11–4.14). These peroxisomes react positively to the DAB/H_2O_2 reagent. Plasmodesmata are numerous between younger cells.

4.3a. Cell Division

Cell division in *C. scutata* has been described by Marchant and Pickett-Heaps (1974). In vegetative colonies, growth takes place by divisions confined to the marginal cells, and the plane of division in these is oriented in either of two directions: radial to (Fig. 4.33) or parallel to (Fig. 4.34) the circumference of the colony. The latter divisions are hereafter termed "circumferential" divisions.

Incipient mitosis is often marked by a deep cleavage developing in the chloroplast. As usual, extranuclear microtubules proliferate during the prophase condensation of chromosomes and dispersal of the nucleolus. The nuclear envelope

Figures 4.36, 4.37
MITOSIS AND CYTOKINESIS IN COLEOCHAETE SCUTATA

4.36. Metaphase (cf. Fig. 4.33). The spindle is open and permeated by endoplasmic reticulum. Microtubules *(t)* run from the pair of polar centrioles *(cn)* to the chromosomes *(cb)*. × 20,000.

4.37. Cytokinesis, radial division. The cell plate *(cp)* is forming the new wall *(w)* in the longitudinally oriented microtubules of the phragmoplast, between widely separated daughter nuclei *(n)*. × 9,200.

SOURCE. Drs. H. J. Marchant and J.D. Pickett-Heaps, published in *J. Phycol. 9*, 461 (1973).

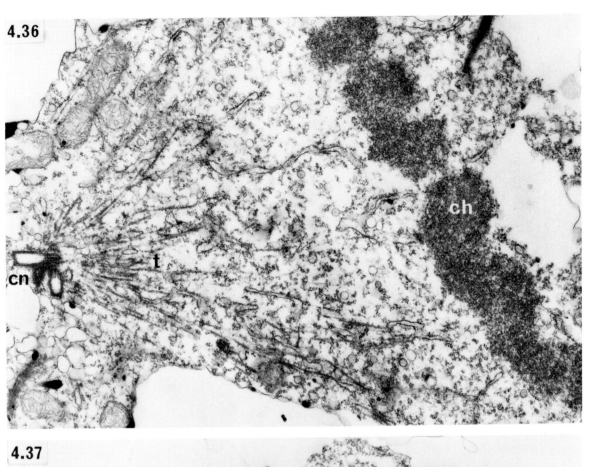

4.36

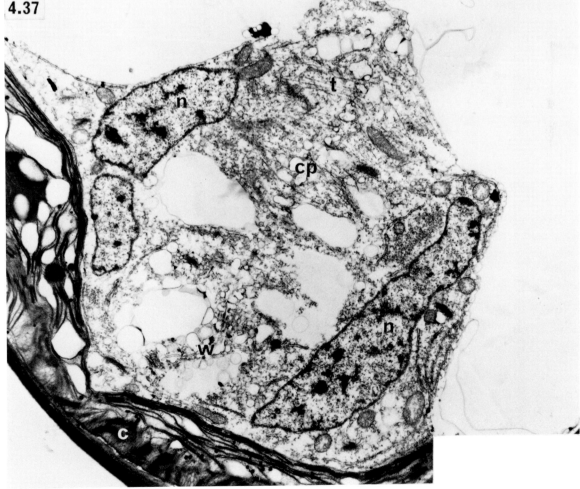

4.37

fragments during prometaphase, and by metaphase it has completely disappeared. The metaphase spindle is centric and open (Fig. 4.36). Anaphase is rapid.

Cytokinesis is particularly interesting in this organism. It is one of the very few algae so far described which uses a phragmoplast for cross-wall formation. The nuclei stay widely separated at telophase, and between them longitudinally oriented microtubules proliferate, and soon vesicles collect in the central region between the nuclei, and these fuse to give the cross wall (Fig. 4.37). However, the course of cytokinesis is subtly different if the cells divide circumferentially (Fig. 4.34). In this event, cytokinesis appears to commence at the walls, and the forming wall grows inward toward the center of the cell. As far as we can tell, the cross wall is still formed entirely by fusion of vesicles. Thus in radial divisions, the cell plate grows from the outside inward.

4.3b. The Hair Cell

Hair cells of *Coleochaete* are fascinating structures for several reasons. Unfortunately, no detailed ultrastructural description of their formation has yet been published, and we have to rely on the microscopic description given by Wesley (1928). She says that a pore forms in the wall, equivalent to that which enables the zoospore to escape. The chloroplast rolls into an incomplete cylinder under the pore, and immediately under the opening, inside the chloroplast, appears a dense granule. From this granule, a "dense granular stream" issues forth through the pore. A thick wall is secreted around the base of the hair (Figs. 4.38–4.40) as a "sheath" grows outward from the pore. When the sheath stops growing, a fine hair (sometimes several) extends from it; from Wesley's account, this hair appears to be a filament of cytoplasm. An ultrastructural investigation of hair differentiation would obviously be of interest.

The hair is also interesting for another reason. It has been demonstrated (Geitler, 1960, 1961a) that the chloroplast and adjacent cytoplasm in these particular cells undergo steady rotation, a complete revolution taking about 1–3 minutes. The cytoplasm near the wall by contrast is stationary. The axis of rotation is about the axis of the base of the hair. Since this rotation is not

Figures 4.38, 4.39
HAIR CELLS OF COLEOCHAETE SCUTATA

4.38. Longitudinal section. The chloroplast *(c)* and nucleus *(n)* are partially suspended away from the wall by vacuoles *(v)*. The wall at the base of the hair *(h)* is thickened. × 5,400.

4.39. Transverse section near the top surface of the hair cell. The thickened base of the hollow hair is visible *(h)*. As in Figure 4.38, the chloroplast is largely separated from the wall by vacuoles. × 9,800.

SOURCE. Dr. H.J. Marchant, unpublished micrographs.

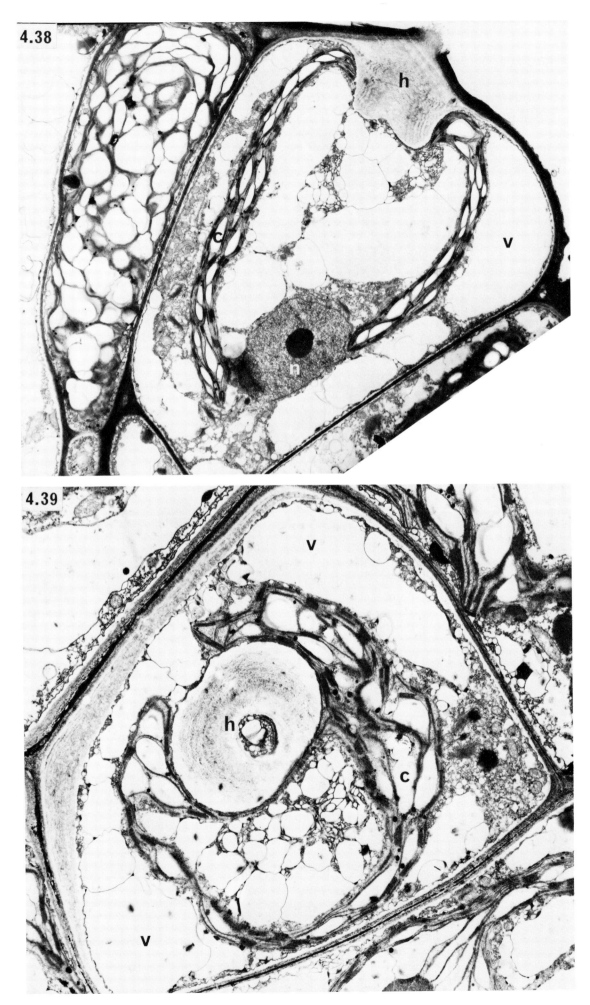

evident in other cells of the organism, it is a most striking phenomenon, particularly when visualized using time-lapse cinematography (Brown, personal communication). Oscillation of cytoplasm in these cells also occurs, but this motion is probably representative of a somewhat unhealthy state, induced, for example, by treatment of the organism with dilute acetic acid. The physiological significance of this rotation remains totally obscure.

The ultrastructure of these cells may be relevant to understanding their unusual behavior. In contrast to surrounding cells, their chloroplast is in the form of a cylinder (as described by Wesley, 1928), containing the nucleus (Fig. 4.38). This cylinder is usually separated to a large extent from the peripheral cytoplasm by an extensive system of vacuoles. The Golgi bodies in these cells are conspicuously active in some way, contrasting to those in neighboring cells. Furthermore, the chloroplast is tightly appressed at one end of the bulbous wall at the base of the hair (Figs. 4.38–4.40). Just inside the surface of the cytoplasm at this wall is a series of microtubules, just visible in Figure 4.40 (small arrows), apparently circumferentially arranged around the axis of the hair. Colchicine experiments indicate that these tubules play no part in generating rotation. Finally, the chloroplast almost always has a characteristic flattened projection appressed to the base of the hair (Fig. 4.40) and often extending into its lumen.

The form of the chloroplast and its separation from the cell wall by vacuoles could obviously be important in achieving rotation. However, the mechanism by which this motion is brought about remains, for the moment, quite obscure.

4.3c. The Zoospore

Unfortunately, zoosporogenesis in *Coleochaete* has not yet been studied ultrastructurally, but the structure of the zoospore is of considerable interest. Wesley (1928) gives a detailed account of the zoosporogenesis observed with a light microscope. She notes that the chloroplast is pressed against the site of the future pore through which the zoospore will escape (this reminds one of *Klebsormidium*; Fig. 4.23). The mature zoospores are round or egg-shaped and they emerge in ameboid fashion through this pore. Wesley says

Figure 4.40
BASE OF A HAIR CELL OF COLEOCHAETE SCUTATA

This micrograph shows the thickened, layered wall at the base of the hair (cf. Figs. 4.38, 4.39). Microtubules are consistently present near the wall in this region (small arrows), as is a peculiar characteristic protrusion from the chloroplast (large arrow) which often extends right into the lumen of the hair. × 12,000.

SOURCE. Dr. H.J. Marchant, unpublished micrograph.

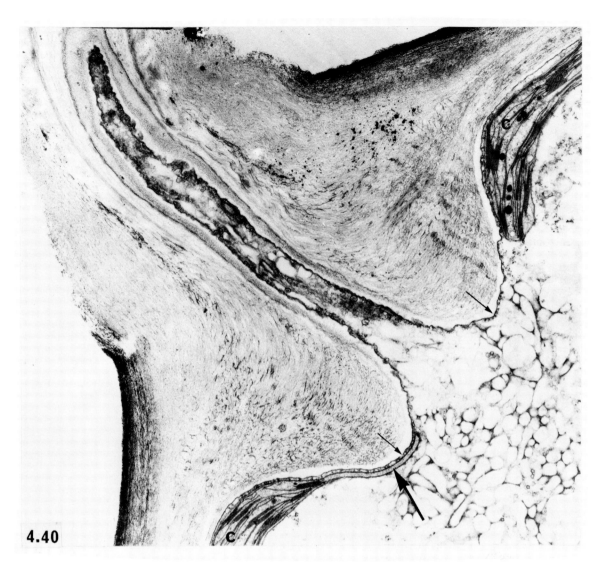

4.40

that they are quiescent for between 5 and 20 minutes before moving, and she also states that there is no doubt that in zoospores of *C. scutata*, the flagella arise during the quiescent period outside the wall. This contrasts with our own observations (Fig. 4.41), which clearly reveal that the zoospore has a pair of flagella while it is still within the parental wall. After the zoospore has undergone a period of motility, it settles and germinates. The first division in this cell is unequal, and the smaller upper cell thus arising will form a hair. The subsequent course of cell division which forms a mature colony is described in some detail in Wesley's paper.

The small amount of ultrastructural work done on the zoospore quite clearly reveals that it is unlike most other ulotrichalean zoospores so far described. The flagella (Fig. 4.42) are covered both by hairs and by a close-packed pattern of tiny, diamond-shaped scales, similar to those covering the sperm of the Charales (Fig. 7.47). The body of the zoospore is also covered by scales (Fig. 4.41, 4.43), but these are quite different from the flagellar scales, being long and somewhat conical (McBride, 1968; Pickett-Heaps and Marchant, 1972). Finally and most importantly, there is a microtubule band running along one side of the zoospore. One end of this band is embedded in a multilayered structure (Fig. 4.43) reminiscent of a similar structure found in the sperm of bryophytes (McBride, 1971; Pickett-Heaps and Marchant, 1972). Indeed, the occurrence of this multilayered structure in *Coleochaete* provided the stimulant for us to investigate carefully the zoospore of *Klebsormidium*; on the basis of the similarity in spindle structure between these two organisms, we predicted that *Klebsormidium* zoospores might have a similar structure. Our discovery of a multilayered structure in *Klebsormidium* (Figs. 4.26–4.28; Marchant, Pickett-Heaps, and Jacobs, 1973) was most exciting and has strenghtened our conviction that ultrastructure of dividing cells and zoospores may have considerable phylogenetic significance (Chapter 8).

4.4. Ulothrix

Many species of *Ulothrix* are common, particularly in freshwater habitats, where they usually are attached to a substrate in a stream or else floating free in still water. The filaments are sim-

Figures 4.41–4.43
THE ZOOSPORE OF COLEOCHAETE SCUTATA

4.41.   Zoospore, almost mature, still inside the parental wall. Note the pair of flagella *(f)*. × 7,100.

4.42.   Diamond-shaped scales and flagella hairs *(fb)* on the flagella. × 50,000.

4.43.   Multilayered structure *(mls)* at the end of the flat band of microtubules *(t)* that runs the length of the zoospore (cf. Figs. 4.26–4.28). The densely stained body scales (visible also in Fig. 4.41) are quite different from those on the flagella. × 72,000.

SOURCE. Dr. H.J. Marchant, Figure 4.43 published in *Cytobios 6*, 255 (1972).

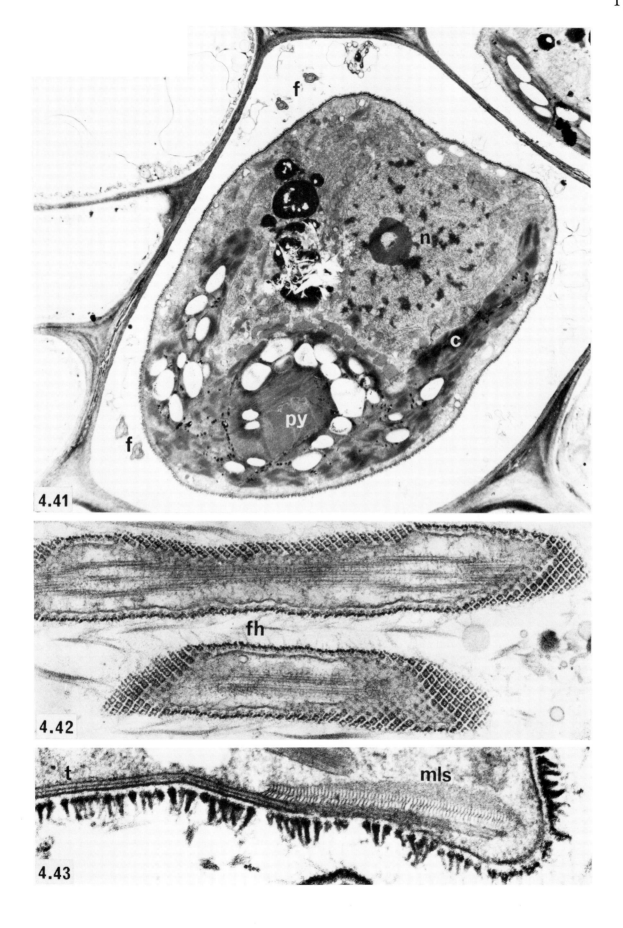

4.41

4.42

4.43

ple in form, occasionally branched (Floyd, Stewart, and Mattox, 1972a), often having a basally differentiated holdfast.

Vegetative reproduction in *Ulothrix* can occur by fragmentation of the vegetative filaments or by release of zoospores, of which 2–16 may be formed per parental cell. Sexual reproduction involves small biflagellate isogametes, which fuse to give a quadriflagellate zygote, which in turn soon forms a thick-walled, resistant zygospore. Meiosis apparently occurs during germination of the zygote, and so, as expected, the vegetative life cycle of *Ulothrix* is haploid. Some species of *Ulothrix (U. zonata)* reportedly produce three types of swarmers (macrozoospores, microzoospores which may be biflagellate, and the gametes; Fritsch, 1935; p. 203).

4.4a. Cell Division

Floyd, Stewart, and Mattox (1972a) have described interphase and dividing cells of *U. fimbriata.* The interphase cell contains persistent centrioles, several pyrenoids, and numerous small microbodies or peroxisomes (cf. *Klebsormidium;* Section 4.2). Plasmodesmata traverse the end walls of the cells (Floyd, Stewart, and Mattox, 1971). During prophase, the chloroplast begins to cleave and the nucleolus disperses. By prometaphase, gaps appear in the nuclear envelope, and microtubules invade the nucleus. The chromosomes are very small and form a typical metaphase plate (Fig. 4.44*A*) inside a highly vesiculated nuclear envelope (Fig. 4.45; cf. *Microspora;* Section 4.5a). Centrioles, which are persistent throughout the cell cycle, are situated at the poles. In living cells, the spindle rocks or "twitches" back and forth from the longitudinal axis before anaphase commences. Floyd et al. have an elegant set of micrographs showing anaphase movement of chromosomes which begins with the spindle tilted to the cell axis (Fig. 4.44 *A–E*). Telophase follows as normal.

Cytokinesis involves a phycoplast/cell plate system. The nuclei return together at the center of the cell and flatten against one another (Fig. 4.46), and between them appear the typical phycoplast microtubules. Vesicles, probably derived from the Golgi bodies, collect among the microtubules and fuse to give a cell plate that usually grows from the center outward (Fig. 4.47).

Figures 4.44, 4.45
MITOSIS IN ULOTHRIX FIMBRIATA

4.44A–E. Metaphase to early telophase in a live cell. The spindle is pointed and oriented slightly skew. A possible centriole is indicated by the arrowhead *(A)*. × 2,200.

4.45. Metaphase spindle. The nuclear envelope *(ne)* is still present, although extensively vesiculated. Spindle microtubules *(t)* converge on polar centrioles *(cn)*. × 33,000.

SOURCE. Drs. G.L. Floyd, K.D. Stewart, and K.R. Mattox, published in *J. Phycol. 8,* 68 (1972).

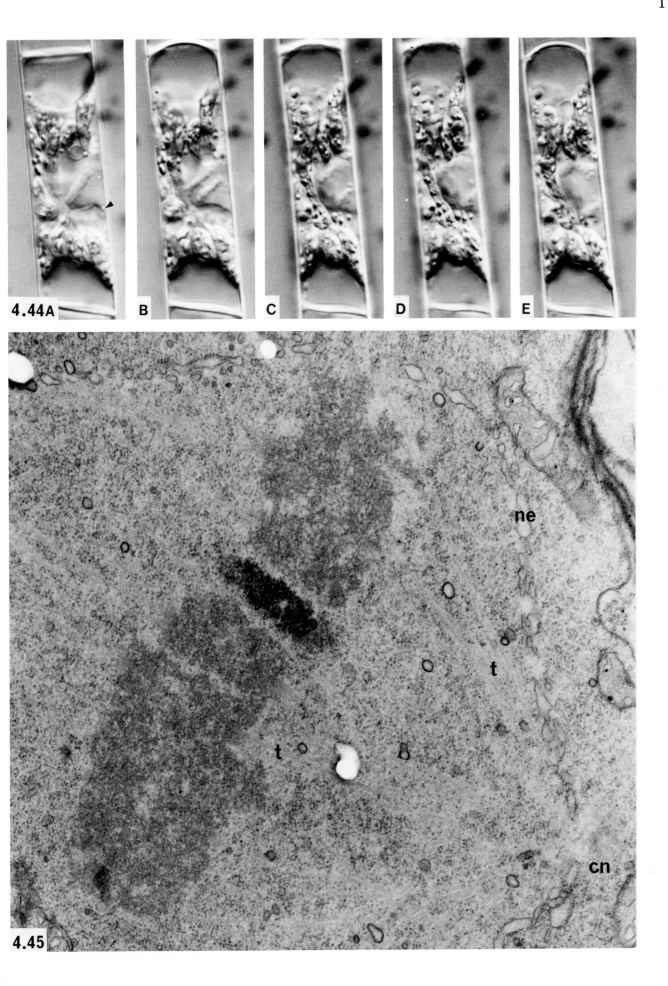

Two interesting points can be noted concerning cytokinesis. First, Floyd, Stewart, and Mattox (1972a) showed that one daughter cell during cytokinesis contains four centrioles adjacent to the nuclear envelope (Fig. 4.47), suggesting perhaps that centriole replication accompanies cytokinesis (unless the daughter cell in this case was abnormal or about to form zoospores). Second, the centrioles remain at the polar regions, in contrast to most other centric, phycoplast-forming cells, where the centrioles move around adjacent to the phycoplast during its formation (cf. *Microspora;* Section 4.5a, or the Chlorococcales; Chapter 3).

Formation of zoospores has not yet been studied in detail, but some information on an unidentified species of *Ulothrix* (Pickett-Heaps, unpublished data) shows that the cell cleaves using a phycoplast/membrane furrow system; no cell plate is apparent (Figs. 4.48, 4.49). This difference is not really surprising, as the zoospores are apparently naked (Fig. 4.48) and do not need a cell wall around them. It is rather striking that *Oedogonium* also reverts from the phycoplast/cell plate to a phycoplast/membrane furrowing system when accomplishing the final cleavage in spermatogenous cells; the spermatozoids to be formed in these cells are also naked (Section 5.1d). The phylogenetic implications of this change in cytokinetic mechanisms are discussed in Chapter 8 (see also Pickett-Heaps, 1972a). In a recent paper, Stewart, Mattox, and Floyd (1973) report, however, that even during vegetative division, *U. zonata* undergoes cytokinesis by furrowing.

4.5. Microspora

"The genus *Microspora* has suffered many vicissitudes," says Fritsch (1935; p. 208), reflecting upon its varied taxonomic history. It is now placed in the Ulotrichales, a reasonable assignment despite some features it possesses distinct from other members of this group. *Microspora* is an unbranched filamentous form which may have a very slightly differentiated basal cell. The parietal chloroplast encloses the whole cell and is uneven and often reticulate. Plasmodesmata and pyrenoids are absent (Figs. 4.50–4.53); the absence of the latter is an unusual feature in algal cells. The cell wall is also rather unusual as it can often be dissociated into distinctively shaped seg-

Figures 4.46, 4.47
CYTOKINESIS IN ULOTHRIX FIMBRIATA

4.46. Live cell at telophase with the cell plate between closely appressed daughter nuclei. × 1,500.

4.47. Cell plate just completed across cell. Phycoplast microtubules can still be detected near the cross wall (arrows; cf. Fig. 4.60). Four centrioles *(cn)* are visible at one pole, and two at the other. × 15,000.

SOURCE. Drs. G.L. Floyd, K.D. Stewart, and K.R. Mattox, published in *J. Phycol. 8,* 68 (1972).

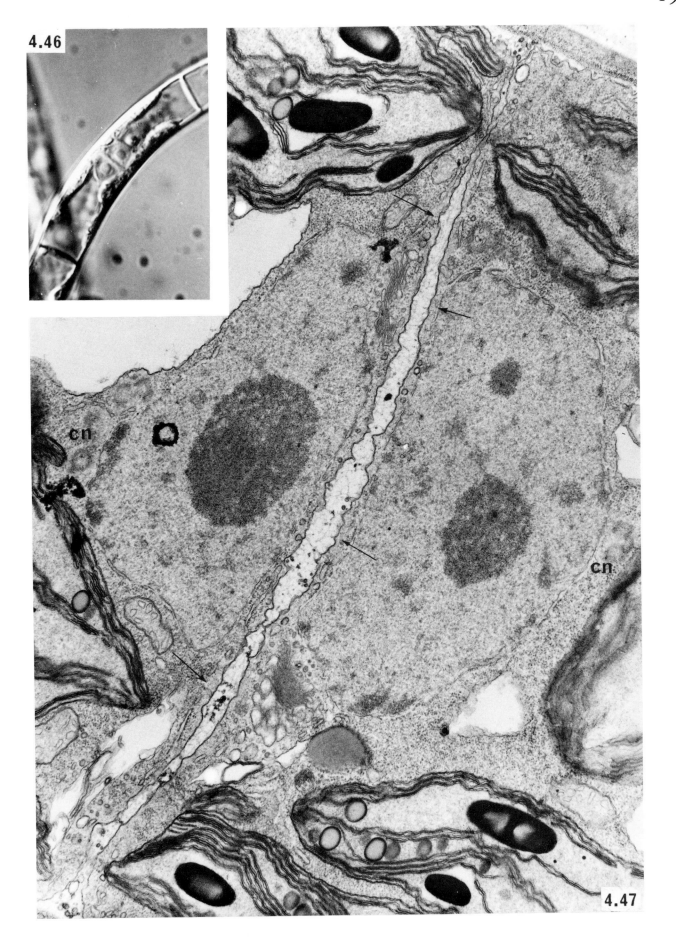

4.46

4.47

ments, H-shaped in optical section, which contain the end wall and half the outer wall of two adjacent cells (Fig. 4.51). The way these segments arise has been set out in Figure 8.56. They result when the cell goes through a distinct phase of wall deposition during interphase, followed by another during cross-wall formation at cytokinesis. The wall segments are in fact closely appressed to one another in the vegetative filament (Figs. 4.50–4.53). Cell elongation, accompanied by a new phase of wall deposition after division (Figs. 4.50, 4.51), probably causes the newer and older wall layers to remain slightly discontinuous and separable under certain conditions.

As in *Ulothrix,* several types of swarmers are known. Two biflagellate zoospores are usually formed per vegetative cell, but quadriflagellate zoospores have also been recorded (Fritsch, 1935). Sexual reproduction is poorly documented, but seems to involve small biflagellate isogametes, 2–16 being released per cell.

Ultrastructurally, the cells are quite simple (Pickett-Heaps, 1973b). The wall is thick and shows evidence of its segmented construction (Figs. 4.50–4.53). The single, flattened nucleus accompanied by Golgi bodies is centrally situated, and the vacuoles on each side are traversed by cytoplasmic strands. The nucleus seems to be appressed to the wall at one point where the persistent centrioles are usually located (Fig. 4.51; small arrow).

4.5a. Cell Division

Cell division in *Microspora* (Pickett-Heaps, 1973b) closely resembles that described already for *Ulothrix* (Section 4.4a), and so need be discussed only briefly. In particular, the spindle is semiopen and enclosed by a vesiculated (Fig. 4.54) nuclear envelope, similar to that of *Ulothrix* (Fig. 4.45). The polar centrioles are often partially enclosed by fragments of membranes lying between them and the spindle. At metaphase, the spindle orientation is most variable; in some cases its axis is perpendicular to the cell axis (Fig. 4.54). Spindle microtubules are attached to chromosomes at distinct kinetochores. The spindle is evidently skew during anaphase, and so it becomes considerably distorted and twisted by confinement within the cell. Consequently, I have found it almost impossible to obtain micro-

Figures 4.48, 4.49
ZOOSPOROGENESIS IN ULOTHRIX SP.

4.48. The cytoplasm of this cell has cleaved to form four zoospores. There is no wall material in the furrows around the zoospores. × 9,000 approx.

4.49. Cleavage, forming zoospore. The furrow (*cf*) passes through the divided chloroplast and is associated with a few phycoplast microtubules (invisible at this magnification). × 18,000 approx.

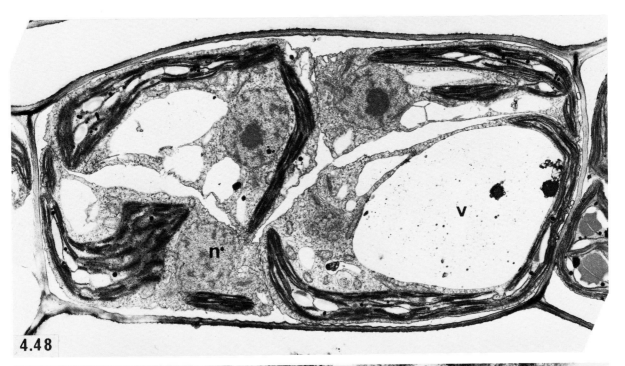

4.48

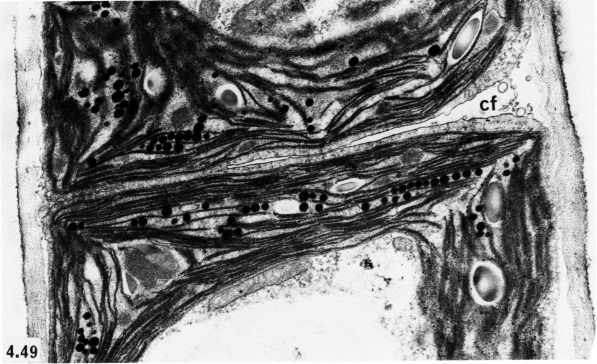

4.49

graphs of late anaphase figures which show both daughter nuclei. Although the nucleolus disperses at prophase, granular aggregates, possibly nucleolar remnants, remain in the spindle and are expelled from it by telophase.

Cytokinesis is mediated by the phycoplast system and, as expected, is preceded by disappearance of the elongated spindle and flattening of the nuclei against one another. The way the cross wall is formed is not completely clear. Some furrowing can always be detected, but vesicles and Golgi bodies also abound in the region around the forming cross-wall (Pickett-Heaps, 1973b). Thus, both cell plate and furrowing may occur, perhaps an intermediate stage in the evolution of a true cell plate (and perhaps analogous to what happens in *Spirogyra*, Section 6.2a, which, however, uses a quite different set of microtubules for the same purpose). The lack of plasmodesmata can also be equated with such an intermediate evolutionary position (Stewart, Mattox, and Floyd, 1973). A notable difference from *Ulothrix* is the disposition of the centrioles during the appearance of the phycoplast microtubules. In *Microspora*, the centrioles invariably move around each daughter nucleus until they are close to the phycoplast. They are situated within an indentation of the nucleus, and pairs in daughter cells are often adjacent to one another.

These results confirm a close affinity of *Microspora* with several of the Ulotrichales. The distinctive construction of the cell wall may be of only minor importance and, indeed, may indicate a common method of cell wall construction and expansion that is normally masked by the layered wall components that are more firmly integrated with one another in other genera. Fritsch (1935; p. 209) mentions that H-shaped pieces can be recognized in the wall structure of other Ulotrichales, including both *Klebsormidium* and *Ulothrix*, and this has been confirmed by Jane and Woodhead (1941). The variable orientation of the spindle at metaphase (Fig. 4.54) is also intriguing and suggests that the spindle moves about even more than that of *Ulothrix* and *Klebsormidium*. Unfortunately, I have been unable to discern the nucleus in living cells because of the thick and uneven parietal chloroplast and thus cannot confirm this possibility directly. The significance of this spindle movement is unknown.

Figures 4.50–4.53
WALL STRUCTURE IN MICROSPORA SP.

4.50. High magnification of the wall region between the arrows in Figure 4.51. The wall is obviously discontinuous, and the two overlapping portions of the H-shaped segments can be discerned. The new cylindrical inner layer (arrows; see also Fig. 4.52) is just beginning to be secreted. × 13,000.

4.51. Interphase cell soon after division. The chloroplasts *(c)* do not possess pyrenoids. The cell is enclosed by interlocked portions of adjacent H-shaped wall segments. The region of wall between the large arrows is shown in Figure 4.50. The interphase centriole is indicated by the small arrow. × 6,200.

4.52. Cell just after mitosis and prior to cytokinesis. It has elongated considerably during interphase. Careful inspection of the micrograph, in comparison with Figure 4.51, shows that elongation has been achieved by the sliding apart of the interlocked H-shaped segments concurrent with the deposition inside them of a tapering cylindrical segment. After cytokinesis, this too is converted into another H-shaped segment. The region of wall between the arrows is shown in Figure 4.53. × 5,500.

4.53. Detail of the wall indicated by the arrows in Figure 4.52. The two portions of the H-shaped segments (previously overlapped; Fig. 4.51) can now be seen to be almost separated (large arrows). The inner layer of the cylindrical segment (small arrow) has a bulge in it indicating the place of cytokinesis. × 12,000.

SOURCE. Author's micrographs, published in *New Phytol.* 72, 347 (1973).

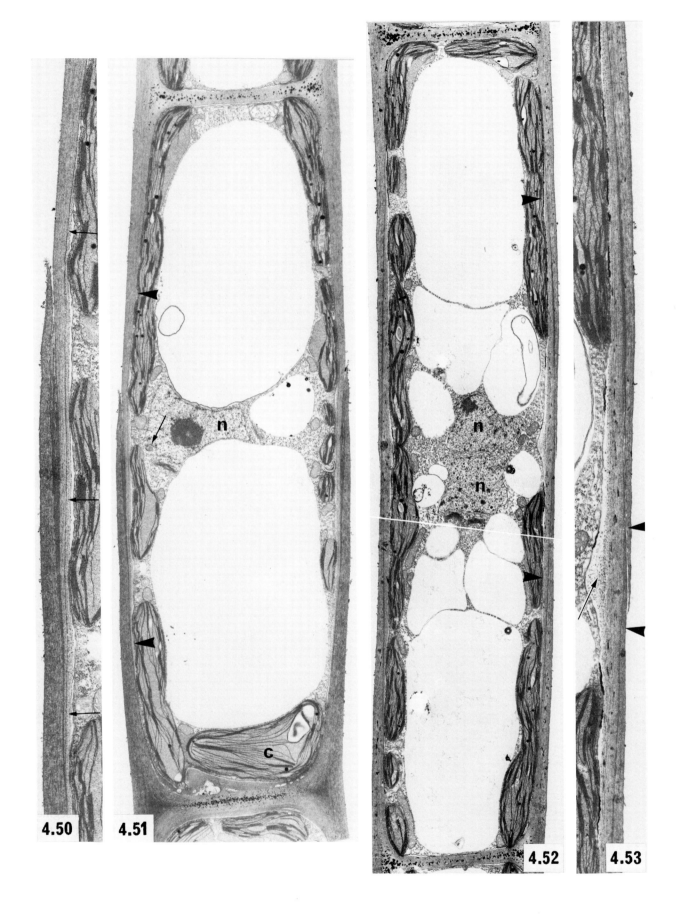

4.50

4.51

4.52

4.53

4.6. Stigeoclonium

This genus, placed in the group Chaetophorales by Fritsch (1935) and others, is obviously closely related to *Ulothrix;* yet equally obvious is the considerable increase in vegetative specialization it has achieved. *Stigeoclonium* is almost always branched, and some species are heterotrichous. The branched filaments are usually tapered and end with a single extremely fine and elongated terminal hair-like cell. The gross morphology of any given species (e.g., degree of branching, heterotrichous habit) varies widely depending upon environmental conditions. Vegetative cells are typically ulotrichalean, with a single parietal chloroplast containing one or more pyrenoids enclosing the central nucleus (Fig. 4.57). In many species, the chloroplast is quite narrow and girdle-shaped, so that the ends of the cells are colorless.

Vegetative propagation is effected by the prolific release of large, quadriflagellate zoospores which contrast with the smaller isogametes, which are also quadriflagellate. The latter are sometimes capable of giving rise to thickwalled resting cells or new plants directly, and so possibly gametes and microzoospores may be easily confused.

4.6a. Cell Division

The ultrastructure of interphase in dividing cells of *S. helveticum* has been described by Floyd, Stewart, and Mattox (1972a). Interphase cells are virtually identical in morphology to those of *Ulothrix,* and cell division is also very similar. The only appreciable difference is that the nuclear envelope in *Stigeoclonium* remains intact around the spindle (Fig. 4.58). The centrioles evidently move into the polar fenestrae, and membranes close over them so that they become intranucleate during mitosis. Not all spindle microtubules focus upon the polar centrioles. Cytokinesis involves the usual phycoplast/cell plate system (Figs. 4.59, 4.60) as in *Ulothrix* (see also Pickett-Heaps, 1972a).

It is a little surprising that *Stigeoclonium*, a more "advanced" organism than *Ulothrix,* has a closed spindle, which is perhaps more primitive (Pickett-Heaps, 1969a, 1972a). However, since the nuclear envelope is also present in *Ulothrix,*

Figures 4.54–4.56
MITOSIS AND CYTOKINESIS IN MICROSPORA SP.

4.54.   Metaphase spindle, oriented across cell. Note kinetochores (arrows). The nuclear envelope is still intact although extensively vesiculated (as in *Ulothrix;* Fig. 4.45). × 12,000.

4.55.   Initiation of cytokinesis. Nuclei are coming together in the center of the cell prior to phycoplast formation. The centrioles are close together at the future phycoplast, in contrast to what occurs in *Ulothrix* (Fig. 4.47). Note also the bandlike structure (arrow) that interconnects the paired centrioles briefly at cytokinesis. This structure may represent some component of a flagellar apparatus that is not now used for cytokinesis, a stage in more primitive cells when zoospores are often formed. × 15,000.

4.56.   Phycoplast containing the typical transversely oriented microtubules *(t)*. Cross-wall formation seems to involve fusion of vesicles as well as furrowing of the cell membrane (arrow). × 30,000 approx.

SOURCE. Author's micrographs, published in *New Phytol.* 72, 347 (1973).

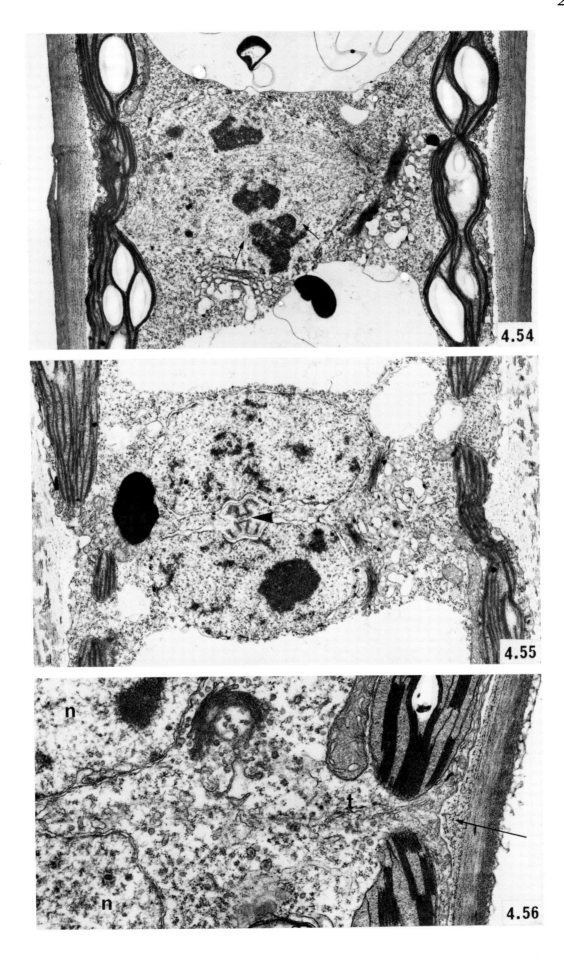

although vesiculated, the difference between these two states may not be highly significant.

4.6b. The Zoospore

Manton (1964b) has described in considerable detail the structure of zoospores appearing in a culture of *Stigeoclonium.* Osmium fixation alone was used in this early work and probably some loss of microtubules resulted, but otherwise the preservation achieved was excellent. It seems likely that this zoospore is fairly representative of the structure of zoospores of other phycoplast-forming ulotrichalean algae.

The naked zoospore resembles *Chlamydomonas* in many respects. The four flagellar bases are arranged as two V-shaped pairs not quite symmetrically arrayed in the cell. Four sets of rootlet microtubules radiate from between the flagellar bases, two containing two microtubules (the two-stranded roots) and the other two containing five microtubules (the five-stranded roots). Although the number of microtubules is consistent in each of the roots, their arrangement in the five-stranded roots is somewhat variable. A striated fiber is associated with, and lies parallel to, each two-stranded root; this fiber is similar to that near the rootlet microtubules in the zoospore of *Oedogonium* (Fig. 5.61). There are two contractile vacuoles near the flagellar bases associated with "hairy" or bristle-coated vesicles. The chloroplast in these strongly phototactic organisms contains the usual eyespot. In other respects, the cells are fairly typical of other unicellular Green Algae.

Upon germination, the cell rapidly withdraws its flagella, changes shape, and becomes sticky all over. Later, a wall is secreted and the cell divides. The flagellar microtubules are found intact in the cytoplasm of germinating zoospores, as is the case with *Hydrodictyon* (Section 3.7a) and zoospores of another Green Alga, *Enteromorpha* (Evans and Christie, 1970). However, contrary to what happens in the zoospores of *Hydrodictyon,* Manton says that in the *Stigeoclonium* zoospore the flagellar bases shorten considerably, losing the transitional structures between them and the flagella, and then sink into the cell coming to rest on the nuclear surface. They then probably become associated with the spindle as normal; this was not confirmed by direct observation of dividing cells, but centrioles were shown at the polar

Figures 4.57, 4.58
STIGEOCLONIUM HELVETICUM, INTERPHASE AND METAPHASE

4.57. Live cell showing large central nucleus and peripheral chloroplast. × 1,500 approx.

4.58. Metaphase spindle. The metaphase plate of chromosomes is tilted (cf. Fig. 4.44A–E). A pair of centrioles *(cn)* is located at one pole within a pocket of the intact nuclear envelope *(ne).* × 41,000.

SOURCE. Drs. G.L. Floyd, K.D. Stewart, and K.R. Mattox. Figure 4.58 published in *J. Phycol. 8,* 68 (1972).

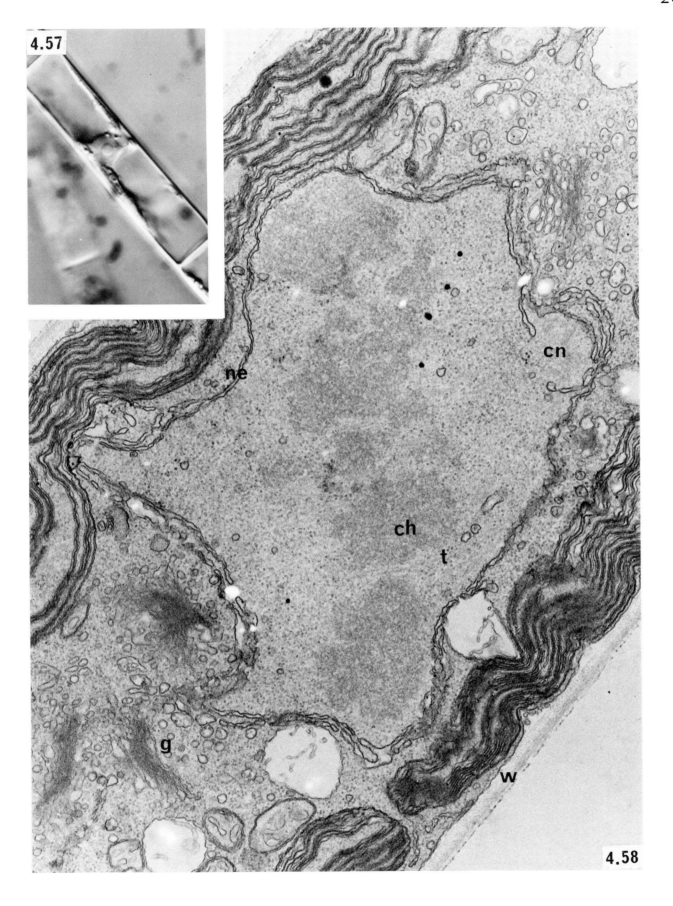

4.57

4.58

region of two daughter nuclei at cytokinesis, just as in vegetative cells. (A phycoplast microtubular system is almost certainly present, although not preserved by Manton's fixation with osmium alone.) Thus, Manton says that the basal bodies are persistent and are passed on to daughter cells. She also noted that when the zoospores became sticky, a population of characteristic vesicles disappeared from the cytoplasm, and she inferred that these were discharged through the cell wall (cf. the zoospores of *Oedogonium;* Section 5.1c). While germination proceeds, the cellular contents undergo considerable reorganization.

Further cell division gives rise to a new vegetative thallus. As in many other Green Algae, the zoospore itself forms the holdfast of this new thallus.

4.7. Ulva

As mentioned earlier, thalloid plant forms arise when cells in a plant body divide in more than one plane. A good example is provided by the Ulvaceae. The common genus *Ulva,* usually marine, consists of a thin, undulating layer of tissue (Fig. 4.62) that essentially is a bilayer of typical ulotrichalean cells containing the usual chloroplast and a single pyrenoid. The cell axis is usually oriented perpendicular to the plane of the thallus; the chloroplast can move about inside the cell to some extent, dependent upon lighting conditions. The thallus is attached for much or all of its life to a substrate by a holdfast which is originally a single basal cell derived from a germinating zoospore and (Fig. 4.61) which subsequently is strengthened by numerous multinucleate rhizoids growing from many of the cells downward through the center of the bilayer.

Vegetative reproduction can take place by fragmentation of the thallus or by quadriflagellate zoospores formed four to eight at a time in any of the vegetative cells. When zoospores germinate, they first form a filament much like that in *Ulothrix,* providing a good example of the principle that "ontogeny recapitulates phylogeny" (Fritsch, 1935) and strongly suggesting the origin of the Ulvaceae from filamentous ulotrichalean algae (Fig. 4.61). Sexual reproduction is isogamous with a slight tendency toward anisogamy in related genera. The biflagellate gametes are produced in the same manner as zoospores. *Ulva* is especially interesting because

Figures 4.59, 4.60
CYTOKINESIS IN STIGEOCLONIUM HELVETICUM

4.59. Live cell. Note the nuclei typically appressed to each other on either side of the developing cell plate. $\times$ 1,250.

4.60. Early stage of cytokinesis. Numerous vesicles are collecting among the transversely oriented microtubules (arrows) of the phycoplast. $\times$ 39,000.

SOURCE. Drs. G.L. Floyd, K.D. Stewart, and K.R. Mattox, published in *J. Phycol. 8,* 68 (1972).

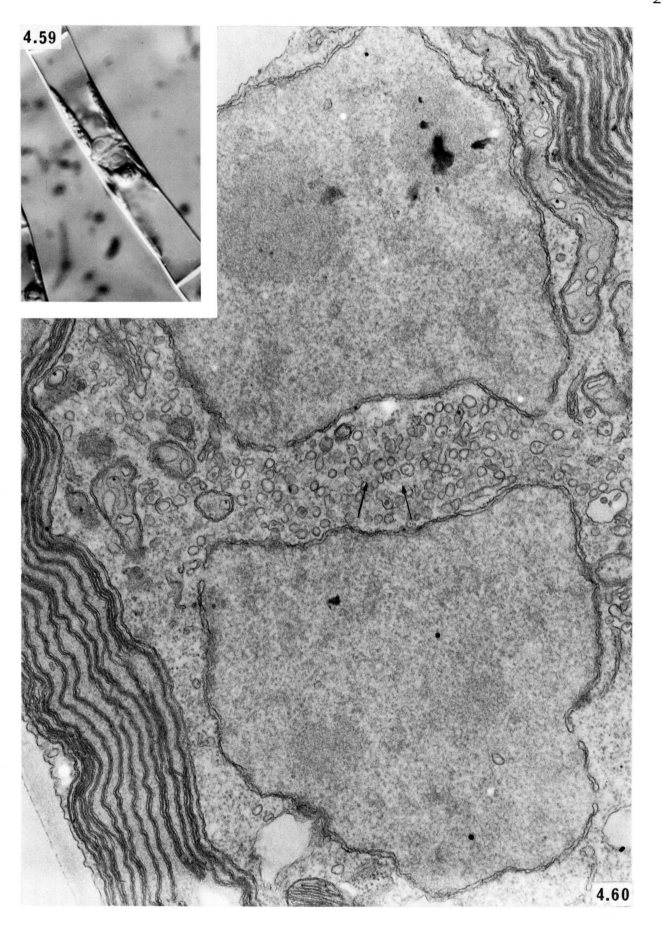

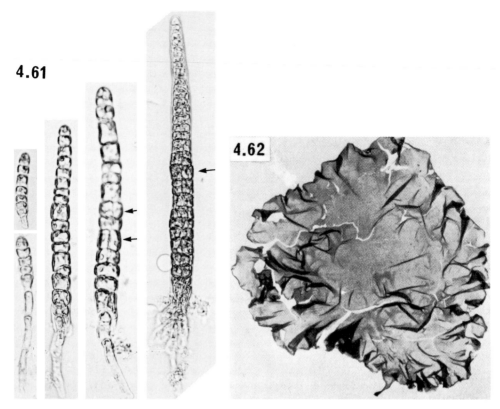

4.61

4.62

Figures 4.61–4.63
ULVA MUTABILIS

4.61.   Young plants of *U. mutabilis*. They commence growth as filaments (much like *Ulothrix*), with a basal holdfast. Later, cell division becomes extensive in two planes (arrows). × 300.

4.62.   Mature plant (gametophyte) of *U. mutabilis*. × 0.5.

4.63.   Comparison between the wild-type *U. mutabilis* organism (arrow) and the slender mutant, both the same age and grown in identical conditions.

SOURCE. Dr. A. Lovlie. Figures 4.61, 4.62 published in *Arch. Protistenk. 104*, 238 (1959).

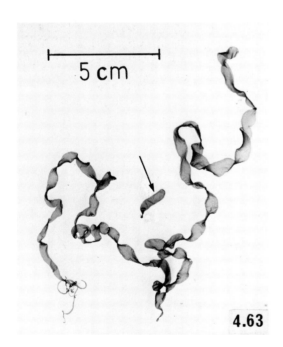

5 cm

4.63

Figures 4.64, 4.65
MITOSIS IN ULVA MUTABILIS

4.64.   Midanaphase. The spindle is closed, with polar fenestrae. × 30,000.

4.65.   Telophase. The cleavage furrow *(cf)* is growing between the daughter nuclei *(n)*. Centrioles *(cn)* are not situated at the poles and are off the spindle's axis. × 37,000.

SOURCE. Drs. A. Løvlie and T. Bråten, published in *J. Cell Sci. 6*, 109 (1970).

it has a regular alternation between haploid sexual and diploid asexual generations. Moreover, vegetative individuals in the two generations are morphologically quite similar in all respects. Thus, the organism undergoes an "isomorphic" alternation of generations; from this state can de-

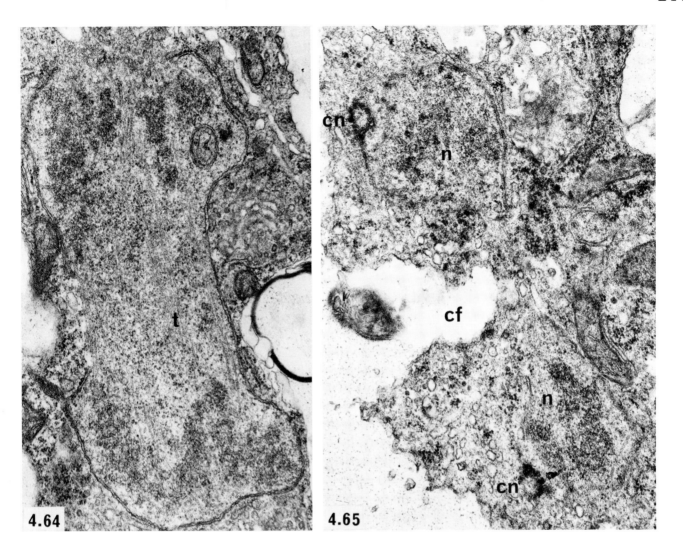

4.64

4.65

velop the "heteromorphic" alternation of generations, characteristic of many advanced Brown Algae (but not of the Green Algae), where the haploid and diploid phases become increasingly distinct morphologically from one another (see Chapter 8; Fritsch, 1942b). The alternation of generations represents another advance achieved by the Ulvaceae over their simpler ulotrichalean predecessors.

Løvlie (1969a,b) has used *Ulva* in studies on the cell cycle. For example, in comparing the "slender" mutant with the wild type, he found that the amount of protein and cell wall material synthesized per unit of DNA was similar in both types, and the efficiency of utilization of incident light energy was also similar (Løvlie, 1969a). However, these slender mutants grow (in the morphological sense) much faster, and the amount of protein per cell is considerably less, than in the wild type (Fig. 4.63). This mutation, therefore, appears to shorten the cell cycle considerably without achieving any greater efficiency in synthesis of cytoplasmic constituents, and the gene affected may be concerned with the events that prepare cells for division. The mutation may be in the chloroplast. A number of other spontaneous mutants have been isolated which also radically alter the morphogenesis of wild-type organisms (Løvlie and Bråten, 1968). Løvlie (1969b) has also been able to measure the changes that take place in photosynthesis during the cell cycle using single cell zygotes (i.e., effectively measuring a single chloroplast). The zygote was contained within a small Cartesian diver suspended in a concentration gradient. The buoyancy of the diver was dependent upon oxygen evolution and could be compared with that of a standard. West and Pitman (1967) have also used *Ulva* in their investigation of ion fluxes (particularly of sodium and potassium) through a cell. The large and active Golgi bodies in these cells could have some unspecified role in ion transport; direct evidence of such an involvement, however, was not obtained.

4.7a. Cell Division

Cell division in *Ulva*, described by Løvlie and Bråten (1968, 1970), is preceded by movement of the nucleus and Golgi bodies away from the chloroplast to the opposite side of the cell. A

Figures 4.66–4.69
FERTILIZATION IN ULVA MUTABILIS

4.66.   Median longitudinal section through gamete. × 21,000.

4.67.   Pair of gametes before cytoplasmic fusion has been initiated, held together by their flagella. × 7,150.

4.68.   Gametes having just established cytoplasmic fusion. Note the apparent lack of cytoplasmic differentiation at the site of fusion. × 16,000. *Inset* shows similar cells under the scanning electron microscope. × 6,200.

4.69.   Gametes (fixed 30 seconds after copulation) in which later fusion of the protoplasts is well under way. × 18,000. *Inset* shows similar cells visualized by the scanning electron microscope. Note that flagellar insertion is at the pointed end of the paired cells (see Fig. 4.66) and one flagellum (arrow) is beginning to fold back along the protoplast. The large spherical extrusion on the surface is reported to be a vacuole. × 8,350.

SOURCE. Dr. T. Bråten, published in *J. Cell Sci. 9*, 621 (1971).

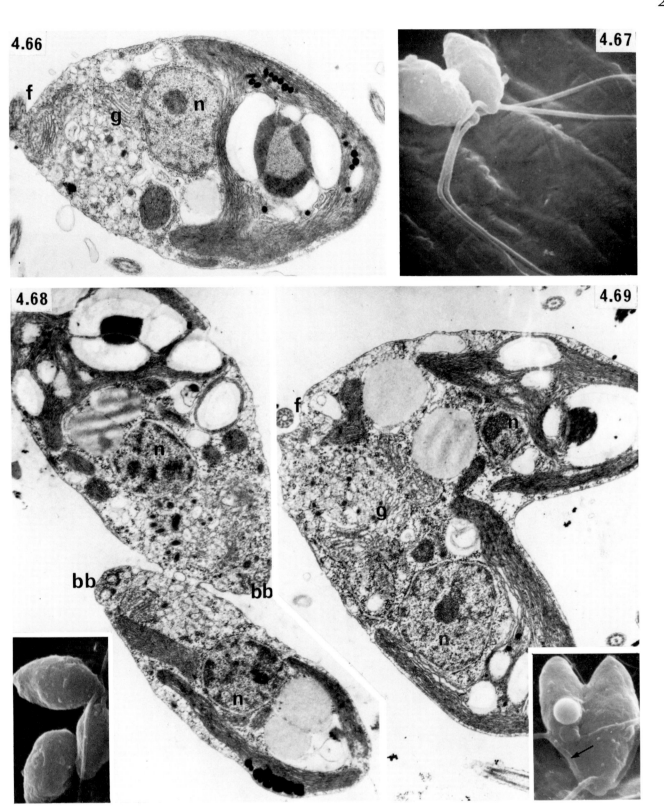

vacuole becomes interposed between them. The nucleus is oval at interphase and prophase, but becomes flattened by metaphase, when the nucleolus has disappeared. The authors do not say whether the centrioles are persistent and when they duplicate; likewise, although they say that no microtubules are in the prophase nucleus, the existence of the usual extranuclear spindle is not mentioned. During prometaphase, the nuclear envelope develops small polar fenestrae, where intranuclear microtubules are first detected. The latter become numerous by metaphase. Polar centrioles, however, appear to be situated to one side of the polar fenestrae (as in Fig. 4.65). The nuclear envelope remains essentially intact during anaphase (Fig. 4.64), although breaks appear in it at the interzone as it stretches. Spindle microtubules do not converge toward the polar fenestrae, and the centrioles are not situated on the spindle axis (Fig. 4.65). Cytokinesis occurs by a symmetrical growth of a furrow of the cell membrane across the cell from the wall between the daughter nuclei. The furrow appears to constrict and then cut the chloroplast in two, and the pyrenoid seems to break down during division.

Unfortunately, a number of interesting points remain unclarified by their description of cell division in this organism, which is obviously difficult to work with. In the discussion, however, they mention that the chromosome-to-pole distance does *not* decrease and indeed appears actually to increase under the light microscope; they believe that the two groups of chromosomes are separated mainly by spindle elongation. An involvement of microtubules with cytokinesis is not mentioned in their paper. From these results, I suspect that cell division in this organism may be more closely related to that in *Klebsormidium* than to that in *Ulothrix,* even though the spindle is largely closed. The fact that the centrioles lie off the spindle axis suggests that either the axis is bent (as in *Tetraspora;* Section 2.3a) or else this spindle resembles that described in sperm packets in *Volvox* (Section 2.2c). It would obviously be of considerable significance to ascertain definitively whether microtubules are involved in cytokinesis (i.e., that a phycoplast is not present).

4.7b. Sexual Reproduction

Bråten (1971) has described fertilization

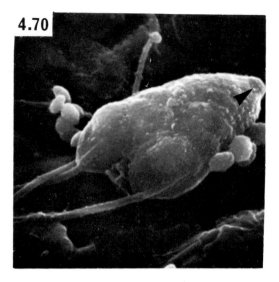

Figure 4.70
FUSED GAMETES OF ULVA MUTABILIS

Scanning electron micrograph. The flagella, inserted at the pointed, anterior end of the cells (arrow; Fig. 4.66), have now folded back along the protoplast and are already partly absorbed into the cell (Fig. 4.71). × 11,600.

SOURCE. Dr. T. Bråten, published in *J. Cell Sci. 9,* 621 (1971).

Figures 4.71, 4.72
ZYGOTE FORMATION IN ULVA MUTABILIS

4.71. Young zygote, 20 minutes after copulation. Note the two distinct sets of chloroplasts and the four complements of flagellar tubules within the cytoplasm (arrows). × 27,000.

4.72. Young zygote, 30 minutes after copulation. The two nuclei *(n)* and parts of the flagellar apparatus are clearly visible. Note particularly that one of the chloroplasts *(c)* is disintegrating; the other is normal. × 38,000.

SOURCE. Dr. T. Bråten, published in *J. Cell Sci. 9,* 621 (1971).

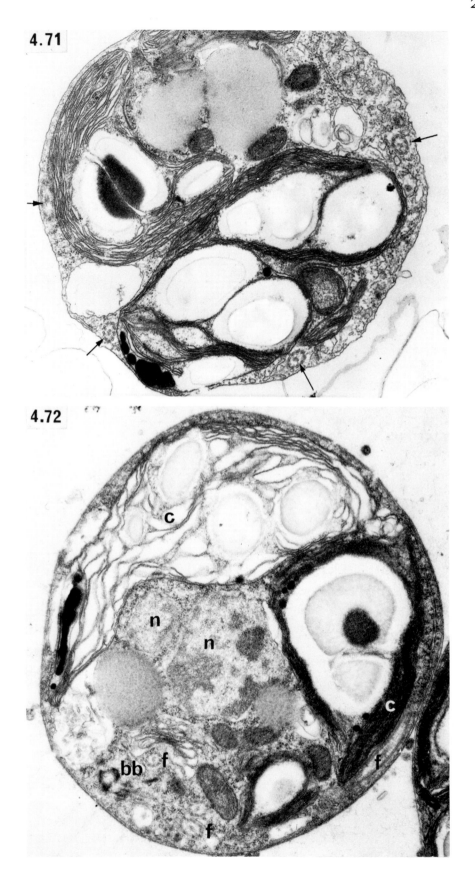

4.71

4.72

and zygote formation in *U. mutabilis.* The process is a remarkably rapid compared, for example, with fertilization in *Chlamydomonas.* This may be a reflection of the exposed conditions (e.g., swirling water in littoral zones) inhabited by *Ulva.* The conjugants are isogamous and indistinguishable from one another even at the ultrastructural level, in contrast to the isogametes of *Chlamydomonas* and *Hydrodictyon,* for example (Sections 2.1b, 3.7b). However, these and most other species of the Ulvaceae are diecious, and as complementary strains are required for conjugation, some form of chemical differentiation between strains must exist. The gametes appear oval in longitudinal sections (Fig. 4.66). One end of the cell is occupied by the large chloroplast and pyrenoid and the other contains the nucleus, flagella, and Golgi apparatus. Conjugation commences with the cells clustering, attached to one another by their flagellar tips (Fig. 4.67); as with *Chlamydomonas,* these clusters break up into pairs of gametes held together by the flagella. However, this process is much more rapid than that in *Chlamydomonas.* Within a few seconds, the cells themselves make contact near the flagellar bases (Fig. 4.68). No specific organelle or differentiation of the plasmalemma was detected at the sites where the membranes of the gametes first fuse. Within 30 seconds, the gametes are partially fused as they jackknife together (Fig. 4.69), and fusion is complete within 3 minutes (Fig. 4.70). Zygotes remain motile for a short period, but soon become attached to a substrate; the attachment is apparently mediated by secretion of the contents of electron-dense vesicles through the plasmalemma, as is the case also with settling zoospores of *Stigeoclonium* and other organisms. The flagella are folded back along the cell, and the proximal part (i.e., nearest the basal body) is absorbed directly into it as the flagella and cell membranes fuse (Figs. 4.70, 4.71). This process is common in algal (and fungal) cells which retract their flagella (Section 3.7a). The rest of the flagella may be drawn into the cell by some sort of contractile process. The flagellar microtubules slowly disperse in the cytoplasm over the next 12 hours.

The settling zygote begins a quite rapid phase of wall secretion. The nuclei fuse within 30 minutes, and Bråten (1971, 1973) provides clear

Figures 4.73–4.75
FRITSCHIELLA TUBEROSA

4.73, 4.74.   Cultured plants of *F. tuberosa.* Note the heterotrichous complexity of the plant body, with colorless rhizoids (arrows) arising from the prostrate portion. Figure 4.73, × 140 approx; Figure 4.74, × 190.

4.75.   Cytokinesis in *F. tuberosa.* The cell plate *(cp)* grows outward from between the daughter nuclei *(n),* which are close together in the center of the cell. × 14,000.

SOURCE. Dr. G. E. McBride. Figure 4.74 published in *Arch. Protistenk. 112,* 365 (1970). Figure 4.75 published in *Nature (Lond.) 216,* 939 (1967).

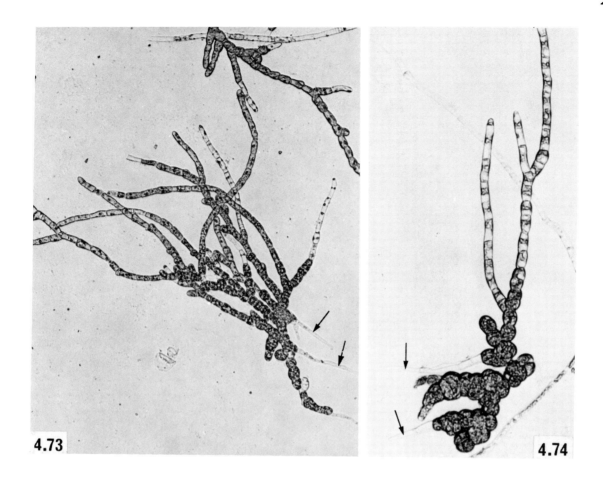

4.73

4.74

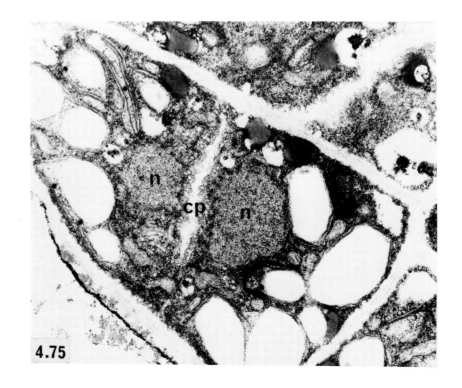

4.75

evidence that one conjugant's chloroplast disintegrates while the other's remains intact (Fig. 4.72). The latter behavior contrasts with observations on other algae in which chloroplasts fuse after conjugation (e.g., *Chlamydomonas;* Cavalier-Smith, 1970).

4.8. Fritschiella

Fritschiella tuberosa (Iyengar, 1932; Figs. 4.73, 4.74) is a highly evolved heterotrichous relative of *Ulothrix* and *Stigeoclonium.* The erect thallus is branched and often differentiated into two, the primary and secondary, systems (Singh, 1947). The prostrate portion of the plant body may grow under the surface of the substrate and has a rhizoidal system emanating from it (Figs. 4.73, 4.74). *F. tuberosa* is markedly terrestrial and can grow in remarkably inhospitable habitats—on alkaline, dry soils low in organic matter (Singh, 1941).

As in *Ulothrix,* reproduction is brought about by three kinds of swarmers—quadriflagellate macrozoospores, quadri- or biflagellate microzoospores, and biflagellate gametes—all of which are formed only in the reserve-laden prostrate portion of the plant. After gametes conjugate, the zygotes give rise directly to a new plant. *Fritschiella,* like *Ulva,* may undergo an isomorphic alternation of generations (Singh, 1947), although this observation was not confirmed by Varma and Mitra (1964). The prostrate system under the soil's surface can also give rise to new erect systems directly, and the club-shaped tissue in which swarmers will eventually arise is often enclosed by a thick, cuticle-like secretion. As will be seen, the nature of this alga makes it preeminently important in discussion concerned with the origin of higher land plants (Chapter 8).

Under the electron microscope, interphase cells appear typically ulotrichalean (McBride, 1970). Plasmodesmata are present, and microtubular structures (larger than, and not equivalent to, cytoplasmic microtubules) occur frequently in chloroplasts of actively growing cells, as is the case with many other Green Algae (cf. Hoffman, 1967; Pickett-Heaps, 1968c). The aerial, erect cells contain the usual band-shaped chloroplast, but in parenchymatous cells, the chloroplast is more reticulate.

4.8a. Cell Division

Owing to numerous problems associated with culturing and handling these cells for electron microscopy, we have no information concerning mitosis. McBride (1967, 1970) has, however, described cytokinesis (Fig. 4.75), which proceeds via cell plate formation with the new wall growing from between the nuclei outward to the outer wall, as in *Ulothrix,* apparently from the fusion of vesicles derived from nearby Golgi bodies. McBride says that no microtubules are involved in cytokinesis; however, since the position of the daughter nuclei after mitosis is typical of phycoplast-forming cells, I remain a little sceptical of this particular observation. McBride fixed his material under conditions that should have preserved microtubules and he also considers their absence perplexing; his sections were dense and heavily stained, a phenomenon which may have obscured these organelles, and one commonly encountered in all types of algal cells prepared for electron microscopy.

McBride (1967, 1970) emphasizes the similarities of the cytokinetic system of *Fritschiella* with that of higher plants. However, for reasons discussed in Chapter 8, I cannot agree with his interpretation, particularly if *Fritschiella* does turn out to contain a phycoplast. The differences, in my view, may be significant, and McBride (1970) draws attention also to *Coleochaete,* which indeed now turns out to have a typical phragmoplast (Section 4.3a). So far as I can tell, it seems appropriate to regard *Fritschiella* as one of the most advanced members of the phycoplast-containing ulotrichalean algae, closely related, for example, to *Ulothrix* and *Stigeoclonium,* but not to *Klebsormidium* or *Coleochaete,* and thus not on the line of algal advance that led to the evolution of land plants. My feeling is that although McBride is on the right track, he and other phycologists have placed undue emphasis upon the *direction* of cell plate growth (i.e., whether it extends centripetally or centrifugally), a factor I believe to be of only secondary or even minor importance. While the possession of a cell plate undoubtedly is a sign of evolutionary advancement, a number of other Green Algae besides the Ulotrichales (e.g., *Chara, Oedogonium*), which are not all related to higher plants, have evolved this structure.

5

THE OEDOGONIALES

This group of organisms has many fascinating and distinctive features which set it quite apart from other Green Algae. Its taxonomic situation in the Green Algae is uncertain in view of these features (e.g., unique method of cell division, unusual zoospores, and complex sexual reproduction). However, since these uninucleate and filamentous organisms all contain phycoplasts, they probably have some distant phylogenetic affinity with one of the ulotrichalean lines of evolution (Chapter 8). I shall be concerned here mainly with the genus *Oedogonium* and its relative, *Bulbochaete*.

5.1. Oedogonium

5.1a. Vegetative Structure

Oedogonium is an unbranched, filamentous alga, usually attached by a differentiated basal cell (Section 5.1c) to some substrate (Fig. 5.1). A conspicuous diagnostic feature of the order is the existence of "caps" or scars on the wall, resulting from cell division (Section 5.1b). The uninucleate vegetative cells, in some species long and highly vacuolate, are rather simple in structure, having a parietal, reticulate chloroplast which contains pyrenoids. Plasmodesmata with distinctive substructure interconnect the cells (Fraser and Gunning, 1969), which show a marked apical/basal polarity along the filament. Wall microtubules are oriented longitudinally, in contrast to their circumferential distribution in many other algae and higher plants. As in many other algae, the "forming" face of Golgi bodies is invariably intimately associated with an element of endoplasmic reticulum, and vesicles with a characteristic surface coat probably pass between these two membrane systems, as is illustrated in Figure 5.2.

Hoffman (1961) has made a detailed study of the morphology, cytology, and reproduction of *Oedogonium* and its close relative, *Oedocladium*.

5.1b. Cell Division

The events occurring before and during cell division in the Oedogoniales are unique and quite extraordinary. The principal features of division in *Oedogonium* described below are probably common to other members of this group, whose ultrastructure has yet to be described in detail. Early accounts of cell division in *Oedogonium* include those of Strasburger (1880), Tuttle (1910), Ohashi (1930), Kretschmer (1930), and Conard (1947); see also the references given by Hill and Machlis (1968).

RING FORMATION

Division in this alga is unique because of the way the cells elongate (Fig. 5.3). Most plant cells grow steadily in length during interphase, and

Figures 5.1, 5.2
OEDOGONIUM

5.1. Whole mount of fixed specimens of *Oedogonium* sp. in their typical situation in nature, epiphytic upon some larger object—here, a leaf of a water weed.

5.2. Golgi body in *O. cardiacum,* apparently replicating by binary fission. In these as in many other algae, the forming face of the cisternal stack is invariably intimately associated with an element of endoplasmic reticulum *(e)*. Furthermore, the micrographs usually suggest an active transfer of small coated, vesicles between the two membrane systems. × 44,000.

SOURCE. Figure 5.2: Drs. J. D. Pickett-Heaps and L. C. Fowke, published in *Aust. J. Biol. Sci. 23,* 71 (1970).

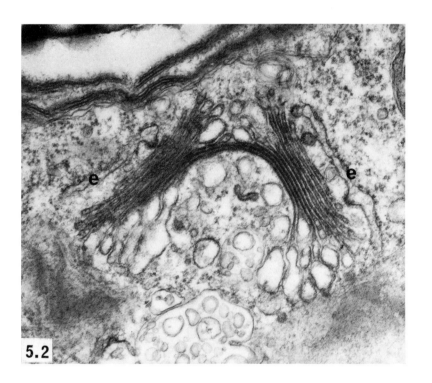

5.3

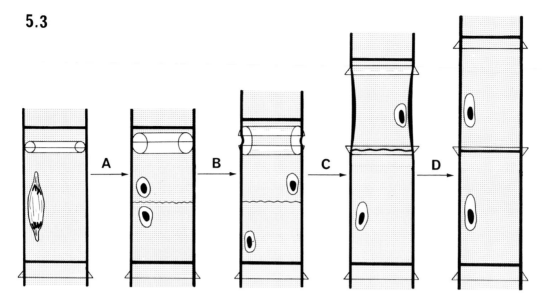

Figure 5.3
CELL DIVISION IN OEDOGONIUM

Diagrammatic representation. Before and during mitosis, a doughnut-shaped apical ring of wall material grows in size, attached to the parental wall. At telophase *(A)*, a phycoplast extends across the cell between daughter nuclei, but a cross wall is not yet formed in it. The parental wall now ruptures violently and precisely at the ring *(B)*. The cells then expand quite rapidly in volume. The material of the ring is drawn out into a cylinder enclosing the apical daughter cell, and the cytoplasmic layer or septum containing the phycoplast moves apically like a diaphragm, until it comes to lie just past the edge of the ruptured parental wall *(C)*, whereupon the new cross wall forms in it *(D)*. A new secondary wall is also laid down inside the stretched ring material to enclose the apical daughter cell *(D)*. As a consequence of ring expansion, the rims of the original parental wall are bent outward *(B, C)* to give rise to the caps characteristic of all members of the Oedogoniales.

when they have roughly doubled their size, cell division cuts the parent cells in two. (The desmids are one notable exception.) In contrast, cell expansion in *Oedogonium* is remarkably abrupt and rapid, primarily occurring during a short period after mitosis is complete, and it makes a fascinating subject for detailed study.

The cell wall in *Oedogonium* contains two major components: (1) an outer layer, derived from the ring (see below), which has an open, fibrous

Figures 5.4–5.6
CELL DIVISION IN OEDOGONIUM

5.4.   Part of the division ring *(dr)* that has formed below a series of five caps arising from five previous divisions in an apical daughter cell. The wall itself contains two thick layers, the outer one derived from the stretched material of the ring in the earlier divisions. The ring in this cell is sited at the wall weakening a precise distance below the other caps, and rupture of the wall at this site obviously would have added another cap to the preexisting sequence. Note the lips of dense secondary wall material (arrows) attaching the ring to the wall on either side of the rupture site. × 10,000.

5.5.   A tiered cap, whose three tiers indicate that this basal daughter cell has undergone three earlier divisions. The wall weakening is clearly visible below the third tier (arrow). The ring would have been located at this precise position (cf. Fig. 5.4), and rupture of the parental wall would have added another tier to the series. × 23,000.

5.6.   As for Figure 5.4. This cell was just about to undergo wall rupture, and the wall weakening is distinctly distended. Note particularly the faint striation in the *basal* lip (arrow) that attaches the ring to the wall. This striation becomes the wall weakening in the basal daughter cell (see Fig. 5.22) and so is instrumental in forming the tiered type of cap (Fig. 5.5). × 35,000.

SOURCE. Drs. J. D. Pickett-Heaps and L. C. Fowke, published in *Aust. J. Biol. Sci. 23*, 93 (1970).

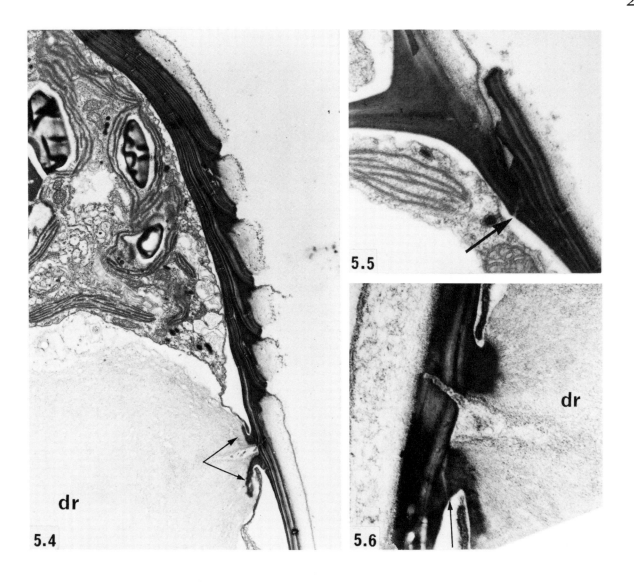

texture and which probably contributes little if any mechanical strength to the wall; and (2) a thick, even inner layer which is almost certainly tough and rigid (Fig. 5.4). In all cells so far examined, there is a circumferential discontinuity in the inner layer of the wall at the apical end of the cell which usually shows up in electron micrographs as a lighter-staining, diagonal striation in the denser wall material (Figs. 5.4–5.6). This is a weakened region of wall, since during cell division and for release of the zoospore (Section 5.1c), the wall invariably ruptures circumferentially and precisely at this site. How this weakened region is built into the wall unfortunately remains quite mysterious (Pickett-Heaps and Fowke, 1970b).

Before cell division, the cell begins depositing rather amorphous material adjacent to the wall weakening (a process illustrated in more detail for the related *Bulbochaete;* Section 5.2b). This activity continues steadily during mitosis, building up a substantial, doughnut-shaped ring of material attached at its outer edge to the cell wall (Figs. 5.3, 5.4, 5.7). The appearance of such apical "rings" in living material is an infallible sign that the cells are entering or undergoing division (Fig. 5.7*A–H*). The ring material stains in various ways, displaying a loose, fibrillar texture under the electron microscope. Two distinct lips of material, similar in appearance to the inner layer of the wall, bracket the wall weakening and can be considered to attach the ring to the wall (Figs. 5.4, 5.6). Material is almost certainly added to the ring via vesicles derived from Golgi bodies (Hill and Machlis, 1968; Pickett-Heaps and Fowke, 1970b; see also Section 5.2b).

MITOSIS

We have followed the time course of mitosis in the haploid male strain of *O. cardiacum* at 20° C (Fig. 5.7*A–H*). The following times are perhaps unreliable since nuclear division is noticeably variable and slowed under the light microscope by the illumination and/or the flattening of the specimen needed to visualize the spindle fibers. The main stages blend into one another, and preprophase is quite prolonged. With a clear-cut prophase nucleus defining time zero, prometaphase and metaphase follow at 5 and 10 minutes, respectively. When Nomarski optics are used, the kinetochore fibers are striking features (Figs. 5.7*A*, 5.7*B*, 5.28) running up

Figure 5.7
MITOSIS IN LIVE CELLS OF OEDOGONIUM CARDIACUM

A. Early prophase, chromosomes condensing. The arrowheads indicate the division ring. *B.* Metaphase. Cord-like chromosomal fibers run from kinetochores to the poles of the closed spindle. *C.* Early anaphase. Chromatids are just separating. *D.* Midanaphase. Note the pointed poles. *E.* Late anaphase. Spindle is elongated. *F.* Telophase. Collapse of the interzonal spindle. *G.* Phycoplast, extending from between closely appressed daughter nuclei, has reached the cell periphery (arrows). *H:* Separation of the daughter nuclei after phycoplast (arrows) has been fully formed, prior to wall rupture. All × 1,200 approx.

SOURCE. Dr. R. Coss, paper in preparation.

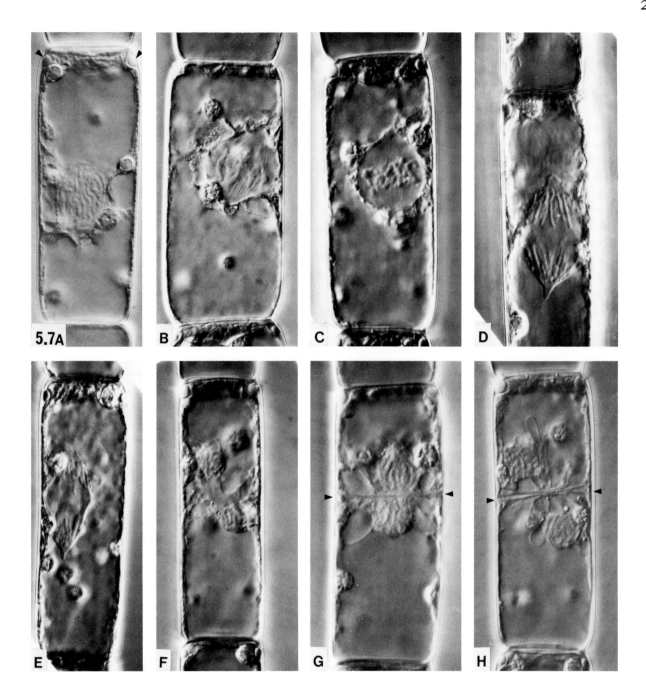

5.7A B C D E F G H

into the pointed poles of the closed spindle. Anaphase movement of chromosomes starts at 12 minutes and at 17 minutes is virtually complete. By 19 minutes the nuclear envelopes have constricted around daughter nuclei, and 30 seconds later the nuclear envelope around the interzonal nucleoplasm vanishes. The separated nuclei now quickly move back together, and by 24 minutes they are flattened against one another. By 32 minutes the phycoplast (Fig. 5.7*G*) is clearly formed, and at 50 minutes the daughter nuclei have separated again, leaving the phycoplast partitioning the cell (Fig. 5.7*H*). Wall rupture takes place at about 75–90 minutes after prophase (Figs. 5.19, 5.20).

The spindle in two species of *Oedogonium* has been described ultrastructurally (Pickett-Heaps and Fowke, 1969, 1970a). The nuclear envelope remains essentially intact throughout mitosis, although there are numerous small gaps in it. If the cell is highly vacuolated, the premitotic nucleus often moves slightly toward the ring and becomes spindle-shaped, with attenuated and finely drawn-out ends. By prophase, proliferating longitudinally oriented microtubules form an extranuclear sheath. At the poles of the spindle, the nuclear envelope becomes characteristically extended into doubled profiles extending into the cytoplasm (as in Fig. 5.17); their significance is not obvious. The nucleolar structure loosens up at this stage (Fig. 5.8); in *O. cardiacum* the granular remnants of the nucleolus remain cohesive, resembling loosely organized strands or lamellae. These often are moved toward the poles during metaphase, but during anaphase are always seen in the middle of the spindle, eventually being eliminated as the daughter nuclei re-form. In a second species of *Oedogonium* studied (Pickett-Heaps and Fowke, 1969), the nucleolar material is more dispersed than in *O. cardiacum* during mitosis until telophase, when it also is eliminated as a large granular body situated between the daughter nuclei and traversed by spindle microtubules.

The kinetochores are conspicuous, containing up to seven layers in *O. cardiacum* (Figs. 5.10, 5.11); they become detectable on chromosomes at prophase, scattered throughout the nucleus (Fig. 5.8). During prometaphase (Fig. 5.9), nuclei become invaded by microtubules (the extranuclear microtubules essentially vanish by this stage, but how they enter the nucleus is undeter-

Figures 5.8, 5.9
EARLY MITOTIC STAGES IN OEDOGONIUM CARDIACUM

5.8. Prophase. The nucleolus *(nc)* is dispersing. Several scattered kinetochores can be discerned on the condensing chromosomes (arrows). × 9,500.

5.9. Prometaphase. Six sets of kinetochores can be seen in this micrograph. Although still scattered within the nucleus (cf. Fig. 5.8), they have all become aligned along the spindle axis (arrows) as microtubules become associated with them (see Fig. 5.17). × 8,300 approx.

SOURCE. Drs. J. D. Pickett-Heaps and L. C. Fowke, published in *Aust. J. Biol. Sci. 23*, 71 (1970).

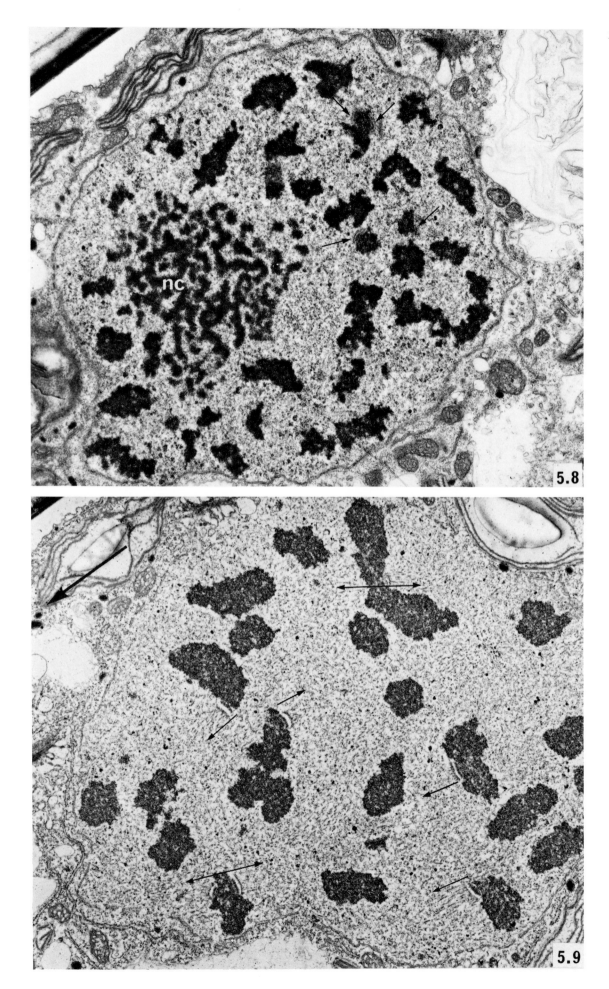

5.8

5.9

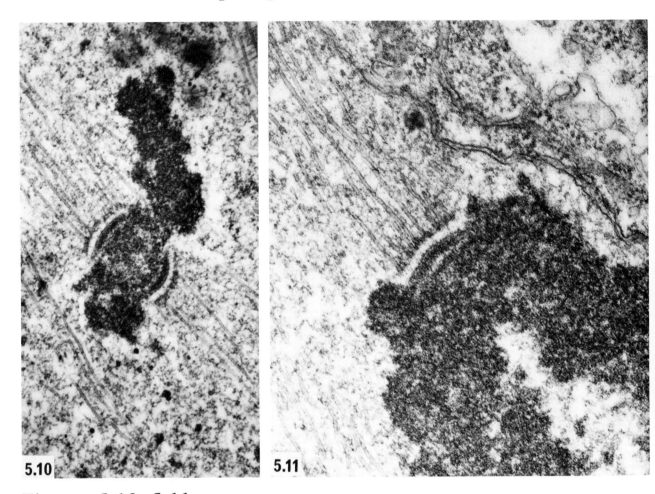

5.10

5.11

Figures 5.10, 5.11
KINETOCHORES IN OEDOGONIUM CARDIACUM

5.10. Paired metaphase kinetochores. × 25,000.

5.11. Single anaphase kinetochore. Six or seven separate layers can be discerned in it. × 39,000.

SOURCE. Drs. J. D. Pickett-Heaps and L. C. Fowke, published in *Aust. J. Biol. Sci. 23*, 71 (1970).

mined). Soon tufts of microtubules are seen attached to the kinetochores, and the former then become aligned with the future spindle axis. By metaphase, all kinetochores have moved into a roughly planar array. These tufts of tubules attached to the kinetochores run up into the long polar extensions of the nuclear envelope. It appears likely that elongation of the kinetochore tubules, confined within the nuclear envelope, could have given rise to these evaginations. This suggests how metakinesis (movement of chromosomes into the metaphase plate configuration) occurs. First, as the scattered kinetochores start

Figures 5.12–5.14
ANAPHASE IN OEDOGONIUM CARDIACUM

5.12. Midanaphase. Only one chromosome in one of the two separating groups can be seen. Its kinetochore microtubules extend into an evagination of the closed nuclear envelope (arrow; see Fig. 5.13). Some of the nucleolar remnants (*nr*) have collected at the middle of the spindle. × 7,000.

5.13. Detail of Figure 5.12, showing the microtubules (*t*) from the kinetochore (*k*) running into the evagination of the nuclear envelope. × 23,000.

5.14. Later anaphase. Bundles of kinetochore tubules are converging into the polar evagination in the nucleus (cf. Fig. 5.7*D, E*). × 11,000.

SOURCE. Drs. J. D. Pickett-Heaps and L. C. Fowke, published in *Aust. J. Biol. Sci. 23*, 71 (1970).

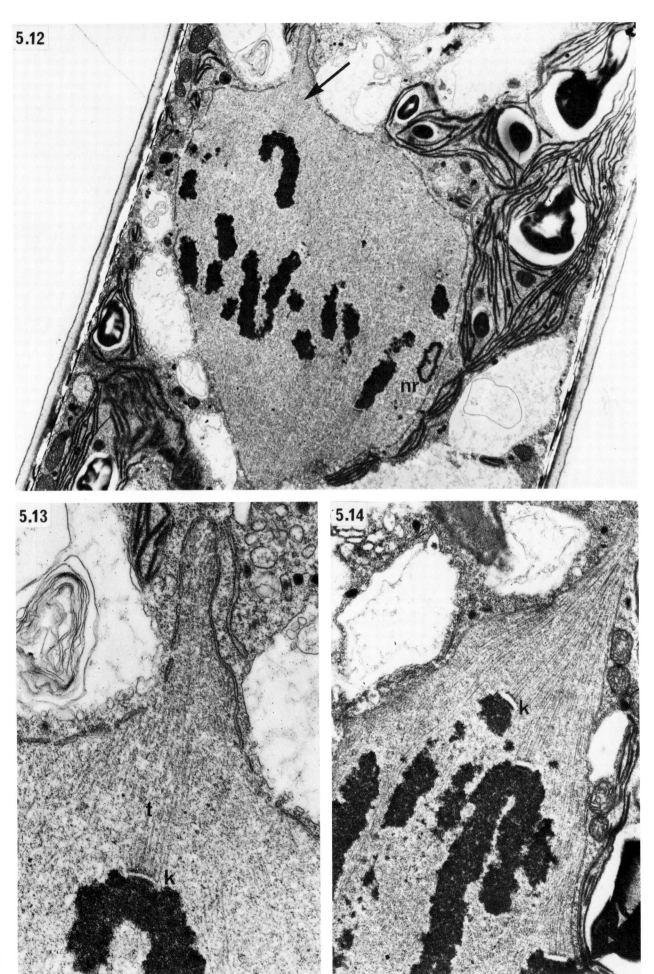

5.12

5.13

5.14

nr

t

k

k

polymerizing their tubular systems, the latter interact with longitudinal microtubules traversing the nucleus, thereby orienting the kinetochores across the spindle. Second, as the fibers elongate at an equal rate, the kinetochores tend to be pushed into the middle of the spindle; i.e., if a kinetochore starts out closer to one pole than the other, one tuft of kinetochore tubules pushes against the nuclear envelope before the other. These ideas are summarized in Figure 5.17.

Anaphase (Figs. 5.12–5.15) is characterized by a considerable elongation of the spindle (Fig. 5.7 B–E), during which the kinetochore tubules shorten and eventually disappear (Fig. 5.15; Pickett-Heaps and Fowke, 1970a). What happens at the poles of the spindle is rather unclear as all these tubules terminate inside the narrowly pointed tip of the nucleus (Figs. 5.7 B–E, 5.12, 5.13, 5.28). High-voltage electron microscopy of thick sections points to the existence of a discrete but rather diffuse polar organizing center in this polar invagination (Fig. 5.36). By telophase, the stretched nuclear envelope contracts tightly around each nucleus, thereby eliminating the spindle microtubules, nucleolar remnants, and other cytoplasmic components from the daughter nuclei.

As with various other Green Algae, the elongated spindle collapses completely after telophase (Fig. 5.7 E–G), and thus the daughter nuclei approach and then flatten against one another. Between them, a typical phycoplast appears (Fig. 5.16), with proliferating microtubules extending transversely in all directions across the cell. These seem to cut through the peripheral chloroplast, and thus the phycoplast effectively divides the cell in two. In vacuolate species of *Oedogonium* (Pickett-Heaps and Fowke, 1969), the cytoplasmic strand containing the phycoplast or "septum" is seen under the light microscope as a distinct layer growing across and transecting the cell. Numerous small vesicles and membranous elements collect among the phycoplast microtubules. There is no clear evidence that these could be derived from Golgi bodies, nor do they contain polysaccharide detectable with the silver-hexamine technique (Pickett-Heaps and Fowke, 1970b). However, such components, if present and of low molecular weight, might well be extracted during processing. It is most important to realize that the cross wall is not yet formed in the phycoplast at this

Figures 5.15, 5.16
LATE MITOTIC STAGES IN OEDOGONIUM CARDIACUM

5.15. Late anaphase. The considerable spindle elongation has apparently stretched the nuclear envelope so much that gaps appear in it. × 2,700.

5.16. Phycoplast formation, with daughter nuclei appressed to each other (cf. Fig. 5.7 G). Numerous vesicles (*vs*) have collected among the microtubules of the phycoplast (*ph*), which have not yet cut the peripheral chloroplast. × 7,500.

SOURCE. Drs. J. D. Pickett-Heaps and L. C. Fowke, published in *Aust. J. Biol. Sci. 23,* 71 (1970).

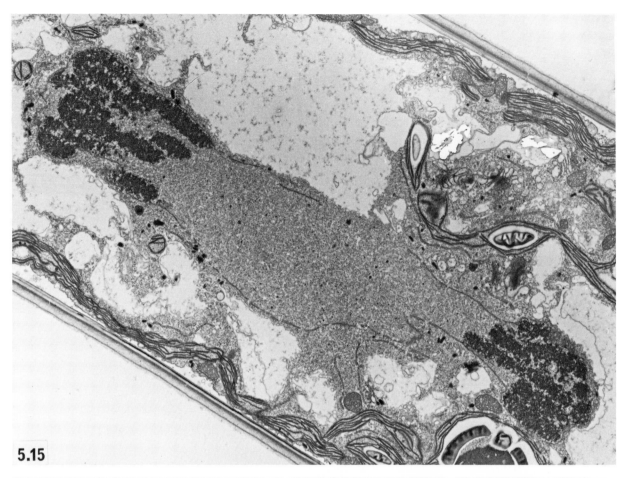

5.15

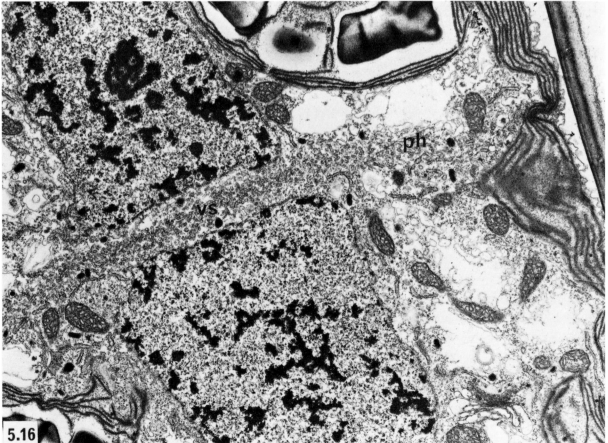

5.16

stage (Hill and Machlis, 1968; Pickett-Heaps and Fowke, 1969, 1970a).

WALL RUPTURE AND CELL ELONGATION

Once mitosis is complete and the septum has been formed between daughter nuclei, the inner layer of the wall ruptures at the weakening, often quite violently in living material. Immediately the material of the ring begins to stretch out evenly into a cylinder which eventually delineates the new daughter cell (Figs. 5.19, 5.20). During cell expansion, the septum moves along inside the older wall like a semirigid diaphragm, until it reaches just beyond the lower point on the rim of the wall's rupture. Then a cross wall appears in the septum among the tubules from a series of linearly arranged, disconnected and flattened cisternae, which fuse together to separate the cells. Polysaccharide (localized by the silverhexamine reagent) is now clearly detectable between the new plasmalemmas (Fig. 5.21). The end wall then thickens considerably; plasmodesmata traverse this wall from its inception. When fully extended, the material of the ring forms the new outer wall layer of one of the two new daughter cells. Inside it is soon deposited the tough, cylindrical inner wall layer, complete with its own weakened region. The other cell, of course, is contained within the older portion of the wall. These events are illustrated for *Bulbochaete* in Section 5.2b.

A direct consequence of this rupture of the outer wall and ring expansion is the formation of the caps so characteristic of this group of algae. The rims of the wall at the rupture site are both deformed outward during cell extension. Thus, soon after rupture of the wall, the basal cell of the dividing pair develops a cap facing "upward" at its apical end; this is complemented by a "downward" facing cap at the apical end of the apical daughter cell. These two types of caps are then precisely related to the position of the new end wall and the weakening in the new walls (Fig. 5.22).

When cells have a downward-facing cap, the new wall weakening appears at a certain, rather precise distance below (i.e., basal to) it. The next ring is formed at this weakening as usual, so that after a subsequent cell division, another separate and distinct cap is added to the apical cell below the first one (Fig. 5.23E). This sequence of events is repeated every division, thereby adding

Figures 5.17, 5.18
SUMMARY OF MITOSIS IN OEDOGONIUM CARDIACUM

5.17. Diagrammatic representation of the main events of the prophase/metaphase transition. During prophase, complex membranous outgrowths arise at the poles of the spindle (see reference given below). Kinetochores seem to be randomly situated and oriented at prophase (Fig. 5.8). However, as microtubules grow from the kinetochores, these tubules probably become first aligned with the other microtubules of the spindle *(A, B)*, thereby reorienting the scattered kinetochores (Fig. 5.9). Then increasing elongation of these kinetochore fibers (Fig. 5.7B) causes the appearance of the polar evaginations of the nuclear envelope. If each set of kinetochore fibers from the one chromosome grows at the same rate, the kinetochores will probably move to the center of the spindle *(D)*, causing metakinesis.

5.18. Diagrammatic representation of the events of anaphase. During anaphase, microtubules (i.e., spindle fibers) run into evaginations of the nuclear envelope. As the kinetochore fibers shorten, the spindle elongates *(A, B)*. The depolymerized microtubule subunits from the kinetochore fibers are probably reassembled to form the elongating continuous spindle.

SOURCE. Drs. J. D. Pickett-Heaps and L. C. Fowke, published in *Aust. J. Biol. Sci. 23,* 71 (1970).

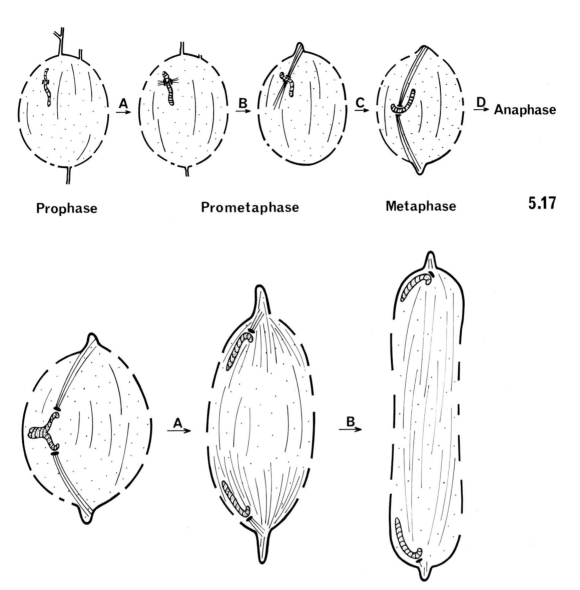

Prophase   Prometaphase   Metaphase   D→ Anaphase

5.17

Anaphase

5.18

one cap to the apical wall per division. Conversely, in the basal cell of a newly divided pair, wall weakenings are situated immediately adjacent to upward-facing caps, and curiously enough, these weakenings can be traced to a faint dislocation visible in the lower of the two lips (see earlier) that attach the ring originally to the cell wall (Figs. 5.4–5.6; Pickett-Heaps and Fowke, 1970b). Subsequent rupture of the wall in this case adds one "tier" to the single cap already present (Fig. 5.23*F, G*). Each subsequent division adds a further tier to this original cap, which thereby remains single but grows in size. Furthermore, we can for our amusement construct a scheme showing the pattern of caps that will form in a filament derived from one cell whose progeny continue to divide synchronously (Fig. 5.24). That such a sequence of caps can be found in nature is revealed by Figure 5.39; the apical cell in this series has acquired extra caps from earlier divisions, and so forms part of an even larger pattern—too large to present here.

How does the cell control this whole process of wall rupture and cell expansion? An obvious answer is by control of turgor pressure. The actual wall rupture is quite violent (Pickett-Heaps and Fowke, 1969), apparently a result of an increase in pressure within the cell. Hill and Machlis (1968) found that expansion ceased when expanding cells were placed in media with increasing osmotic pressure, and we have confirmed their observation. The turgor pressure is probably generated by the vacuoles, which effectively double in volume during cell expansion (see Section 5.2b). We have presented evidence that in *O. cardiacum* Golgi vesicles are discharged into the vacuoles, but it must be appreciated that such events are difficult to demonstrate unequivocally. Much more striking is the appearance of the Golgi vesicles themselves. They become hypertrophied during mitosis and cytokinesis (as in Fig. 5.2), and remain so in both daughter cells until the septum reaches its final position (Figs. 5.22, 5.23). Then all the Golgi bodies in the basal cell of a dividing pair revert to normal interphase size, while those in the apical cell remain conspicuously hypertrophied. This may be significantly related to the fact that the basal cell at this stage is full-sized, whereas the apical cell still has to expand further before reaching its own full size (equivalent roughly to that of the basal cell). Tschermak (1943) revealed a

Figures 5.19–5.21
WALL RUPTURE AND CELL ELONGATION IN OEDOGONIUM CARDIACUM

5.19*A, B.* Wall rupture and cell elongation in live cells of *O. cardiacum*. In both micrographs, the position of the septum is indicated by arrowheads; the edge of the ruptured parental wall by the paired arrows. × 1,000 approx.

5.20. Scanning electron micrograph showing wall rupture and ring stretching in *O. cardiacum*. Note the series of caps above the wall rupture site. × 3,000 approx.

5.21. Section stained with the silver-hexamine technique for the ultrastructural localization of polysaccharide. The septum of this slightly plasmolyzed cell has moved up the cell after wall rupture and has reached its final position, near the lower lip attached to the stretched material of the division ring (*dr;* an equivalent cell of *Bulbochaete* is shown in Figs. 5.145–5.147). The vesicles in the phycoplast are unreactive to this stain, but when the new cross wall *(xw)* starts to form, the wall reacts strongly, as here. The arrow indicates the wall weakening in the lower lip (see arrow in Fig. 5.6). × 18,000.

SOURCE. Figure 5.21: Drs. J. D. Pickett-Heaps and L. C. Fowke, published in *Aust. J. Biol. Sci. 23,* 93 (1970).

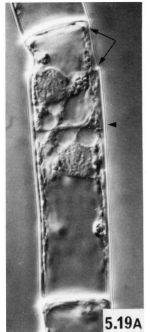

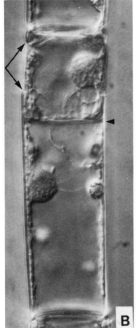

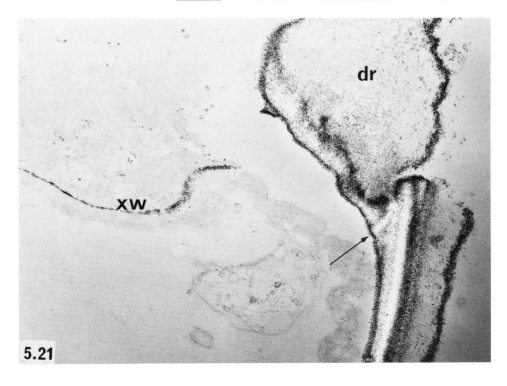

physiological difference between daughter cells before ring breakage. She placed live cells in solutions (1:100,000) of the dye neutral red and found that the apical daughter cell becomes much more heavily stained than its basal counterpart.

One inevitably wonders how this extraordinary behavior associated with cell division originally arose. The actual deposition of the ring probably represents an adaptation of a common method of mucilage secretion whereby Golgi vesicles, instead of being discharged through the wall, concentrate their content into a clearly defined extracellular region. The ring material is certainly very plastic and easily deformable, and Hill and Machlis (1968) showed that one major component in the ring stains with ruthenium red and so may contain pectic material. Incidentally, this phase of "wall" deposition involving Golgi vesicles is quite distinct from that which gives rise to the inner wall layer. The latter is probably secreted directly by the plasmalemma; there is no direct evidence that vesicles are involved in its formation. Indeed, it is probably a valid principle in many plant cells that Golgi bodies may be involved in secretion of mucilage or the expandable primary wall, whereas they are not involved in secretion of the tough inner wall. For example, in the desmid *Cosmarium* (Section 6.4b), the expanding primary wall is also probably secreted via vesicles, but the secondary wall is not; the primary wall is also eventually shed.

But how did this whole complicated and unique system originate? I have suggested that it represents a modified wound-response mechanism (Pickett-Heaps, 1972e). Thus, the cell detects the weakening in its own apical wall before and during division, and secretes a plugging type of wall material around it; then a carefully timed increase in cell turgor ruptures the wall to permit the accumulated material to expand. The evidence for this suggestion is set out in detail elsewhere (Pickett-Heaps, 1972e) and can be summarized as follows. First, inadvertent breaks are often found in the inner wall, mended by the deposition of both amorphous material and then a new inner wall layer (Fig. 5.26). The inner wall layer, incidentally, always contains numerous minor dislocations within it, but these are not continuous around the cell as is the case with the apical wall weakening. Second, when the holdfast of a germling attaches itself to another filament of *Oedogonium,* the latter always, in my experience,

Figures 5.22, 5.23
CAP FORMATION IN OEDOGONIUM

5.22. The wall weakening or rupture site *(rs)* is always located a precise distance below the last of a series of downward-facing caps *(dfc₁* in this case). Before and during mitosis, a division ring is formed at this site *(A)*, attached to the tough inner wall layer by two lips. The upper lip is integrated into the parental wall structure, whereas the lower lip has a faint dislocation in it (see Fig. 5.6). The wall then ruptures at the preweakened site, and the material of the ring is stretched into a cylinder as the original daughter cells double in volume *(B, C)*. As a result of this cell expansion, the rims of the parental wall are bent outward to form a new series of caps. The apical daughter cell acquires a second cap *(dfc₂)* below the first. The septum *(s)* moves up the basal daughter cell until it is past the newly formed upward-facing cap *(ufc₁)*, at which point, a cross wall is formed in it (Fig. 5.21). Notice particularly that the faint dislocation built into the basal lip of the wall ring *(A)* has now become the wall rupture site *(rs)* in the basal daughter cell.

5.23. Tiered cap formation, a sequel to the events outlined in Figure 5.22. Following cell expansion *(D)*, the material of the ring is converted into the outer, nonstructural layer of the wall as a new tough secondary wall is secreted inside it around the apical daughter cell. This wall in some mysterious way acquires the next wall rupture site *(rs)* below the series of caps, and ring formation *(E)*, wall rupture, etc. will later follow as set out in Figure 5.22. In the *basal* daughter cell, the next ring is laid down adjacent to the wall weakening *at* the single tiered cap. Subsequent wall rupture, ring expansion *(F)*, etc., now give rise to two daughter cells, the apical one of which possesses a single cap *(dfc₁)*, and the basal one, the *single* cap *(tc)* with two distinct tiers. Each time this basal daughter cell divides, it adds a tier to its existing series.

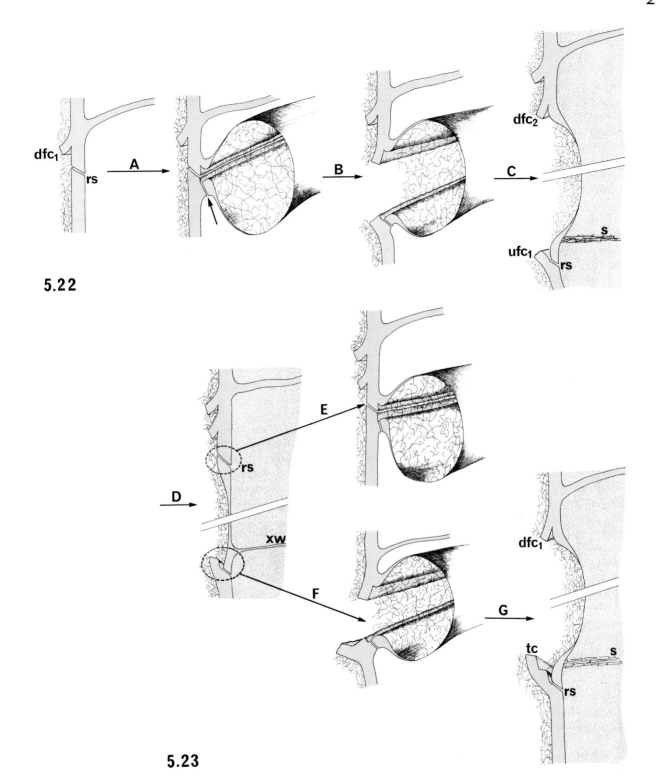

5.22

5.23

5.24

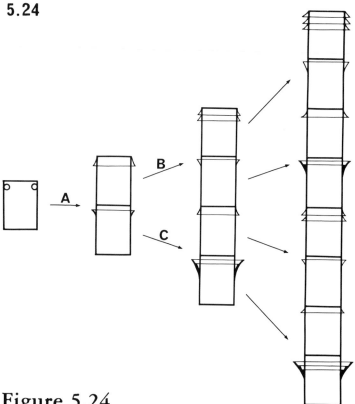

Figure 5.24
PATTERN OF CAPS GENERATED BY ONE CELL OF OEDOGONIUM UNDERGOING THREE DIVISIONS

Diagrammatic representation. The first division *(A)* results in a downward-facing cap in the apical cell and an upward-facing cap in the basal cell (Fig. 5.22). When these two cells divide *(B, C)*, the most apically situated cell acquires another cap below its first, whereas the most basal cell acquires a second tier to its single upward-facing cap, and so on into the third and subsequent divisions. The filament in Figure 5.39 has a pattern almost exactly equivalent to that shown in this diagram, with some additional caps in the apical cell from earlier divisions.

SOURCE. Drs. J. D. Pickett-Heaps and L. C. Fowke, published in *Aust. J. Biol. Sci. 23,* 93 (1970).

responds by thickening the wall immediately adjacent to the holdfast (Figs. 5.25, 5.81, 5.91), again with the two layers of wall material. Apparently, the growing holdfast constitutes a threat to the integrity of the host's cell wall. In the cases cited above, the two layers deposited by the cell could easily be analogous to the ring material and the inner wall layer formed every division.

Figures 5.25–5.27
POSSIBLE WOUND RESPONSE IN OEDOGONIUM

5.25.   Detail of Figure 5.81, showing attachment site of the rhizoid of a germinating zoospore to another filament of *Oedogonium.* The host cell has reacted by thickening its wall immediately beneath the rhizoid, first with an amorphous layer (equivalent perhaps to the material of the ring), and then with a secondary wall layer. × 22,000 approx.

5.26.   Repair of a wall fracture in a germling. There are two layers sealing the hole, the outer layer rather amorphous and the inner layer a typical secondary wall. × 11,000.

5.27.   The division ring of a cell adjacent to an empty cell *(ec)* recently vacated by a zoospore. The material of the ring extends over the whole end wall, which is bulging outward in the absence of a protoplast in the next cell. Presumably this end wall is somewhat liable to rupture. × 10,000.

SOURCE. Author's micrographs, published in *Protoplasma 74,* 195 (1972).

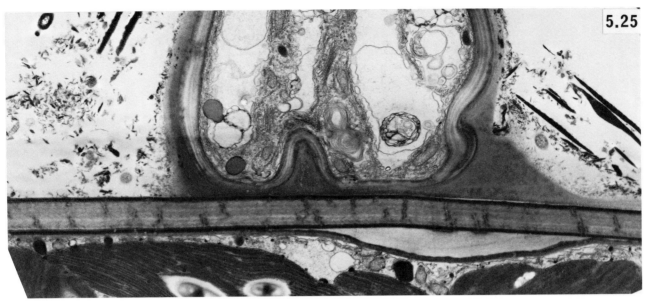

5.25

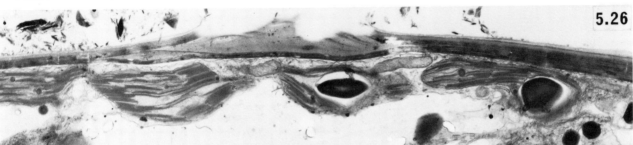

5.26

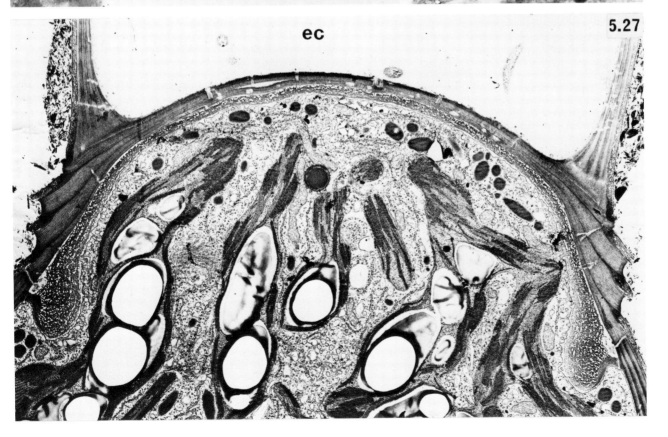

ec

5.27

Third, one cell examined that was about to divide, was situated immediately basal to an empty cell, recently vacated by a zoospore. In this particular case, the ring material was deposited all over the end wall of the cell as well as at the wall weakening (Fig. 5.27). Presumably, the absence of a neighboring cell rendered this end wall more liable to rupture, and so the cell reinforced it.

Since the spindle of *Oedogonium* is large and easily visualized with the light microscope in living cells (Figs. 5.7, 5.28), this organism provides ideal experimental material for investigating the nature of the spindle. The spindle fibers are highly birefringent, and as would be expected (Pickett-Heaps, 1967a), they disappear when the cell is treated with the drug colchicine (Fig. 5.29) and other agents (e.g., low temperature) that cause dissolution of microtubules. The herbicide IPC (isopropyl-N-phenylcarbamate) has been shown to cause mitotic abnormalities in higher plants, in particular, causing the spindles to become multipolar (Hepler and Jackson, 1969). When appropriate concentrations are applied to dividing cells of *Oedogonium,* mitosis is blocked (Coss and Pickett-Heaps, 1974). If, however, the dividing cells are chilled to 0° C, placed in culture medium containing IPC, and then warmed to room temperature with the IPC washed out, the results are intriguing. Chilling dissolves the spindle, which can re-form after the IPC is removed and the cells warmed again. However, these new spindles are now oriented across the cell and are often multipolar (Fig. 5.30). Tripolar anaphases are not infrequent (Fig. 5.31), and three daughter nuclei may be formed. Phycoplast formation is also severely disturbed, with the microtubules often becoming oriented parallel to the cell axis (i.e., at right angles to their normal axis; compare Fig. 5.32 with Figs. 5.7, 5.19). Sometimes both nuclei end up on the same side of the phycoplast (Fig. 5.33), in which event subsequent wall rupture and cell elongation give rise to enucleate and binucleate daughter cells (Figs. 5.33, 5.34).

Such cells are also fascinating when viewed as thick sections with a high-voltage (1,000kv) electron microscope. Figure 5.35 shows a very early anaphase spindle after the IPC treatment outlined above. The bundles of microtubules emanating from the kinetochores are most striking. Similar high-voltage electron microscopy of normal (untreated) spindles reveals a rather diffuse polar body in the tip of the drawn-out

Figures 5.28–5.34
THE SPINDLE IN OEDOGONIUM CARDIACUM

5.28.   Normal metaphase spindle, live cell of *O. cardiacum.* × 1,300.

5.29.   Similar metaphase spindle after treatment with 0.1% colchicine. All spindle fibers have disappeared, and the nucleus is spherical. × 900.

5.30.   *O. cardiacum* cell chilled (0° C) at metaphase and then treated with the herbicide IPC. Later the IPC was washed out, and the cell returned to room temperature in normal medium. The spindle has re-formed, but at right angles to the cell axis, and it is now tripolar (arrows). × 1,300.

5.31.   *O. cardiacum* treated as for the cell in Figure 5.30, showing a tripolar anaphase spindle (arrows). × 1,000.

5.32.   As for Figure 5.31. Telophase, with the phycoplast (arrow) forming abnormally parallel to the cell axis. This spindle was presumably bipolar during anaphase. × 900.

5.33.   As for Figure 5.31. In this cell, the phycoplast (arrowheads) has divided the cell into binucleate and enucleate daughter cells. The wall has just ruptured at the ring (paired arrows). × 900.

5.34.   The final result of the division shown in Figure 5.33, with a binucleate cell (double arrow) next to an enucleate cell (single arrow). × 400.

SOURCE. Dr. R. A. Coss, unpublished micrographs.

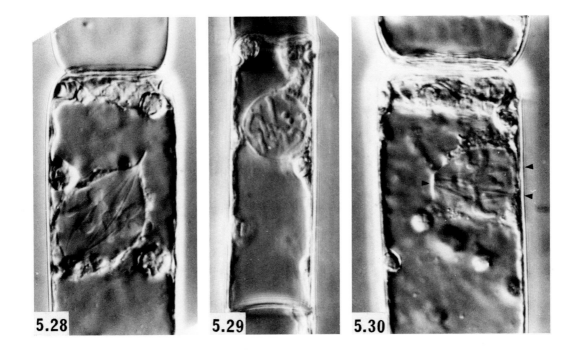

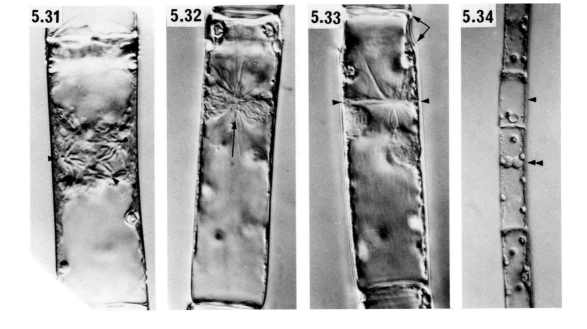

nuclear membrane (Fig. 5.36; cf. Fig. 5.28), into which the spindle microtubules insert. This polar structure is not discernible with thin sections, although the organization of the spindle strongly hints at the existence of such a center. This polar body (MTOC?) is even more conspicuous at the focal points of microtubules in multipolar spindles of IPC-treated cells (Fig. 5.36).

5.1c. The Zoospore Cycle

Vegetative reproduction in *Oedogonium* occurs by the frequent formation of large (one per vegetative cell), multiflagellate, very active zoospores (De Bary, 1854; Gussewa, 1931; Kretschmer, 1930; Ohashi, 1930). These naked, motile cells emerge from vegetative filaments, swim about for some rather variable period, and then lose their flagella as they differentiate back into thick-walled germlings that constitute the basal cells of a new vegetative filament (Fig. 5.38). While cultures in the laboratory usually contain a few cells undergoing zoosporogenesis (particularly following subculturing), considerable numbers of these motile cells can easily be induced by exposing an old culture to a higher than normal concentration of carbon dioxide. Using this method, Retallack and von Maltzahn (1968) found that zoosporogenesis is completed within 15–20 hours of induction.

ZOOSPOROGENESIS

Light microscopy of living cells reveals little about the morphogenetic events that give rise to zoospores, partly because the cells usually become quite dense during differentiation. Several clear signs, however, indicate that zoosporogenesis is in progress. Most conspicuous is a marked basal and much less obvious apical contraction of the protoplast away from the cell wall (Fig. 5.40). Soon a clear region of cytoplasm also appears on one side of the central region of the cell, around which the flagella may be detected later on. This cytoplasm will form the "dome" of the mature zoospore (see below).

Electron microscopy allied with light microscopy of thicker sections (Pickett-Heaps, 1971b) provide a much more detailed picture of the fascinating morphological events of zoospore differentiation. The "basal contraction" of the cytoplasm (Figs. 5.39, 5.40) can be ascribed to the localized secretion of "basal mucilage" (Fig.

Figures 5.35–5.37 HIGH-VOLTAGE (1,000 kV) ELECTRON MICROSCOPY OF THICK (0.5µ) SECTIONS OF SPINDLES OF OEDOGONIUM CARDIACUM

5.35. Early anaphase, cell treated with the drug IPC as for Figure 5.30. The spindle axis has become rotated, now lying almost perpendicular to the cell axis. Thick sections show clearly the tufts of microtubules extending from the kinetochores and converging at the poles. This spindle was multipolar, and the kinetochore tubules inserted into several dense polar bodies (MTOCs?), as shown in Figure 5.37. × 9,000.

5.36. Pole of a normal (untreated) late anaphase spindle. The kinetochore tubules appear to insert into a dense body located in the polar invagination. This structure is not discernible in ordinary thin sections. × 9,500.

5.37. One of the dense bodies at the poles of a multipolar, IPC-treated spindle, such as that shown in Figure 5.35. × 8,000.

SOURCE. Drs. R. A. Coss and J. D. Pickett-Heaps, paper in preparation.

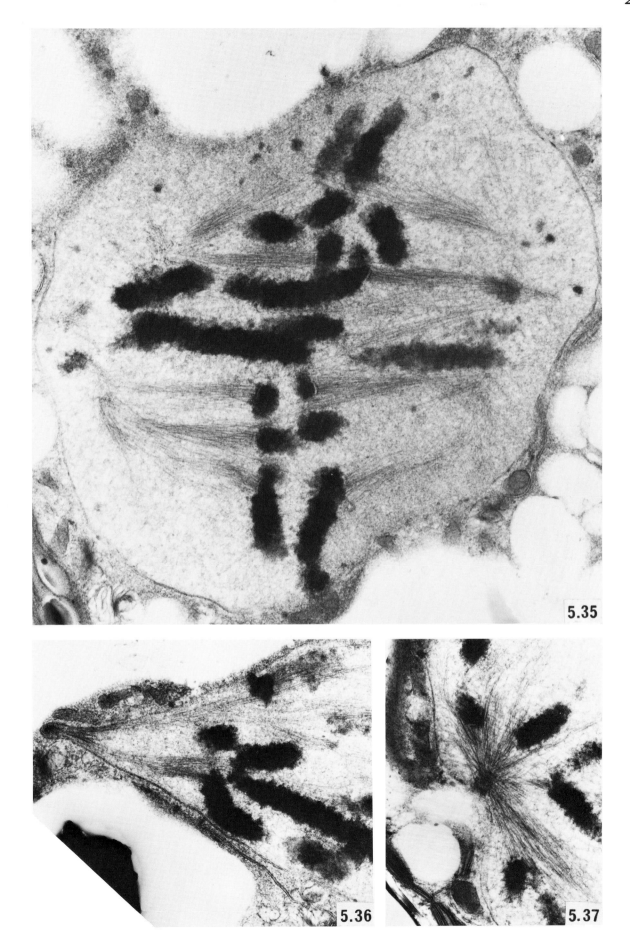

5.35

5.36

5.37

5.47), a material derived from Golgi vesicles and containing polysaccharide detectable by the PAS and PAS/H reactions (i.e., using light and electron microscopic histochemistry; Figs. 5.41, 5.48). The methods are described elsewhere (Pickett-Heaps, 1967d, 1968d). The Golgi bodies in the basal cytoplasm become hypertrophied early in zoosporogenesis, a change soon followed by the accumulation of basal mucilage; in contrast, the Golgi elsewhere in the same cell are conspicuously different in appearance. Indeed, throughout zoosporogenesis the cell contains various different populations of Golgi bodies with apparently diverse functions (see later).

While basal mucilage is being secreted, another distinctly different, polysaccharide-containing layer (Figs. 5.41, 5.45) is also being evenly laid down enclosing the apical portion of the cell. As it thickens, the plasmodesmata traversing the end wall stretch and finally break. I have called this the "hyaline layer," since it is transformed into the expanding "hyaline vesicle" that encloses the fully developed zoospore during and just following emergence (Figs. 5.66, 5.68, 5.71–5.73). At least part of this layer (Figs. 5.45, 5.46) seems also to be derived from the contents of quite small Golgi vesicles secreted through the plasmalemma. I also have evidence that some elements of the endoplasmic reticulum are continuous with the plasmalemma in the apical region of the cell during this secretion. However, this most unusual association of these two membranous systems could easily be artifactual (Pickett-Heaps, 1971b).

The mature zoospore is multiflagellate (Fig. 5.67); Hoffman and Manton (1962) have counted approximately 120 flagella in some zoospores of *O. cardiacum*. As far as we know, formation of flagella is totally dependent upon the prior existence of basal bodies in the cell. The well-known spatial configuration of the basal body's nine triplet tubules is matched precisely by the configuration of the nine doublet tubules that run the length of the flagellum (see Section 2.1). It is generally accepted that the basal body acts as a template or "seed" that determines the pattern and possibly the assembly of the tubular system that is presumably vital for flagellar motion and behavior. However, vegetative cells of *Oedogonium* have never been found to contain any basal bodies or detectable centriolar equivalent. Where then do those of the zoospore come from?

Figure 5.38
THE ZOOSPORE CYCLE IN OEDOGONIUM

Diagrammatic representation. The whole contents of one cell differentiates into a zoospore *(A)* whose dome, surrounded by a ring of flagella, is invariably situated on one side of the cell. The mature zoospore emerges *(B)* after the parental wall ruptures circumferentially at the wall weakening, and it is temporarily enclosed by an expanding hyaline vesicle. During its actively motile phase, it undergoes some elongation *(C)*. It then sheds its flagella, and the dome differentiates into a holdfast *(D)*, usually attached to a substrate. The young germling thus formed secretes a wall, much thickened at the holdfast, and later undergoes normal division *(E)* to initiate a vegetative filament.

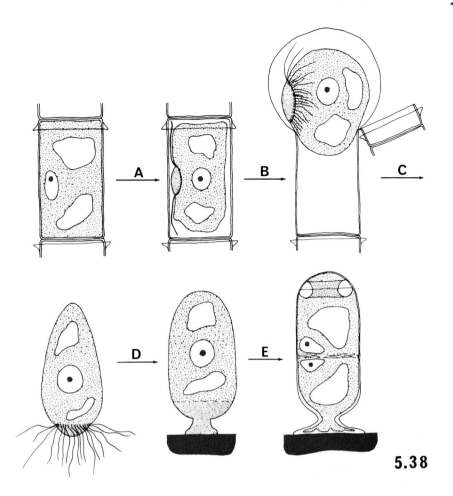

5.38

I agree with Hoffman's (1966) assertion that they arise de novo. This is refuted by Retallack and von Maltzahn (1968), who, on the basis of light microscopical evidence alone, maintain that basal bodies, visualized as fluorescent particles in living cells, collect from around the cytoplasm to be incorporated into the flagellar apparatus. In the two species I have studied, the genesis and rearrangement of centrioles follow a remarkably precise and reproducible pattern of events, quite different from that described by Retallack and Von Maltzahn.

The earliest cytological evidence of impending zoosporogenesis I have ever detected is the appearance of a rather nondescript ball of amorphous material in a small basal invagination of the nuclear envelope (Fig. 5.49). Its significance becomes obvious when cells are examined at a slightly later stage of differentiation: several short and incompletely formed centrioles are apparently condensing from or forming within this material (Fig. 5.50). Within a short time, a considerable number of centrioles is created in a highly ordered array. Following their biogenesis they are always arranged into two adjacent rows; the centrioles in each row are essentially parallel to one another and slightly inclined to the parallel centrioles of the adjacent row. The two rows themselves lie in an arc close to and curving around the nucleus (Fig. 5.51). Between each of the centrioles in a row, a short set of linearly arranged triplet tubules embedded in a densely staining matrix is also formed (Figs. 5.52, 5.53). These structures I have termed "rootlet templates" for reasons set out below.

The two parallel rows of centrioles then move to the lateral cell wall, accompanied by the nucleus. Preceding them, an array of microtubules appears to cleave the peripheral chloroplast (Fig. 5.54). During this move, the nucleus becomes deformed so that the rows of centrioles surmount an increasingly pronounced ridge in the nuclear membrane (Fig. 5.55A, B). When the rows of centrioles are close to the wall, the centers of the rows then move apart while the two ends stay together (Figs. 5.55B, 5.55C, 5.56). Following some rearrangement of centrioles at each end of the rows, a circular array of centrioles becomes appressed to the side of the cylindrical cell wall. During these maneuvers, the nucleus remains close to the centrioles (see also

Figures 5.39–5.44
ZOOSPOROGENESIS IN OEDOGONIUM

5.39. This filament contains cells midway through zoosporogenesis. All exhibit basal contraction of the cytoplasm following secretion of the basal mucilage (Figs. 5.47, 5.48), and the nucleus in several cells has migrated to one side while forming the ring of basal bodies (cf. Figs. 5.55A). The caps are labeled (1T for a single-tiered cap, 2T for a two-tiered cap, etc.; the arrowheads indicate the downward-facing caps). This sequence is exactly that predicted in Figure 5.24 except for the apical cell, which has seven extra caps acquired from earlier divisions. × 460.

5.40. Whole mount of differentiating cells. The dome (d) is visible on one side of the cells, whose cytoplasm has contracted away from the wall, particularly where the basal mucilage has been secreted (arrow). × 660.

5.41. Section of plastic-embedded filament (the same one as that from which Figs. 5.45–5.48 were also taken), stained with the standard PAS technique for the detection of polysaccharide. The starch grains are particularly reactive, the wall much less so. Note the strong reactivity of the basal mucilage (large arrows) and the hyaline layer (small arrows). × 1,600.

5.42–5.44. Golgi bodies from three sequential cells in the same filament. In Figure 5.42, elements of the endoplasmic reticulum (e) are collecting at the forming face, and the Golgi vesicles are large. In Figure 5.43, the endoplasmic reticulum seems to be stacking up on the forming face, and in Figure 5.44, these elements appear to have become new cisternae in the Golgi body, whose vesicles are altered in appearance. × 26,000 approx.

SOURCE. Author's micrographs, published in *Protoplasma* 72, 275 (1971).

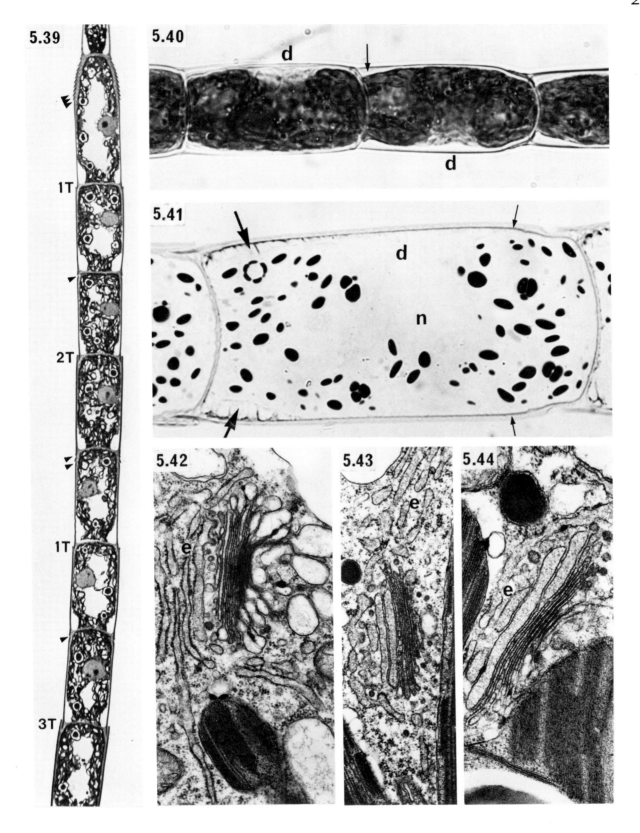

Fig. 5.39). Thus, the ridge on it splits into two smaller ridges which separate as the two rows separate. Finally, the nucleus returns to its central position in the cell, and the region it had occupied thereafter remains distinctively differentiated from the rest of the cytoplasm, filling up with smaller organelles and particularly a marked concentration of endoplasmic reticulum (Fig. 5.65). This region, now circumscribed by the array of centrioles, will become the highly refractile dome of the mature zoospore (see Figs. 5.68–5.76), and it is analogous to the clear area that appears on one side of live cells undergoing zoosporogenesis (Fig. 5.40).

The centrioles next begin extruding their flagella and, by definition, thereby become basal bodies. Concurrently, the rootlet templates also begin assembling their own specific microtubular system, the rootlet microtubules. These radiate outward from the dome and lie just under the plasmalemma (Figs. 5.60, 5.61, 5.64*B*). In mature zoospores, they form a symmetrical cytoskeletal framework presumably transmitting the stress of flagellar motion throughout the zoospore. I believe it is most significant (Pickett-Heaps, 1971a) that these two organelles, the centrioles and the rootlet templates, arise concurrently and de novo in strict spatial relation to one another from amorphous material, and that later they both assemble specific microtubular systems. They would both fit my definition (Pickett-Heaps, 1969a) of highly structured and therefore conspicuous microtubule-organizing centers (MTOCs) in the cell. This example, among many others (see Chapter 8) divests the centriole of part of its classic aura of mystery, as I believe it can best be visualized as only one of several different forms that the MTOCs of a cell can take on, as necessary (Pickett-Heaps, 1971a).

While the flagellar and rootlet microtubules are being assembled, the flagellar apparatus around the dome soon acquires increasingly complex additional components. The way these are formed is quite unclear, and so I shall describe mainly some aspects of their completed structure. Most prominent is the rather massive "fibrous ring" (Fig. 5.59; Hoffman and Manton, 1962), which connects all the basal bodies in a highly specific manner through two of their nine triplet tubules (Figs. 5.57, 5.58). The complexity of its structure can best be appreciated by inspection of

Figures 5.45–5.48
EXTRACELLULAR POLYSACCHARIDE SECRETION DURING ZOOSPOROGENESIS IN OEDOGONIUM

5.45. Section stained with the silver-hexamine reagent after peroxidation. The wall rupture site is conspicuous near the base of the tiered cap, and the hyaline layer (*hl*) reacts strongly, as do the contents of vesicles near the Golgi (*g*). × 24,000.

5.46. As for Figure 5.45, but without silver-hexamine staining, and showing the hyaline layer. The wall rupture site here, of course, is below the last of the series of caps. Plasmodesmata traverse the end wall. × 10,000 approx.

5.47. Basal mucilage (*bm*) accumulating and causing the cytoplasm to shrink from the wall (see Figs. 5.39–5.41). × 6,000 approx.

5.48. As for Figure 5.47, but the section was peroxidized and then stained with silver-hexamine. The basal mucilage and the contents of Golgi vesicles have reacted strongly. × 13,000 approx.

SOURCE. Author's micrographs, published in *Protoplasma* 72, 275 (1971).

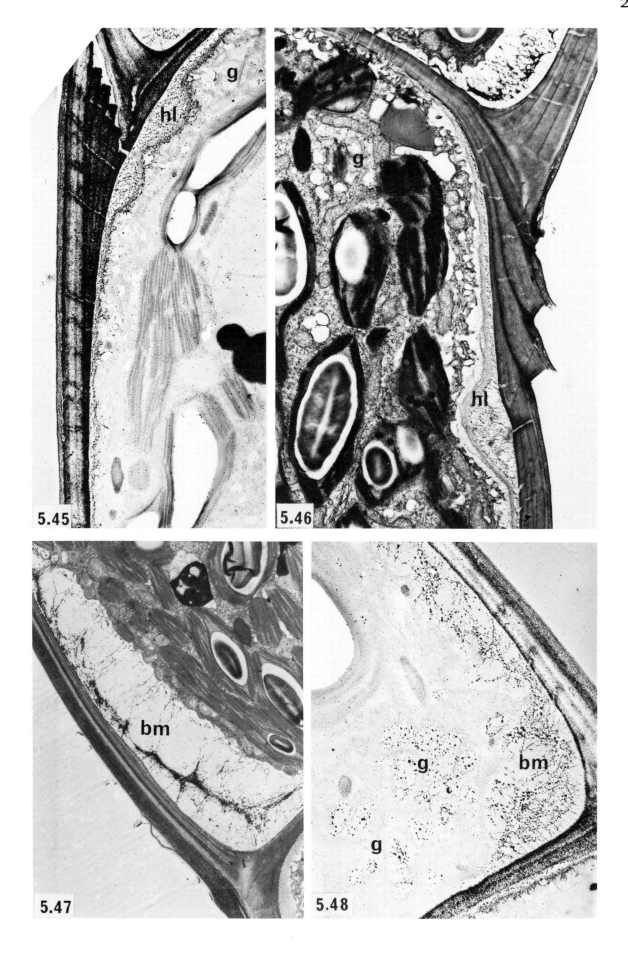

micrographs. When transected, it appears to contain thin, densely staining lamellae. The distal end of some of the triplet microtubules of the basal bodies becomes capped by characteristic, localized amorphous material. Also, each set of rootlet microtubules now acquires a long, thin, tapering "striated fiber" (Figs. 5.61, 5.63; Hoffman and Manton, 1962; Hoffman, 1970) connected to the rootlet templates. Here again, a picture of this structure serves better than any description of it. Note, too, the extra individual microtubules, variable in number and disposition, that also appear near the triplet rootlet microtubules (Figs. 5.60, 5.61). The striated fibers are circular in cross section and are clearly visible between the basal bodies (Figs. 5.59, 5.60). Filaments also apparently connect the caps of amorphous material on the basal bodies to the striated fibers (Fig. 5.57). The integrity of this incredibly complex flagellar apparatus is best shown by the work of Hoffman and Manton (1962, 1963), who succeeded in isolating the ring of basal bodies intact from both zoospores and spermatozoids (Fig. 5.62), the latter having roughly a quarter the number of basal bodies as the former. Similar results obtained later by Hoffman (1970; his Figs. 1, 2) show clearly the fine, straight—almost rigid looking—striated fibers between basal bodies, and their associated rootlet microtubules splitting into three individual tubules further away from the ring (Fig. 5.62).

As I mentioned earlier, two separate populations of Golgi bodies secrete the basal mucilage and the hyaline layer. Those elsewhere in the cell vary predictably in appearance as differentiation proceeds. Sometimes their vesicles are small; sometimes large and hypertrophied. In the almost mature cells of one species I have studied, some Golgi vesicles always contain distinctive reticulate material. Earlier on, when the nucleus is returning to the center of the cell away from the basal bodies, Golgi collect around the internal surface of the dome as it fills with endoplasmic reticulum. They then begin forming considerable numbers of large vesicles whose amorphous contents upon release seem to condense and gain in electron opacity (Fig. 5.65). These vesicles appear to move through the massive concentration of endoplasmic reticulum, collecting at the dome's surface in considerable numbers (Figs. 5.68–5.74). These I have termed the "basal particles" (Pickett-Heaps, 1972d);

Figures 5.49–5.52
INITIATION OF FLAGELLAR APPARATUS DURING ZOOSPOROGENESIS IN OEDOGONIUM

5.49. The earliest cytological sign of impending zoosporogenesis yet discovered, a small indistinct mass (arrow) in a basal invagination of the nucleus (n). × 19,000.

5.50. Centrioles forming in the same position as the dense mass arrowed in Figure 5.49, a little later in zoosporogenesis (two cells along the same filament as that from which Fig. 5.49 was obtained). × 42,000.

5.51. Proliferation of centrioles around the nuclear envelope. One of the two rows can be seen here. × 15,000.

5.52. Detail of Figure 5.51, showing the rootlet templates (arrow) associated with an element of endoplasmic reticulum (e) which have formed between the centrioles. × 46,000.

SOURCE. Author's micrographs, published in *Protoplasma* 72, 275 (1971).

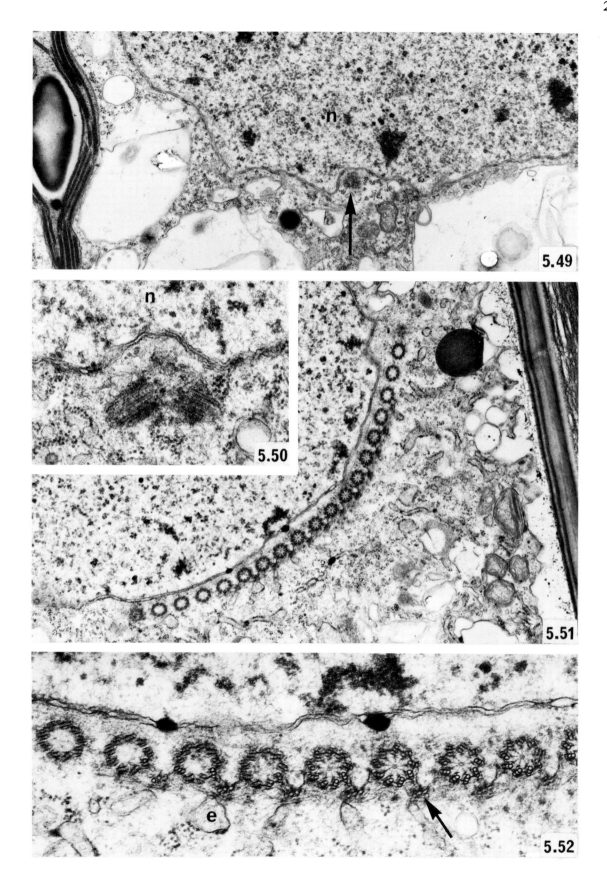

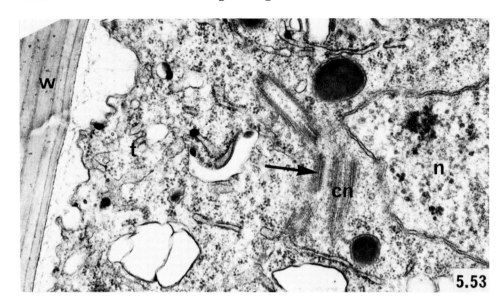

Figure 5.53
ROOTLET TEMPLATE

This micrograph shows a rootlet template (arrow) embedded in dense material in one of the two rows of centrioles (*cn*; cf. Fig. 5.52). × 29,000.

SOURCE. Author's micrograph, published in *Protoplasma 72*, 275 (1971).

their significance will soon become obvious. They are not PAS/H-positive but stain densely with toluidine blue. Before the zoospore is released, the Golgi around the dome undergo yet another change; they cease forming the basal particles and invade the dome while giving rise to large vesicles with diffuse, fibrillar contents.

The Golgi bodies in these cells then are heterogeneous in appearance and seem to be performing several specialized functions. Clearly they are versatile organelles. Many workers have reported that in algae (and in many other cells) a transfer of coated vesicles between the endoplasmic reticulum (or nuclear envelope) and Golgi bodies probably occurs (as in Fig. 5.2). Such a coupling of the two membranous systems immediately suggests that the endoplasmic reticulum could control Golgi functions (i.e., by supplying enzymes, etc., to be included in the vesicles). One species of *Oedogonium* undergoing zoosporogenesis takes this process one step further: whole cisternae of endoplasmic reticulum

Figures 5.54, 5.55
REARRANGEMENT OF THE TWO ROWS OF CENTRIOLES INTO A RING IN FORMING ZOOSPORES OF OEDOGONIUM

5.54.   The two rows of centrioles (*cn*) in end view here, accompanied by the nucleus (*n*), moving to the side wall (*w*). Ahead of them, the chloroplast (*c*) is cleaving, perhaps under the influence of the numerous microtubules (*t*) associated with the centrioles. × 19,000.

5.55 *A–C.*   The light micrograph (*A*) shows two nuclei appressed to the side wall. *B* and *C* show the ridge on these same nuclei under the electron microscope at different stages during the separation of the two rows of the centrioles into the ring configuration. *A;* × 1,400; *B;* × 19,000; *C;* × 16,000.

SOURCE. Author's micrographs, published in *Protoplasma 72*, 275 (1971).

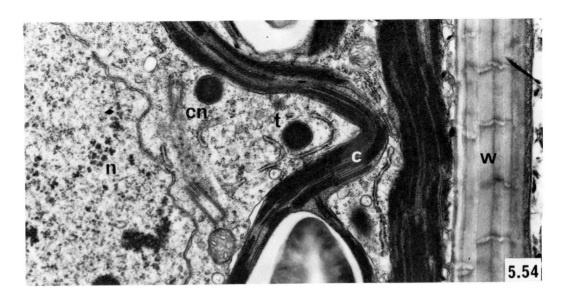

5.54

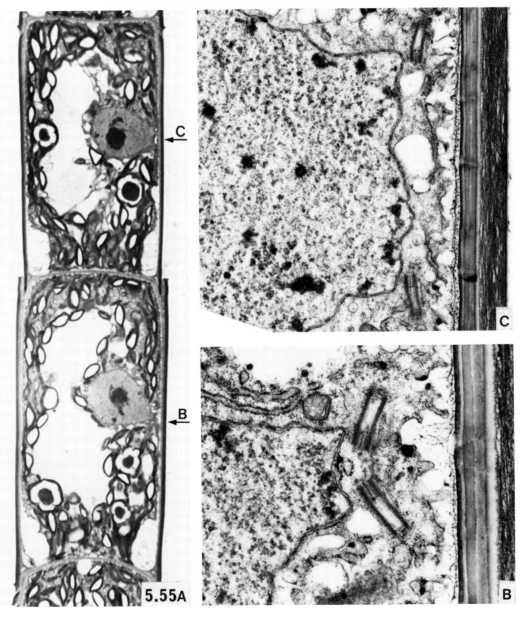

5.55A

C

B

lose their ribosomes and accumulate, two or three layers deep, on the forming face of the Golgi (Figs. 5.42–5.44; Pickett-Heaps, 1971b). They then condense down into typical Golgi cisternae. The nature of the vesicles formed by the Golgi is noticeably different before and after this takes place. These events lead one to suspect that Golgi bodies can be regarded as highly differentiated regions of endoplasmic reticulum, a concept which ties in with ultrastructural work on certain fungi, where "Golgi" vesicles seem to be derived from a "Golgi" system consisting of a single membranous cisterna (of differentiated endoplasmic reticulum?) (Bracker, 1967, p. 349).

There are several other more subtle changes detectable in cytoplasmic organelles whose significance is not obvious. One conspicuous feature acquired, however, is the eyespot (Fig. 5.72), a single layer of uniform, spherical inclusions which accumulates in one small outward-facing region of chloroplast near the dome (Fig. 5.64).

MOTILE PHASE OF THE ZOOSPORE

Release of the zoospore in vivo from the filament is initiated by a circumferential rupture of the cell wall (Fig. 5.69) at the apically weakened region used in cell division (see earlier). Immediately the basal cytoplasm rounds up and the protoplast swells outward from the confines of the wall (Figs. 5.66, 5.68, 5.70). Once it has completely emerged, it becomes quite spherical (Figs. 5.66, 5.71), and almost immediately, the thin membranous hyaline vesicle detaches from its surface, ballooning out steadily while remaining attached to the inside of the empty cell. Meanwhile, the flagella around the zoospore's dome become increasingly active, and soon the zoospore is swimming around inside the ever more tenuous hyaline vesicle. Eventually the hyaline vesicle ruptures and disappears, allowing the zoospore to rush forth and commence its brief period of motility.

The hyaline vesicle is derived directly from the hyaline layer secreted around the apical portion of the zoospore during differentiation (Fig. 5.68). Its behavior during release of the zoospore suggests that it must be an extensible, semipermeable membrane. Presumably it takes up water continuously from the surrounding medium, becoming inflated until it ruptures. The role of the basal mucilage in zoospore release is not clear. This polysaccharide-containing

Figure 5.56
RING OF CENTRIOLES

The ring of centrioles has now been established on the side wall of the cell, delineating the site of the future dome. Note how the peripheral chloroplast has been cleared from this region of the wall. These are light and electron micrographs of the same cell. *A;* × 1,800; *B;* × 30,000.

SOURCE. Author's micrographs, published in *Protoplasma 72,* 275 (1971).

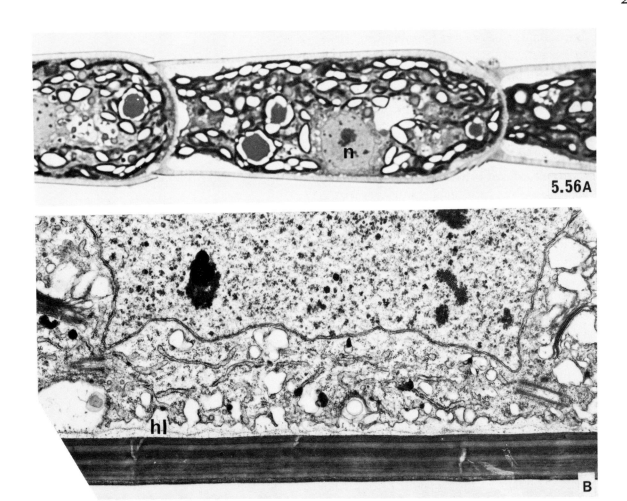

material could take up water too and thereby help eject the zoospore from the parental wall, a conclusion suggested by Steinecke (1929) and also by Fritsch (1902a). It certainly becomes more disperse as the zoospore emerges (Figs. 5.68, 5.71). It is important to realize that although rupture of the parent cell wall is necessary for both cell division and zoospore release, the increase in pressure inside the wall required to effect this rupture presumably must be achieved in different ways. Following mitosis, the volume occupied by daughter cells approximately doubles; the vacuoles increase in size, and therefore an internal increase in osmotic pressure of the cell is probably responsible. However, the volume of the zoospore does not increase markedly after it is freed from the wall. Indeed, upon release, contractile vacuoles rapidly appear all over the zoospore's surface (Fig. 5.74), indicating, as might be expected, that they must continuously expel water taken up by the cytoplasm whose high internal osmotic pressure is no longer balanced by confinement within a rigid cell wall. This all means that the pressure required to effect wall rupture for zoospore release must be generated inside the wall, but outside the protoplast if possible. In view of the behavior of the hyaline vesicle after wall rupture, I conclude that it may take up water, increasing the pressure inside the cell wall—but not inside the protoplast. We may also surmise that the cell could easily trigger this event by secreting through its plasmalemma some osmotically active material at the appropriate moment.

While the zoospore is essentially spherical immediately after emergence (Fig. 5.66, 5.71), the highly refractile dome soon begins to protrude from a slight constriction that develops around the cell where the circular array of flagella is inserted (Fig. 5.73). The rest of the cell now also undergoes a slow but steady elongation during the motile phase (Figs. 5.66E, 5.67, 5.72, 5.74), which is highly variable in duration, lasting in my experience between a couple of minutes to an hour or longer. The end of the motile phase is marked by several dramatic changes to the cell. The first of these commences when the zoospore comes to rest and then begins an unusual shivering or vibratory motion, during which all the flagella are violently shed. Their membrane usually balloons out, rendering them conspicuous as they give a few final twitches

Figures 5.57–5.60
FLAGELLAR APPARATUS OF MATURE ZOOSPORES OF OEDOGONIUM

5.57. Edge of the fibrous ring connected to the basal bodies, which have a complex series of additional components associated with them. × 42,000.

5.58. At this level the fibrous ring has two dense striations connected to two specific triplet tubules of the basal bodies. The rootlet tubules and striated fiber between the basal bodies have been sectioned obliquely. × 45,000.

5.59. Part of the fibrous ring and the dense material capping one end of the basal bodies. × 42,000.

5.60. This section passes through the edge of the flagellar apparatus. The basal bodies are sectioned obliquely, and their flagellum (f) passes out of the cell. In contrast, the rootlet templates (the row of three adjacent tubules), striated fibers, and variable numbers of additional tubules are sectioned transversely. These extend around the cell periphery, under the plasmalemma (arrow). × 35,000.

SOURCE. Author's micrographs, published in *Protoplasma* 72, 275 (1971).

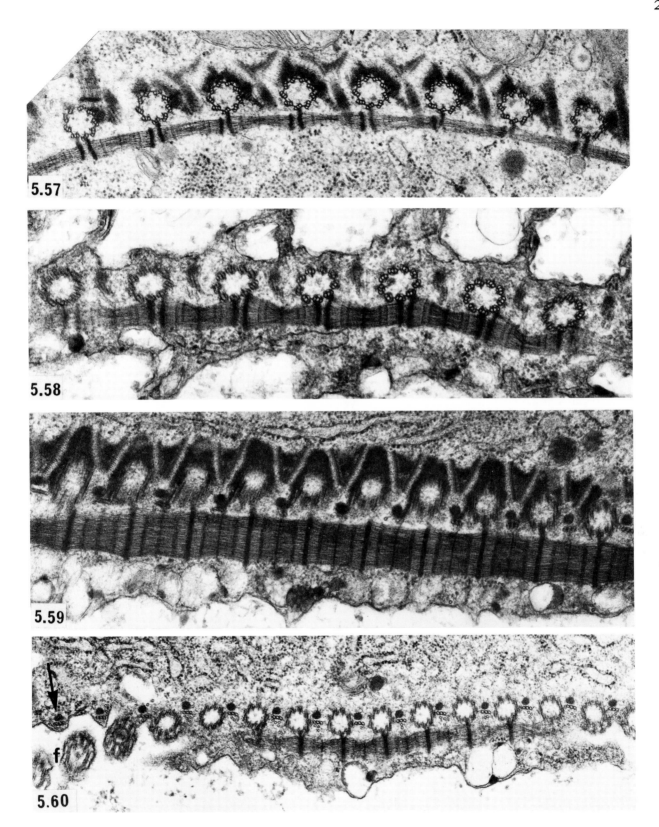

5.57

5.58

5.59

f

5.60

before floating away (Fig. 5.75) from the now quiescent zoospore. This shedding of the flagella appears a great waste of cellular protein. I have observed it repeatedly under the light microscope and it probably occurs in nature too, as I have never, for example, come across flagellar microtubules in the cytoplasm of the germinating zoospore, as often happens in other algae which retract their flagella (Chapters 3 and 4).

DIFFERENTIATION OF THE GERMLING

Soon after the flagella have been shed (Pickett-Heaps, 1972c), the constriction around the dome disappears and the cell begins quite rapidly to elongate further (Fig. 5.80A), often becoming cylindrical as it increases in volume (Pickett-Heaps, 1972d). This enlargement soon ceases, but meanwhile a protrusion of cytoplasm appears from the dome and grows continuously over the next few hours (Fig. 5.76). The rhizoid outgrowth thus forming is to become the holdfast of the germling, the future basal cell of a new vegetative filament. If zoospores are frustrated in their attempts to find a substrate for the holdfast to attach to, the cells will usually collect at the surface of the medium they are growing in (Fig. 5.77). The rhizoid will then continue to grow, often becoming long and attenuated or else bifurcated (Figs. 5.77–5.79, 5.84A; Fritsch, 1902b; Pickett-Heaps, 1972d). The cytoplasm of unattached rhizoids remains quite transparent for some time, and within it numerous cytoplasmic particles and vacuoles can be seen (Fig. 5.79). Such unattached germlings often undergo cell division later to form new floating vegetative filaments. Others instead differentiate back into zoospores to try again. For example, in *O. capillare*, Fritsch (1902b) records several successive cycles of zoosporogenesis from single-celled, unattached germlings; each generation of zoospores was less well developed than that preceding it, until in that case, the cells finally expired.

Rhizoids are covered by a diffuse mass of presumably sticky material, and those that manage to attach to some substrate differentiate rather differently. They appear to ramify into a radiating array of finger-like outgrowths of cytoplasm that envelope other filaments of algae or else spread out radially over flat surfaces (e.g., floating leaves; Fig. 5.90). Their ameboid activity is strikingly demonstrated if the zoospores are allowed to attach to Millipore filters. Cytoplasmic

Figures 5.61–5.63
FLAGELLAR APPARATUS OF OEDOGONIUM

5.61. The striated fiber in longitudinal section, closely associated with rootlet *(rt)* and additional tubules at the cell periphery (see Fig. 5.60). × 83,000.

5.62. Flagellar apparatus isolated from the spermatozoid of *O. cardiacum*. Some flagella *(f)* are still attached to the ring of basal bodies from which also radiate the arrays of the striated fiber/rootlet microtubule complex. × 12,000.

5.63. Negatively stained rootlet microtubule *(rt)* and striated fiber from the sperm of *O. cardiacum*. × 60,000.

SOURCE. Figure 5.61: Author's micrograph, published in *Protoplasma 72*, 275 (1971). Figure 5.62: Dr. L. R. Hoffman, published in *Amer. J. Bot. 50*, 455 (1963). Figure 5.63: Dr. L. R. Hoffman. Reproduced by permission of the National Research Council of Canada from *Can. J. Bot. 48*, 189 (1970).

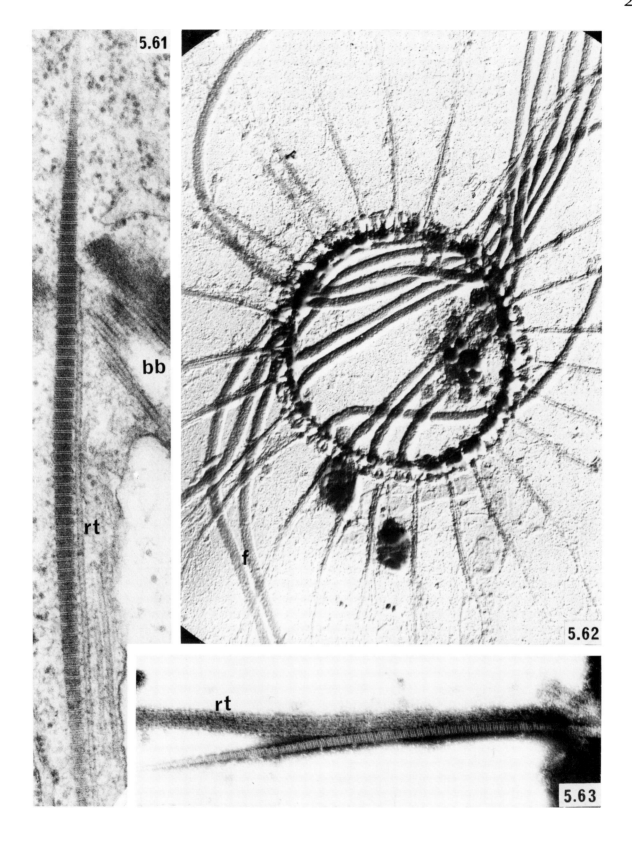

5.61

bb

rt

f

5.62

rt

5.63

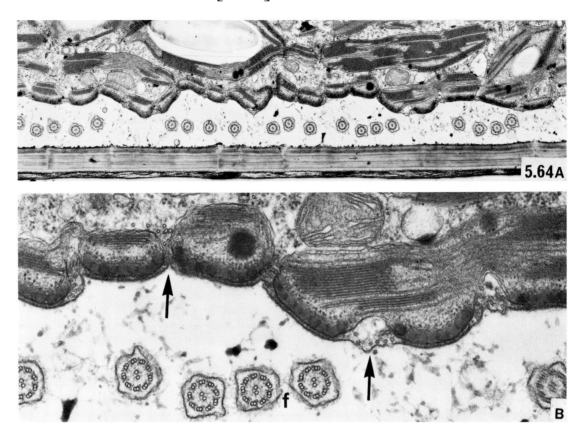

Figure 5.64
EYESPOT ON THE ZOOSPORE OF OEDOGONIUM

The eyespot consists of a layer of granules at the periphery of the chloroplast, the latter being indented apparently by the sets of cytoskeletal microtubules emanating from the flagellar apparatus (arrows). *A;* × 14,000; *B;* × 44,000.

SOURCE. Author's micrographs, published in *Protoplasma* 72, 275 (1971).

processes, covered with the diffuse mucilaginous material, penetrate quite deeply into the interstices of the filter pad (Fig. 5.83). Likewise, these outgrowths can attach and partly penetrate into the outer wall layer of other cells of *Oedogonium* (Fig. 5.81), thereby provoking the wound response in the host cell that I have already mentioned (Fig. 5.25). While the holdfast is forming, the typical two-layered, even wall is being secreted around the rest of the cell. A similar wall is also laid down around the holdfast, much more irregularly at first, and later massively thickened (Figs. 5.85–5.87, 5.90–5.94). Furthermore, lo-

Figure 5.65
DOME OF A MATURE ZOOSPORE OF OEDOGONIUM BEFORE RELEASE

The dome is circumscribed by the ring of flagella *(f),* and it contains much endoplasmic reticulum *(e).* The Golgi bodies *(g)* at the inside edge of the dome apparently give rise to a population of vesicles which condense and move through the endoplasmic reticulum, collecting at the surface of the dome (see Figs. 5.68–5.73). The peripheral chloroplast remains totally excluded from this region. × 7,200.

SOURCE. Author's micrograph, published in *Protoplasma 72,* 275 (1971).

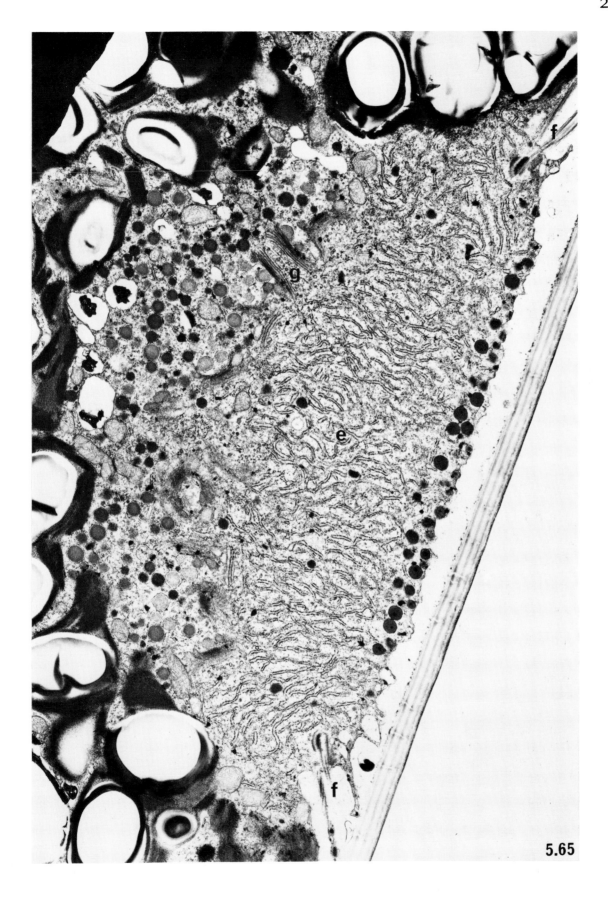

5.65

calized areas of wall deposition create partitions that divide up the cytoplasm in the holdfast into a complex labyrinth (Figs. 5.81, 5.82). By the time the holdfast is fully formed on a flat surface, it resembles a flattened cone (like the base of a large vase) penetrated by small radiating cytoplasmic channels.

Examination with the electron microscope shows clearly that although the flagella are shed from germinating zoospores, the basal bodies and other components of the flagellar apparatus remain inside the cell (Fig. 5.80B). Once the constriction disappears from around the edge of the dome, the entire flagellar apparatus (including the fibrous ring) can be seen further inside the cell, and it soon becomes increasingly dislocated and disrupted. During subsequent differentiation of the germling, all components of the flagellar apparatus steadily lose their form and stainability while apparently floating free in the cytoplasm. By the time the holdfast is partly developed, no trace of this complex system can be seen anywhere in the cell. Such observations indicate that the fibrous ring probably has considerable mechanical strength, being responsible for the constriction around the dome of the zoospore —not a surprising conclusion in view of Hoffman and Manton's ability to isolate the flagellar apparatus intact (Fig. 5.62). The rootlet template microtubules may become dissociated from the fibrous ring as the flagella are shed, but this is far from conclusive (Pickett-Heaps, 1972c). There is no evidence to suggest that either the fibrous ring or the striated fibers could be contractile. They could perform other functions in the cell which, for want of evidence, I shall not discuss at this stage. The eyespot, incidentally, also slowly disappears during differentiation of the germling.

While the growing rhizoids are attaching to a substrate, at least two large, conspicuous, and different types of vesicles appear to be discharging into this region of expanding wall (Fig. 5.80 C, E). The numerous Golgi bodies in the dome are producing hypertrophied vesicles containing finely divided, PAS/H-positive fibrils (Fig. 5.80 D); these vesicles form one of the populations of vesicles thus secreted. The other class of vesicles is derived from the numerous basal particles, preformed during zoosporogenesis by the Golgi bodies and retained at the surface of the dome. These basal particles apparently are altered

Figures 5.66, 5.67
ZOOSPORES OF OEDOGONIUM CARDIACUM

5.66A–F. Sequence showing emergence of two live zoospores. The hyaline vesicle enclosing each swells continuously after wall rupture, until it disappears and the zoospore swims off. The arrows indicate flagella. × 350 approx.

5.67. Scanning electron micrograph, showing the ring of flagella at the dome. × 800 approx.

SOURCE. Author's micrographs. Figure 5.66 published in *Protoplasma 74*, 149 (1972).

263

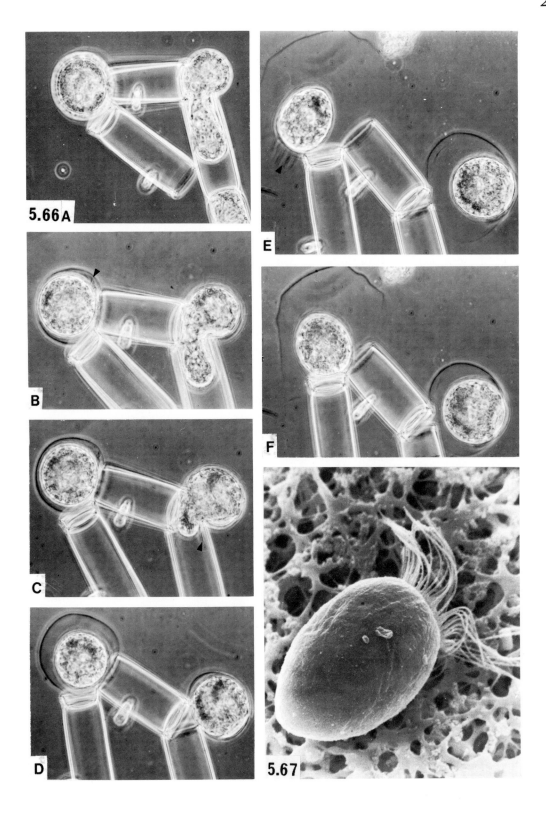

before being discharged into the wall; they become noticeably less electron-dense and their outer membranes become distended (Fig. 5.80 B, C). Moreover, they are now rendered PAS/H-positive. The discharge of both these types of vesicles into the wall is observed only at the holdfast, and the material in the vesicles surely contributes to the adhesive properties of this end of the cell. How the two wall layers are secreted around the rest of the cell (e.g., Fig. 5.81) is not understood, and the apical discontinuity or weakening in the inner wall layer appears early on. Longitudinally oriented microtubules are always present near the wall, and these are particularly conspicuous in elongating rhizoids of unattached germlings (Fig. 5.84). Just inside the layer of microtubules in the holdfast is usually found a sheet of reticulate, smooth endoplasmic reticulum, similar to that near the cell wall of some higher plant cells (Fig. 5.84 B; O'Brien, 1972).

The differentiating holdfast of *Oedogonium* always contains a massive accumulation of endoplasmic reticulum on whose surface some polysomes can be found in twisting, parallel arrays. This rather striking feature is not uncommon in developing zoospores and has been seen in other algae. I have also observed in some unattached germlings (Pickett-Heaps, 1972d) an apparent breakdown of this endoplasmic reticulum into masses of smooth membranes or else homogeneous cytoplasm relatively free of membranes but containing large numbers of free ribosomes. The significance of these observations remains to be established, but they could possibly represent a massive reprogramming of cytoplasmic systems, previously set up for germling formation, prior to a renewed cycle of zoospore re-formation.

One other intriguing point can be briefly mentioned here. I unexpectedly encountered (Pickett-Heaps, 1972f) in several fully developed germlings of one species of *Oedogonium*, numerous particles of extremely regular size (ca. 230 nm across) and shape (hexagonal in cross section). Furthermore, the existence of other, apparently incompletely formed particles nearby indicated that an icosahedral "coat" was being formed which was then filled with a dense core of fibrillar material. These observations immediately suggested that one of the species of *Oedogonium* being investigated had been infected by a large virus which replicated in the moribund germling. If this is true—and the particles have

Figures 5.68–5.70
ZOOSPORE RELEASE IN OEDOGONIUM

5.68. Zoospore beginning to emerge. The hyaline layer (arrows) enclosing the apical end of the cell is clearly turning into the hyaline vesicle (Fig. 5.66). The basal mucilage *(bm)* is becoming disperse. × 3,400.

5.69. Wall rupture (arrow) at the weakened site below the caps. × 1,600.

5.70. Zoospore half out. × 800.

SOURCE. Author's micrographs, published in *Protoplasma 74,* 149 (1972).

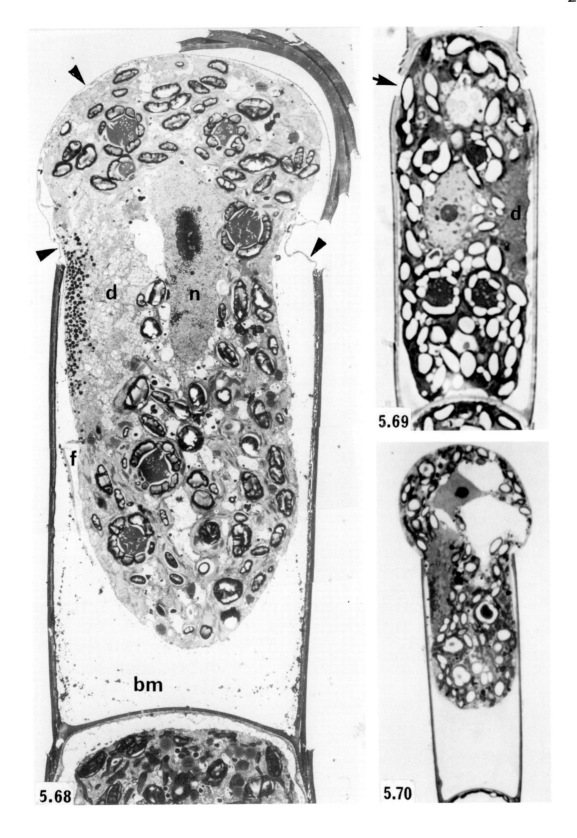

5.68

5.69

5.70

not been isolated or their infectivity demon-strated—then it would appear that the presumed virus infected the cells while they were naked (and perhaps, therefore, vulnerable) zoospores. This interesting possibility hints at why viral in-fections of Green Algae are rarely reported. Viruses may have difficulty infecting host cells enclosed by a rigid wall, and so many algal cells are resistant throughout almost all their life cy-cles.

CELL DIVISION IN THE GERMLING

Essentially, cell division occurs as expected in mature, single-celled germlings. However, the ring, which is normally laid down at the apical wall weakening (Fig. 5.85), may be hypertro-phied in some species to cover the whole apical end of the cell (Fig. 5.91; Fritsch, 1904; Pickett-Heaps, 1972e). Such a variation in behavior is consistent with the hypothesis that the formation of rings may represent a modified wound re-sponse (Pickett-Heaps, 1972e), in this case one that reinforces the entire end wall of the cell that is perhaps more liable to fracture than normal. Rupture of the wall in the germling (Figs. 5.85, 5.88) and expansion of the ring (Fig. 5.86) fol-low as normal. The cap of secondary wall derived from the basal cell often becomes incorporated into the wall of the apical cell (Figs. 5.86, 5.87), in which event, the ring in the next division of the apical cell is formed below this cap (Fig. 5.89). Sometimes the cap is not integrated into the wall (Fig. 5.93; Fritsch, 1902a), in which case the next ring in the apical cell may cover the entire end wall. In some species, the basal cell does not undergo more than one division (Fig. 5.87); in other species, it can divide at least twice (Fig. 5.92; Pickett-Heaps, 1972e). Further cell divisions then give rise to the new filament (Fig. 5.94).

5.1d. Sexual Reproduction

Sexual reproduction in *Oedogonium* is ooga-mous. The large spherical oogonium (Figs. 5.95, 5.117) is fertilized by a small, multiflagellate sperm (Figs. 5.115, 5.117) which bears some resemblance to a miniature zoospore. As usual, species of *Oedogonium* may be either monecious (Fig. 5.95) or diecious. Two interesting types of sexual behavior are known for this order (Fritsch,

Figures 5.71–5.73
EMERGENT ZOOSPORES OF OEDOGONIUM

5.71.   Zoospore rounded up after emergence, still enclosed by the hyaline vesicle (cf. Fig. 5.66 B, D). The flagella (f) are starting to erect as they be-come functional. × 1,700.

5.72.   Swimming zoospore. Note flagella (arrows) and eyespot (es). This zoospore is appreciably longer than those in Figures 5.71 and 5.73; com-pare with Figure 5.67. × 1,200.

5.73.   Zoospore still within the hyaline vesicle (ar-rows). The dome has begun to protrude. × 1,000.

SOURCE. Author's micrographs, published in *Protoplasma* 72, 275 (1971) and 74, 149 (1972).

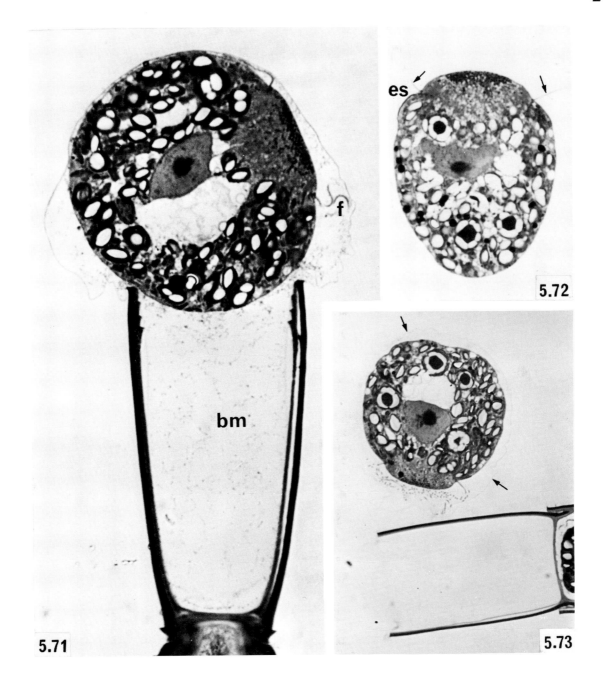

1935; Ohashi, 1930), termed "macrandrous" and "nannandrous" reproduction.

MACRANDROUS REPRODUCTION

The diecious *O. cardiacum* provides an excellent example of macrandrous reproduction. Antheridia arise as a series of small, discoidal cells at the apical end of a parental cell (Figs. 5.95, 5.96; Ohashi, 1930), and in each of these cells differentiate two or four sperm. The sperm, upon release from the antheridia, are attracted to the oogonia chemotactically (Hoffman, 1960), and one fertilizes each oogonium directly (as in Fig. 5.124).

NANNANDROUS REPRODUCTION

In the more complicated nannandrous reproduction (summarized in Fig. 5.102), differentiation of male cells commences with the release of an "androspore" from a parental filament. The androspore arises in a small disc-like cell (Ohashi, 1930), but it is appreciably larger than a sperm cell, although smaller than a zoospore. It does *not* fertilize the oogonium directly; instead it is attracted to an oogonial mother cell and then attaches to this filament (Fig. 5.102*A*), germinating as would a miniature zoospore. It is noteworthy that zoospores almost never attach to a vegetative filament of the *same* species of *Oedogonium*, whereas androspores almost always do so (Tiffany, 1957). Upon germination, the androspore gives rise to an epiphytic "dwarf male" filament following a series of cell divisions (Fig. 5.102*B*). Then the apical cells in these small filaments differentiate sperm, which when released (Fig. 5.102*C*) fertilize the adjacent oogonium, now fully differentiated following a concurrent division of the mother cell (Fig. 5.102*B*). Both androspores and dwarf male filaments can revert to the vegetative condition (Rawitscher-Kunkel and Machlis, 1962).

Hormonal integration of reproduction in some nannandrous species may be complex and subtly organized (Rawitscher-Kunkel and Machlis, 1962). The mother cell (Figs. 5.98, 5.99) attracts androspores chemotactically as before (Fig. 5.97). This attraction was demonstrated by soaking a thread of cotton in the culture medium containing female cells and then placing it in a suspension of androspores, which immediately swarmed to the thread. Once attached to the female filament, the germinating androspores

Figure 5.74
MATURE ZOOSPORE OF OEDOGONIUM

This zoospore had been free-swimming for some while and is quite elongated. Note the protruding dome bounded by the flagella *(f)* and the numerous contractile vacuoles (arrows) on the surface. × 7,100.

SOURCE. Author's micrograph, published in *Protoplasma 74*, 149 (1972).

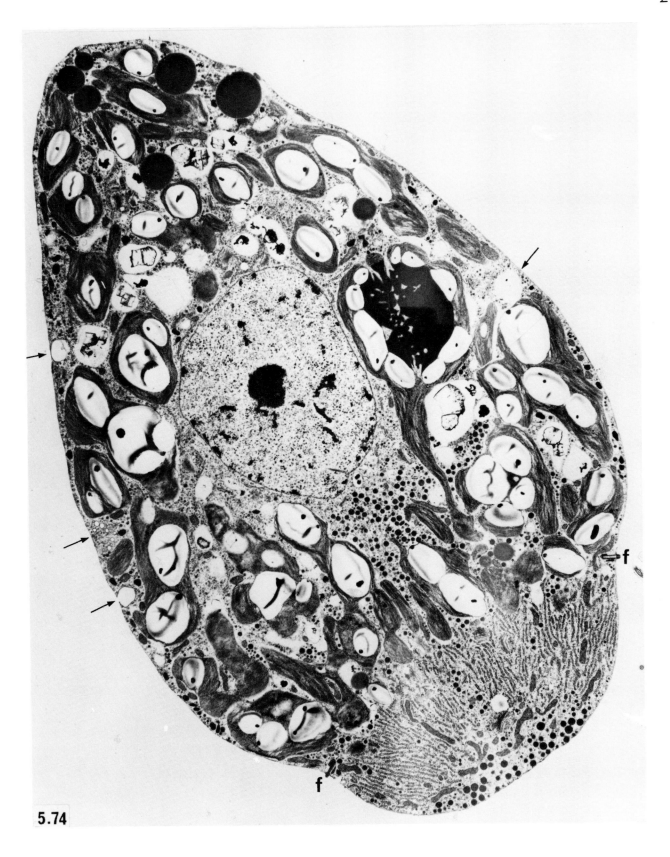

5.74

grow in well-defined directions, presumably in response to the chemotactic agent (Fig. 5.100). Division of the mother cell into the oogonium and the basal "suffultory" or "supporting" cell is delayed until attachment and subsequent germination of the androspores (Fig. 5.99), which indicates some specific inductive effect of the males upon the mother cells. This supposition was ingeniously confirmed experimentally by encasing the female cells in a thin jacket of agar (a sort of algal contraceptive!). Androspores were attracted to the mother cell for the duration of the experiment (21 days) and they attached themselves to the agar; however, the mother cells did not divide and differentiate, presumably since the androspores were not attached directly to their wall. Once formed, the maturing oogonium secretes a large amount of mucilaginous material (Fig. 5.101), which envelopes it and the dwarf males. This secretion, also stimulated by the presence of the males, serves to trap released sperm, preventing their dissemination in the culture medium. The sperm are attracted to the oogonial pore presumably again by chemotaxis. Fertilization, involving entry of one sperm into the pore and fusion with the oogonium, is rapid. Rawitscher-Kunkel and Machlis (1962) mention the extremely rapid appearance of 'a protoplasmic papilla' through the oogonial wall, to which the sperm attach; I interpret this observation to be attributable to the emergence under pressure of the polysaccharide plug responsible for pore formation (see below).

We can summarize the advantages of such a pattern of reproductive behavior (Fig. 5.102) as follows:

1. Differentiation of the oogonium is contingent upon the proximity of males, thereby ensuring that the sperm will be produced concurrently. In macrandrous species, oogonia develop even when males are not nearby, and consequently, fertilization is not always achieved.
2. The sperm are close to the oogonium when they are formed and cannot move far from it.
3. Differentiation of the androspore does not involve a final commitment to sexual reproduction, since these cells can, in the absence of females, give rise to vegetative filaments.

Bulbochaete hiloensis undergoes nannandrous reproduction, to be described in more detail in Section 5.2e.

Figures 5.75–5.79
GERMINATION OF LIVE ZOOSPORES OF OEDOGONIUM

5.75. Next to a young germling, a zoospore recently emerged from the empty cell *(ec)* is undergoing germination almost immediately (i.e., without passing through an extended phase of motility). It is elongating and shedding its numerous flagella (arrow). × 580.

5.76A–C. Successive stages, photographed in a interval of a few minutes, in the elongation of a germinating zoospore and the initiation of a rhizoid (arrow) from the dome *(d)*. × 540.

5.77. Young single-celled germlings collected on the surface of culture medium. × 300.

5.78, 5.79. Germlings of two species, showing the vacuolated rhizoids, often extensively lobed when the germlings are unattached to a substrate. Both × 680 approx.

SOURCE. Author's micrographs, published in *Protoplasma 74*, 149 and 169 (1972).

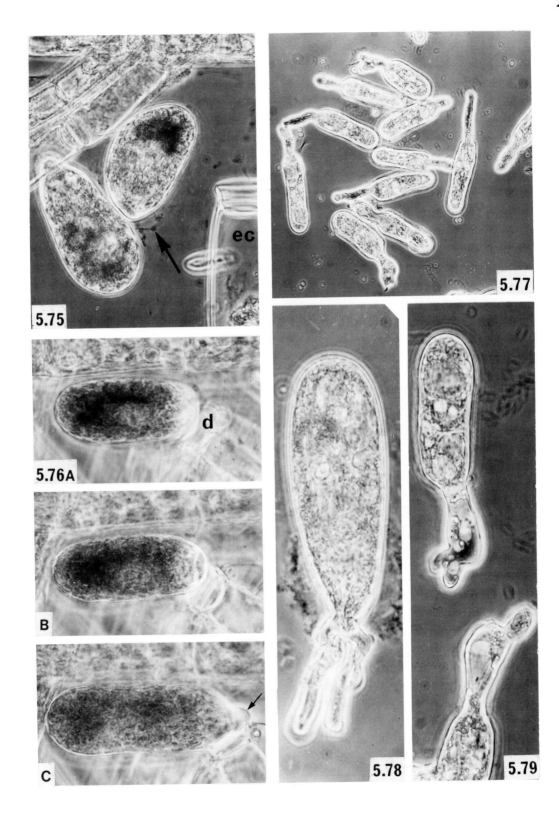

5.75

ec

5.76A

d

B

C

5.77

5.78

5.79

SPERMATOGENESIS (MACRANDROUS
SPECIES)

Spermatogenesis in macrandrous species is
preceded by the creation of the series of disc-like
antheridial cells (Figs. 5.95, 5.96), a result of a
modified type of the cell division so characteristic
of the Oedogoniales, in an antheridial mother
cell. These divisions and the subsequent differentiation of two to four sperm inside each antheridial cell, has been described by Ohashi
(1930) and others (Fritsch, 1935) and more recently at the ultrastructural level by Coss and
Pickett-Heaps (1973, 1974). Spermatogenesis is
summarized in Figure 5.103.

The antheridial mother cell appears indistinguishable from other vegetative cells, but since
we have never observed it to undergo normal
vegetative division after it has formed antheridial
cells, we suspect it has been terminally differentiated and cannot revert back to the vegetative
condition. The cell division that gives rise to the
antheridial cells is highly asymmetrical. A ring is
formed, but it is much smaller than normal. The
premitotic nucleus migrates right up to the apical
transverse wall (Figs. 5.104–5.106), and the
spindle appears to be so appressed to this wall
that it suffers considerable deformation (Fig.
5.105). Following telophase, the phycoplast cuts
off a small apical region of cytoplasm (Figs.
5.106, 5.107), and after wall rupture and expansion of the ring (Fig. 5.108), the discoidal antheridial cell is formed. Further similar divisions
give rise to a series of such cells.

The antheridial cell then undergoes at least
one (Figs. 5.105–5.107) and sometimes (probably rarely) two more divisions, to form the sperm
mother cells. In *O. cardiacum,* a normal ring is not
formed for these divisions, and instead a vestigial
ring or circumferential layer of diffuse material is
secreted at the wall weakening (Figs. 5.107–
5.109); neither is there wall rupture after cytokinesis. However, Ohashi (1930) reports
"stretching of the ring" after this division in antheridial cells of *O. grande.* The spindle is always
oriented across the cell during this division, presumably because of the restricted volume of the
antheridial cell. The nucleoplasm is quite pale
ultrastructurally, allowing us (Coss and Pickett-
Heaps, 1973) to discern two polar bodies

Figure 5.80*A–C.*
GERMINATING ZOOSPORE OF OEDOGONIUM

5.80*A.* Light micrograph showing the elongated cell
and dome *(d).* × 800.

5.80*B.* Dome region, same cell. The broken up remains of the flagellar apparatus are still present
(arrows). A thin wall has been secreted around
the cell, and the contents of the dome—mainly
endoplasmic reticulum, dense vesicles, and Golgi
bodies *(g)*—have become intermingled. ×
7,500.

5.80*C.* Wall secretion at the surface of the dome.
Two types of vesicles seem to be discharged into
this adhesive end of the cell. One type has quite
diffuse contents, perhaps derived directly from
the nearby Golgi bodies (see Fig. 5.80*B*). The
other apparently is derived from the mass of
dense particles that accumulates at the surface of
the dome (Figs. 5.65, 5.68.) and which has subsequently undergone some slight modification,
becoming less dense. × 30,000.

SOURCE. Author's micrographs, published in *Protoplasma
74,* 169 (1972).

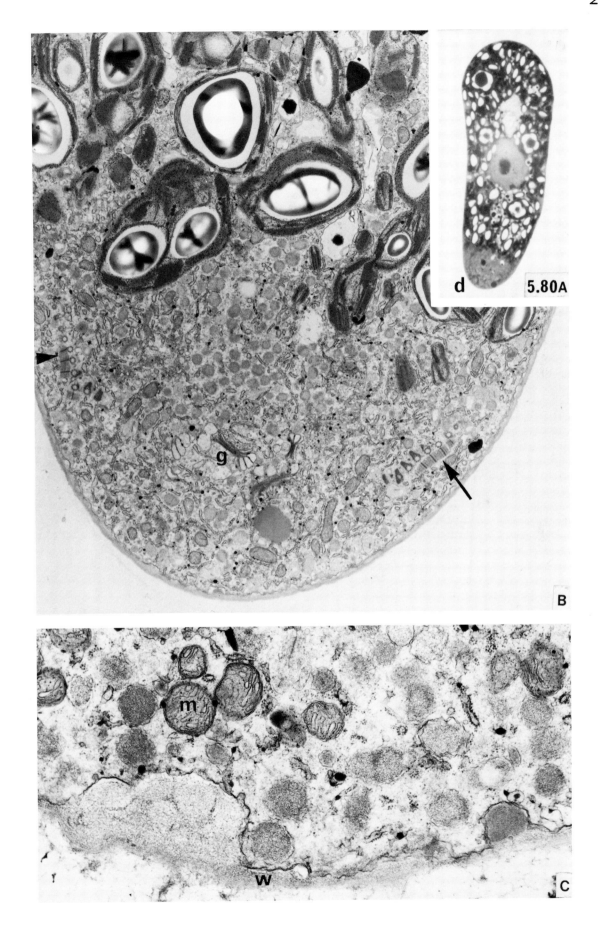

(MTOCs?) in which many spindle microtubules are embedded (cf. Figs. 5.36, 5.37).

Apparently, cytokinesis is transverse after this division, following rearrangement of daughter nuclei within the cell (Ohashi, 1930; cf. Figs. 5.108, 5.109). However, in our experience, the new cross wall has a quite variable orientation when visualized in thin sections. On several occasions (e.g., Fig. 5.108), we encountered four sperm in one antheridial cell, and so another division must have taken place right after the first. Hoffman (personal communication) says that the latter behavior is most unusual in *O. cardiacum.*

One most intriguing variation was encountered in the final division in the antheridial cells. Apparently cytokinesis is effected by a membrane furrow (Fig. 5.109*A, B*), rather than a cell plate, although a phycoplast is present as usual. This observation is again consistent with the general notion that during the formation of motile cells, more primitive cytological features may appear. Furrowing is undoubtedly a primitive form of cytokinesis, but it would probably be difficult to use with the method of cell elongation employed by extant members of the Oedogoniales, where the septum (defined by the phycoplast) has to move up the cell some time after telophase. However, in the absence of wall rupture and cell expansion (i.e., the situation during cytokinesis in antheridial cells), this problem does not arise, and so the more primitive cytokinetic feature can reappear (Pickett-Heaps, 1972a).

Sperm now differentiate in the sperm mother cells (Coss and Pickett-Heaps, 1974), and spermiogenesis resembles zoosporogenesis in several ways. Centrioles and rootlet templates (Fig. 5.110) form de novo, and they are reorganized into the typical flagellar ring, smaller than that of the zoospore (Hoffman and Manton, 1962, 1963) and containing about a quarter the number of basal bodies. The flagellar apparatus acquires rootlet tubules (Fig. 5.111), striated fibers, a fibrous ring, etc., but many of these components are less well developed than in zoospores. Golgi bodies secrete a mucilaginous sheath around the cells, which merges structurally with the cross wall between sperm mother cells and the vestigial ring formed at the first division of the antheridial cell (Fig. 5.108). The dome of the sperm cell is quite small; it contains some vesicles, apparently derived from the Golgi bodies, but these are not equivalent to the nu-

Figure 5.80*D, E*
GERMINATING ZOOSPORE OF OEDOGONIUM

Further micrographs of the zoospore shown in Figure 5.80*A–C.* Both these sections have been peroxidized and stained with the silver-hexamine reagent for the detection of polysaccharide. The two types of vesicles discharged into the wall can be recognized, one of them coming directly from the Golgi bodies (*g*). Note particularly the fibrous (adhesive?) material outside the wall in *E. D;* × 27,000; *E;* × 35,000 approx.

SOURCE. Author's micrographs, published in *Protoplasma* 74, 169 (1972).

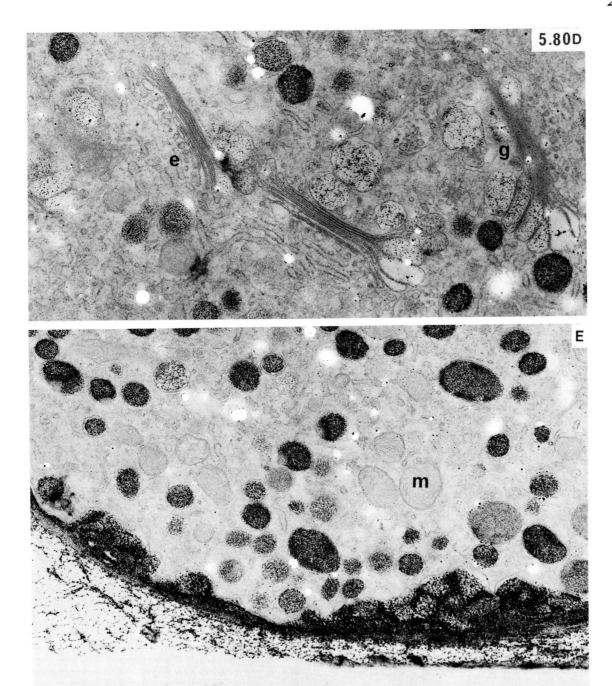

5.80D

E

merous, densely stained basal particles of the zoospore. The function of these vesicles in the sperm is uncertain.

The mature sperm is released in much the same fashion as the zoospore following rupture of the wall at the weakened region (Fig. 5.114). The material of the vestigial ring, in particular, is transformed into the swelling hyaline vesicle that briefly encloses the emerging sperm cell. The newly released sperm is almost colorless and markedly spindle-shaped (Figs. 5.113, 5.115; see also Hoffman, 1973a). Numerous microtubules run its length into an apical protrusion (Fig. 5.112). Sperm have a quite limited life, whose brevity seems at least partly due to their inefficient or short-lived osmoregulatory system. If they do not achieve fertilization, the slender cells begin to swell after a comparatively short period of motility. They become quite spherical, and a few minutes later they burst. Hoffman (1973a) notes that their competitiveness in achieving fertilization falls off with their age, and spherical sperm achieve plasmogamy only with much difficulty.

OOGENESIS, FERTILIZATION, AND GERMINATION OF OOSPORES

The oogonium of *Oedogonium* is a large spherical cell with a fertilization pore near its apical end (Fig. 5.116). The pore appears to consist of a split in the oogonial wall. When the oogonium is ready for fertilization, its cytoplasm contracts somewhat away from the wall, and the sperm enters through this pore (Fig. 5.117).

Oogenesis in macrandrous species such as *O. cardiacum* commences with a highly asymmetrical division (Coss and Pickett-Heaps, 1973) of the oogonial mother cell which, prior to division, is dense and filled with food reserves (Fig. 5.118; Ohashi, 1930). A large division ring appears in the mother cell, and the premitotic nucleus migrates to the basal cell wall (Fig. 5.119). Following mitosis (Fig. 5.120), cytokinesis cuts off the small suffultory cell (Fig. 121). Wall rupture and ring expansion follow as normal; however, the suffultory cell, undergoing about a sevenfold increase in volume, becomes almost colorless and devoid of cytoplasm, whereas the apical daughter cell, containing most of the mother cell's cytoplasm, becomes the swollen oogonium (Figs. 5.122, 5.123). These events are summarized in Figure 5.124. Thus, highly asymmetrical divi-

Figure 5.81
YOUNG GERMLING OF OEDOGONIUM ATTACHED TO ANOTHER FILAMENT OF A DIFFERENT SPECIES OF OEDOGONIUM

The germling is now quite elongated, with a thickening wall. The holdfast is also developing. Note the debris collected around the sticky holdfast. The attachment of this zoospore has elicited a wound response in the host cell (see Fig. 5.25). × 4,200 approx.

5.81

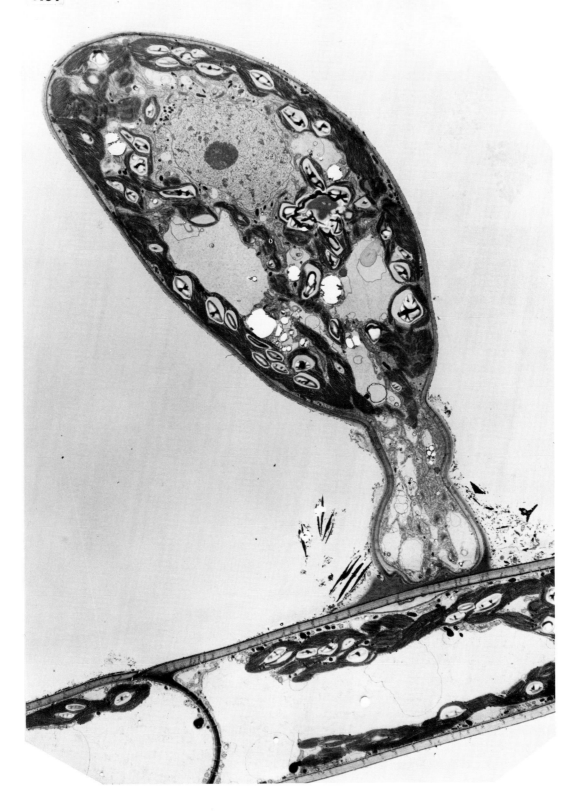

sions precede both oogenesis and spermatogenesis in *Oedogonium*, but the polarity of division is opposite, depending upon whether tiny, almost nonphotosynthetic sperm, or a large and long-lived oogonium is to be formed. Ohashi (1930) notes that in some species of *Oedogonium*, a true suffultory cell is not evident, since this basal cell may undergo a further division to form a second (and even third and fourth) oogonium.

Hoffman (1971) has described the structure of the oogonium prior to fertilization (Fig. 5.125). The egg cytoplasm is depressed at the region of the pore, and an accumulation of hyaline material at this region also tends to be extruded through the open pore. Such extruded material is clearly not cytoplasm. Using the electron microscope, Hoffman showed that the hyaline material is fibrous (Fig. 5.128) and says it is "clearly associated with the chemical breakdown of the oogonial wall during pore formation;" earlier he says that "pore formation is apparently associated with a localized chemical alteration of the wall" (see also Hoffman, 1973a).

I feel I must disagree with Hoffman's otherwise beautifully documented papers in this matter. Earlier stages of oogenesis are not illustrated, and Hoffman could have misinterpreted his results owing to a lack of developmental studies, vital in this instance, since he has not yet established that the oogonial wall does in fact become degraded. Reference to both hair cell formation and oogenesis in *Bulbochaete* (Sections 5.2c, 5.2e) permits an entirely different interpretation, quite consistent with Hoffman's observations. I suspect that the oogonial wall is already thinned out and contains a future rupture site long before the pore is formed and that localized polysaccharide secretion takes place at the pore site (as also occurs during hair cell formation and oogenesis in *Bulbochaete*). Uptake of water by polysaccharide could easily be visualized as rupturing the pre-weakened wall (and concurrently causing depression of the nearby egg cytoplasm) to form the pore, and this mucilage would then tend to emerge through the pore precisely as described by Hoffman and others. This explanation has also been proposed by Retallack and Butler (1973) to account for pore formation during oogenesis in *Bulbochaete*, and Hoffman (1973a) notes a relief of the pressure exerted by the pore substance after the wall has ruptured. The pore material is lined by a thin layer attached to the oogonial wall

Figure 5.82
MATURE GERMLING OF OEDOGONIUM

5.82A. Germling with well-developed holdfast in a typical situation, attached to the surface of a water weed (cf. Fig. 5.1). × 820.

5.82B. Detail of Figure 5.82A, showing the thick wall secreted around the rhizoids of the holdfast. Secretion of material into this wall is evident at arrows. The Golgi bodies (g) are now no longer hypertrophied (cf. Fig. 5.80D). A thin, electron-transparent line separates the cuticle of the host and the holdfast wall. × 26,000.

SOURCE. Author's micrographs, published in *Protoplasma* 74, 169 (1972).

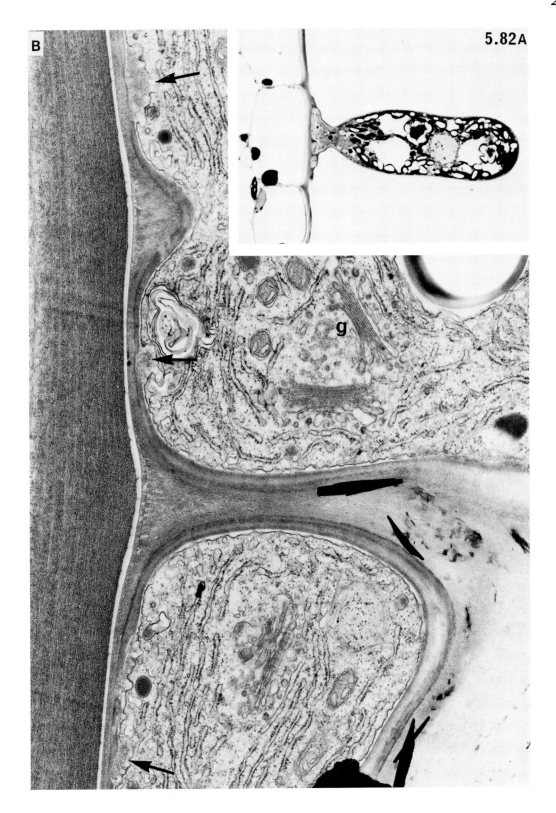

(this layer is just visible in Fig. 5.125) which persists in the oogonium even after the pore material has been dissolved away during fertilization (Hoffman, 1973a). The polysaccharide secreted at the hair cell site in *Bulbochaete* is likewise composed of two layers, the first amorphous and the second quite persistent (Section 5.2c).

The mature egg (Fig. 5.125) contains the usual organelles, including a prominent nucleus, a large central vacuole, much lipid and starch, a few pyrenoids, and extensive ramifications of the nuclear envelope into elements of endoplasmic reticulum. When the egg cytoplasm becomes shrunken (Figs. 5.117, 5.125), Hoffman (1973a) reports the existence of contractile vacuoles in its surface cytoplasm (Fig. 5.126), as might be anticipated.

Hoffman (1973a) has provided one of the most comprehensive and beautifully illustrated papers ever published concerning fertilization in algae (or any other organism, for that matter) using *O. cardiacum*. One of his sequences, taken from a cine film, is reproduced here (Fig. 5.127). As mentioned before, chemotaxis attracts the sperm to the oogonial pore region (Hoffman, 1960), sometimes even before the pore has opened. Once near the pore, the active sperm make vigorous thrusting movements to enter it. If hyaline material blocks the pore, it soon disappears once a sperm is nearby, apparently being digested by some material released by the sperm (and perhaps contained within the vesicles at the dome). The sperm alters shape appreciably during these movements, with its elongated tip constantly changing length. The flagellar end also elongates and moves flexibly and rapidly, sweeping over the egg surface. Plasmogamy coincides with the rapid loss of the sperm's flagella, shed with the same vibratory motion that characterizes loss of the flagella in a germinating zoospore (Section 5.1c). This process is apparently just preceded by fusion of the membrane of the egg and sperm. Within a few seconds, the tail of the sperm collapses and its whole body disappears into the egg (Fig. 5.127).

Ultrastructural examination of recently fertilized, binucleate oogonia show the flagellar apparatus, minus flagella, of the sperm inside the egg cytoplasm (Fig. 5.129) undergoing dissolution (cf. Fig. 5.80B). The sperm nucleus is quite different in appearance from the oogonial nucleus (Fig. 5.130), allowing Hoffman (1973b) to

Figures 5.83, 5.84
GROWTH OF RHIZOIDS IN GERMLINGS OF OEDOGONIUM

5.83. This germling had germinated on a Millipore filter pad whose surface is indicated by the accumulation of fine debris. (The pad was subsequently dissolved out during dehydration of the specimen for electron microscopy, leaving electron-transparent holes.) Note how the rhizoidal outgrowth penetrated into the filter pad. × 11,000.

5.84A. This germling was unattached, a situation that usually provokes considerable extension of the rhizoid(s) as here (see Figs. 5.77–5.79.). Figure 5.84B was taken from the region indicated by the arrow. × 1,500 approx.

5.84B. Detail of Figure 5.84A. Microtubules *(t)* are longitudinally oriented in the rhizoid, and there is usually a characteristic layer of reticulate membrane (endoplasmic reticulum?) close to the wall, seen here in surface view. × 35,000.

SOURCE. Author's micrographs, published in *Protoplasma* 74, 169 (1972).

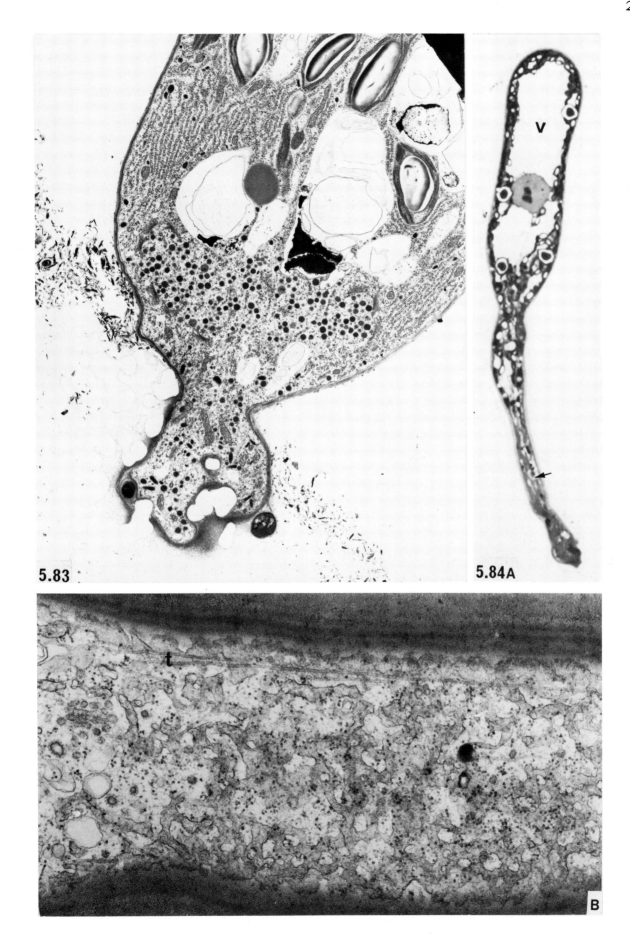

5.83

5.84A

B

confirm that rare cases of polyspermy do occur. Shortly after fertilization, a thin layer is rapidly secreted around the cell, and it may serve to prevent such multiple fertilizations. Unfertilized oogonia do not form parthenospores in *O. cardiacum,* but do so in many other species of *Oedogonium.*

As they mature, oospores (Fig. 5.131) usually acquire a red or brownish coloration as they secrete their thick, ornamented, and resistant cell wall. Germination (Hoffman, 1965) in *O. foveolatum* occurs more readily if the oospores are aged, and it is stimulated in the laboratory by light and transfer of the oospores to fresh culture medium. Two meiotic divisions give rise to a quadrinucleate cell which cleaves to release four haploid zoospores, at first contained within a vesicle derived from the inner layer of the oospore's wall (Fig. 5.132). These zoospores give rise to vegetative filaments as normal. Diploid strains of *Oedogonium* are not unusual, following upon failure of the meiotic reduction divisions.

5.2. Bulbochaete

The genus *Bulbochaete* is closely related to *Oedogonium,* but its greater morphological complexity makes it quite distinctive (Fig. 5.133). Tiffany (1928) has described this genus in considerable detail, and the life cycle of *B. hiloensis,* the species described in the next sections, has been detailed by Cook (1962).

5.2a. Vegetative Structure

The filament, arising from a basal holdfast cell derived from a zoospore (as in *Oedogonium*) is always branched to a varying extent, and the branches themselves are further similarly ramified. Most conspicuous on the plant body (Fig. 5.133) are numerous colorless and very fine hair cells with their characteristic bulbous base (from which the genus derives its name), attached apically and slightly laterally to many vegetative cells. Terminal cells on a branch often bear two such hairs, which usually die soon after they are formed, often breaking off and floating away. As we shall see, the formation of these hairs is directly linked to the ability of the filament to become branched. Indeed, hair cell formation initiates branching (Sections 5.2b, 5.2c), and over short distances along the filament, both

Figures 5.85–5.87
CELL DIVISION IN GERMLINGS OF OEDOGONIUM

5.85. The wall has just ruptured at the ring (arrows) in this binucleate germling (see Fig. 5.88). The septum containing the phycoplast is labeled *(s).* × 1,400.

5.86. Ring expansion. The septum has moved almost to its final position. Note the cap (arrows) on the apical cell derived from the basal cell. × 31,000.

5.87. The apical cell of this pair is about to divide. The ring (arrows) is sited just below the cap (see Fig. 5.89). × 1,000.

SOURCE. Author's micrographs, published in *Protoplasma 74,* 195 (1972).

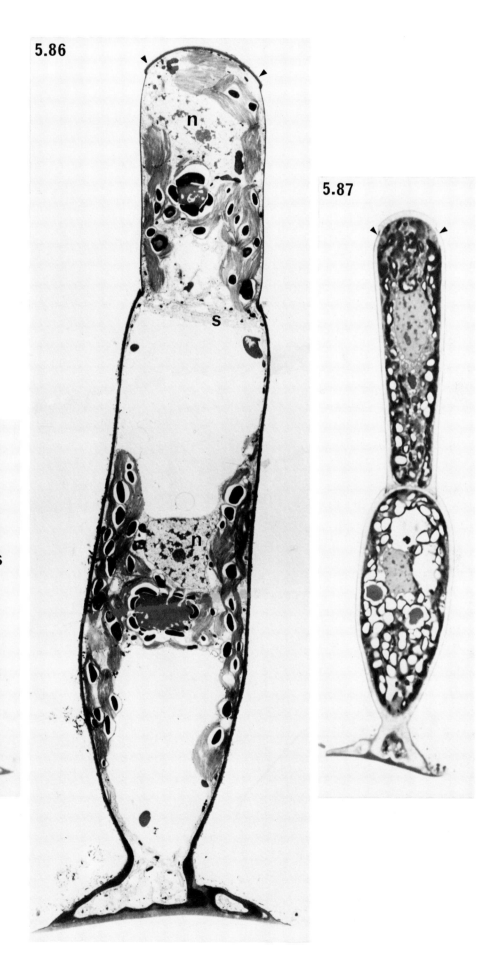

hairs and branches tend to arise from one side (Fig. 5.134). Some species of *Oedogonium* may bear hairs also, but these are always single and terminal (e.g., Geitler, 1961b). Because the plant body is highly branched and possesses numerous hair cells, these plants often form flocculent masses in culture.

The vegetative cells are smaller and shorter than those of most species of *Oedogonium* and are usually somewhat swollen; their overall shape, however, is quite variable. The peripheral chloroplast contains numerous small pyrenoids (Retallack and Butler, 1970a). The caps formed in the wall during cell division are usually quite difficult to discern with the light microscope. Ultrastructural examination reveals that they are entirely equivalent to the caps of *Oedogonium* (Fig. 5.135). However, as will be seen, the number of single, downward-facing caps is usually limited (in *B. hiloensis*) by the tendency of apical daughter cells to form hairs. The plasmodesmata in *Bulbochaete* are unusual (although probably typical of the Oedogoniales); their wall is lined with helically arranged particles, and they reportedly contain no connecting strand (Fraser and Gunning, 1969).

5.2b. Cell Division

Vegetative cell division in *B. hiloensis* (Pickett-Heaps, 1973d) resembles that already described for *Oedogonium* in most important respects. The wall always contains the weakening below single caps or adjacent to tiered caps (Figs. 5.135–5.138) that precisely determines the siting of the division ring. The latter is almost certainly composed mainly of the contents of Golgi vesicles (Figs. 5.136–5.138), although its electron microscopical appearance is somewhat inhomogeneous. The lips of secondary wall attaching the ring to the parental wall are also equivalent to those in *Oedogonium* (Fig. 5.138) except that the basal lip, complete with its own weakening, is much larger.

In the case of cell division at the branch point, or in a cell that has just previously formed a hair, the division ring is *invariably* sited under the branch or hair (Figs. 5.140, 5.142). Careful inspection of the micrographs reveals a circular wall discontinuity again in precisely the position taken up by the ring. The origin of this discontinuity will be described in Section 5.2c. Con-

Figures 5.88–5.90
GERMLINGS OF OEDOGONIUM

5.88. Wall rupture at the ring *(dr)* of the cell shown in Figure 5.85. × 9,500.

5.89. Ring formation at the wall weakening below the cap in the cell shown in Figure 5.87. × 6,400.

5.90. Edge of the holdfast of a germling, showing the radiating rhizoidal outgrowths with the thick holdfast wall secreted around them. × 1,300.

SOURCE. Author's micrographs, published in *Protoplasma 74*, 169 and 195 (1972).

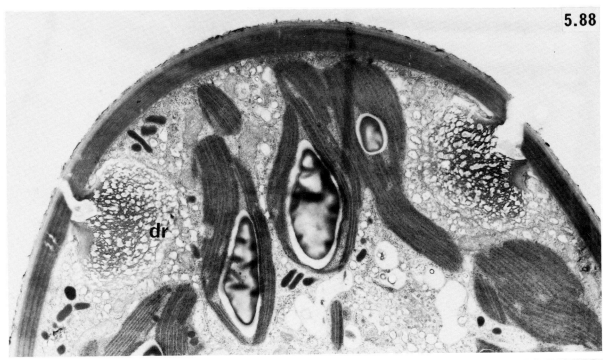

5.88

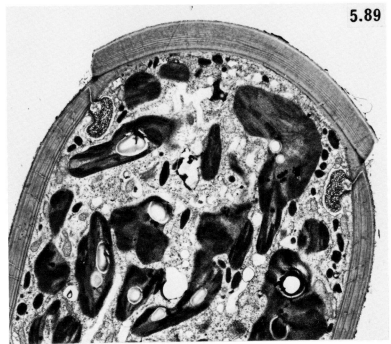

5.89

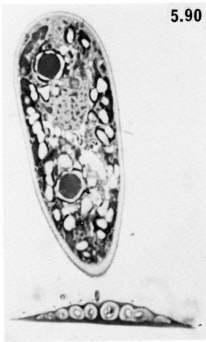

5.90

versely, there is no trace of the discontinuity in the wall under the next apical cell along the main filament axis. This observation confirms that cell expansion following division at the base of a branch or hair must necessarily be in the direction of that branch or hair, and such a cell can *never* thereafter intercalate a new daughter cell in the original filament axis (see also Cook, 1962). The plane of cell division is thus controlled by the siting of the wall weakening.

Mitosis proceeds as in *Oedogonium* (Pickett-Heaps, 1973d). The spindle is closed, the kinetochores are complex, and the nucleolar material persists as a rather large and apparently inert body in the spindle, being eliminated from daughter nuclei at telophase. In cells undergoing the asymmetrically oriented division at the base of branches or hairs, the spindle is aligned with the main filament axis (Fig. 5.140); only during phycoplast formation (Fig. 5.142) does the division apparatus become oriented toward the direction of future cell expansion.

After mitosis, the daughter nuclei approach one another, and the phycoplast forming between them cuts through the peripheral chloroplast as usual (Figs. 5.139, 5.141). However, in cells dividing to form or elongate a branch, the diagonal phycoplast (Fig. 5.142) tends not to grow all the way across the cell (Fig. 5.143), presumably because during subsequent cell expansion it has to move into the rather restricted space at the base of the branch.

Wall rupture (Figs. 5.141, 5.143) and expansion of the ring follow, and as expected, the rupture is quite violent, with cell expansion being initially rapid (Fig. 5.144). Figure 5.145 shows two cells at late stages of elongation, with two septa containing the phycoplast at or near their final position. The ring material remains somewhat inhomogeneous during expansion (Fig. 5.146), but the secondary wall secreted inside the ring material and continuous with its lower lip is much more even (Fig. 5.147). The cross wall forms in the phycoplast as normal.

Abnormal siting of the division ring is not uncommon in *B. hiloensis,* particularly following abortive attempts at hair cell formation (Pickett-Heaps, 1973d). Apparently in such cells the disposition of the wall weakening is abnormal and incorrect, and as a result, cell expansion cannot be achieved properly. Many such cells then seem to secrete a secondary wall over the malformed,

Figures 5.91–5.94
DIVISION IN GERMLINGS OF ANOTHER SPECIES OF OEDOGONIUM

5.91.   The division ring here is continuous over the end wall (as in Fig. 5.27). Furthermore, the germling of another very small species of *Oedogonium* has attached itself to this cell, invoking a wound response (cf. Fig. 5.25) consisting of an extension of the ring material under the holdfast (arrow). × 1,100.

5.92.   *Both* cells of this germling have rings (arrows) and are about to divide. × 400.

5.93.   In this germling, the upper cell has its ring spread over the end wall (as in Fig. 5.91), and the germling is surmounted by a cap derived from the basal cell wall (as in Figs. 5.85–5.89), but which had become detached from the wall of the new daughter cell. × 400.

5.94.   Continuing cell divisions in the germling. Both upper cells have rings (arrows). × 300.

SOURCE. Author's micrographs, published in *Protoplasma 74,* 195 (1972).

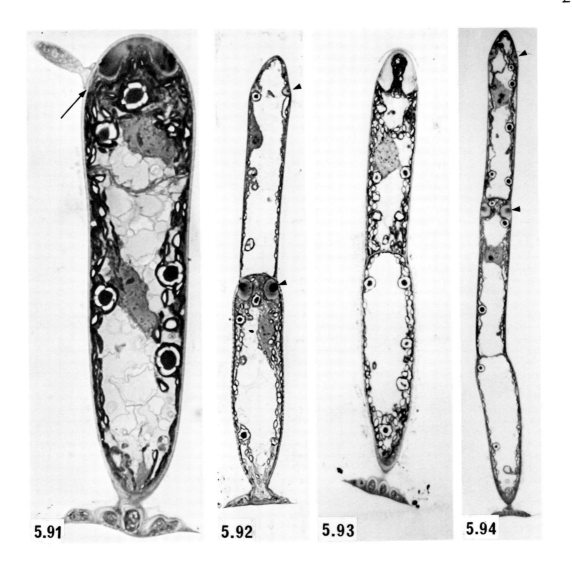

5.91 **5.92** **5.93** **5.94**

unexpanded division rings, and they subsequently die. An example of an abnormal division ring is shown in Figure 5.148.

5.2c. Hair Cell Formation

The extremely long (Fig. 5.133), colorless hair cells, entirely lacking a chloroplast, are a most distinctive feature of *Bulbochaete.* Fraser and Gunning (1973) describe these unusual cells, noting a lack of a nucleolus (an organelle I have seen in several hairs; Fig. 5.160) and a complement of active Golgi bodies and massed endoplasmic reticulum. This appearance suggests a secretory function of some sort, but the significance of these cells, which never undergo division but soon senesce and die, is not at all obvious. Numerous longitudinally oriented microtubules line the shaft of the hair, mostly near the wall, and others near the basal nucleus are suggested as having a role "anchoring" it (Fraser and Gunning, 1973), although it is not clear to me where else the nucleus could really move.

Formation of the hair cell (Pickett-Heaps, 1974e) takes place following a highly modified form of cell division, thus offering a fascinating comparison with normal cell division. The physical nature of the wall is vital in understanding how and why the hair forms. Following normal cell division, a certain percentage of *apical* daughter cells secretes a secondary wall inside the stretched division ring that is different from that previously described for both *Bulbochaete* and *Oedogonium. No* apical circumferential weakening appears in it. Instead, it is markedly thinned out on one side at the apical end of the cell, and in the middle of the bulging wall thus created, a wall discontinuity is now apparent, extending only part way around the cell (small arrow in Fig. 5.150). This different form of the secondary wall is unmistakably present immediately after its secretion; there is no evidence that the wall in such a cell has been altered later in the cell cycle. If it divides again, any cell with this different secondary wall necessarily has to give rise to a hair cell and then subsequently form and elongate a branch under that hair.

The course of hair cell formation in live cells in *B. hiloensis* is shown in Figure 5.149. The bulge on the wall becomes filled with refractile cytoplasm, devoid of chloroplast (Fig. 5.149*B*). I have not been able to observe mitosis in live

Figures 5.95, 5.96
SEXUAL REPRODUCTION IN MACRANDROUS SPECIES OF OEDOGONIUM

5.95.   Sexually induced *O. foveolatum,* a homothallic species, showing the large oogonium just below a cell that has formed two antheridial cells, each of which contains two maturing sperm. The oogonium is almost mature, having rounded off inside its wall (cf. Fig. 5.125). × 800 approx.

5.96.   Series of antheridial cells in the heterothallic *O. cardiacum.* × 720.

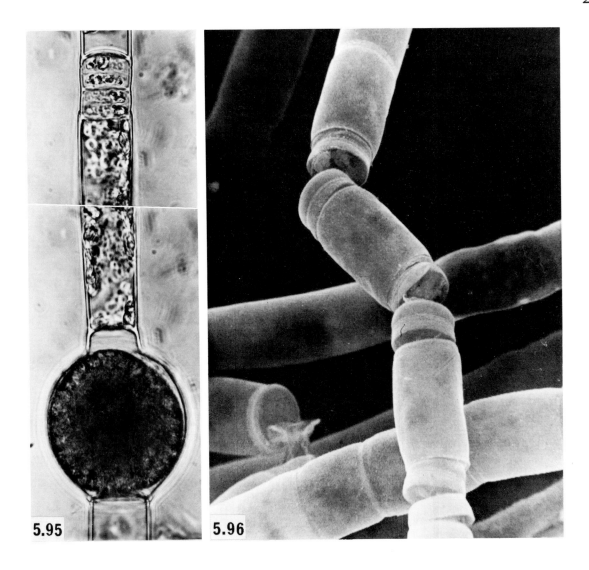

5.95

5.96

cells, but the wall soon afterward splits at the middle of the bulge (Fig. 5.149*A*, lower cell; compare with the normal wall rupture in the upper cell). Then the colorless and bulbous hair cell expands through the split in the wall and quite rapidly (0.25µm/minute; Fraser and Gunning, 1973) pushes out the long, narrow shaft of the hair itself (Fig. 5.149*C–E*).

Ultrastructurally (Pickett-Heaps, 1974e), the first sign of imminent hair cell formation is the secretion of initially scattered lumps of wall material at the bulge (Figs. 5.150, 5.151) and particularly at the wall weakening (Fig. 5.151). Soon, a thin, fairly even pad of this primary wall layer has been secreted (Fig. 5.152), of limited extent but always covering the thinned-out wall region. As might be expected, it is again probably derived from the nearby population of Golgi bodies, which become hypertrophied only during this limited phase of secretion (Fig. 5.151). Next, an initially thin, even wall layer is secreted inside the pad of primary wall (Fig. 5.153). I have termed this the "secondary" wall, although it undergoes some expansion during later stages of hair extension. The Golgi bodies do not apparently play any role in secretion of this layer, which thickens during division and which possesses a very important property. Its structure is *not* continuous with the secondary wall around the remainder of the cell (small arrows, Fig. 5.153). As the secondary wall layer thickens during and after mitosis and cytokinesis, the circular discontinuity that has arisen (Fig. 5.157) becomes the wall weakness that later invariably locates the future division ring under the hair (Figs. 5.142, 5.143) and later at the base of the branch (Fig. 5.140). Careful inspection of Figures 5.153–5.157, 5.159, and 5.160 will clearly establish this important sequence of events, vital to any understanding of the basic morphological difference in vegetative structure between *Oedogonium* and *Bulbochaete.*

As this secondary wall layer is being secreted, the cell becomes polarized prior to its highly asymmetrical division. The chloroplast, previously lining the entire wall including the bulge (Figs. 5.151, 5.152) now moves away as the nucleus migrates toward the bulge (Fig. 5.153). Mitosis follows (Fig. 5.154), and cytokinesis commences with the formation of a slightly curved phycoplast (Fig. 5.155), which

Figures 5.97–5.101
SEXUAL REPRODUCTION IN HETEROTHALLIC, NANNANDROUS SPECIES OF OEDOGONIUM

5.97. Low-power micrograph showing large numbers of androspores being powerfully attracted to several oogonial mother cells. × 40.

5.98. Oogonial mother cell in an India-ink suspension, showing the faint sheath surrounding it. Note how the cytoplasm of this cell is apically polarized (cf. Figs. 5.118–5.121). × 250.

5.99. Oogonial mother cell with some dwarf males attached, about to undergo division (see the prominent division ring). An adjacent androspore is probably about to attach to the mother cell. × 400.

5.100*A, B.* Two micrographs, taken approximately 60 seconds apart, showing sperm clustering about the region where the fertilization pore will form on the maturing oogonium. These sperm have just been released from the nearby dwarf males. × 380.

5.101. Oogonium and suffultory cell with dwarf male attached, in an India-ink suspension. The mucilaginous gel around the oogonium is obvious. × 160.

SOURCE. Drs. E. Rawitscher-Kunkel and L. Machlis, published in *Am. J. Bot. 49,* 177 (1962).

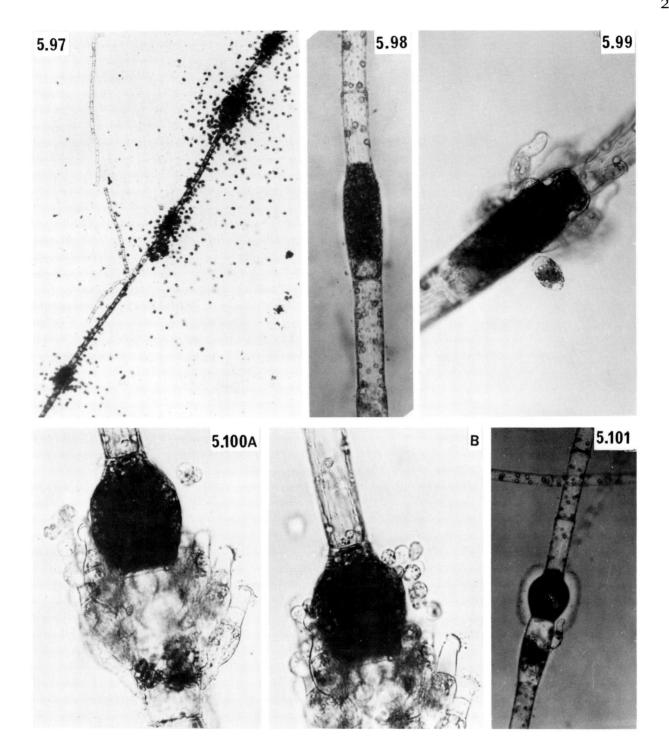

thereafter totally excludes the chloroplast from the future hair (Fig. 5.156). The parental wall then ruptures (Fig. 5.149A) and the hair cell begins protruding through the rent formed (Figs. 5.156, 5.158). The phycoplast moves in the direction of the hair cell's expansion through the slit until it is a little distance past the circular discontinuity in the secondary wall layer, whereupon a cross wall is formed in it (Fig. 5.159). The new cross wall is continuous with the secondary wall, and both later thicken considerably (Fig. 5.160). The hair cell's cytoplasm becomes noticeably paler than that of its parental cell, and many of its organelles (particularly the Golgi bodies) begin to look quite different (Pickett-Heaps, 1974e). Two flaps of the thin parental wall originally enclosing the bulge thereafter enfold the base of the hair or branch (Figs. 5.149A, 5.156, 5.158–5.161).

Further elongation of the hair shaft occurs by tip growth (Fraser and Gunning, 1973). Numerous longitudinally oriented microtubules line the wall of the hair. Many of these seem to embed in an amorphous region just behind the tip (Fig. 161B), whose organization resembles that of the spitzenkörper found in the tips of certain fungal hyphae (Grove and Bracker, 1970). Terminal cells often bear two hairs (Fig. 5.161A). Apparently the older hair begins to act as an apical cell, and then its parental cell undergoes formation of a secondary hair as outlined above. The differences between normal cell division and hair cell formation are summarized in Figure 5.162.

Hair cell formation in *Bulbochaete* may broaden our understanding of cell division in the *Oedogionales* generally and our appreciation of its subtleties (Pickett-Heaps, 1974e). It is clear, of course, that the wall weakening is of key importance in the division process, and that its type and positioning in *Bulbochaete* profoundly affect the type and orientation of subsequent cell division. Hairs can form only in apical cells, as the required form of wall weakening has to be created anew in the newly secreted secondary wall. In contrast, basal cells invariably inherit a circular discontinuity at the tiered cap, a legacy from the morphology of the division ring (i.e., the weakening in its basal lip; Figs. 5.138, 5.147). In the process of hair cell formation, an apical daughter cell becomes transformed into a *basal* daughter cell. It has two systems of caps demonstrating

Figure 5.102
COURSE OF SEXUAL REPRODUCTION IN NANNANDROUS SPECIES OF OEDOGONIUM

Diagrammatic representation. An oogonial mother cell remains inactive until a male androspore *(as)*, resembling a small zoospore, attaches to it. This attachment and subsequent germination of the epiphytic androspore *(A)* stimulates the oogonial mother cell to divide into the oogonium and suffultory cell *(B)*. This division is presumed to be asymmetrical (see Figs. 5.118–5.123). The oogonium and dwarf male filament *(dm)* differentiate concurrently *(C)*. The mature oogonium contracts slightly from its wall, and the fertilization pore is formed as the oogonium secretes an envelope of mucilage, into which the sperm are released from the dwarf males. Fertilization gives rise to a diploid zygote *(D)*.

SOURCE. Redrawn after Rawitscher-Kunkel and Machlis (1962).

Figure 5.103
COURSE OF SPERMIOGENESIS IN MACRANDROUS SPECIES OF OEDOGONIUM

Diagrammatic representation. An antheridial mother cell secretes a small division ring and undergoes highly asymmetrical division to cut off a small discoidal antheridial cell. This sequence of events is repeated several times *(A)*. The antheridial cells also undergo one (or sometimes two) more divisions *(A, B)* and then differentiate sperm, two or four per original antheridial cell. These sperm are later released in somewhat the same fashion as zoospores.

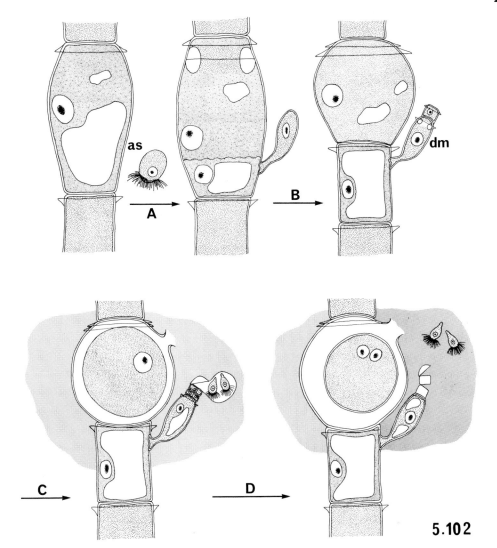

as

dm

C   D

5.102

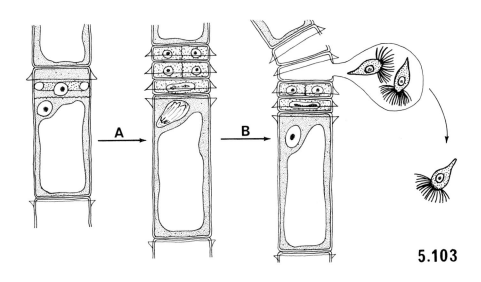

A   B

5.103

this genealogy: single downward-facing cap(s) around the axis of the original filament, and later (if it forms a branch) an increasingly large tiered cap at the base of the branch (Fig. 5.140). If we want to push the homology further, we can regard the whole secondary wall layer initially lining the primary wall pad inside the bulge in the parental wall (Fig. 5.153) as equivalent to the *basal* lip of a normal division ring; like the basal lip, it irreversibly locates the future division ring and thereby gives rise to tiered caps. The apical wall type is obviously less "stable" than the basal type—hardly surprising as the latter does not alter significantly after division, being maintained intact to enclose the basal daughter cell. The Golgi bodies seem to perform homologous functions during normal division and hair cell formation, secreting the material of the ring and its homologue, the pad of primary wall. They are apparently not involved in either case in secondary wall formation, which commences a little earlier in the case of hair cell formation unless we regard this latter secondary wall as equivalent to the secondary wall of the basal lip of the division ring.

How could such a modified form of division have arisen? Hair cell formation could also be considered in terms of a wound response mechanism (Section 5.1b) in which the resultant new cell is different from that arising from a more normal division. It would, of course, help our speculation considerably if we could assign some function to these cells! There is, however, one point I consider most important. The type of wall discontinuity needed for hair cell formation (a partial one, situated in a thinned region of apical wall, some distance below the caps), is also, I believe, found in one other type of cell, the oogonial mother cell of *Bulbochaete* (and *Oedogonium*), where it leads to the formation of the fertilization pore. As already mentioned, there is a distinct similarity between the morphology of the wall in oogonial mother cells and hair cells and in the material secreted under the rupture site, and this fertilization pore in *Oedogonium* and *Bulbochaete* always forms in the apical cell wall following ring expansion. The significance of these considerations in explaining the morphological differences between *Bulbochaete* and *Oedogonium* will be emphasized at the end of this chapter.

Figures 5.104–5.106
SPERMATOGENESIS IN OEDOGONIUM CARDIACUM

5.104. Initial divisions in antheridial mother cells. The premitotic and anaphase nuclei are highly polarized, and the division rings are tiny. × 900.

5.105A. Both antheridial mother cells here have already cut off one discoidal antheridial cell. Both these antheridial cells are dividing a second time. The upper mother cell is also dividing and displays considerable distortion of its late anaphase spindle (see Fig. 5.105B). × 900.

5.105B. Same cell as that in Figure 5.105A, showing the severely bent spindle and the typical small division ring (arrow). × 5,200.

5.106. Both antheridial mother cells here are undergoing their third asymmetrical division (the small arrows mark the division rings). The upper mother cell has formed its phycoplast (large arrow). Two of the antheridial cells have divided and differentiated sperm, and the other two are at prophase of their division. × 900.

SOURCE. Drs. R. A. Coss and J. D. Pickett-Heaps, published in *Protoplasma 78*, 21 (1973).

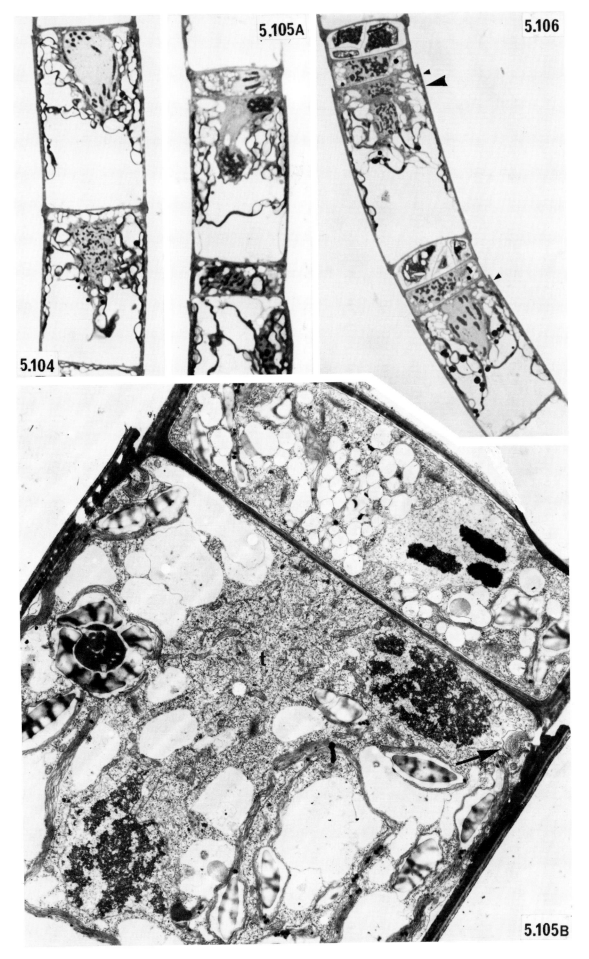

5.104

5.105A

5.106

5.105B

5.2d. The Zoospore

Just as in *Oedogonium,* zoospores are frequently differentiated from vegetative cells, usually following changes in the external medium (e.g., following subculture). The zoospore has a prominent reddish eyespot and is positively phototactic (Cook, 1962), unlike the spermatozoids.

The course of zoosporogenesis—the structure and release of the mature zoospore and its germination—to the best of my knowledge closely resembles what has already been described for *Oedogonium* (Section 5.1c) and can be dealt with briefly. One difference between these two genera lies in the staining properties of the hyaline layer. The zoospore of *Bulbochaete* does not display the marked basal "contraction" of the zoospore of *Oedogonium,* and basal mucilage is not secreted as it is in *Oedogonium.* Instead, the differentiating zoospore is enclosed in a bilayered sheath, markedly thickened at the future wall rupture site (Figs. 5.163, 5.168; Retallack and Butler, 1970b). I interpret the outer dense layer of this sheath to be equivalent to the basal mucilage of *Oedogonium,* which it closely resembles in appearance, and the inner, more finely fibrous layer to be equivalent to the hyaline layer. Retallack and Butler (1970b) say that this vesicle may serve to isolate the zoospore from the rest of the filament, allowing it to develop its own polarity; Retallack and von Maltzahn (1968) also suggest that severance of intercellular connections between successive cells of filaments in *Oedogonium* "is considered to be the primary event during induction of zoosporogenesis and subsequent events are considered to result from it." In both these genera, my own results clearly show that the plasmodesmata remain intact throughout most of zoosporogenesis, and I have no reason to believe that such connections are nonfunctional.

The zoospore of *Bulbochaete* (Retallack and Butler, 1972) seems to have a flagellar apparatus similar to that of *Oedogonium.* It is released, as in *Oedogonium,* by rupture of the enclosing wall, either at the circular weakening that also locates the division ring or else at the partial discontinuity characteristic of those certain apical daughter cells which would otherwise give rise to

Figures 5.107–5.109
CELL DIVISION PRECEDING SPERMATOGENESIS IN OEDOGONIUM CARDIACUM

5.107. The antheridial cell is in metaphase. Note the absence now of a true division ring and the typical conspicuous kinetochore (arrow). The antheridial mother cell has formed its phycoplast *(ph)* so as to cut off the small antheridial cell. × 4,700.

5.108. This micrograph shows the rather unusual differentiation of four (rather than two) sperm in one of the antheridial cells. The next cell has formed its phycoplast following division. Expansion of the division ring (arrow; *cf.* Fig. 5.107) is occurring in the antheridial mother cell. × 4,800.

5.109*A.* Apparent *cleavage* (rather than cell plate formation) during cytokinesis in an antheridial cell. The furrow seems to have grown from the side of the cell (arrow). See Figure 5.109*B.* × 10,000.

SOURCE. Drs. R. A. Coss and J. D. Pickett-Heaps. Figures 5.107, 5.108 published in *Protoplasma 78,* 21 (1973). Figure 5.109*A* published in *Cytobios 5,* 59 (1972).

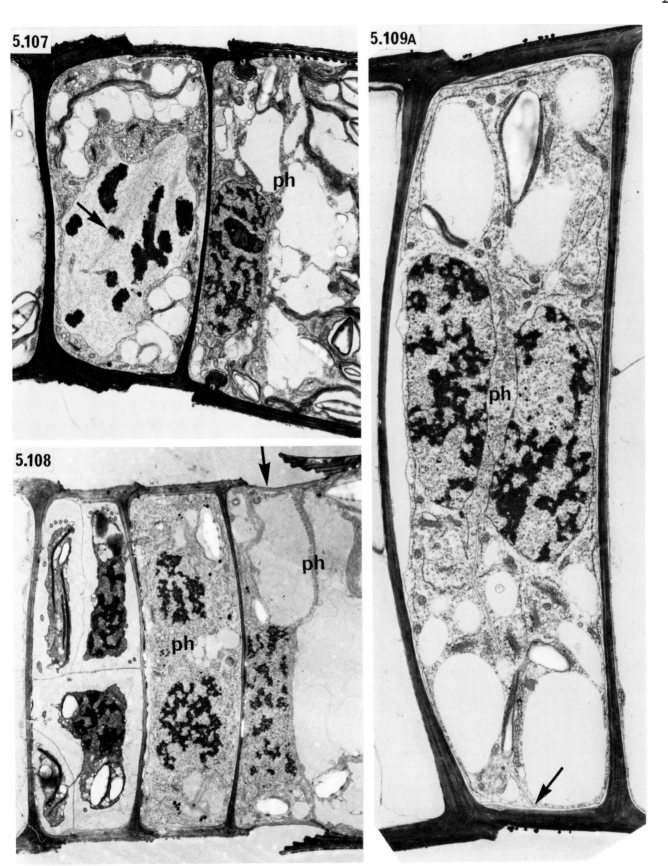

hairs. An example of the latter is shown in Figure 5.163; the thickening of the hyaline layer at the thin, bulging wall region indicates clearly the site where the wall would have split for release of this zoospore. Figure 5.164 shows another equivalent filament after zoospore escape; the pore is in precisely the analogous position to that of the hair in the next cell along the filament. Clearly, the terminal cells in Figures 5.163 and 5.164 would have formed a second hair (as in Fig. 5.161*A*) had they proceeded with cell division instead of zoosporogenesis.

Cook (1962) reports that the first division in a germling typically results in the creation of a hair cell, but in my cultures, this sequence of events has been relatively unusual.

5.2e. Sexual Reproduction

Bulbochaete hiloensis is homothallic and nannandrous. Sexual reproduction does not appear to be as tightly controlled as it is in the species of *Oedogonium* described by Rawitscher-Kunkel and Machlis (1962). In particular, development of the oogonium does not appear to be dependent upon the prior attachment of dwarf male filaments to the oogonial mother cell. Cook (1962) has described sexual reproduction in *B. hiloensis* in considerable detail, noting the occasional tendency for dwarf male filaments to revert to the vegetative condition, giving rise to vegetative cells after having formed several antheridial cells. Retallack and Butler (1973) have more recently illustrated ultrastructurally some aspects of sexual reproduction in this species. A filament from a sexually active culture is shown in Figures 5.133 and 5.165.

Just as in nannandrous species of *Oedogonium*, the dwarf male is formed from an androspore, much like the zoospore although a little smaller (Fig. 5.168). A live androspore still within its hyaline vesicle is seen in Figure 5.166; it is obviously far larger than the nearby spermatozoid. The androspore germinates as normal (Fig. 5.169), but it then undergoes a series of highly asymmetrical cell divisions to give rise to the dwarf male. These divisions are all strongly apically polarized (Figs. 5.170, 5.174). The division ring is small, and subsequent ring expansion (Figs. 5.171, 5.172) and secondary wall secretion (Fig. 5.173) give rise to a small, discoidal,

Figures 5.109*B*–5.112
SPERMATOGENESIS IN OEDOGONIUM CARDIACUM

5.109*B*. Detail of Figure 5.109*A*, showing the apparent edge (arrow) of the cleavage furrow *(cf)* growing through the phycoplast *(ph)*. × 63,000 approx.

5.110. Section through the two rows of recently formed centrioles, with one rootlet template visible (arrow). × 32,000.

5.111. Dome of differentiating sperm. The basal bodies *(bb)* have already extruded their flagella, and the rootlet microtubules *(t)* have also grown from the rootlet templates around the cell's periphery. × 31,000.

5.112. High-voltage (1,000 kV) electron micrograph of a thick (0.5μ) section of a sperm, showing the numerous cytoskeletal microtubules which run the length of the cell and up into its projecting tail (cf. Figs. 5.113–5.115). × 20,000 approx.

SOURCE. Drs. R. A. Coss and J. D. Pickett-Heaps. Figure 5.109*B* published in *Cytobios 5*, 59 (1972). Figures 5.110–5.112 published in *Protoplasma 81*, 297 (1974).

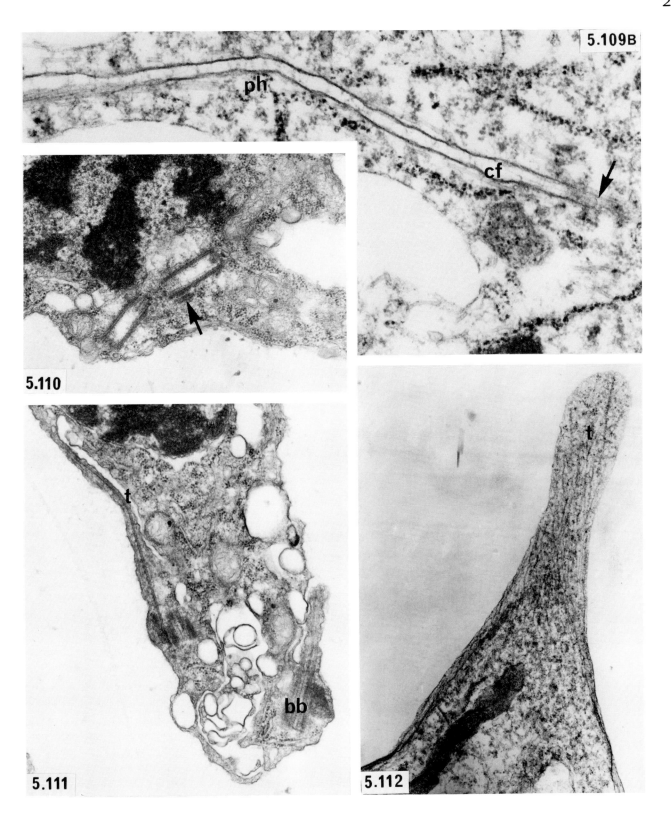

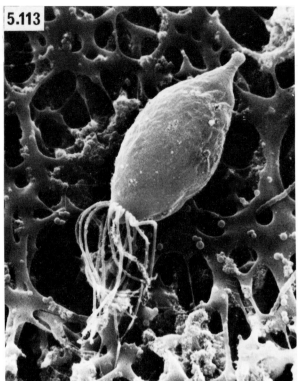

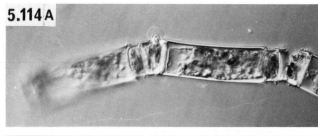

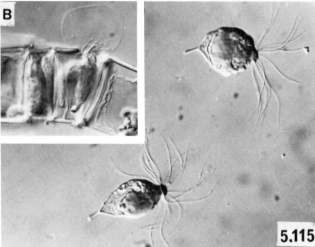

Figures 5.113–5.115
SPERM OF OEDOGONIUM CARDIACUM

5.113. This scanning electron micrograph can be compared with Figure 5.67 of the much larger zoospore. × 2,400.

5.114*A, B.* Live sperm emerging from an antheridial cell. *A,* × 480; *B,* × 950.

5.115. Two sperm fixed briefly before being photographed. × 1,100.

SOURCE. Drs. R. A. Coss and J. D. Pickett-Heaps, published in *Protoplasma 81,* 297 (1974).

and almost colorless antheridial cell, again in a manner similar to what has already been described for macrandrous species of *Oedogonium* (Section 5.1d). The basal cell of the dwarf male is, therefore, analogous to the antheridial mother cell of macrandrous species of *Oedogonium.* Successive divisions of the basal cell then create the typically segmented male filament (Figs. 5.165, 5.167, 5.175). Each antheridial cell divides once more, as might be expected. No ring is laid down for this secondary division, although

Figures 5.116, 5.117
FERTILIZATION IN OEDOGONIUM CARDIACUM

5.116. Oogonium of *O. cardiacum* with fertilization pore. Note caps at the apical end of the oogonium. × 1,700.

5.117. Fertilization in *O. cardiacum.* The sperm is entering through the fertilization pore, and the cytoplasm of the mature oogonium is typically spherical and slightly shrunken in the wall. × 789, approx.

SOURCE. Figure 5.117: Dr. R. A. Coss, unpublished micrograph.

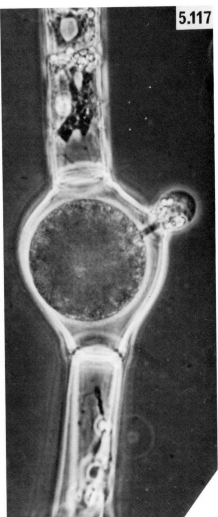

some fibrous material is secreted at the future wall rupture site (Fig. 5.174). The two sperm mother cells soon differentiate into spermatozoids. In my experience, many pairs of mature sperm apparently accumulate (Fig. 5.175) before they are released. Centrioles appear de novo in the sperm mother cells (Fig. 5.176), whose active Golgi apparently secrete the fibrous layer enclosing each cell (Fig. 5.177). A small ring of flagella is extruded from the dome, which contains a collection of small vesicles, mitochondria, and the complex flagellar apparatus characteristic of the Oedogoniales.

In my preparations of material fixed for electron microscopy, the dwarf male soon became segmented as the mother cells differentiated. The wall split apart (Fig. 5.175), and the integrity of the filament was maintained by the fibrous vesicle enclosing the cell. The sperm are often briefly enclosed by this vesicle upon release (Fig. 5.167), but it soon swells and ruptures.

OOGENESIS

The course of oogenesis is different in *B. hiloensis* and *O. cardiacum* (Section 5.1d). The difference is most clearly indicated by close inspection of the scanning electron micrographs reproduced in Figures 5.179–5.181. The mature oogonium and fully differentiated oospore are large and quite elongated, and contain a matching set of single caps around them; they appear to be some type of double cell. Equivalent caps are not found in the oogonia of *Oedogonium;* they arise in *Bulbochaete* because the oogonial mother cell undergoes *two* asymmetrical divisions before it differentiates into an oogonium. The events described in detail below are outlined in Figure 5.182.

The oogonial mother cell commences differentiation by undergoing an asymmetrical division in which the spindle is strongly basally polarized (*inset,* Fig. 5.183). Interestingly and perhaps significantly, if this asymmetrical division takes place under a hair cell, the spindle is oriented toward the wall ring (*inset,* Fig. 5.183). In contrast, the spindle of a vegetative cell in this situation is aligned along the spindle axis (Fig. 5.140) until the phycoplast is formed (Figs. 5.142, 5.143). After a highly asymmetrical cytokinesis, ring expansion gives rise to a basal, "primary" suffultory cell, almost devoid of cytoplasm (Figs.

Figures 5.118–5.123
FORMATION OF OOGONIUM IN OEDOGONIUM CARDIACUM

5.118. Oogonial mother cell before division, full of starch, lipid, etc. × 1,000 approx.

5.119. Basal migration of the premitotic nucleus. The ring is large. × 1,000 approx.

5.120. Anaphase at the basal end of the cell. This is the diploid strain, hence its larger size. × 1,000 approx.

5.121. The septum (*s*) has divided the oogonial mother cell asymmetrically. × 1,000 approx.

5.122. Wall rupture and partial expansion of the ring. × 1,000 approx.

5.123. The basal suffultory cell is now almost devoid of cytoplasm. The apical cell is differentiating into the oogonium. Note the mucilage secreted under the future pore site, where the wall is thinned out. × 1,000 approx.

SOURCE. Drs. R. A. Coss and J. D. Pickett-Heaps, published in *Protoplasma 78,* 21 (1973).

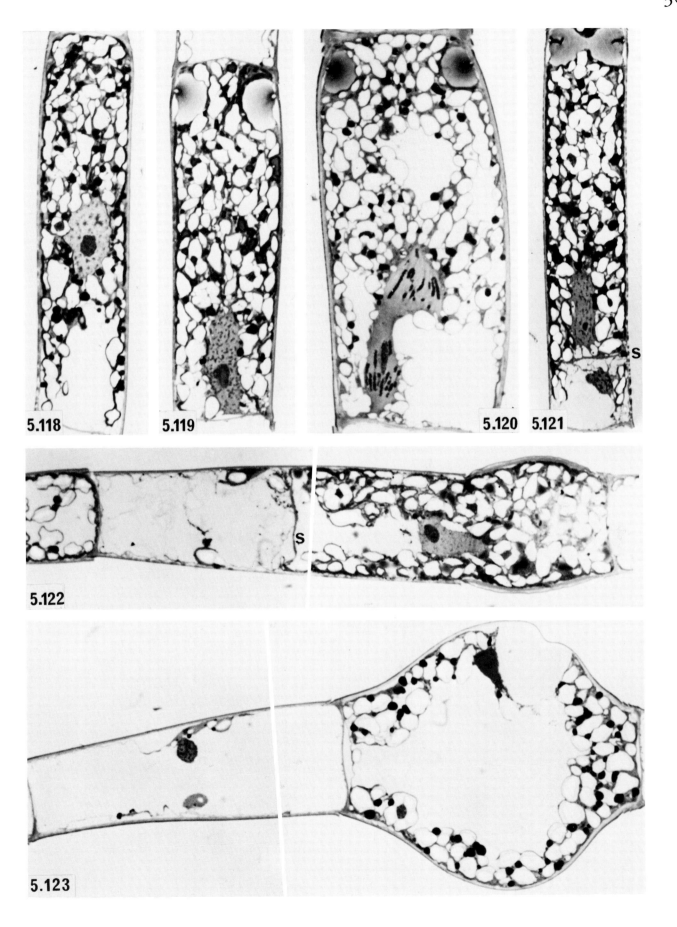

5.118

5.119

5.120

5.121

5.122

5.123

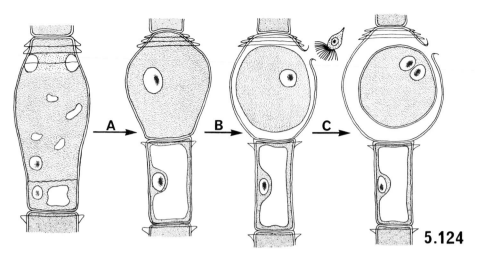

5.124

Figure 5.124
FORMATION OF OOGONIUM AND OOSPORE IN MACRANDROUS SPECIES OF OEDOGONIUM

Diagrammatic representation. The oogonial mother cell divides asymmetrically *(A)*, and subsequent cell expansion gives rise to a swollen apical oogonium and (usually) a basal suffultory cell that is almost devoid of cytoplasm, chloroplast, etc. After the oogonium has matured and formed its fertilization pore, its cytoplasm shrinks slightly prior to fertilization *(B)*. Later, the oospore secretes a thick, resistant wall *(C)*.

5.182*A*, 5.183) and an apical oogonial mother cell, which becomes quite rotund. So far, these events resemble what happens in *Oedogonium* (Figs. 5.118–5.123). Unlike *Oedogonium*, this mother cell in *Bulbochaete* undergoes a second highly asymmetrical division (Fig. 5.182*C*). Another ("secondary") suffultory cell is cut off. However, upon expansion of the division ring, the septum cutting off this secondary suffultory cell *does not move.* The second asymmetrical division, therefore, results in the creation of another, very small, flat suffultory cell and an oogonium which is considerably larger than the other cells in the filament (Figs. 5.165, 5.182*C*). The caps arising from the second division give the oogonium and the oospore of *Bulbochaete* the appearance shown in Figures 5.179–5.181.

Figure 5.125
OOGONIUM AFTER FERTILIZATION

Oogonium soon after fertilization, similar to an unfertilized egg. The sperm nucleus is not visible in this section. The large vacuolated cell contains much reserve material. The mucilaginous material previously blocking the fertilization pore (Fig. 5.128) has disappeared. A membrane-like layer remains (arrow), perforated by the sperm although it looks intact in this section. × 2,650.

SOURCE. Dr. L. R. Hoffman, published in *Contributions in Phycology.* Lawrence Kansas, Allen Press, 1971, p. 97.

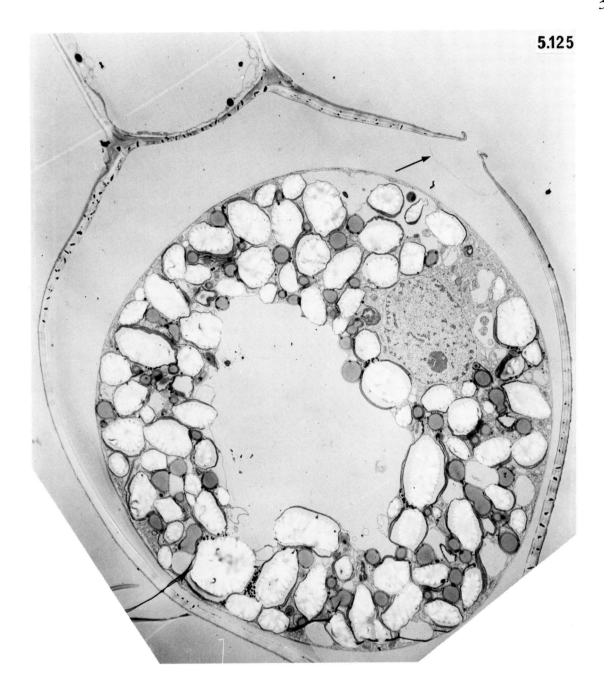

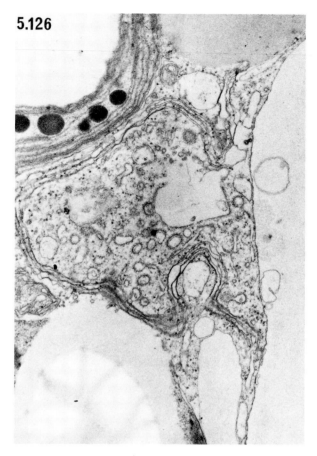

Figure 5.126
CONTRACTILE VACUOLE IN OOGONIUM

A contractile vacuole in the peripheral cytoplasm of a mature oogonium that has become shrunken inside its wall (as in Figs. 5.117, 5.125).

SOURCE. Dr. L. R. Hoffman, published in *J. Phycol.* 9, 62 (1973).

The second asymmetrical division is interesting for several reasons. The division ring does not form under the caps as might be expected; instead, when the mother cell's secondary wall is secreted after the first asymmetrical division, the wall weakness appears about a third of the way down from the apical end of the cell. The ring is later located at this site (Figs. 5.183, 5.184). The premitotic nucleus (Fig. 5.183), spindle, and phycoplast (Fig. 5.184) are all strongly basally polarized, as would be expected. Although the phycoplast does not move during ring expansion,

Figure 5.127
FERTILIZATION IN LIVE CELLS OF OEDOGONIUM CARDIACUM

Sequence taken from a cine film (16 fps). In the first four frames *(A–D),* dissolution of the exuded pore substance can be followed as the elongated sperm first makes contact with the oogonium. The next three *(E–G)* show entrance of the flagellar end of the sperm through the pore. At *G,* the sperm has contacted the oogonial cytoplasm, 57 seconds after the filming started. At *H,* the sperm has become momentarily quiet, prior to shedding its flagella, occurring at *J,* 60.25 seconds after filming commenced (the cast-off flagella are only visible in the projected film). During *K–M,* the sperm's tail rapidly collapses, and absorption of the sperm *(N–P)* is complete within another 27 seconds. All × 1,200.

SOURCE. Dr. L. R. Hoffman, published in *J. Phycol.* 9, 62 (1973).

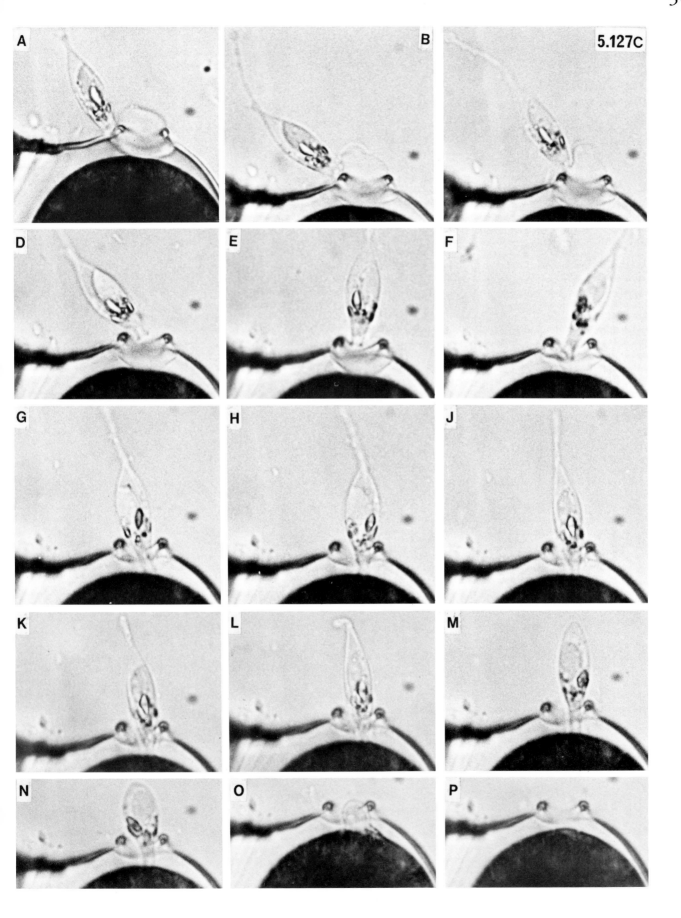

the new cross wall does not appear in it until after ring expansion is complete (Fig. 5.185). Expansion of the ring is uneven, since one side of the cell expands considerably more than the other (Fig. 5.185). Thus, the two caps arising from this division are not parallel (Figs. 5.181, 5.185). At the site undergoing greatest expansion, the new secondary wall layer secreted inside the ring material is markedly thinner than that elsewhere. Furthermore, in the apical region of this thinner wall is a dislocation that will give rise to the fertilization pore (Figs. 5.186, 5.187). It is emphasized that this wall morphology is present right from the inception of secondary wall secretion (cf. the comments in Section 5.ld). The oogonium is initially quite vacuolate, but it soon becomes dense, accumulating starch and lipid. Incipient pore formation is signaled by the secretion of a fibrous (polysaccharide?) material under the wall dislocation, presumably derived from the nearby active Golgi bodies (Figs. 5.186, 5.187). The nucleus migrates to this region of the cell, and the cytoplasm near the nucleus becomes filled with microtubules oriented toward the pore (Fig. 5.188). Then the wall ruptures, and the plug of material originally lining the wall bulges through the fertilization pore. Once the pore has been formed, the cytoplasm of the oogonium contracts somewhat from the wall (Cook, 1962), just as in *Oedogonium*. Fertilization of the mature oogonium has not, to my knowledge, been described in detail; presumably it resembles that described in Section 5.1d. Cook (1962) reports observing fertilization on one occasion.

The oospore then matures, forming a massive and highly complicated wall. First, a series of ridges of fibrous material is laid down inside the original oogonial wall (Fig. 5.189). Golgi bodies are active at this stage, but whether they contribute to the secretion of this material is unknown. This material apparently is instrumental in giving the oospore its characteristic ridged surface (Figs. 5.179, 5.181), since several different layers are next secreted inside this material, following the contours established by it. A section through a fairly mature cell wall is shown in Figure 5.191, and I need not emphasize its thickness or complexity. During this process of maturation, the cytoplasm of the oospore becomes dense (Fig. 5.190) and accumulates much lipid and starch. In live cultures, maturing oospores soon become

Figures 5.128–5.130
OOGONIA OF OEDOGONIUM CARDIACUM BEFORE AND AFTER FERTILIZATION

5.128.   The pore and material exuded through it in a mature oogonium prior to fertilization. × 6,000.

5.129.   Part of the disintegrating flagellar apparatus of a sperm, inside the cytoplasm of a fertilized oogonium. The numerous nearby mitochondria *(m)* are also derived from the sperm. × 40,000.

5.130.   The nucleus *(n)* of the sperm in the oogonial cytoplasm soon after plasmogamy. It appears quite different from the oogonial nucleus (see Fig. 5.128). One cast-off flagellum *(f)* belonging to the sperm cell can still be seen. × 5,000.

SOURCE. Dr. L. R. Hoffman, published in *J. Phycol. 9,* 62 (1973).

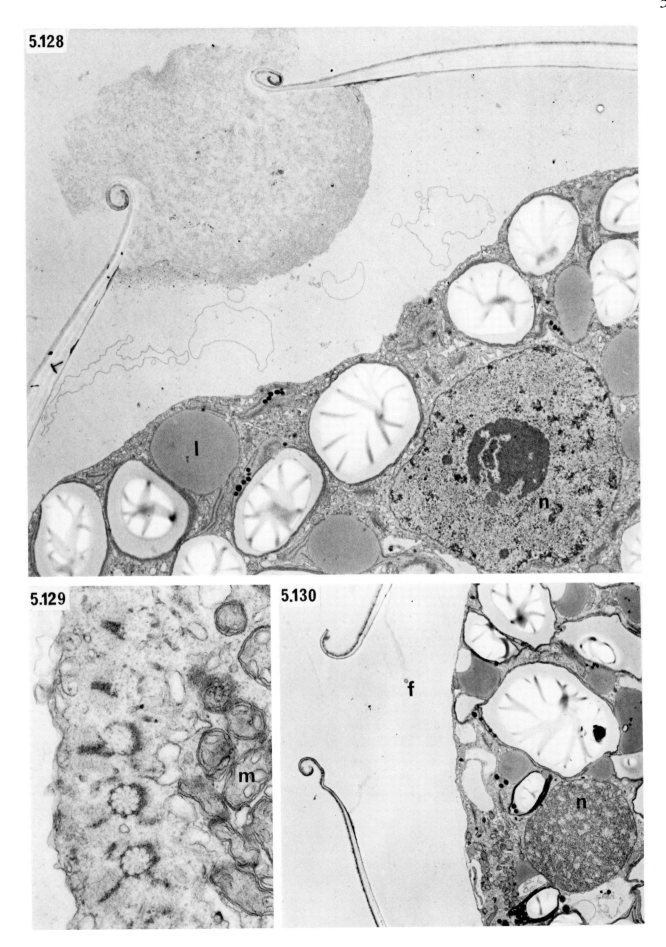

5.128

5.129

5.130

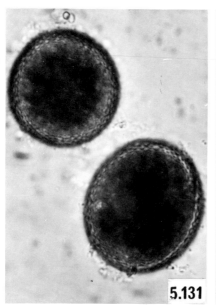
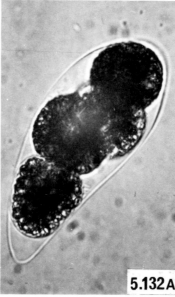
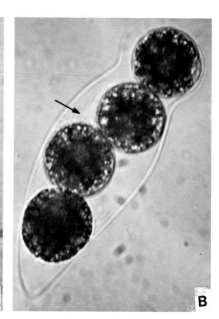

Figures 5.131, 5.132

OOSPORES OF OEDOGONIUM FOVEOLATUM

5.131. Two oospores of *O. foveolatum,* with their thick patterned wall. × 625.

5.132*A, B.* Germination of an oospore of *O. foveolatum.* A. Tetrad of zoospores within the vesicle released from the oospore. B. Release of the zoospores from this vesicle. Note flagella (arrow). × 625.

SOURCE. Dr. L. Hoffman, published in *Am. J. Bot. 52,* 173 (1965).

bright orange. Cook (1962) has described germination of oospores. They become greener, increase in size, and split their outer wall. The contents of the cell cleave and differentiate into four dense zoospores, each of which germinates to give rise to a new filament.

Cook also notes that following its first asymmetrical division (Fig. 5.182*A*), the oogonial mother cell apparently still possesses the potential to differentiate into an enlarged vegetative zoospore, but this reversion is apparently not possible after the second asymmetrical division. Several oogonia I encountered showed no sign of having formed the secondary suffultory cell, even though they had undergone expansion of the ring from the second division (e.g., the middle

Figure 5.133

BULBOCHAETE HILOENSIS

This organism consists of a highly branched filament, derived from a basal holdfast cell. Numerous characteristic hairs arise laterally from vegetative cells. These are colorless, and several are seen detached. This specimen is undergoing sexual reproduction. The empty cells have released androspores (or perhaps some zoospores), which have germinated near the larger oogonia, forming dwarf male filaments.

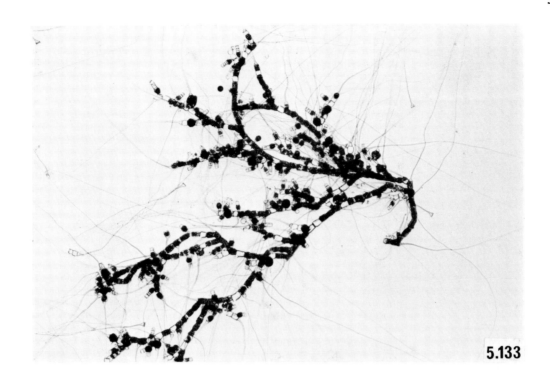

5.133

oogonium in Fig. 5.192). Neither have I found in my cultures any binucleate oospores, the likely situation following fertilization, which may indicate a tendency for the oogonia to have differentiated into parthenospores (at least in my cultures).

The sequence of events described above clearly indicates that pore formation is a direct consequence of the morphological alteration of the oogonial cell wall so as to split in the right region. The mechanism of pore formation in the Oedogoniales has already been discussed (Section 5.1d). The resemblance between the formation of this fertilization pore and the initial sequence of events involved in hair cell formation (particularly the secretion of the polysaccharide and the polarization of the nucleus) is unmistakable. I have obtained a fortunate sequence of cells in one filament which documents even more forcibly the ontogenic relation between hair cell formation and the oogonial pore. Figure 5.192 shows, first, an oogonial complex (i.e., an oogonium and two suffultory cells) forming a branch on the main filament axis. The next pair of cells is another oogonial complex, this time with all the cells in the axis of the original filament. The oogonium clearly reveals the site of its future fertilization pore in the thinned-out wall on one side of the filament (large arrowhead, Fig. 5.192). Next along the filament is a pair of vegetative daughter cells from a very recent division; in this pair, the apical cell has the modified wall morphology that would have led later to hair cell formation (large arrowhead, Fig. 5.192). The incipient wall rupture site is on the same side of the filament axis in both this future hair cell and in the oogonial cell, these in turn are on the same side as the branch represented by the most apically situated oogonial complex. It has already been emphasized (Sections 5.2a, 5.2c) how hair cells and consequently branches tend to arise on one side of a filament (Fig. 5.134; Pickett-Heaps, 1973d, 1974e). This sequence of cells unequivocally ties in the siting of the fertilization pore with the siting of the future hair cell. Thus, the wall rupture sites in these cells are almost certainly homologous structures. Also revealing (Pickett-Heaps, results in preparation) is the occasional tendency for new hair cells to arise through the fertilization pore; this phenomenon is puzzling and under investigation, but I have several exam-

Figures 5.134–5.137
VEGETATIVE CELLS OF BULBOCHAETE HILOENSIS

5.134. This scanning electron micrograph shows the hair cells *(h)* emerging through a lateral rupture of the parental cell wall. The hairs and also the branches that arise underneath them tend over short distances to form on one side of a filament. × 1,750.

5.135. An eight-tiered cap with its wall weakening, exactly analogous to that in *Oedogonium.* × 15,000.

5.136. A division ring just beginning to form at the wall weakening of a single-tiered cap. × 18,000.

5.137. A division ring *(dr)* forming below a downward-facing cap, at a wall weakening, obscured here by contamination on the section. The ring, like that in Figure 5.135, appears to be formed from the contents of vesicles derived from the nearby hypertrophied Golgi bodies *(g)*. × 18,000.

SOURCE. Author's micrographs, published in *J. Phycol. 9,* 408 (1973).

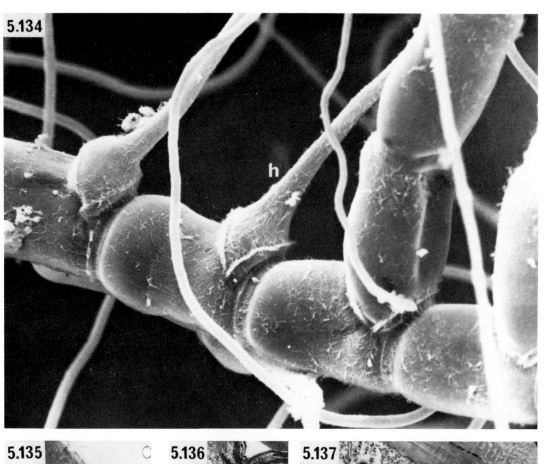

5.134   h

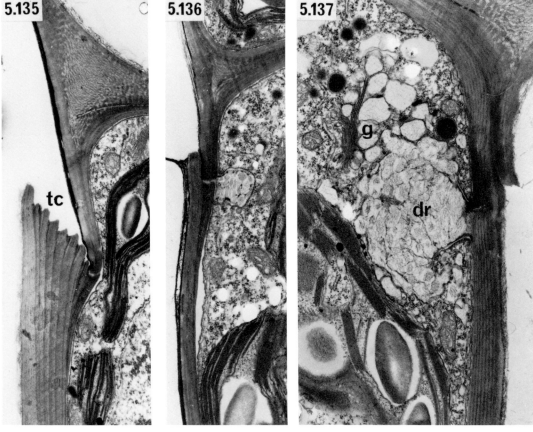

5.135   tc

5.136

5.137   g   dr

ples of hairs arising in this manner. Apparently, if the wrong type of wall weakening arises after the first asymmetrical division, the oogonial mother cell responds by forming a hair cell upon its second division.

As I see it, the more complex vegetative morphology of *Bulbochaete* compared with *Oedogonium* is fairly easily explicable. The differentiation of hair cells and the subsequent creation of branches in *Bulbochaete* appear to be direct consequences of the appearance in *vegetative* cells of a wall morphology that is primarily sexual in origin and involved in giving rise to the fertilization pore. *Oedogonium* is not branched because this wall feature appears only for sexual reproduction. Thus, the formation of hair cells in *Bulbochaete* and its characteristically branched morphology need be considered not merely as an isolated peculiarity, but as a result of highly evolved modifications to cytoplasmic and morphological features that are characteristic of all members of this extraordinary group of organisms.

Figures 5.138, 5.139
CELL DIVISION IN BULBOCHAETE HILOENSIS

5.138. Fully formed division ring. While similar in most respects to the ring in *Oedogonium* (note, for example, the weakening in the basal lip; arrow), it is more inhomogeneous in texture, and the basal lip is much larger. × 22,000.

5.139. The septum of a cell after telophase (e.g., Fig. 5.141), with the microtubules *(t)* of the phycoplast cutting through the peripheral chloroplast *(c)*. Note the banded tubule in the chloroplast (arrow). × 40,000.

SOURCE. Author's micrographs, published in *J. Phycol. 9*, 408 (1973).

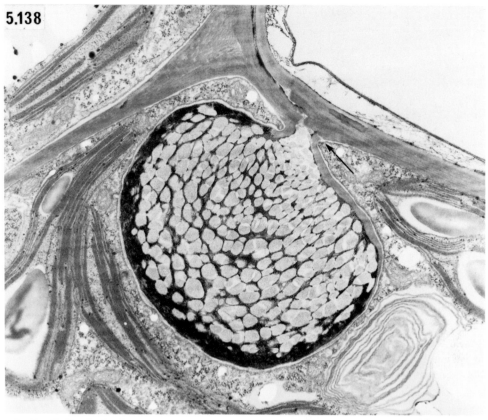

5.138

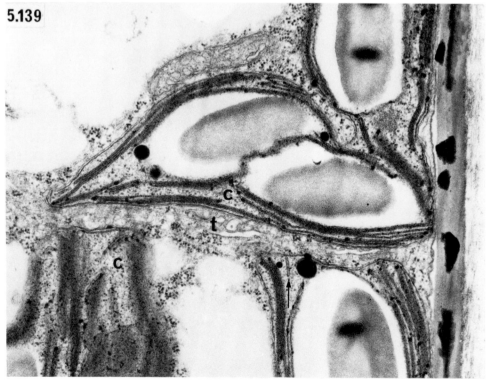

5.139

Figures 5.140, 5.141
CELL DIVISION IN BULBOCHAETE HILOENSIS

5.140. Typical metaphase figure. The spindle axis is parallel to the axis of the filament (arrow), even though the division ring *(dr)* is located at the wall weakening near the tiered cap *(tc)*. This ring is at the base of a branch *(br)*. Once a hair has been formed, any division thereafter in the cell subtending the hair must necessarily be in the direction of the hair, forming (Figs. 5.142, 5.143) and then elongating (as here) a branch. $\times$ 6,700.

5.141. Wall rupture and the first stage of expansion of the division ring *(dr)*, with division in this case along the axis of the filament (cf. Fig. 5.143). The phycoplast *(ph)* extends across the cell (arrows), through the chloroplast (see Fig. 5.139). $\times$ 8,600.

SOURCE. Author's micrographs, published in *J. Phycol. 9*, 408 (1973).

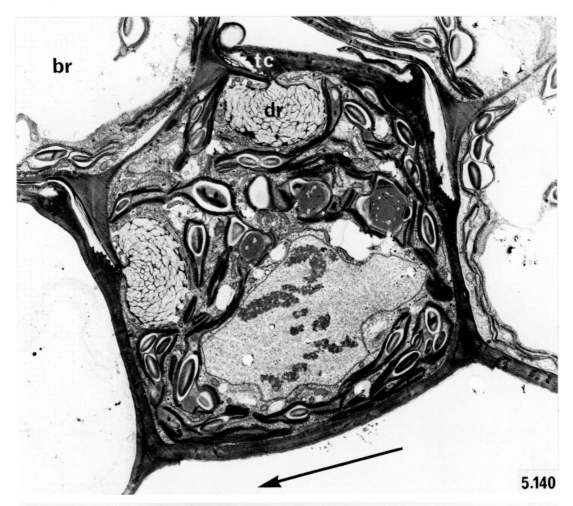

5.140

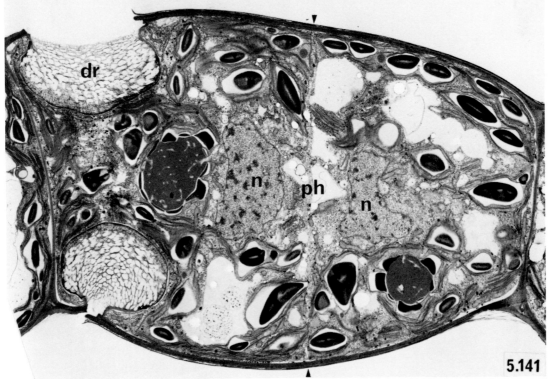

5.141

Figures 5.142, 5.143
CELL DIVISION FORMING A BRANCH IN BULBOCHAETE HILOENSIS

5.142.   The division ring *(dr)* is sited under the base of the newly formed hair cell *(h)*. The phycoplast *(ph)* between daughter nuclei is now tilted toward the hair cell, even though the spindle had previously oriented along the axis of the filament (Fig. 5.140). The next cell in the filament has a normal division ring. $\times$ 4,300.

5.143.   As for Figure 5.142, but the wall has now ruptured, and expansion of the division ring has commenced. Note the flaps of wall (arrows), visible also in Figure 5.142 (arrows), that remain as a result of hair cell formation. $\times$ 5,600.

SOURCE. Author's micrographs, published in *J. Phycol.* 9, 408 (1973).

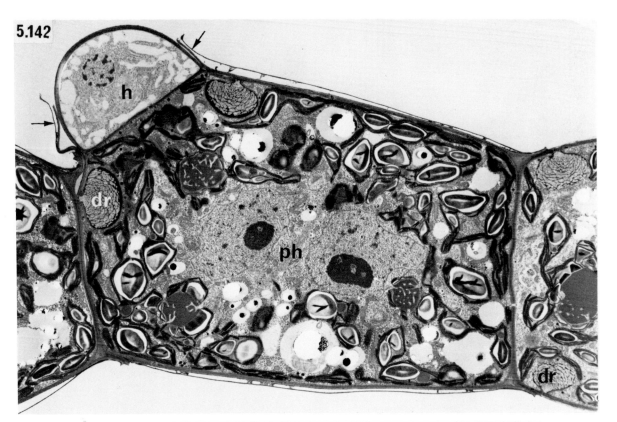

5.142

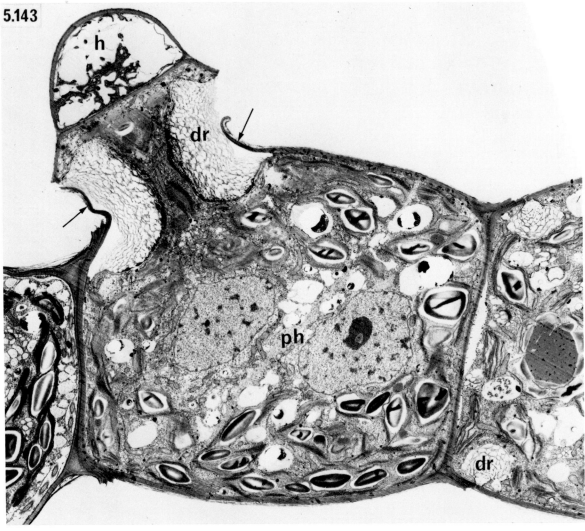

5.143

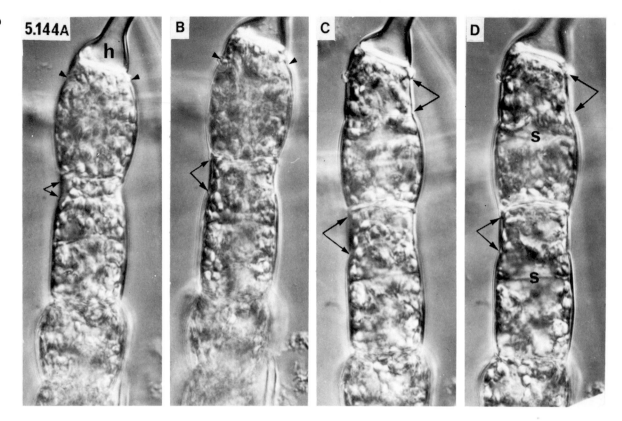

Figure 5.144
CELL EXPANSION FOLLOWING WALL RUPTURE IN LIVE CELLS OF BULBOCHAETE HILOENSIS.

In *A*, the wall has just ruptured in the lower cell, with the edges of the old wall indicated by the paired arrows. In the upper cell, arrowheads mark the wall ring at the base of the hair *(h)*. Further expansion of the lower cell has taken place in *B–D*. Meanwhile, the wall has also ruptured in the upper cell, which then also undergoes expansion (upper paired arrows, *C, D*). The septum is labeled *(s)*. × 1,800 approx.

Figures 5.145–5.147
CELL EXPANSION AND SECONDARY WALL FORMATION IN BULBOCHAETE HILOENSIS

5.145. Two cells at slightly different stages of expansion of their division ring. In the upper cell, the phycoplast *(pb)* has not yet quite reached its final position, indicated by the phycoplast in the lower cell, which is just initiating the new cross wall (see Fig. 5.147). × 2,500.

5.146. Appearance of an extended division ring *(dr)*. Note the weakening in the basal lip (arrow). The phycoplast was situated below the edge of this micrograph. × 4,300.

5.147. Secretion of the new secondary wall (large arrows) as an even layer inside the extended material of the ring and continuous with the basal lip of the ring whose weakening is still clear (small arrow). The new transverse cross wall *(xw)* has just started to form in the microtubules *(t)* of the phycoplast. × 9,000.

SOURCE. Author's micrograph's, published in *J. Phycol. 9*, 408 (1973).

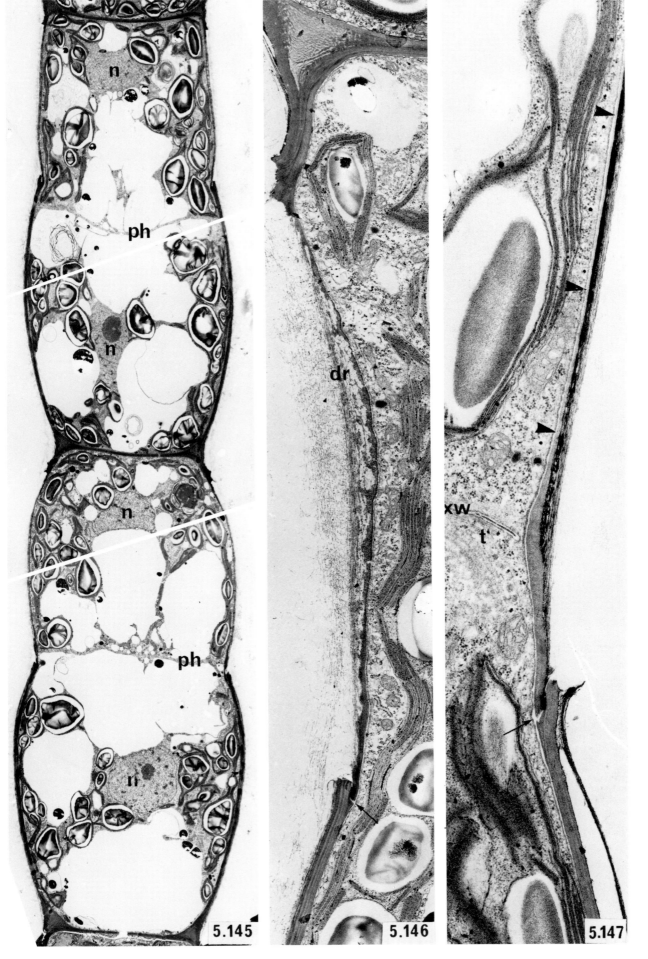

5.145

5.146

5.147

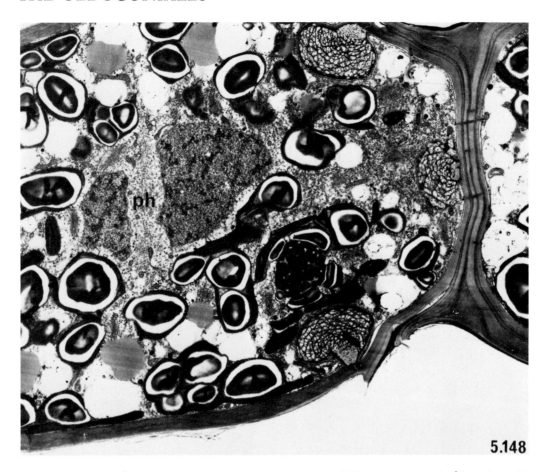

5.148

Figure 5.148
ABNORMAL DIVISION RING IN BULBOCHAETE HILOENSIS

Part of the ring is located in the middle of the end wall of the cell. × 11,000.

Figures 5.149, 5.150
HAIR CELL FORMATION IN BULBOCHAETE HILOENSIS

5.149*A–E.* Live cells. *A:* Wall rupture during normal division (paired arrows, upper cell) and hair formation (lower cell; arrows mark the edge of the rupture site). × 2,200. *B:* Premitotic or mitotic cell, with the bulge in the wall filled with refractile cytoplasm. × 2,000. *C:* Early emergence of the shaft of the hair after mitosis and cytokinesis. Flaps of parental wall are indicated by arrows. × 1,000. *D, E:* Successive stages in the growth of the hair. *D,* × 2,000; *E,* × 1,700.

5.150. Premitotic cell, about to form a hair. The apical region of wall is devoid of the weakening below the caps (large arrow) that would lead to normal division rings (*dr* in the next basal cell). Instead, the wall is thinned out on one side and contains a discontinuity (small arrow) extending only part way around the cell. The Golgi in this cell are hypertrophied, and the first trace of the primary wall pad is being secreted under the bulging region of the wall. × 10,000.

SOURCE. Author's micrographs, published in *J. Phycol. 10,* 148 (1974).

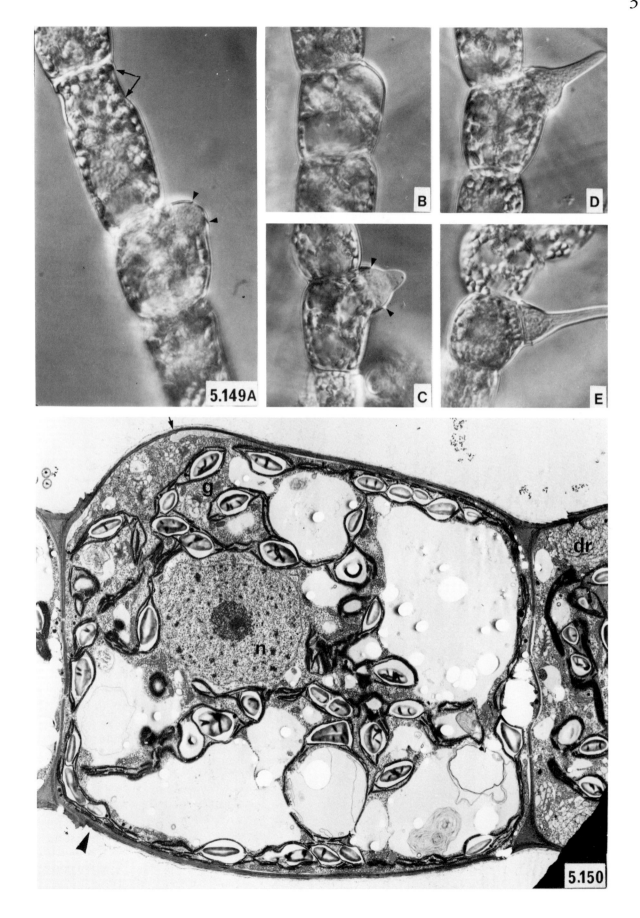

5.149A

B

C

D

E

g

n

dr

5.150

Figures 5.151–5.153
PREMITOTIC WALL SECRETION PRIOR TO HAIR FORMATION IN BULBOCHAETE HILOENSIS

5.151.   Very early stage (cf. Fig. 5.150) of primary wall secretion under the thinned, bulging region of parental wall. The nearby Golgi bodies *(g)* are hypertrophied, and the wall rupture site is indicated by the arrowhead. $\times$ 24,000.

5.152.   As for Figure 5.151, but the primary wall layer is almost complete. The chloroplast *(c)* is moving away from this region of wall as the nucleus *(n)* approaches. $\times$ 18,000.

5.153.   Secretion of the secondary wall layer inside the primary wall has been initiated. This thin layer is *not* continuous with the secondary wall of the parental cell. The discontinuity thus created between the two secondary walls (small arrows) will later (cf. Figs. 5.156, 5.157, 5.159, 5.160) position the division ring and orient cell expansion so as to initiate a branch under the newly formed hair (as in Fig. 5.143). The weakening in the parental wall is again denoted by the large arrow. The nucleus has almost taken up its premitotic position at the bulge, from which the chloroplast has now been excluded. $\times$ 20,000 approx.

SOURCE. Author's micrographs, published in *J. Phycol. 10.* 148 (1974).

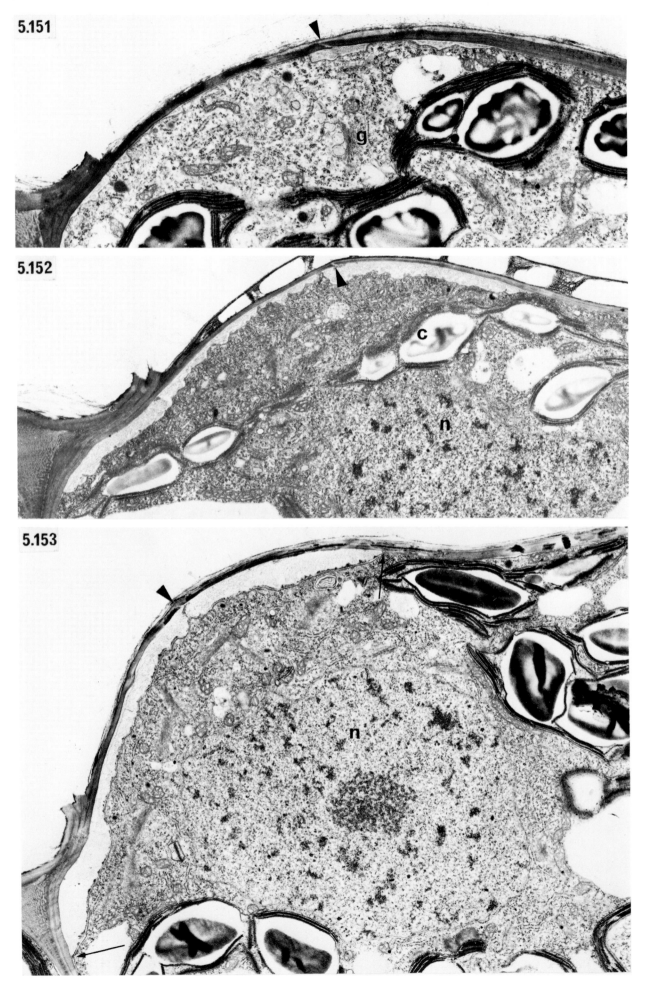

5.151

5.152

5.153

Figures 5.154, 5.155
ASYMMETRICAL MITOSIS PRECEDING HAIR FORMATION IN BULBOCHAETE HILOENSIS

5.154.   Late anaphase, with one pole of the spindle directed toward the bulge in the wall. × 10,000.

5.155.   Telophase. The daughter nuclei have come back together again, and the usual phycoplast (ph) is forming between them. This micrograph clearly demonstrates how the thick secondary wall of the parental cell (large arrow) is thinned out at the site of the future hair. The site of imminent rupture is indicated by the small arrow. × 18,000 approx.

SOURCE. Author's micrographs, published in *J. Phycol. 10,* 148 (1974).

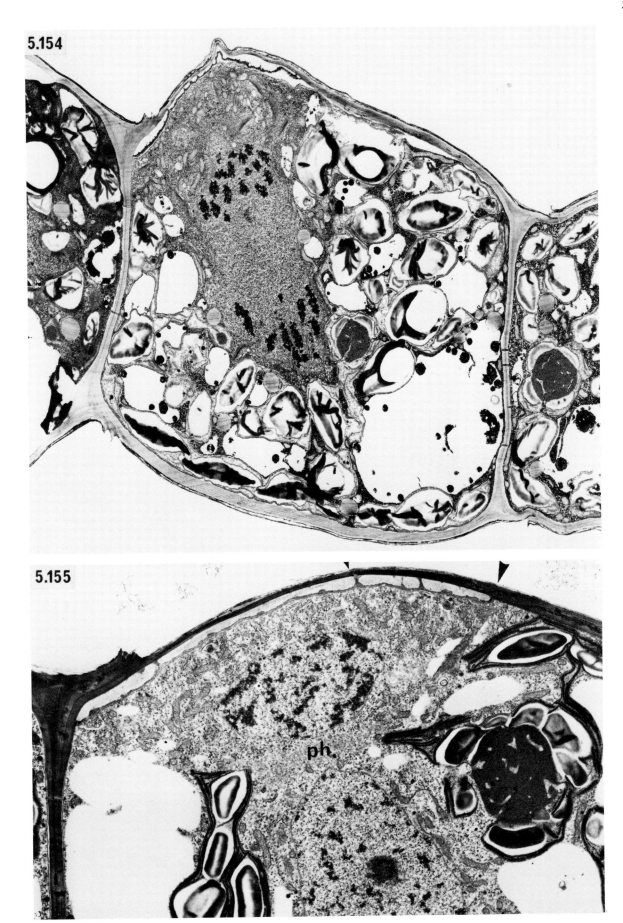

5.154

5.155

ph.

Figure 5.156
WALL RUPTURE PRECEDING EMERGENCE OF HAIR CELL IN BULBOCHAETE HILOENSIS

The flaps of ruptured parental wall (cf. Figs. 5.149A, C) enclosing the base of the emerging hair are particularly obvious. The phycoplast (ph) is moving slightly in the direction of the cell expansion, and it entirely prevents the chloroplast (arrows) from getting into the future hair. × 14,000.

SOURCE. Author's micrograph, published in *J. Phycol. 10,* 148 (1974).

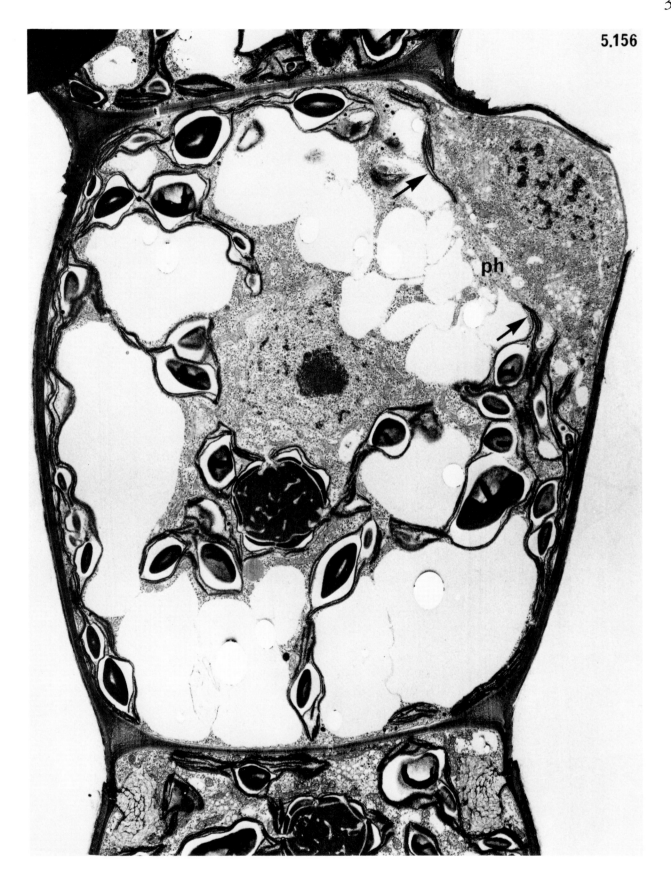

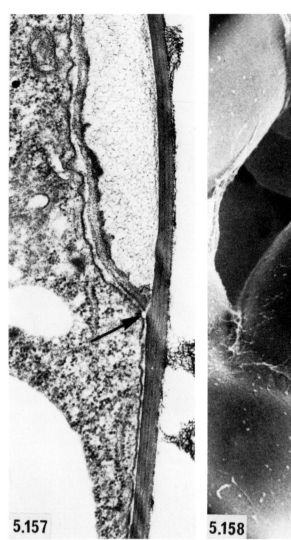

5.157

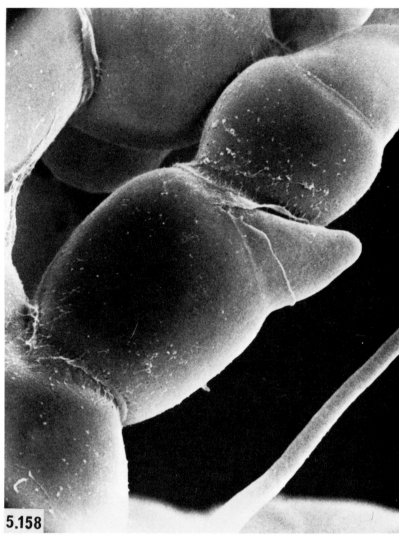

5.158

Figures 5.157, 5.158
EMERGENCE OF HAIR CELL IN BULBOCHAETE HILOENSIS

5.157. The secondary wall under the primary layer (cf. Fig. 5.153) is thickening, but the discontinuity (arrow) where it abuts the parental wall is increasingly pronounced. This weakening will locate future division rings in the parental cell. × 30,000 approx.

5.158. Young hair cell emerging through the ruptured parental wall. × 5,000.

SOURCE. Author's micrographs, published in *J. Phycol. 10,* 148 (1974).

Figures 5.159, 5.160
YOUNG HAIR CELL IN BULBOCHAETE HILOENSIS

5.159. Cytokinesis. The phycoplast *(pb)* has moved some distance past the circular discontinuity in the secondary wall layer (arrows), and the new cross wall (xw) is just forming in it. The Golgi in the hair cell are quite different in appearance from those in the parent. × 17,000 approx.

5.160. The secondary wall and the new cross wall have thickened around the base of the hair cell as the shaft of the hair (not in the plane of this section) grows. Note the circular discontinuity (arrows) which in future will locate the division ring at the base of the hair (Fig. 5.142) and later, the branch (Fig. 5.140). The cytoplasm of the hair is distinctly paler than that of the parental cell. × 17,000 approx.

SOURCE. Author's micrographs, published in *J. Phycol. 10,* 148 (1974).

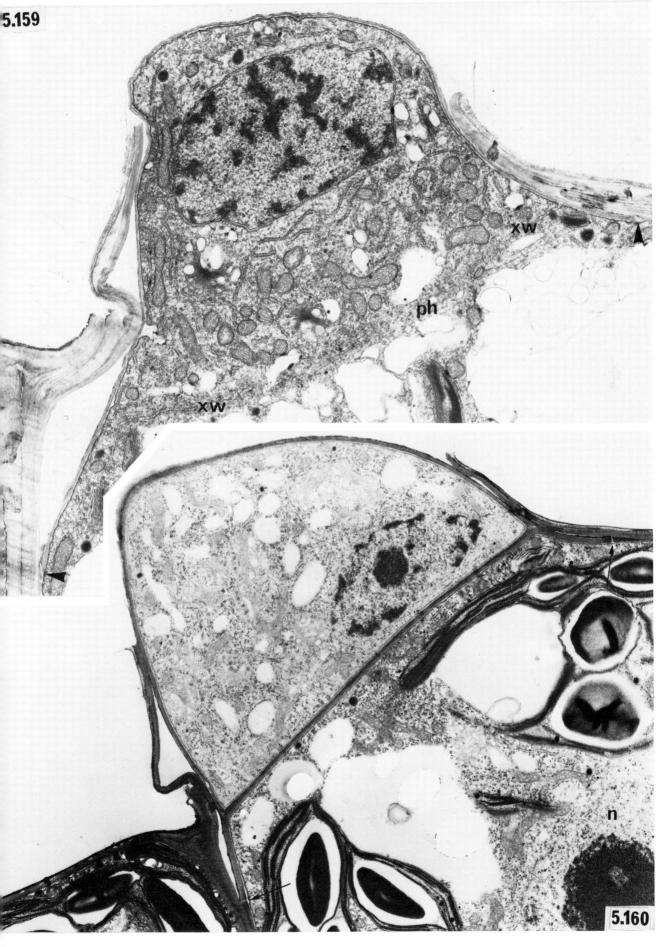

5.159

xw

ph

xw

n

5.160

Figure 5.161
HAIR CELLS IN BULBOCHAETE HILOENSIS

5.161A. Twin hairs on a terminal cell of *B. hiloensis*. One is always much older than the other. × 12,-000.

5.161B. Detail of the tip of the younger hair cell in Figure 5.151A. Numerous longitudinally oriented microtubules *(t)* are embedded in an amorphous region devoid of ribosomes, immediately behind the tip. × 40,000.

SOURCE. Author's micrographs, published in *J. Phycol. 10*, 148 (1974).

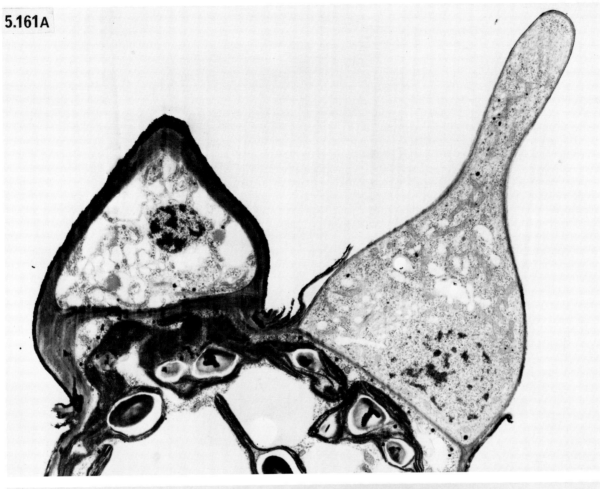

5.161A

B

5.162

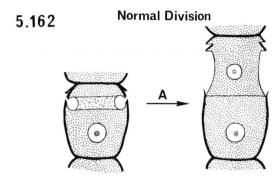

Normal Division

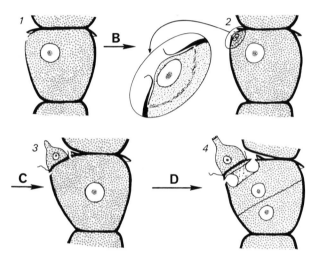

Hair Cell Formation

Figure 5.162
SUMMARY OF NORMAL CELL DIVISION AND HAIR CELL FORMATION IN BULBOCHAETE

In normal division, the circular wall weakening locates the division ring, as in *Oedogonium,* and leads to the same resultant cell expansion *(A)* along the axis of the filament. During *hair cell formation,* no such weakening exists. Instead, a modified type of weakening is situated in a thinned-out region of the parental wall. This type of weakening necessarily leads to asymmetrical division *(B)* and hair cell formation *(C)*. As a further consequence of this asymmetrical type of division, a new circular weakening arises at the base of the hair *(B, C)* which thereafter locates the division ring in the parental cell under the hair and the branch that arises at this site.

Figures 5.163, 5.164
ZOOSPORES IN BULBOCHAETE HILOENSIS

5.163. Zoospore differentiating in a terminal cell of a filament. The dome *(d)* is partly formed. This cell had already formed a hair *(h),* now dead. The bulge on the side wall and the future rupture site (arrow) indicates the way the zoospore would have emerged (see Fig. 5.164). This cell obviously would have formed a second hair (cf. 5.161 *A*) if it had divided again instead of forming a zoospore. × 5,300.

5.164. A terminal cell equivalent to that in Figure 5.163, but now empty following escape of the zoospore. Note that the position of the rent in the wall coincides with the position of the hair formed by the next cell. × 1,300.

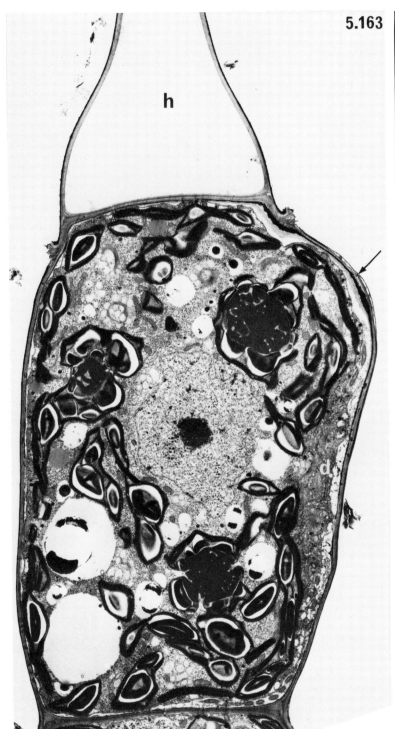

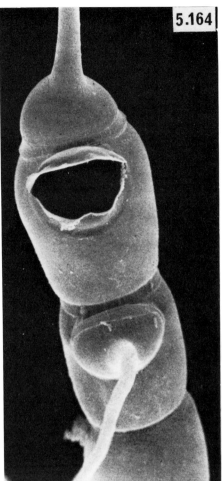

Figures 5.165–5.167
SEXUAL REPRODUCTION IN BULBOCHAETE HILOENSIS

5.165. The empty cells marked with small arrowheads have been vacated by zoospores, or more likely, androspores. Many dwarf male filaments are attached near the maturing, large oogonia (large arrows). Two smaller oogonial mother cells (small arrows) have not yet undergone their second asymmetrical division, although their primary suffultory cell, almost devoid of cytoplasm, can be clearly seen (see Fig. 5.182 for a summary of the formation of oogonia).

5.166. An escaping androspore, still enclosed within the hyaline vesicle (arrow), is seen near a sperm cell (top of micrograph). The disparity in size is obvious. × 700 approx.

5.167 A, B. Micrographs taken about a minute apart, showing a sperm cell emerging from a dwarf male filament (arrow). A, × 900; B, × 1,800.

SOURCE. Author's micrographs, paper in preparation.

337

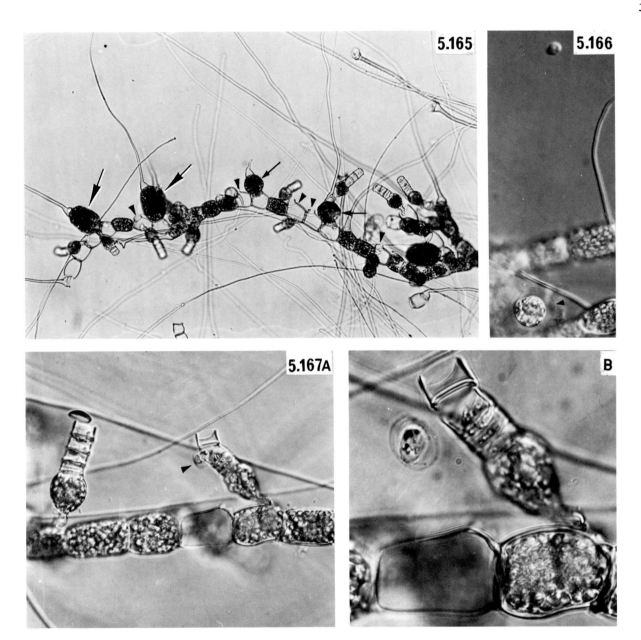

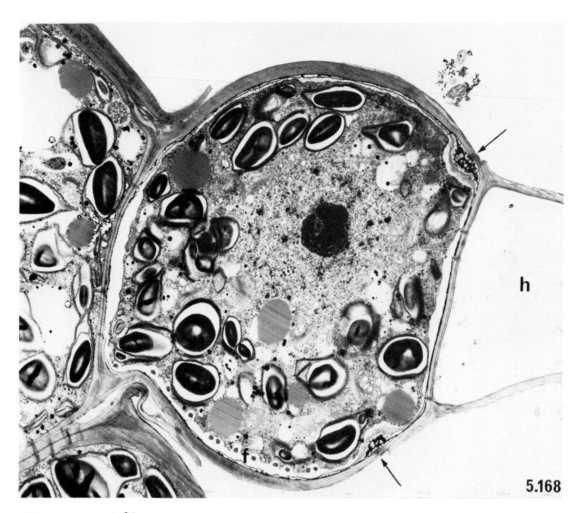

Figure 5.168
DIFFERENTIATING ANDROSPORE OF BULBOCHAETE HILOENSIS

The androspore closely resembles the zoospore, although usually a little smaller. Note the bilayered hyaline vesicle thickened at the future wall rupture site (arrows). × 5,900.

SOURCE. Author's micrograph, paper in preparation.

Figures 5.169, 5.170
GERMINATING ANDROSPORES OF BULBOCHAETE HILOENSIS

5.169. Premitotic dwarf male which at this stage resembles a vegetative germling, although appreciably smaller. × 5,200.

5.170. The first division in a dwarf male. The division ring *(dr)* is much smaller than that of vegetative divisions, and the cytoplasmic partitioning, marked by the site of the phycoplast *(ph)*, is highly asymmetrical (cf. Figs. 5.105–5.108). × 4,700.

SOURCE. Author's micrographs, paper in preparation.

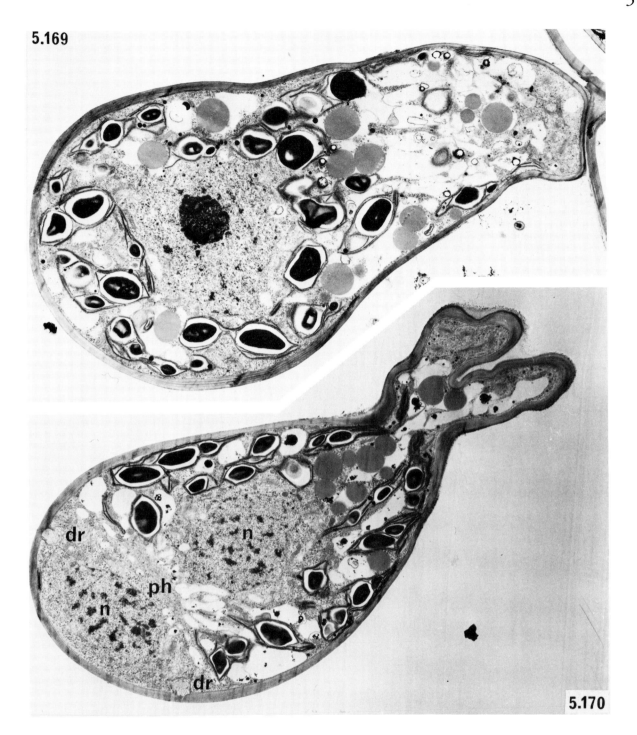

5.169

5.170

Figures 5.171–5.174
CELL DIVISIONS IN DWARF MALES OF BULBOCHAETE HILOENSIS

5.171. Expansion of the division ring, first division. × 2,800.

5.172. As for Figure 5.171. The cell plate *(cp)* is fusing into a new cross wall in the phycoplast. × 7,000.

5.173. Dwarf male after the first division, with the new secondary walls thickened. × 3,700.

5.174. The basal cell is undergoing its second highly asymmetrical division. Note one group of late anaphase chromosomes *(cb)* appressed to the end wall, near the small division ring *(dr)*. The antheridial cell has just completed its division, with a new cross wall forming in the phycoplast *(pb)*. As in *Oedogonium,* no division ring or cell expansion features in this second division, but some material is secreted at the wall rupture site (arrows). × 6,500.

SOURCE. Author's micrographs, paper in preparation.

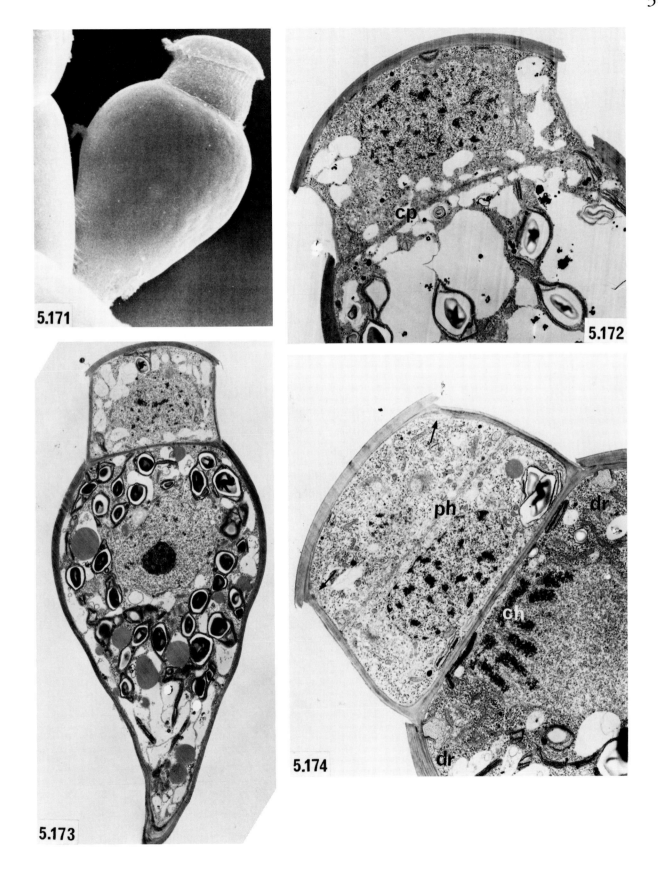

5.171

5.172

5.173

5.174

Figures 5.175–5.178
SPERMATOGENESIS IN DWARF MALES OF BULBOCHAETE HILOENSIS

5.175.   A series of five antheridial cells arising from repeated divisions of the basal cell. The lowest antheridial cell has not yet undergone its second division. The next contains two partially differentiated sperm cells; the flagella are just being extruded (arrow). The two sperm are probably fully differentiated in each of the three apical antheridial cells. Note how the wall in these three cells is segmented, having begun to split at the rupture site. The layers secreted early in spermatogenesis clearly form the hyaline vesicle enclosing them (*cf.* Fig. 5.167). $\times$ 4,700.

5.176.   Centrioles *(cn)* arising de novo near the nucleus *(n)*. $\times$ 17,400.

5.177.   Early stage in spermatogenesis. Material has been secreted at the wall rupture site (arrows) in the preceding division. Each spermatid is enclosed by a fibrous hyaline layer, almost certainly secreted by the active Golgi bodies *(g)*. The forming dome *(d)* is visible in one spermatid. $\times$ 11,600.

5.178.   The dome of a mature sperm, with its collection of dense vesicles and mitochondria *(m)*. Part of the striated fiber/rootlet microtubule complex extending from the basal bodies is indicated by the arrows. $\times$ 20,500.

SOURCE. Author's micrographs, paper in preparation.

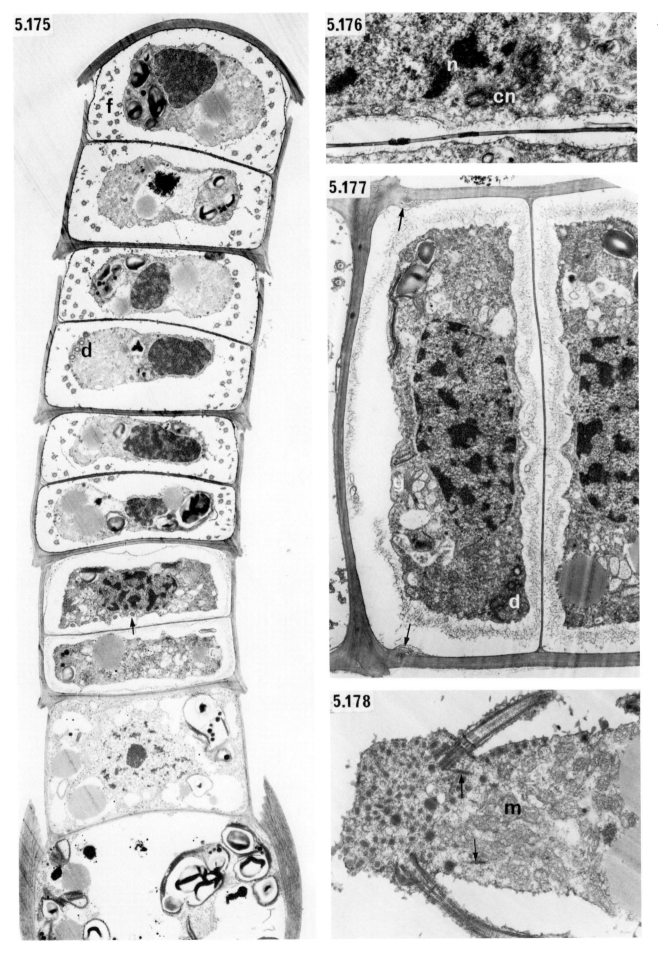

Figures 5.179–5.181
SEXUAL REPRODUCTION IN BULBOCHAETE HILOENSIS

5.179. Oospore and mature oogonium with adjacent dwarf male *(dm)*. The two empty cells had probably produced androspores. $\times$ 1,100.

5.180. Mature oogonium showing the fertilization pore. Note particularly its characteristic shape and wall caps, the basal one running around the middle of the cell. $\times$ 1,400.

5.181. Mature oospore. The wall becomes massively thickened and somewhat convoluted. The fertilization pore and division caps are still evident. The arrow marks the secondary suffultory cell. $\times$ 2,300.

SOURCE. Author's micrographs, paper in preparation.

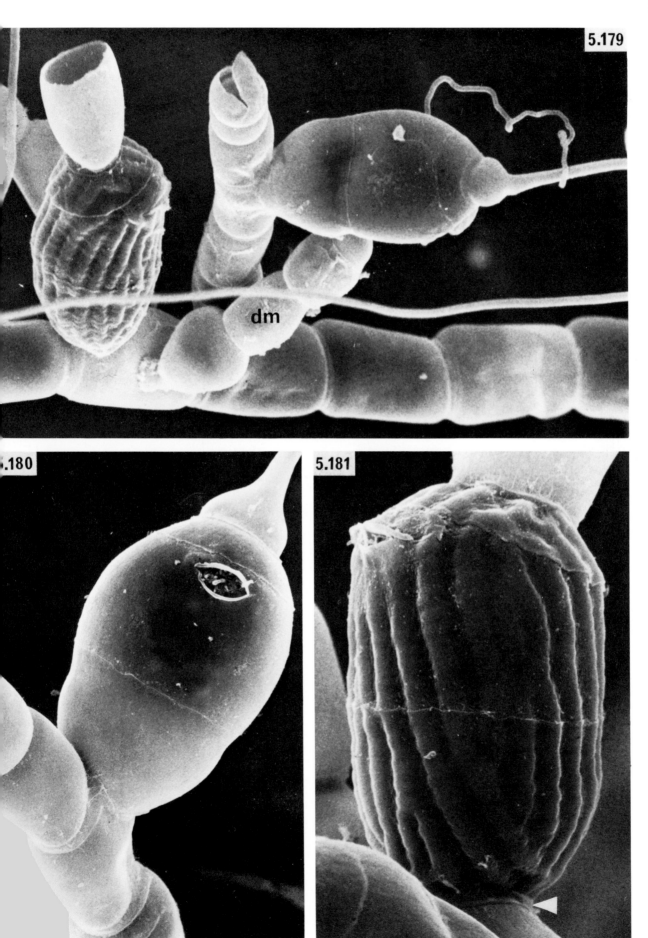

5.179

5.180

5.181

dm

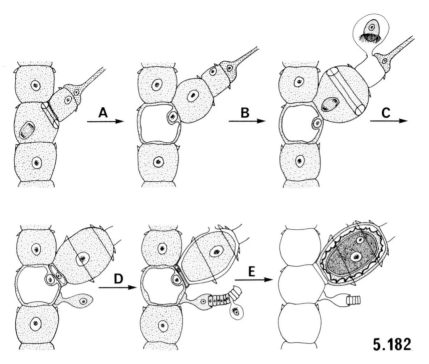

5.182

Figure 5.182
COURSE OF SEXUAL REPRODUCTION IN BULBOCHAETE HILOENSIS

Diagrammatic representation. Formation of an oogonium commences with a highly asymmetrical division of an oogonial mother cell, which gives rise to a basal primary suffultory cell, almost devoid of cytoplasm, and a more rotund oogonial mother cell *(A)*. This mother cell enlarges and then undergoes *another* highly asymmetrical division *(C)*. A tiny secondary suffultory cell is cut off, and during subsequent expansion of the division ring *(C)* this suffultory cell *stays the same size*. Thus the oogonial cell approximately doubles in volume after ring expansion. It matures, forms the pore, shrinks somewhat prior to fertilization, and then after fertilization, differentiates into the orange oospore with its characteristic wall *(D, E)*. This sequence of events often occurs at a branch point, as depicted here (see also Fig. 5.165). Quite often, the cell surmounting an oogonium differentiates into an androspore, as shown here *(C)*. Androspores attach to the filament near the oogonia and differentiate into dwarf male filaments *(D, E)*.

Figure 5.183
OOGONIAL MOTHER CELL AT PREPROPHASE OF SECOND ASYMMETRICAL DIVISION

The premitotic nucleus is strongly basally polarized, and the division ring (arrows) lies close to the middle of the cell. The primary suffultory cell *(sfc)* is almost devoid of cytoplasm. × 6,000. *Inset.* Late anaphase of the first asymmetrical division in a mother cell. Again, the spindle is strongly basally polarized and aligned with the direction of future ring expansion under a hair cell.

SOURCE. Author's micrographs, paper in preparation.

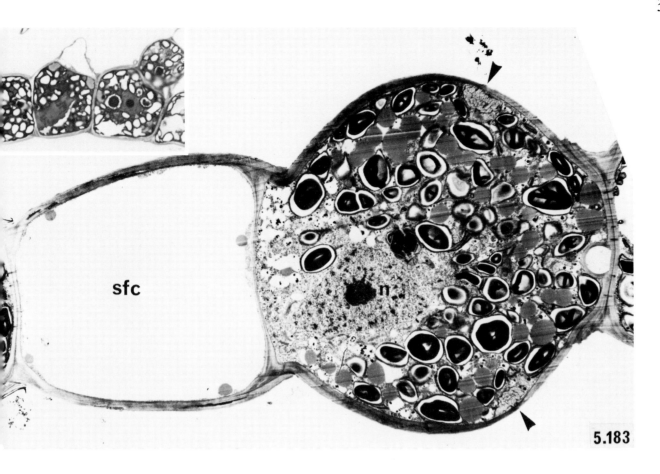

5.183

Figure 5.184

SECOND ASYMMETRICAL DIVISION IN OOGONIAL MOTHER CELL OF BULBOCHAETE HILOENSIS

The phycoplast *(ph)* is basally situated between daughter nuclei. It will *not* move along the cell during expansion of the division ring *(dr)*; see Figure 5.185. The nucleus of the primary suffultory cell *(sfc)* is visible. × 8,000 approx.

SOURCE. Author's micrograph, paper in preparation.

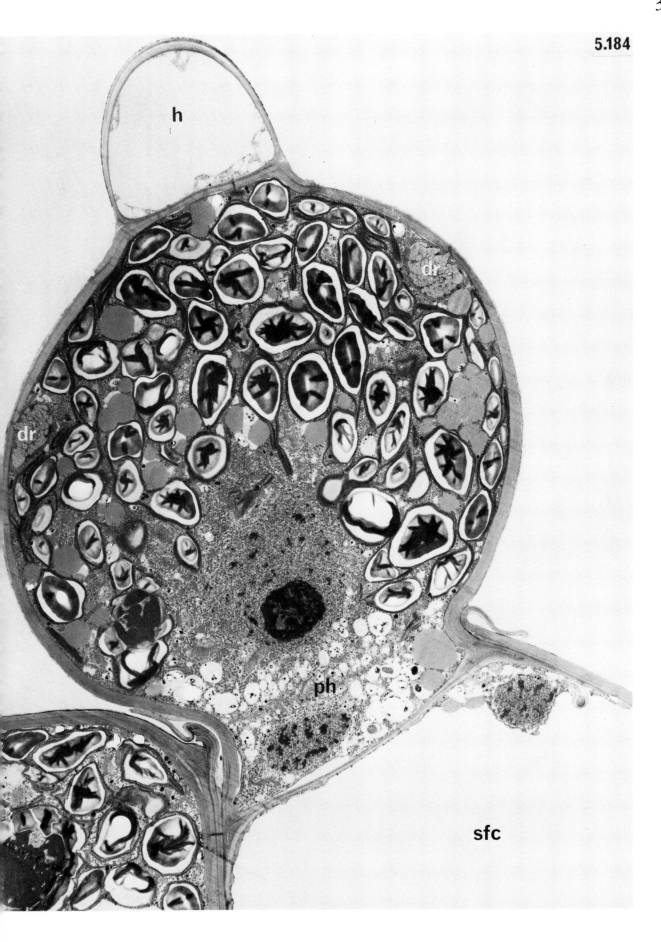

Figures 5.185, 5.186
DEVELOPMENT OF OOGONIA
IN BULBOCHAETE HILOENSIS

5.185. Expansion of the division ring, second asymmetrical division of the oogonial mother cell. The new cross wall is just forming in the phycoplast *(ph)*, which has cut off the second, very small suffultory cell *(sfc)*. Note particularly how the ring has expanded unevenly (the edges of the older wall are indicated by the small arrows). In the middle of the region of greatest expansion (large arrowhead) is already found the wall dislocation that will give rise to the fertilization pore. × 6,000.

5.186. The secondary wall has been secreted under the stretched material of the division ring. This wall, too, is unevenly thickened. At its thinnest region (arrow; see Fig. 5.187) fibrous material has been deposited prior to rupture of the wall and formation of the fertilization pore. × 7,000.

SOURCE. Author's micrographs, paper in preparation.

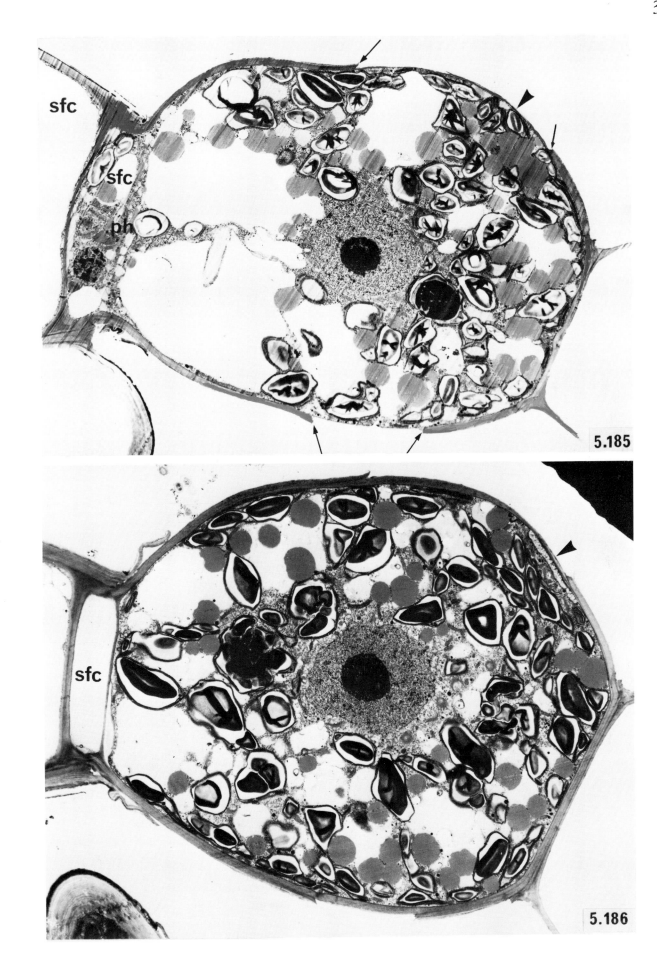

5.185

5.186

Figures 5.187, 5.188
FORMATION OF THE FERTILIZATION PORE IN OOGONIA OF BULBOCHAETE HILOENSIS

5.187. Detail of the wall region in Figure 5.186 indicated by the arrow, showing the fibrous (polysaccharide?) material secreted at the thinnest part of the wall, under the future rupture site (arrow). × 17,000.

5.188. The nucleus *(n)* has moved toward the fertilization pore, which has formed by rupture of the oogonial wall at the preweakened site. The Golgi bodies *(g)* are very active in this part of the cytoplasm, which is also filled with numerous microtubules *(t)* directed at the pore site. × 13,000. *Inset.* The same cell at a lower magnification. × 1,500.

SOURCE. Author's micrographs, paper in preparation.

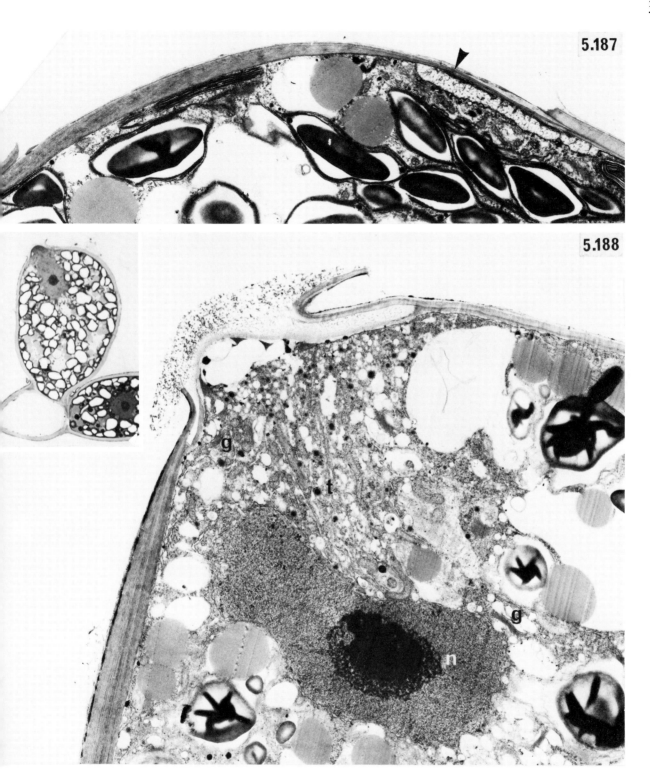

5.187

5.188

Figures 5.189–5.191
MATURATION OF OOSPORE
IN BULBOCHAETE HILOENSIS

5.189. The dense oospore fills with lipid *(l)* and other storage material and begins differentiation of its complex wall by secreting ridges of loosely fibrous material inside the oogonial wall. × 18,000 approx. *Inset.* The same cell at a lower magnification. × 1,500 approx.

5.190. Light micrograph of a maturing oogonium which has already secreted several of its wall layers. The fertilization pore (small arrow) and the two caps (between the set of small arrowheads) are also visible. The secondary suffultory cell *(sfc)* is labeled. × 2,800.

5.191. Detail of the wall layers in a maturing oogonium containing much lipid *(l)*. The arrow indicates the cap of the parental oogonial wall *(pw)*. × 11,000 approx.

SOURCE. Author's micrographs, paper in preparation.

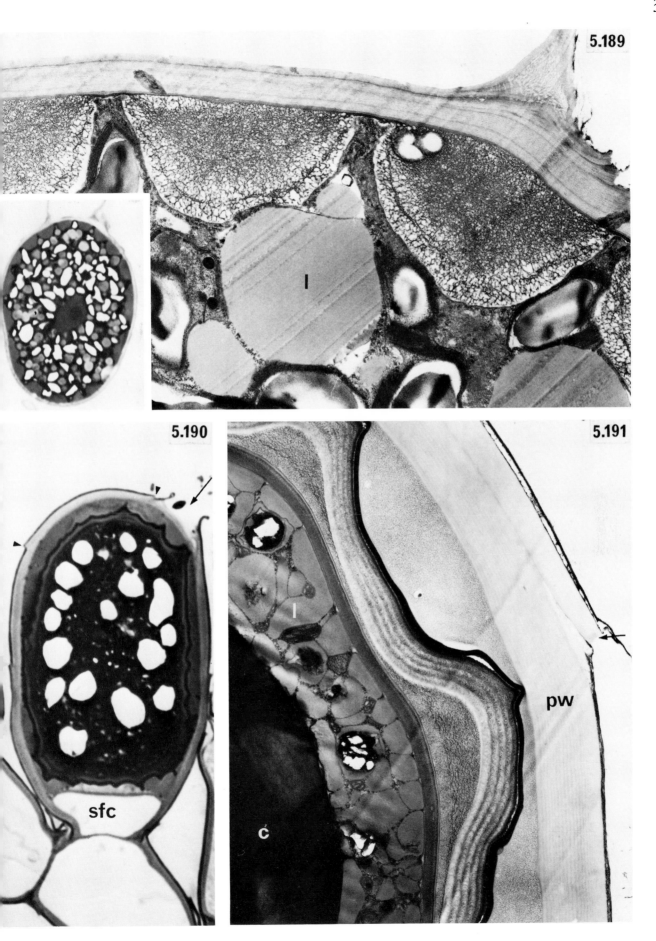

5.189

5.190

5.191

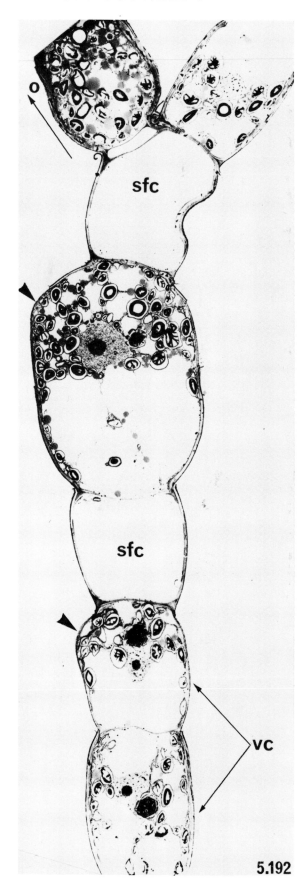

5.192

Figure 5.192
DEVELOPMENTAL RELATION BETWEEN BRANCHING, HAIR CELL FORMATION AND CREATION OF FERTILIZATION PORE

Low power area view. At the top is an oogonium *(o)* forming a branch (arrow) off the main filament. One of its two suffultory cells *(sfc)* is labeled. Basal to this is another oogonium, which has only one of its suffultory cells visible. The site of its future fertilization pore is indicated by the arrowhead. At the bottom is a pair of vegetative daughter cells *(vc)*, arising from a recent cell division. The apical daughter cell already displays the altered wall morphology (Fig. 5.150 *et seq.*) and the partial wall weakness (at arrowhead) that predetermines the site of hair cell formation and subsequent branching. Note that the branch, future fertilization pore, and hair cell site are all on the same side of the filament. Since branches and hairs consistently arise thus (*e.g.,* Fig. 5.134), the pore site is obviously a structural feature homologous with the hair cell site. × 2,600 approx.

SOURCE. Author's micrograph, paper in preparation.

6

THE CONJUGALES (ZYGNEMATALES)

Perhaps the most distinctive feature of the Conjugales, a well-defined group of Green Algae, is the complete lack of any flagellate stages in their life cycle. During sexual reproduction, ameboid gametes, which appear to consist of the protoplast of vegetative cells, fuse in the characteristic process of conjugation. The resultant zygote secretes a tough, resistant wall and can often outlast periods of extreme environmental stress before meiosis and germination take place. Many members of the Conjugales, including filamentous forms (e.g., *Spirogyra;* Ojima and Tanaka, 1970), possess the ability to undergo active, if sometimes rather slow, movement associated with the secretion of mucilage. The chloroplasts are often highly elaborate in architecture. As with the Oedogoniales, the phylogenetic affinities of this order are difficult to determine (Chapter 8). Chapman (1962) suggests that the Conjugales evolved long ago from the Ulotrichales and that the unicellular forms arose by overspecialization of the fragmentative tendencies of early filamentous types.

The order Conjugales can be divided as follows (Fritsch, 1935):

Suborder I. Euconjugatae
  A. The Mesotaenioideae (saccoderm desmids)
  B. The Zygnemoideae
Suborder II. Desmidioideae (placoderm desmids)

In this chapter, I shall describe some aspects of representatives of these groups. Unfortunately, few species have been studied ultrastructurally in detail, and so generalizations may be risky. The results already obtained, however, suggest profitable areas for future investigation.

6.1. Saccoderm Desmids

These predominantly unicellular organisms, simple in form and internal structure compared with other members of the Conjugales, can be considered as primitive desmids, having clear affinities with the Zygnemoideae. However their relation to the placoderm desmids is unclear, and they may have had a separate origin from these latter organisms (Fritsch, 1935).

Their cell wall is a simple structure devoid of pores and obvious ornamentation, and their chloroplasts, although often lobed, lack the elaboration evident in the placoderm desmids. Hardly any work has been done on the ultrastructure of these organisms. This is unfortunate, since cell division and morphogenesis may well prove interesting, particularly in comparison with other members of the Conjugales. My own attempts at studying the ultrastructure of *Netrium* and *Cylindrocystis* have proved difficult so far, but I include some micrographs which may demonstrate why these organisms merit further study.

6.1a. Netrium

Netrium is a typical saccoderm desmid (Fig. 6.1). The central nucleus is flanked by two large

Figures 6.1–6.3
NETRIUM DIGITUS

6.1. Group of cells at interphase and various stages of division. The chloroplast has cleaved in one cell (arrows), which is undergoing nuclear migration after cytokinesis (cf. Fig. 6.6A–D). × 190.

6.2. Metaphase spindle, live cell. The numerous small chromosomes are arranged in a precise plate (large arrow). The remains of the nucleolus appear at one pole of the spindle as a large mass (small arrow; see Fig. 6.4). × 1,000.

6.3A–C. Anaphase. The paired arrows indicate the positions of the daughter chromosomes. This movement is quick, lasting about 4 minutes, during which time the cleavage furrow grows inward appreciably. × 950.

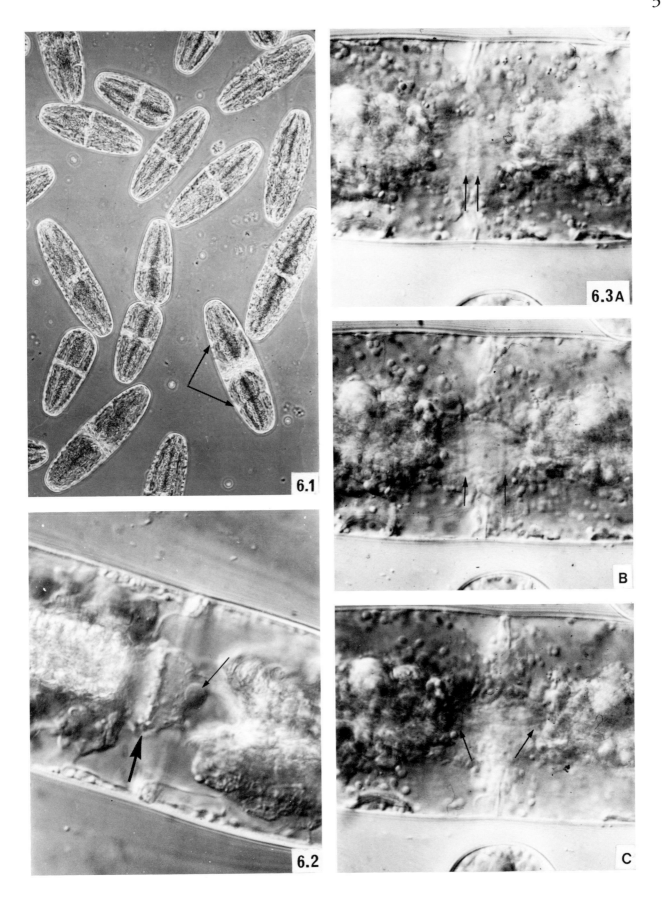

chloroplasts, each filling much of the half cell containing it. The chloroplast is stellate in cross section; its lobes, extending radially to the wall, often exhibit scalloped edges (Carter, 1919a). The large pyrenoid occupies the central portion of the chloroplast. Cytoplasmic streaming is active and clearly visible in live cells, which also secrete copious quantities of mucilage.

CELL DIVISION

Netrium faces the same problems during cell division as do the other desmids described in this chapter. Once the nucleus has divided, an ingrowing furrow cuts the symmetrical cell in half (Fig. 6.1). To restore its interphase symmetry, the chloroplast (and pyrenoid) in each daughter cell cleaves, the daughter nucleus migrates into this cleavage, and a new half cell is formed by the controlled expansion of the cell wall originally derived from the furrow. All these stages can be seen in Figure 6.1 and have been described, for example, by Carter (1920b) and by Biebel (1964). Smith (1950, p. 303) appears incorrect (at least in the case of *Netrium*) when he says that elongation of daughter cells takes place over their entire length. So far as I can tell, the new wall arises entirely from expansion of the wall originally deposited in the cleavage furrow. Division of the chloroplast and postmitotic migration of the daughter nuclei between the resultant halves also occur in genera such as *Closterium* (a placoderm desmid) and *Zygnema* (a member of the Zygnemoideae). In the former, the method of half-cell formation appears similar to that of *Netrium* (see Section 6.3a), while cell elongation in *Zygnema* is (as far as I know) slow and occupies much of interphase, with perhaps the whole wall undergoing the expansion. Detailed study of these processes in *Netrium* and *Zygnema* may yield information of phylogenetic significance when compared with the information we now have on *Closterium* and other placoderm desmids.

Mitosis, cytokinesis, and the restoration of symmetry can easily be followed in suitably chosen live cells. Sometimes the nucleus is clearly visible in the center of the cell; sometimes it is obscured by the chloroplast (as in Fig. 6.1). In the former case, mitotic spindles are easily visible (Figs. 6.2, 6.3). The numerous (over 500; Godward, 1966) tiny chromosomes form a precise metaphase plate in the striated spindle. Conspicu-

Figures 6.4–6.5
METAPHASE IN NETRIUM DIGITUS

6.4. The numerous chromosomes *(ch)* are aligned across the cell among the spindle microtubules *(t)*. At one pole (arrow) is a large body probably derived from the nucleolus (*cf.* Fig. 6.2). × 3,100 approx.

6.5. Metaphase plate of doubled chromosomes which are typically coated with short pieces of membrane (arrows). × 14,000 approx.

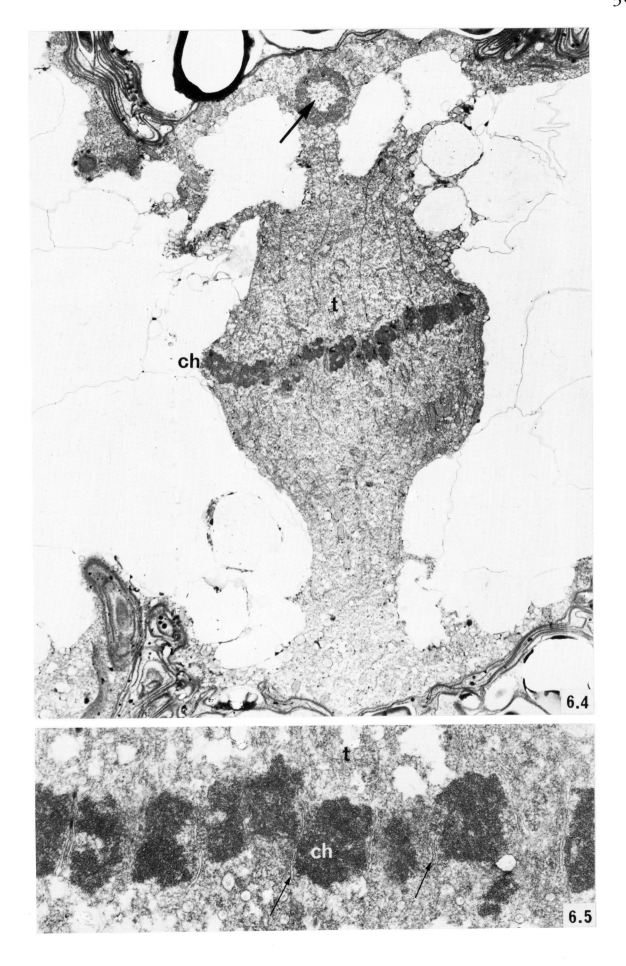

6.4

6.5

ous at one pole of many spindles is a large spherical object (Fig. 6.2) which has been tentatively identified as being derived from the nucleolus (Fig. 6.4). Anaphase separation of the chromosomes (Fig. 6.3) is rapid, lasting less than 5 minutes, during which the septum grows across the interzone, bisecting the cell a little later. Ultrastructurally, the spindle is open (Fig. 6.4) and typical in most respects, except for the mass of presumed nucleolar material at one pole and the numerous short segments of membrane coating the chromosomes (Fig. 6.5).

After the daughter nuclei have re-formed, they migrate along the chloroplast until they reach the cleavage in it (Fig. 6.6A, B). During this time, the cell cleaves in two and the daughter cells begin their expansion. Unfortunately, difficulties with fixation have so far prevented me from making a detailed ultrastructural study of these morphogenetic events, which may prove to be of great interest when compared with the behavior of *Closterium.* Microtubules have been detected near migrating daughter nuclei and near the expanding semicell wall, and colchicine treatment severely affects the migration of daughter nuclei (Tippit and Pickett-Heaps, unpublished data), but it is premature to make too many comparisons of morphogenetic systems in *Netrium* and *Closterium.*

SEXUAL REPRODUCTION (CONJUGATION)
Biebel (1964) has described conjugation in hetero- and homothallic strains of *N. digitus* using the light microscope. Conjugation was reliably induced in axenic cultures and was stimulated by lack of nitrogen in the medium and high population density of cells. Conjugation commences when cells, laden with oil, starch, and other reserves, come to lie side by side (Fig. 6.7A). In the homothallic strain, *N. digitus* var. *digitus,* the conjugants are often daughter cells formed by previous cell divisions. A broad outgrowth appears in the side of each cell, matching that in the other cell, and both conjugants become increasingly bent and pushed apart by its extension (Fig. 6.7B). The protoplasts also begin to shrink away from the cell wall, at which stage contractile vacuoles become active on their surface. The walls separating the outgrowths fuse and then disappear; Biebel describes the protoplasts as moving into their conjugation tube, a structure analogous in my opinion to that observed in other members

Figure 6.6
POSTMITOTIC REORGANIZATION IN NETRIUM DIGITUS

The movement of daughter nuclei along their respective cells before eventual insertion between the halves of the cleaving chloroplast (arrows in *D;* see also Fig. 6.1) can be clearly followed in this series of micrographs. See also postmitotic nuclear movement in *Closterium* (Fig. 6.58A–E). Note also the progressive separation of the daughter cells as the wall material in the septum splits. × 830.

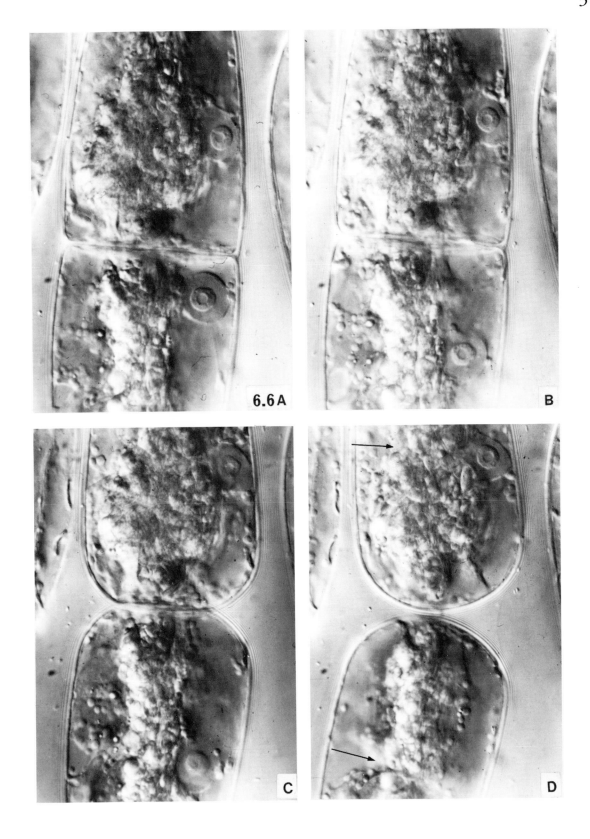

6.6 A

B

C

D

364

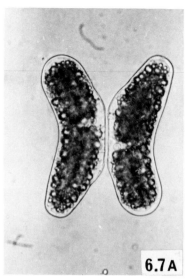
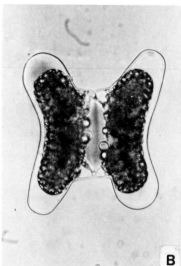
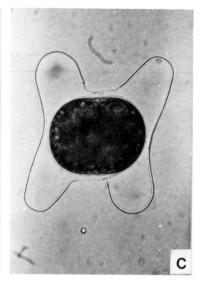

6.7A

B

C

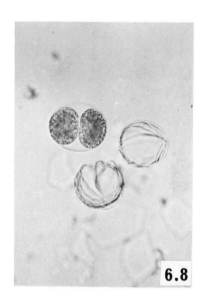

6.8

Figures 6.7, 6.8
SEXUAL REPRODUCTION IN LIVE CELLS OF NETRIUM DIGITUS VAR. DIGITUS

6.7 *A–C.* *A.* Cells paired, beginning to bend. *B.* About 2 hours later. *C.* About 30 minutes later than *B.* Zygote is forming. All × 300.

6.8. Pair of gones released from germinating zygote. × 220 approx.

SOURCE. Dr. P. Biebel, published in *Am. J. Bot. 51,* 697 (1964).

Figures 6.9–6.13
INTERPHASE AND DIVIDING CELLS OF SPIROGYRA SP.

6.9. Fixed whole cells embedded in resin, showing spiral chloroplasts. × 200.

6.10. As for Figure 6.9, showing interphase nucleus suspended in the vacuole by thin cytoplasmic strands. × 350.

6.11. Filament containing prophase, anaphase, and telophase nuclei. × 340.

6.12. Preprophase, enlarged nucleus. The thickened cytoplasmic strands contain numerous microtubules (Fig. 6.14) and will contribute to the polar *(pl)* cytoplasm. Chromosomes are becoming visible (arrow). × 1,600.

6.13. Prophase. The cytoplasmic strands connecting the nucleus and the chloroplasts *(c)* are now thin again, but a large mass of striated cytoplasm has accumulated outside the nuclear envelope (see Fig. 6.15) at the future poles of the spindle. × 1,400.

SOURCE. Drs. L.C. Fowke and J.D. Pickett-Heaps, published in *J. Phycol. 5,* 240 (1969).

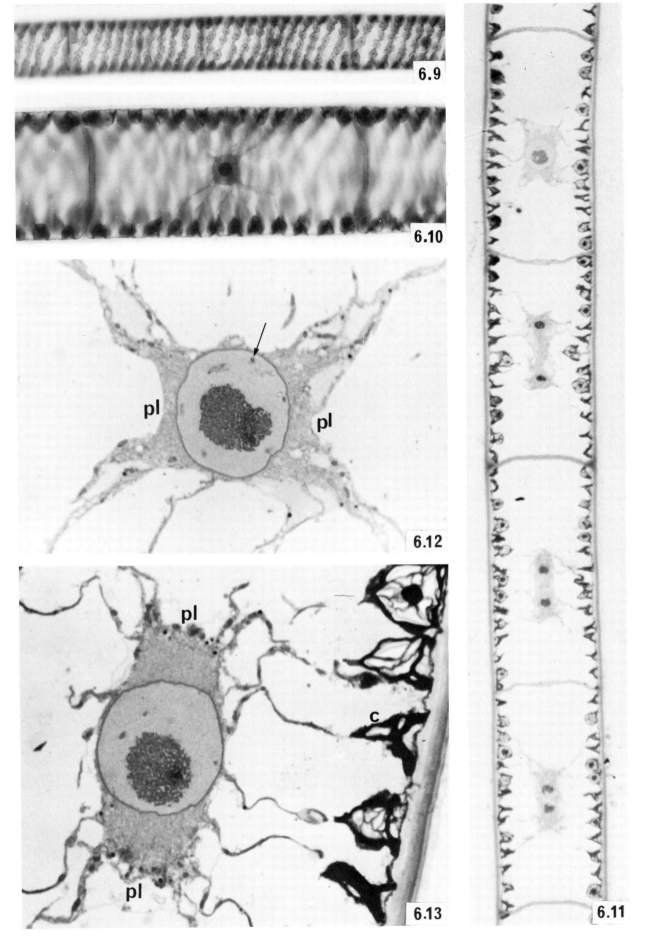

6.9

6.10

6.12

6.13

6.11

of the Conjugales, such as *Spirogyra* (Section 6.2a) and *Closterium* (Section 6.3a). The protoplasts fuse rather suddenly, and then the zygote contracts (Fig. 6.7C). Later, the thick zygospore wall, containing at least three layers, is secreted. The nuclei apparently do not fuse until just before germination. No ultrastructural work has been done on conjugation in these cells, but the events described for live cells quite closely resemble what happens in *Closterium*.

Germination of zygospores takes place after a resting period and is stimulated by their transfer to fresh medium. The zygospores as usual are highly resistent to desiccation. Impending germination is indicated when the brownish-yellow zygospore becomes green. Nuclear fusion probably also occurs some time during this period. The wall of the zygospore ruptures to release the cell contents in a hyaline vesicle (Fig. 6.8). Soon after emergence, the nucleus undergoes the first meiotic division. Presumably, a second meiotic division soon follows, although this was not specifically mentioned by Biebel, who states that upon subsequent cleavage of the protoplast into two, each daughter cell or "gone" can usually be observed to contain both a pycnotic (i.e., degenerating) and a viable nucleus. The two cells are soon released from the hyaline vesicle; in *N. digitus* var. *digitus,* however, up to five "products" may be formed per zygote.

Kies (1967) has described cell division and conjugation in *Roya*. He says that there is no conjugation tube formed around the emerging gametes. As mentioned later, the definition of a conjugation tube is at present vague. Electron microscopic evidence will be needed to confirm whether these gametes are naked; in those members of the Conjugales studied ultrastructurally, the gametes move into the newly secreted, expanding wall (e.g., *Closterium*). I cannot see why this wall should not be designated a conjugation tube, and many light micrographs of conjugating desmids (see later) indicate that such a wall is present.

6.2. The Zygnemoideae

Many of the commonest freshwater algae belong to this group (*Spirogyra* and *Zygnema* are particularly well known to biology students, usually being available in quantity from ponds and lakes everywhere). They are usually found as un-

Figures 6.14–6.17
MITOSIS IN SPIROGYRA SP.

6.14. Preprophase, showing thickening cytoplasmic strands (Fig. 6.12) containing numerous microtubules *(t)*. $\times$ 23,000.

6.15. Prophase (Fig. 6.13). Extranuclear cytoplasm contains numerous highly oriented microtubules. $\times$ 23,000.

6.16. Prometaphase. Extranuclear microtubules are invading the nucleus *(n)*. $\times$ 27,000.

6.17. Very early anaphase. The small diffuse chromosomes *(cb)*, coated with striated nucleolar material, have just started separating. The nuclear envelope is still reasonably intact around the spindle. $\times$ 10,000.

SOURCE. Drs. L.C. Fowke and J.D. Pickett-Heaps, published in *J. Phycol. 5,* 240 (1969).

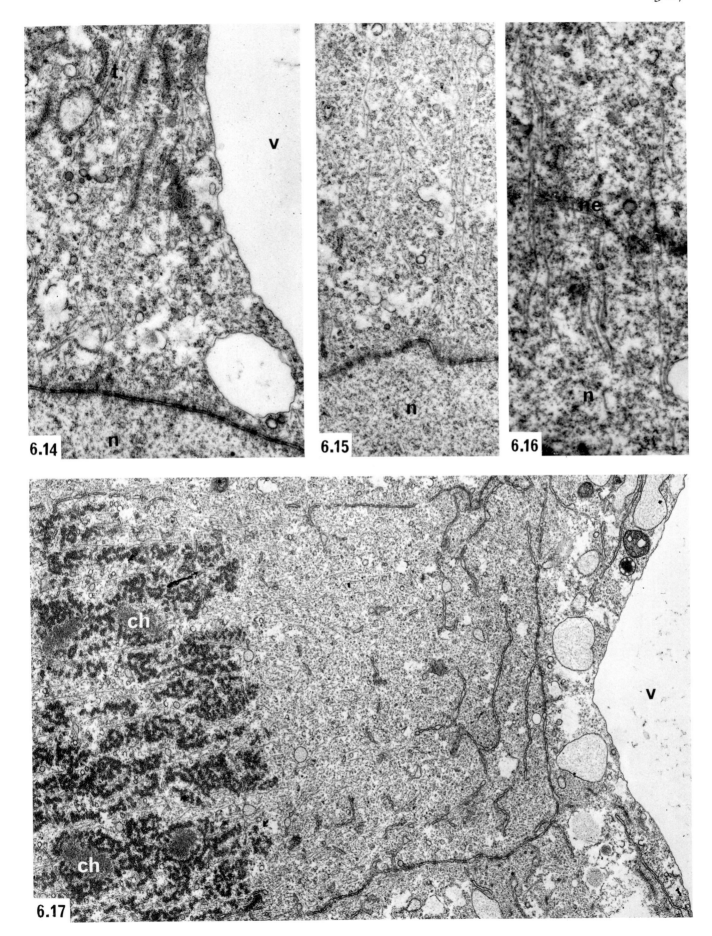

6.14

6.15

6.16

6.17

branched, unattached filaments of cylindrical cells, whose chloroplasts are stellate (e.g., *Zygnema*) or in the form of either a single flat plate (e.g., *Mougeotia*) or one or more spiral bands (e.g., *Spirogyra*). Pyrenoids are conspicuous and often numerous. In most genera, vegetative reproduction can occur by fragmentation of the filaments into shorter segments. In this book, I shall confine my detailed comments to *Spirogyra, Zygnema,* and *Mougeotia,* the only species so far investigated ultrastructurally in detail.

To what extent the following account of *Spirogyra* is typical of other members of the Zygnemoideae is difficult to tell in the absence of published information. *Mougeotia,* for example (Section 6.2c), differs appreciably from *Spirogyra* in certain respects during cell division and conjugation. *Zygnema* contains two stellate chloroplasts whose lobes radiate from two large pyrenoids situated on either end of the nucleus. While mitosis could resemble that in *Spirogyra,* the single pyrenoid/chloroplast in each new daughter cell also divides after cell division, and each daughter nucleus is then inserted between these halves. Equivalent postcytokinetic events are not necessary and do not occur in *Spirogyra,* and they rather resemble the behavior observed in daughter cells of saccoderm desmids (e.g., *Netrium,* described earlier) and the unconstricted placoderm desmid *Closterium.*

6.2a. Spirogyra

The familiar and common *Spirogyra* is easily identified under the light microscope by its one or more elaborate, ribbon-like chloroplasts, containing numerous pyrenoids, which are wound around the inside of the cylindrical cell (Figs. 6.9, 6.10). The nucleus is suspended in the middle of a large vacuole by delicate cytoplasmic strands (Fig. 6.10 et seq.).

Ultrastructurally (Fowke and Pickett-Heaps, 1969a), vegetative cells of *Spirogyra* reveal little of particular interest. The cells are especially prone to cytoplasmic damage following usual methods of tissue preparation. No matter how carefully carried out, this manipulation disrupts the thin peripheral layer of cytoplasm and its membranous components. Nevertheless, the ultrastructure of cell division and conjugation can be examined in spite of evidence of tissue damage. The wall contains at least two layers, the

Figures 6.18–6.22
LATER STAGES OF MITOSIS IN SPIROGYRA SP.

6.18. Midanaphase. The chromosomes are embedded in dense nucleolar material, and remnants of the nuclear envelope are still visible. Note that the chromosome-to-pole distance does not seem to have shortened appreciably (Fig. 6.17). × 2,000.

6.19. Late anaphase/early telophase. The spindle is much elongated, as the nucleolar material rounds up. Vacuoles are appearing in the interzone. × 2,000.

6.20. Early and late telophase. The nucleolar material is seen condensing into new nucleoli. Note the persistence of the interzonal cytoplasm (which contains numerous microtubules; Fig. 6.25). Cleavage, hardly detectable in the upper cell, is quite advanced in the lower. × 1,120.

6.21. Midanaphase chromosomes *(ch)* deeply embedded in nucleolar material *(nc)*. Note numerous continuous spindle microtubules *(t)* around the edge of the spindle. × 14,000.

6.22. Numerous persistent microtubules *(t)* around re-formed daughter nuclei (top cell in Fig. 6.20). × 27,000.

SOURCE. Drs. L.C. Fowke and J.D. Pickett-Heaps, published in *J. Phycol. 5,* 240 (1969).

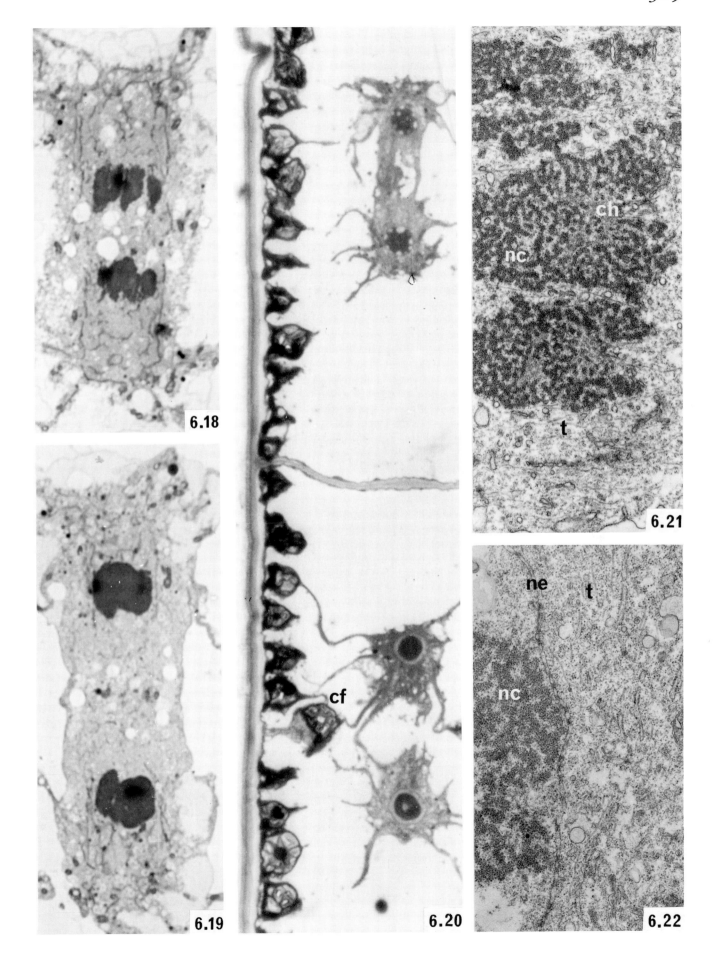

outer one mucilaginous and responsible for the characteristic slimy nature of the filaments. Circumferentially oriented wall microtubules line the inside of the plasmalemma. Plasmodesmata have not been reported in the end wall separating the mature vegetative cells. The large central nucleus contains one (or more) conspicuous dense nucleolus, but during interphase the chromatin is completely dispersed into a fine fibrillar granular matrix.

CELL DIVISION

Cell division in *Spirogyra* (Figs. 6.11, 6.20, 6.23) has been investigated by a number of workers with the light microscope (see references in Fowke and Pickett-Heaps, 1969a). In particular, McAllister's (1931) account of mitosis and cytokinesis has been confirmed in most details using more modern microscopical techniques (Fowke and Pickett-Heaps, 1969a,b). Godward and co-workers have investigated the behavior of the nucleolus during mitosis (see references in Jordan and Godward, 1969).

Mitosis. The interphase nucleus is surrounded by a thin layer of cytoplasm continuous with delicate cytoplasmic strands radiating out to the chloroplasts (Fig. 6.10). The centrally situated nucleolus is spherical and very dense, containing both fibrillar and granular regions. During preprophase, a stage quite distinct from prophase, the nucleus enlarges, and the cytoplasmic strands supporting it thicken markedly (Fig. 6.12). Electron microscopy reveals numerous microtubules in these strands (Fig. 6.14). Furthermore, a count of microtubules per unit length along the cell wall shows that their number decreases approximately fourfold during mitosis and cytokinesis, in comparison with the number along interphase cell walls. From these observations, we conclude (Fowke and Pickett-Heaps, 1969a) that a majority of wall microtubules, many intact, move along with cytoplasm via the cytoplasmic strands into the forming spindle. The apparent transport of cytoplasmic microtubules and their incorporation into the prophase spindle have also been observed in other algal and higher plant cells (Pickett-Heaps, 1973d). By prophase, the cytoplasmic strands have become thin again (Figs. 6.11, 6.13) and virtually devoid of microtubules, but now there is a conspicuous cylindrical mass of cytoplasm capping each end of the

Figures 6.23–6.25
CYTOKINESIS IN SPIROGYRA SP.

6.23. Cytokinesis three quarters complete. The ingrowing cleavage furrow has impinged upon the cylindrical mass of interzonal cytoplasm, which contains numerous microtubules. Note that the daughter nuclei remain widely separated. × 720.

6.24. Cleavage furrow *(cf)* impinging on the dense cylinder of interzonal cytoplasm (Fig. 6.23). Note organizations of microtubules, etc. (Fig. 6.25) at the innermost edge of the furrow. × 27,000.

6.25. Phragmoplast organization (*i.e.,* microtubules embedded in dense material, interspersed with aligned vesicles) at inner edge of cleavage furrow *(cf)*. × 27,000.

SOURCE. Drs. L.C. Fowke and J.D. Pickett-Heaps, published in *J. Phycol. 5,* 273 (1969).

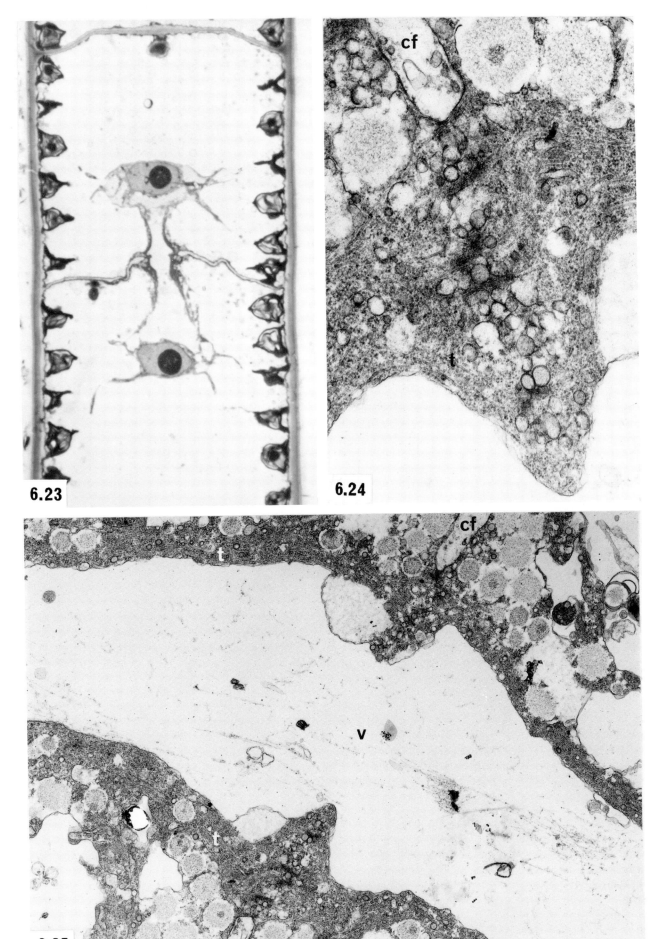

6.23

6.24

6.25

nucleus, filled with thousands of microtubules (Fig. 6.15) and elements of endoplasmic reticulum all oriented along the future spindle axis. Most larger organelles (e.g., mitochondria) have by now been eliminated to the ends of the poles (Fig. 6.13). Commencing during preprophase, the structure of the nucleolus steadily loosens up, becoming reduced to lamellae and strands of granular material. Pale, diffuse chromosomes appear and begin condensing in and around the nucleolus (Figs. 6.12, 6.13).

As prophase advances to prometaphase, the nuclear envelope adjacent to the poles becomes more flattened and develops numerous minor indentations. At prometaphase, gaps appear in the envelope at the poles and microtubules penetrate into the nucleus (Fig. 6.16), which then becomes increasingly striated. These latter changes are clearly detectable with the light microscope (McAllister, 1931), and they are equivalent in many ways to spindle formation in higher plants, where microtubules also penetrate the nucleus at prometaphase. By metaphase, the semi-intact nuclear envelope has expanded poleward, and as a result the prophase caps of cytoplasm are no longer so prominent. The small, rather inconspicuous metaphase chromosomes are arranged in the usual plate configuration; they remain embedded in the large, loosely knit central mass of nucleolar material, which is conspicuously striated by the microtubules, many of which are attached to chromosomes.

During anaphase (Figs. 6.17–6.19), the spindle steadily elongates and remnants of the nuclear envelope completely disperse. The two groups of separating chromosomes, deeply embedded in nucleolar material, become virtually undetectable by late anaphase (Figs. 6.19, 6.21). The microtubules of the continuous spindle seem to be distributed predominantly in a cylindrical array, enclosing the two groups of chromosomes. The nuclear envelope re-forms during telophase as normal (Figs. 6.20, 6.22) around the small daughter nuclei, whose nucleolar material progressively condenses into new nucleoli (Fig. 6.20). Numerous small vacuoles appear briefly in the nucleus at this stage. Between the daughter nuclei the cylindrical array of continuous microtubules persists after telophase, soon becoming involved in cytokinesis (see below), and many microtubules are still seen around reformed nuclei (Fig. 6.22). During the transition

Figures 6.26–6.31
CONJUGATION IN SPIROGYRA SP

6.26.   Very early stage. The two filaments are closely stuck together and the first sign of an incipient conjugation tube can be seen (arrow). Note that the female filament has recently divided. In this and subsequent micrographs showing conjugation, the "male" filament is on the right. × 400.

6.27.   Conjugating filaments are now linked, and pushed apart by fully developed conjugation tubes. Note the layer of adhesive mucilage now suspended between the cells. The pair of cells arrowed is shown in Figure 6.32. × 560.

6.28.   Fertilization. The wall separating the conjugation tubes has been eroded away (Fig. 6.35), and the protoplasts have fused. The male cells are also clearly contracting away from their cell wall (arrows). × 640.

6.29.   Live filaments, showing contraction of the male protoplast. Female protoplasts also are just starting to round up (arrows). Note the debris stuck in the mucilage between the filaments. × 400.

6.30.   Migrating and condensing gametes, fixed and embedded in resin. Note vegetative cells (vc) in one filament. × 400.

6.31.   Section of the cells shown in Figure 6.29. Note much debris between cells. × 400.

SOURCE. Drs. L.C. Fowke and J.D. Pickett-Heaps, published in J. Phycol. 7, 285 (1971).

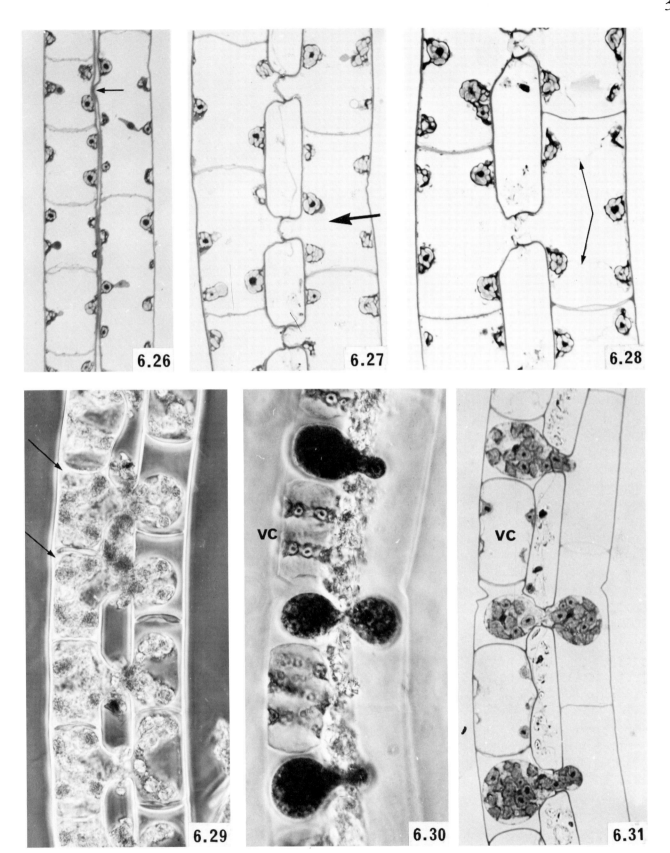

to interphase, the nucleus swells appreciably, acquiring its characteristic fine matrix around the prominent nucleolus. These observations clearly indicate that nucleolar material is persistent throughout division, being distributed along with the chromosomes into the daughter nuclei.

Cytokinesis. The principal features of cytokinesis were described with remarkable accuracy by McAllister (1931) and others; nevertheless, some minor misinterpretations in the earlier literature can now be clarified using electron microscopy. For example, cytokinesis in *Spirogyra* has features common to certain other members of the Conjugales, the desmids, and to higher plants; it is, however, not directly equivalent to cytokinesis in any of these groups.

Cytokinesis is initiated during early anaphase by a narrow annular ingrowth of the plasmalemma at the middle of the cell. This septum grows slowly during the later stages of mitosis (Fig. 6.20, lower cell), often displacing the ribbon-like chloroplasts into the vacuole until they break. We could never detect any microtubules associated with the septum during its early growth (Fowke and Pickett-Heaps, 1969b). Such is the case also with the desmids *Closterium, Cosmarium,* and *Micrasterias,* for example, where cell partitioning by the septum is completed without involvement of microtubules. However, a most intriguing change occurs in *Spirogyra* as the septum impinges upon the cylinder of cytoplasm separating telophase daughter nuclei (Fig. 6.23). This cylinder, it will be remembered, contains numerous persistent, longitudinally oriented microtubules, which now gather and apparently proliferate where contacted by the septum. Dark-staining material, interspersed with vesicles, also begins to collect in a planar array here (Fig. 6.25); soon a small phragmoplast, typical of higher plants, has been organized around the ingrowing edge of the septum (Fig. 6.24). Thus, cytokinesis is completed by the creation of a cell plate.

I have suggested that these observations may have considerable significance, indicating perhaps how a primitive mechanism of cytokinesis (i.e., cleavage) may have been transformed during evolution into the phragmoplast/cell plate method common to higher plants. These ideas have been discussed in detail elsewhere (Fowke and Pickett-Heaps, 1969b; Pickett-Heaps,

Figures 6.32, 6.33
CONJUGATION IN SPIROGYRA SP.

6.32. Detail of two of the cells shown in Figure 6.27. The structure of the conjugation tubes is shown. They are fused at their ends. The adhesive mucilage layer *(a)* is now suspended between the filaments (Fig. 6.27). × 5,000.

6.33. Male protoplast pulling away from the wall after fertilization. Note the massive accumulation of mucilage *(ml)* beginning to build up between the wall and the cell membrane. This material is probably derived from the active Golgi bodies *(g)* in these cells. (The Golgi bodies in the female protoplast appear quite quiescent.) × 5,000.

SOURCE. Drs. L.C. Fowke and J.D. Pickett-Heaps, published in *J. Phycol. 7,* 285 (1971).

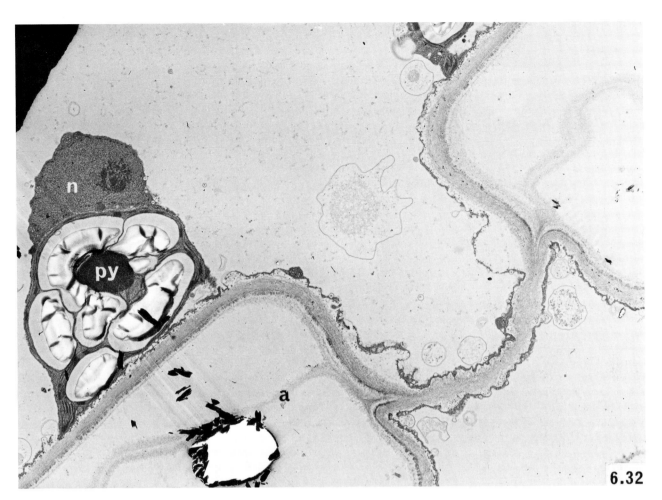

6.32

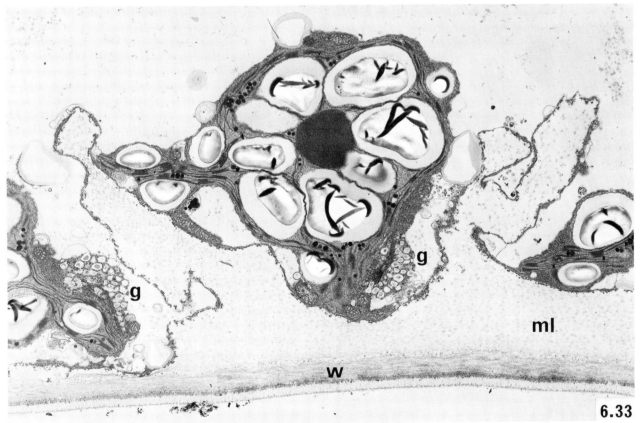

6.33

1969a; Chapter 8). One final point is worth making here. The position of the septum in the cell appears to be determined by the position of the early mitotic nucleus. This relation was neatly demonstrated by Van Wisselingh (1909), who centrifuged a nucleus to one end of the cell, and then (having patiently waited a considerable period) centrifuged it back to the other end as it went into mitosis. While the nucleus completed division in the latter end, the septum was sited at the other end of the cell occupied earlier by the premitotic nucleus.

SEXUAL REPRODUCTION (CONJUGATION)

Sexual reproduction occurs quite readily in many members of the Zygnemoideae, often at specific times of the year, and it involves the characteristic and well-known process of conjugation. Filaments usually come to lie side by side, often following a phase of slow active movement, and the cells from adjacent filaments push out contiguous conjugation tubes which abut one another. The end wall between the two tubes then erodes away, and the protoplasts fuse. In some species, the zygospore may be formed in the conjugation tubes between the filaments (as is also the case, for example, with many desmids; cf. Section 6.4c). In other species, one ameboid protoplast (now actually forming part of the zygote after fertilization) moves into the other cell before the zygote's cytoplasm condenses and a resistent cell wall is secreted around it. In this event, the protoplast that moves is by convention derived from the "male" plant. However, it appears that degrees of sexuality may exist in such species. For example, three or more filaments may be mutually involved in conjugation, and a filament that acts as a male (i.e., donating its protoplast) toward a second filament, may concurrently act as a female toward a third (Fritsch, 1935, pp. 327 et seq.).

In *Spirogyra*, conjugation is usually "scalariform" (as described here) when the filaments lie side by side, linked by conjugation tubes to form a ladder-like array of cells. However, "lateral" conjugation is also known, whereby adjacent cells of the *same* filament form a conjugation tube around their mutually shared end wall, through which one protoplast moves into the other cell, effecting fertilization. Conjugation has been studied by a number of earlier workers (references in

Figures 6.34–6.36 CONJUGATION IN SPIROGYRA SP.

6.34. Zygote cytoplasm is beginning to become dense as contraction and condensation continue. The first sign of wall deposition is detectable around the zygote (arrow). × 5,000.

6.35. Fertilization, showing the two protoplasts having just fused. The end wall *(ew)* previously separating the gametes is frayed and eroded away. × 14,000.

6.36. Row of maturing zygotes within the wall of the female filament. They have rounded up, but their wall is not yet very thick. × 400.

SOURCE. Drs. L.C. Fowke and J.D. Pickett-Heaps, published in *J. Phycol. 7,* 285 (1971).

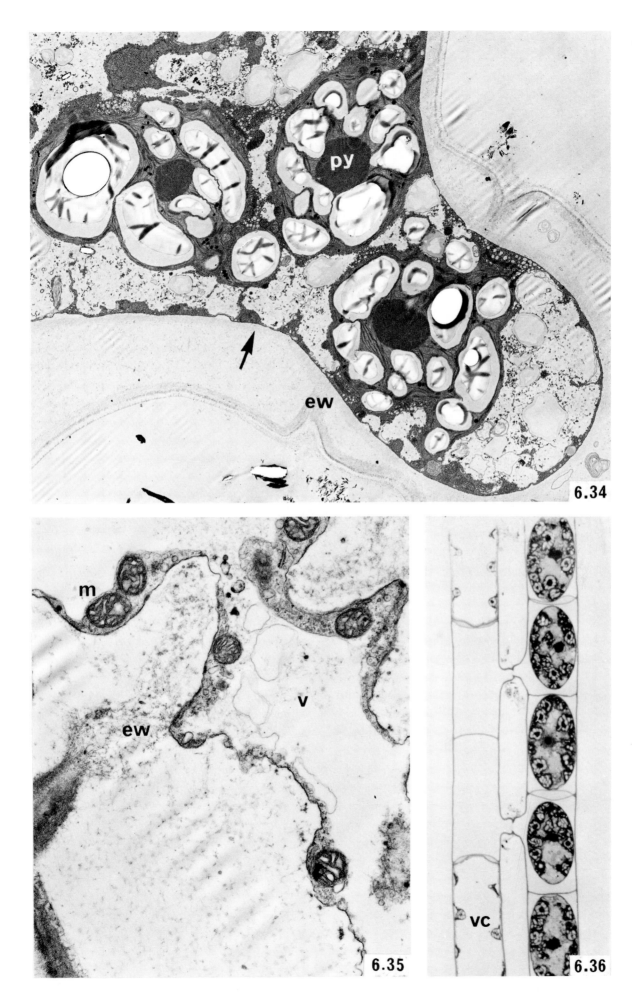

6.34

6.35

6.36

Fritsch, 1935, p. 323; Hoshaw, 1968) and more recently using electron microscopy (Fowke and Pickett-Heaps, 1971).

While many cells have undergone division just prior to conjugation, their partners may sometimes be much larger, and were probably about to divide before being interrupted by conjugation (Fig. 6.26). Consequently, when one large cell conjugates with a smaller partner, numerous unpaired vegetative cells may later be found between maturing zygotes or zygospores (Figs. 6.30, 6.31). There is no evidence in these cases of true anisogamy; i.e., the larger protoplast in such cases could be either male or female.

Filaments about to conjugate are apparently stuck together by mucilaginous material secreted between them. This adhesive tends to collect numerous particles of debris (Fig. 6.31) which render microtomy frustrating and difficult. Conjugation tubes are initiated by a bulge in the wall initially in male cells only (Fig. 6.26), but each bulge is soon precisely matched in position by a similar adjacent bulge in the female cells (Fig. 6.27). The cytoplasmic layer adjacent to this expansion or papilla in the wall is often thickened and contains numerous vesicles as well as other cytoplasmic organelles, but electron micrographs do not indicate what has happened to the wall at this site. The transversely oriented wall microtubules remain near the plasmalemma of the papillae temporarily, and we believe that secretion of new wall material (rather than stretching of the preexisting structure) is responsible for its outgrowth. Continuous extension of the papillae pushes the filaments apart, leaving the layer of adhesive suspended between them (Figs. 6.27, 6.32). Next, the end wall between and shared by pairs of papillae disintegrates (Figs. 6.28, 6.35); how this is accomplished is not indicated by electron micrographs. With their end wall gone, each pair of papillae has now become a conjugation tube, and the protoplasts fuse. As before (Fowke and Pickett-Heaps, 1971), it is convenient to refer to "male" and "female" protoplasts after fertilization, although strictly speaking they both now should be considered a single zygote.

In the male protoplasts only, the Golgi bodies begin to produce numerous vesicles with fibrous contents just before and for some period after syngamy. Then as the male protoplast begins to contract (Figs. 6.29–6.31), large amounts of material, identical to the contents of these vesi-

Figures 6.37–6.41
VEGETATIVE CELLS AND AKINETES OF ZYGNEMA SP.

6.37.   Cell in young vegetative filament. On either side of the central nucleus (rather obscured in this micrograph) are the two large pyrenoids with lobes of the chloroplasts running out to the cell's periphery. × 1,400.

6.38.   Filament from an older culture than that in Figure 6.37. Cells are becoming very dense before forming akinetes, owing partly to an accumulation of reserve material. × 1,400.

6.39.   Mature akinete, separated from the rest of the filament. × 1,400.

6.40.   Young vegetative cell. The nucleus (n) is sandwiched between the large pyrenoids (py). The wall has two distinct layers. × 4,200.

6.41.   Cell from an older culture, beginning to turn into an akinete. The whole cytoplasm is typically dense and filled with inclusions, while starch has accumulated around the pyrenoid (cf. Fig. 6.40). × 3,900.

SOURCE. Drs. R.J. McLean and F. Pessoney, published in *Contributions in Phycology*, Lawrence, Kansas, Allen Press, 1971, p. 145.

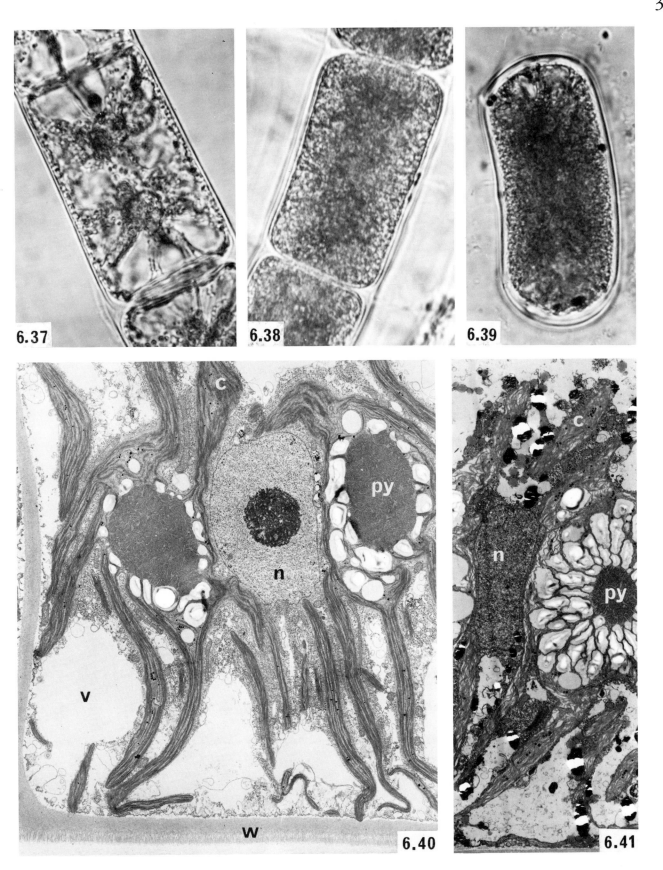

cles (and by inference, derived from them), begin accumulating between the cell wall and the plasmalemma (Fig. 6.33). The female protoplast may also concurrently contract slightly, but it displays no obvious similar cytoplasmic differentiation. As the male begins its journey through the conjugation tube, both protoplasts begin cytoplasmic condensation (Figs. 6.30, 6.31). Lloyd (1928) and others report that contractile vacuoles are now active on the surface of the protoplasts, but we could not detect them in our specimens. The enormous central vacuole (single after fertilization) shrinks steadily, and the cytoplasm generally becomes increasingly densely stained (Fig. 6.34). The zygote starts rounding up, even when part of it is still in the conjugation tube (Fig. 6.31); it eventually becomes ellipsoidal in profile inside the female cell wall (Fig. 6.36). Then a thick, layered wall is secreted to form the resistant zygospore. As with so many other organisms, electron microscopy of maturing zygospores becomes increasingly difficult. The cells' accumulation of starch, lipid droplets, etc. makes processing for electron microscopy less and less rewarding, and the cytoplasm finally becomes so dense that ultrastructural details are often totally obscured. In all cells we examined, the two nuclei remained separate even when appressed together, in agreement with Fritsch's (1935, p. 333) belief that nuclear fusion is delayed for some while after fertilization.

Germination, summarized by Fritsch (1935), is apparently preceded by two meiotic divisions. Three of the resultant nuclei and apparently one of the two complements of original chloroplasts degenerate, while the thick zygote wall ruptures to release its contents as a cell which soon undergoes division to form a vegetative filament (and which may also form a rudimentary holdfast).

Secretion of mucilage by the Golgi bodies in the male protoplast might appear to be responsible for moving that protoplast through the conjugation tube and into the female cell. However, loss of turgor pressure may also be responsible for the failure of mucilage to be pushed through the wall as presumably happens in vegetative cells. Nevertheless, the Golgi bodies are different in the male and female protoplast, and uptake of water by this mucilage in the male cell could at least help move the protoplast in the required direction.

Figure 6.42
PORTION OF AN AKINETE OF ZYGNEMA SP.

Note the typical accumulation of lipid bodies (*l*). Elsewhere, the cytoplasmic components (*e.g.*, the chloroplast) have lost much of their normal substructure and have become dense and heavily stained. $\times$ 5,000.

SOURCE. Drs. R.J. McLean and F. Pessoney, published in *Contributions in Phycology*, Lawrence, Kansas, Allen Press, 1971, p. 145.

Figure 6.43
MOUGEOTIA SP.

The broad, flat single chloroplast in each cell is capable of rotation in response to light. These two micrographs, taken a few minutes apart, show the change in orientation when live filaments were first exposed to weak (*A*) and then intense light (*B*) from the microscope's condenser system. $\times$ 160.

SOURCE. Drs. C.W. Bech-Hansen and L.C. Fowke. Reproduced by permission of the National Council of Canada from [*Can. J. Bot. 50*, 1811 (1972).

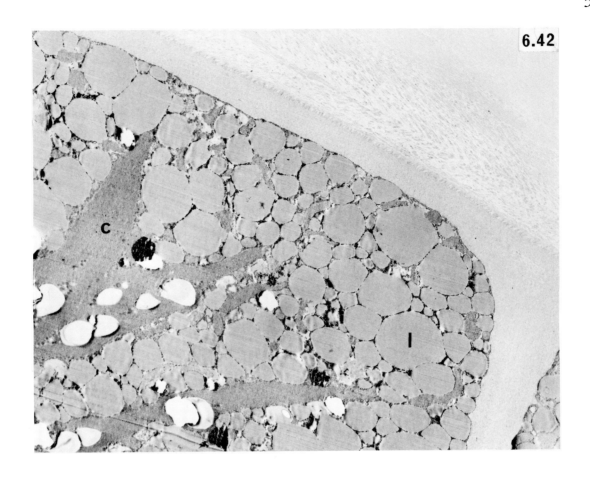

6.42

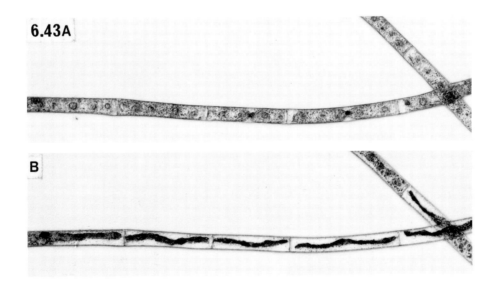

6.43A

B

6.2b. Zygnema

Zygnema, frequently encountered in nature with its close relative *Spirogyra,* is clearly distinguishable from the latter by its two stellate chloroplasts per cell, situated at either end of the central nucleus (Fig. 6.40). Each chloroplast consists of lobes which radiate from a large and complex central pyrenoid. The chloroplasts are often so large that they seem to fill the cell, but usually the two pyrenoids can be discerned even if the nucleus is obscured (Fig. 6.37). During division, the nucleus divides first, and then the cell is partitioned by an annular, ingrowing furrow. Each daughter cell then contains only one chloroplast/pyrenoid complex and one nucleus, eccentrically situated. Consequently, reorganization of the major organelles must occur to reestablish the symmetry of the interphase cell. As in certain desmids (e.g., *Netrium, Closterium;* see Sections 6.1, 6.3), the chloroplast/pyrenoid complex divides in two (sometimes before cytokinesis is over; Fritsch, 1935), and the daughter nucleus migrates into the deepening cleft forming the daughter chloroplasts, so that eventually the symmetry of each daughter cell is restored. However, in contrast to the desmids, where cell expansion is complete shortly after cell division, elongation of these cylindrical cells occurs slowly and steadily throughout the subsequent interphase period. Conjugation in *Zygnema* appears similar to that in *Spirogyra,* with both lateral and scalariform mating reported. However, anisogamy has been reported in a few species, and in others, the zygote may be formed in the conjugation tube rather than in one of the parental cell walls (Fritsch, 1935).

There has been very little ultrastructural work published on Zygnema. The chloroplast is reported to contain an unusual, crystal-like lattice of membranes (McLean and Pessoney, 1970).

FORMATION OF AKINETES

Like many other Green Algae, *Zygnema* may reproduce asexually by the transformation of individual vegetative cells into akinetes—thick-walled cells full of food reserves (oil, starch, etc.) which are able to survive prolonged periods of desiccation and other conditions unfavorable for vegetative existence.

Figures 6.44–6.52
MITOSIS IN MOUGEOTIA SP.

6.44. Early prophase. Prominent nucleolus (arrow) is dispersing. × 1,200.

6.45. Prometaphase. Nuclear envelope is becoming distorted. Numerous chromosomes are visible. × 1,200.

6.46. Metaphase. Nuclear envelope is still relatively intact. Furrowing has commenced. × 1,200.

6.47. Midanaphase. Nuclear envelope is still visible. × 1,200.

6.48. Telophase. Daughter nuclei are re-forming. × 1,200.

6.49. Prophase. Large numbers of extranuclear microtubules *(t)* have not yet penetrated the nucleus (*cf. Spirogyra;* Fig. 6.15). × 18,800.

6.50. Prometaphase. As in *Spirogyra,* the extranuclear microtubules in the polar cytoplasm (Fig. 6.16) have begun penetrating through the indented nuclear envelope. × 14,500.

6.51. Metaphase plate of chromosomes. In contrast to *Spirogyra,* nucleolar material does not coat the chromosomes, and spindle microtubules *(t)* attach to chromosomes at distinct kinetochores (arrows). × 16,200.

6.52. Detail of a metaphase kinetochore (Fig. 6.51). × 47,000.

SOURCE. Drs. C.W. Bech-Hansen and L.C. Fowke. Reproduced by permission of the National Research Council of Canada from *Can. J. Bot. 50,* 1811 (1972).

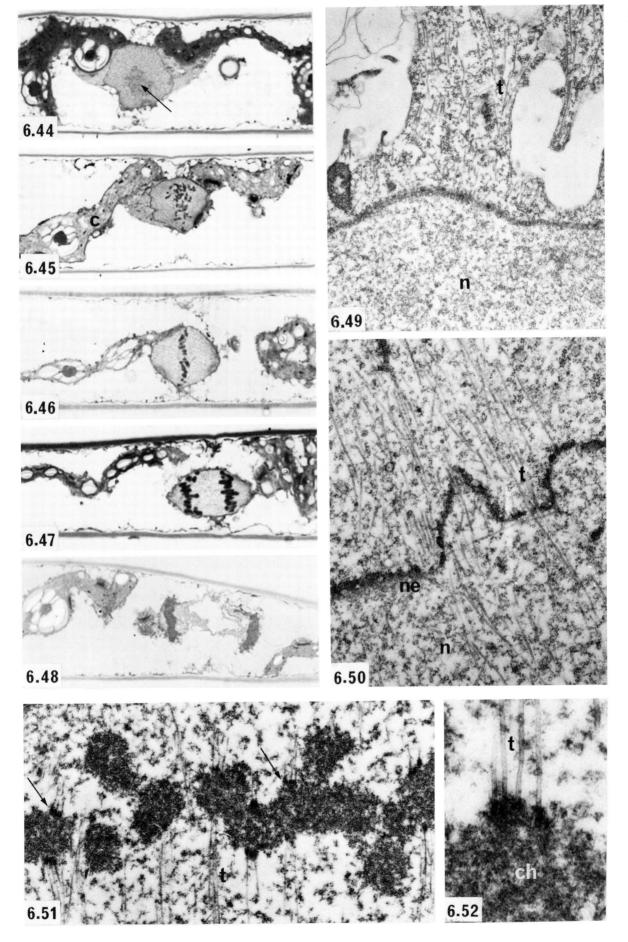

The differentiation of akinetes in *Zygnema* has been studied by McLean and Pessoney (1971) in cultures aging over a period of 6 weeks. After about 4 weeks, cells are denser (Fig. 6.38) than those from younger cultures (Fig. 6.37), having accumulated starch and lipid. The cytoplasm becomes progressively denser when viewed with the election microscope (Fig. 6.41). Golgi-derived vesicles appear to be discharged into the walls of both young and old cells; the wall becomes appreciably thickened during maturation of akinetes. In 5-week-old cultures, filaments begin to fragment into single cells whose cytoplasmic density obscures the pyrenoids and other individual cell constituents (Fig. 6.39). The chloroplasts accumulate starch grains but increasingly lose their internal organization and become homogeneous (Fig. 6.42). By 6 weeks, the abundant mature akinetes containing much reserve material (Fig. 6.42) are capable of surviving at least a year of desiccation before germinating and forming vegetative filaments.

6.2c. Mougeotia

Mougeotia is a distinctive alga because its single plate-like chloroplast lies in the middle of the cell. It is also well known to biology students because this chloroplast can actively rotate in response to changes of light intensity, either to maximize its exposure to weak light or minimize its exposure when the light intensity is too strong (Fig. 6.43; see Virgin, 1968). This movement is not colchicine-sensitive but can be reversibly prevented by treatment with cytochalasin B (Wagner, Haupt, and Laux, 1972), a drug which in some cells stops cytoplasmic streaming and disrupts cytoplasmic microfilaments (Wessells et al., 1971). Conjugation in the group Mougeotiaceae resembles that of the other filamentous Conjugales, with a slight variation. Before fusion, the gametes apparently discard part of their cytoplasm by forming sterile "cells" which are presumably enucleate (Fritsch, 1935, p. 329).

CELL DIVISION

Cell division in *Mougeotia* sp. has been described at the ultrastructural level by Bech-Hansen and Fowke (1972). During prepro-phase, the number of microtubules along the wall decreases, and does not increase again until after

Figures 6.53, 6.54
VEGETATIVE AND MITOTIC CELLS OF CLOSTERIUM LITTORALE

6.53. Vegetative cell. The terminal vacuoles (arrows) often contain small refractile crystals, absent here. The nucleus and its prominent nucleolus are centrally situated, and each of the two semicells is largely filled by a chloroplast. × 1,000.

6.54A–H. Mitosis in a live cell. The central region of the cell has enlarged, and the nucleolus has dispersed, by prometaphase (A). Metaphase (B; cf. Fig. 6.56) is typical. The septum first becomes detectable in this cell during anaphase (C–F), but grows inward mainly during telophase (G, H). × 1,100.

SOURCE. Figure 6.54A–H taken by Jean Linder.

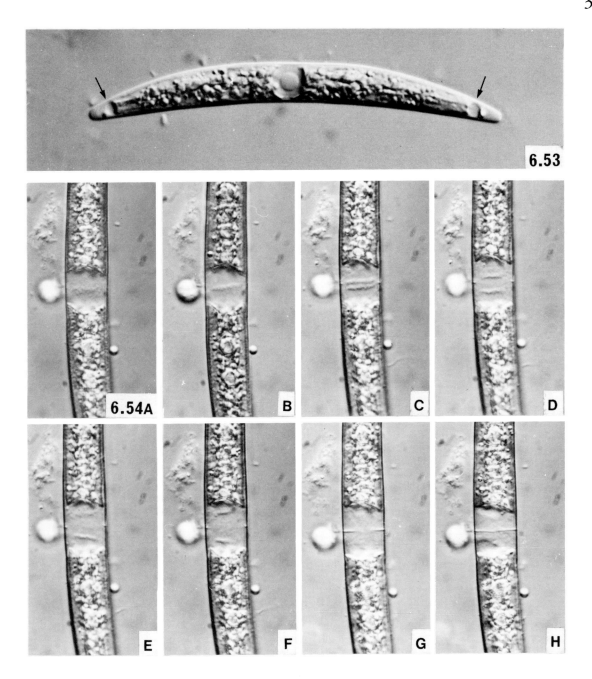

6.53

6.54A
B
C
D
E
F
G
H

mitosis is complete. During this stage, numerous microtubules appear near the chloroplast and around the nucleus, and Bech-Hansen and Fowke (1972) conclude that, as in *Spirogyra,* many wall microtubules move to and are utilized in the spindle. During prophase (Fig. 6.44) and prometaphase (Fig. 6.45), longitudinally oriented extranuclear microtubules accumulate (Fig. 6.49) which invade the nucleus (Fig. 6.50) and organize the typical metaphase spindle (Fig. 6.46). In contrast to *Spirogyra,* however, the prominent interphase nucleolus disperses completely, and furthermore, chromosomes have distinct, densely staining kinetochores (Figs. 6.51, 6.52). During anaphase elongation of the spindle (Figs. 6.47, 6.48), the semiintact nuclear envelope suffers increasing disruption, and interzonal microtubules markedly proliferate. Cytokinesis was not covered in their paper, but Bech-Hansen and Fowke indicate that a phragmoplast/cell plate system, characteristic of *Spirogyra,* is not utilized to complete the cross wall.

These results offer an interesting example of both the similarities and dissimilarities that may exist in mitotic and cytokinetic structures in two quite closely related algae, *Spirogyra* and *Mougeotia.*

6.3. Unconstricted Placoderm Desmids

The majority of the true (placoderm) desmids are characterized by a more or less pronounced constriction in the middle of the cell, which divides it into beautifully symmetrical halves or semicells. These species can conveniently be termed the constricted desmids. Relatively few "unconstricted" placoderm desmids, such as *Closterium* and some species of *Penium* (and perhaps *Hyalotheca*), possess no such central constriction. The little ultrastructural work so far published reveals some important differences, particularly during cell division, between the constricted and unconstricted desmids. (It is possible that cell division in *Closterium* may resemble that in *Netrium* more closely than that in other constricted desmids such as *Cosmarium.*)

6.3a. Closterium

Vegetative cells of *Closterium* (Fig. 6.53) are

Figures 6.55–6.57
MITOSIS IN CLOSTERIUM LITTORALE

6.55.   Prophase. The nucleolus *(nc)* is dispersing as the nuclear envelope *(ne)* fragments, allowing cytoplasmic constituents such as vesicles (arrow) into the nucleus. Chromosomes are just becoming apparent. × 7,200.

6.56.   Metaphase. The acentric spindle is open, and numerous spindle microtubules *(t)* pass by, or are attached to, the small doubled chromosomes *(cb;* see Fig. 6.54B). The septum *(s)* has grown inward appreciably in this cell. × 7,300.

6.57.   Late anaphase. The chromosomes groups *(cb)* care widely separated in the elongating spindle. The septum is still only partially formed. × 4,500.

SOURCE. Drs. J.D. Pickett-Heaps and L.C. Fowke, published in *J. Phycol.* 6, 189 (1970).

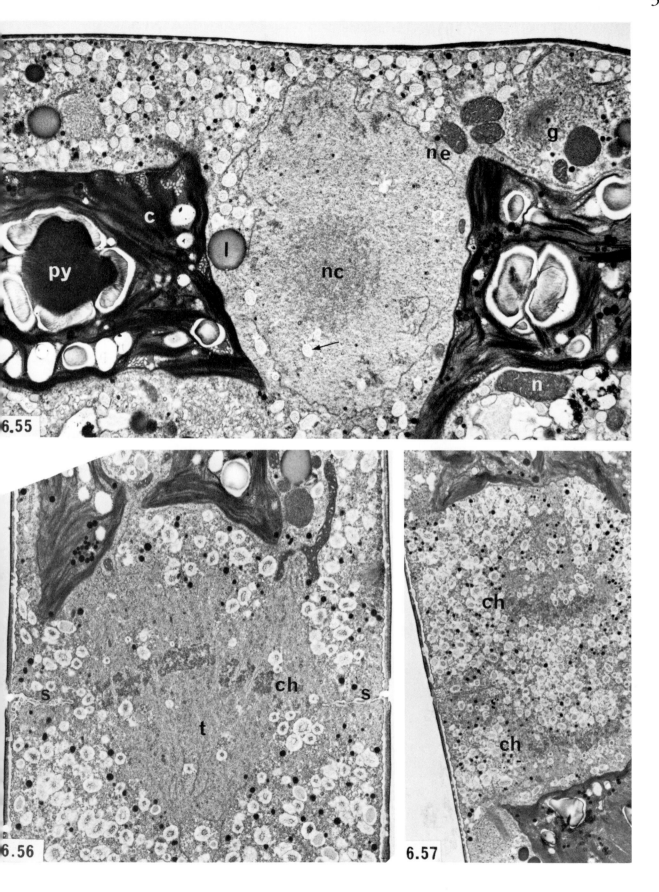

lunate or crescent-shaped in profile and curved to varying extents; some are broadened in the middle of the crescent, and most, if not all, are circular in cross section. The nucleus with its prominent nucleolus (usually single) is situated in the middle of the cell, which also can be thought of as being composed of two semicells (Lutman, 1910). Pressed against the nucleus and extending along each semicell almost to its tip is a chloroplast containing prominent pyrenoids, which may be either centrally situated or scattered. Although the chloroplasts appear to fill the semicells of live organisms, cross sections of fixed tissue reveal that the chloroplasts have regularly spaced, radially oriented ridges which extend to the cell wall (Fig. 6.70; Carter, 1919a). The long grooves between the ridges are filled with cytoplasm containing the usual assortment of cytoplasmic organelles. In the small clear area of cytoplasm at the tip of each semicell is a small, conspicuous vacuole which often (depending upon the growth conditions) contains a refractile crystal or two (Fig. 6.53). The wall contains numerous rows of pores (Mix, 1969) through which presumably is secreted the mucilage involved in cellular movement.

Ultrastructurally, the cytoplasm of vegetative *Closterium* cells (in this case, *Cl. littorale*) is slightly unusual in several respects (Pickett-Heaps and Fowke, 1970c). Large vesicles with diffuse fibrillar contents are numerous (e.g., Figs. 6.55–6.57); I suspect these contain mucilage destined for secretion. Small vacuoles, containing a variety of different inclusions, are also common. Virtually no microtubules are seen near fully expanded, vegetative cell walls, except sometimes for a small number in a band encircling the cell near the nucleus. This lack of wall microtubules during interphase is in striking contrast to the number associated with the expanding wall secreted after cell division (see below). A layer of endoplasmic reticulum often follows the contours of the chloroplasts (Fig. 6.70).

CELL DIVISION

Mitosis and cytokinesis in *Closterium* is quite normal in most respects, but as with other desmids, every time the cell divides its symmetry is destroyed. The new cross wall or septum bisects the crescent-shaped organism, and each daughter semicell, therefore, contains only one chloroplast

Figures 6.58–6.60
POSTMITOTIC MORPHOGENETIC EVENTS IN LIVE CELLS OF CLOSTERIUM LITTORALE

6.58*A–E.* Nuclear movement along semicells after cytokinesis is complete. The course of this movem
ent can be followed by the position of the prominent nucleoli (large arrows), and the nuclei eventually become inserted in the cleavage that cuts each chloroplast in two *(E)*. Note also in *E* that the cells are just beginning to expand and push apart (small arrow). × 1,100.

6.59. Early stage of semicell expansion. The position of the nucleus in one semicell is indicated *(n)*. × 1,600.

6.60. A slightly later stage. The cells are almost separated, but remain ensheathed with an investment of mucilage. Note the new terminal vacuoles appearing at the ends of the forming semicells. × 1,600.

SOURCE. Figure 6.58*A–E* taken by Jean Linder.

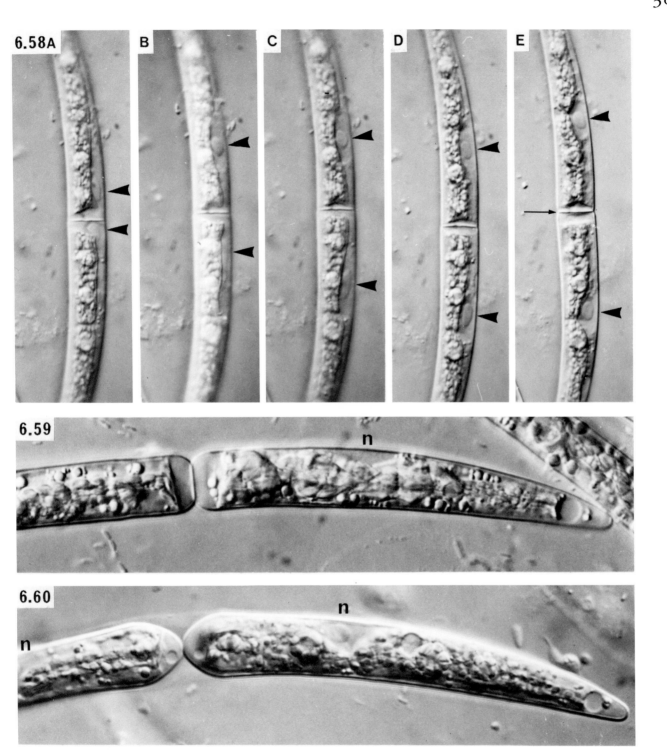

and a nucleus near the septum. To restore its beautiful interphase symmetry (Fig. 6.53), after cytokinesis each daughter cell carries out three distinct but integrated morphogenetic activities: (1) the chloroplast divides, (2) the nucleus moves along the semicell until it is inserted between the halves of the chloroplast, and (3) precisely controlled expansion of a new wall originally forming the septum, concurrent with continuous secretion of material into it, allows the crescent-shaped profile of a new semicell to be established. These events have been investigated by several earlier workers (in particular, Lutman, 1911; see Fritsch, 1935, p. 345), but many of the fascinating details have been revealed only recently by ultrastructural analysis (Pickett-Heaps and Fowke, 1970c).

Mitosis and cytokinesis. In live cells of *Cl. littorale* (Linder and Pickett-Heaps, unpublished data), the onset of prophase is revealed by the rapid dissolution of the nucleolus, which takes about 2–3 minutes (Fig. 6.54*A*). In the next few minutes, the chromosomes condense and move into the metaphase plate configuration, in which they remain for about 10 minutes (Fig. 6.54*B*). The spindle, occupying the region between the chloroplasts, is free of cytoplasmic streaming, but in it, the metaphase plate of chromosomes slowly oscillates and moves around. Anaphase lasts about 2–3 minutes, and the two plates of chromosomes, initially sharply defined as they separate, become more obscure as they reach the poles of the elongating spindle (Fig. 6.54*C–G*). After anaphase is completed, cytoplasmic streaming slowly resumes between the daughter nuclei. Formation of the septum is usually detectable by anaphase (Fig. 6.54*C*) and is completed within 20 minutes (Fig. 6.54*H*). Then follows a variable interval (15–30 minutes) during which the nuclei wander around the cytoplasm adjacent to the septum, and eventually they become more tightly sandwiched between the chloroplast and the cross wall, where they remain for up to about 15 minutes.

The events leading up to restoration of cell symmetry usually commence with the initiation of chloroplast cleavage in the species we studied *(Cl. littorale);* in some species, chloroplast cleavage is initiated by prophase. In the space of 7–15 minutes, the nuclei migrate along the side of the semicells (Figs. 6.58) and become inserted in the

Figures 6.61–6.64
EARLY STAGES IN POSTMITOTIC NUCLEAR MIGRATION IN CLOSTERIUM LITTORALE

6.61. Early telophase. The small nucleus *(n)* at this stage is difficult to discern, and the re-forming nuclear envelope is indicated by arrows. The persistent spindle microtubules *(t)* are already becoming focused upon the microtubule center *(mc)*, next to the chloroplast *(c)*. × 25,000.

6.62. Later stage, with the nucleus now enlarged and clearly visible. × 25,000.

6.63. The microtubule center, the focus of the numerous microtubules, has just started migrating along the cytoplasm lying in a lobe of the chloroplast (see Figs. 6.70–6.72). × 30,000.

6.64. Further migration of the microtubule center along the semicell. The nucleus is becoming deformed. × 32,000.

SOURCE. Drs. J.D. Pickett-Heaps and L.C. Fowke, published in *J. Phycol.* 6, 189 (1970).

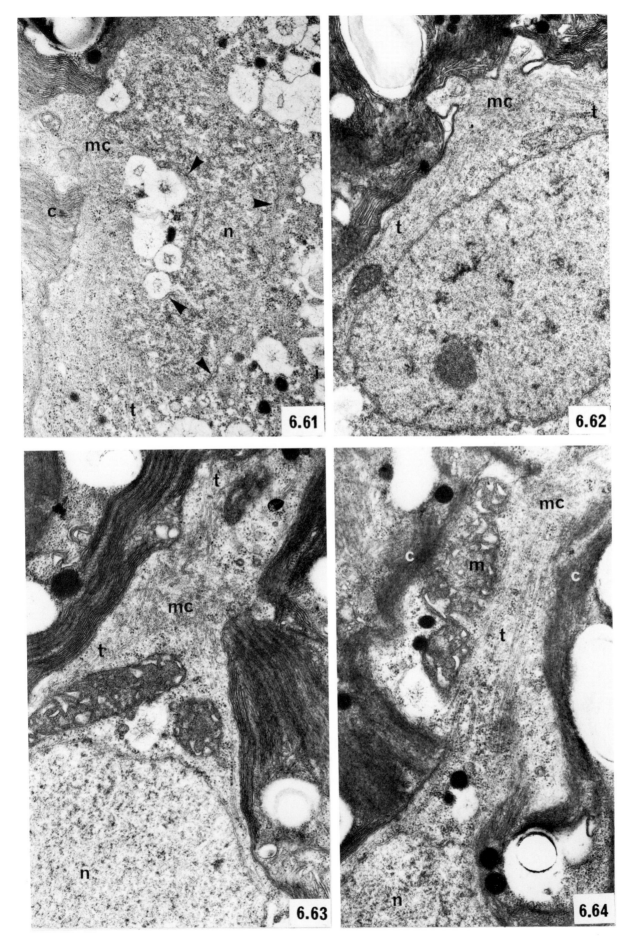

deepening cleavage in the chloroplast. Daughter cells often begin expansion during nuclear movement (Fig. 6.58E). This re-formation of the new semicell is, however, a long process. Within an hour of cytokinesis, the daughter cells often become separated (Figs. 6.59, 6.60), but they may not become completely symmetrical for at least another hour or so.

During prophase, when chromosomes appear in the nucleus, the nucleolus becomes increasingly disperse (Fig. 6.55) until it disappears. Microtubules proliferate outside and then inside the nucleus as its membrane begins to fragment. By metaphase, the numerous tiny doubled chromosomes are aligned across the spindle in a precise plate (Fig. 6.56); typical spindle microtubules pass between or else attach to the chromosomes. The chromatids separate as two distinct plates during anaphase (Fig. 6.57) and soon lose their individual identity as telophase commences (Fig. 6.61). The small daughter nuclei are initially difficult to discern in the cytoplasm (Fig. 6.61), but soon enlarge and acquire nucleoli. Most importantly, and in striking contrast to what is observed in most other spindles, extranuclear (spindle?) microtubules persist around the re-forming nuclei (Figs. 6.61, 6.62); the significance of this phenomenon will soon become clear.

Cytokinesis is achieved by an annular ingrowth of plasmalemma. The forming septum is quite obvious by metaphase, when it is coplanar with the metaphase chromosomes (Fig. 6.56), and it is completed, bisecting the cell, by telophase. Microtubules are not associated with the septum until after cytokinesis has been completed for some time. This new cross wall may grow by incorporation of vesicles into it, a possibility that has not been definitively established.

Reestablishment of cell symmetry. The three main events necessary to reestablish cell symmetry (see earlier) will be described separately here, but it is reemphasized that these are all concurrent and carefully integrated events.

The mechanisms involved in chloroplast division remain quite obscure. We (Pickett-Heaps and Fowke, 1970c) found microtubules in the deepening cleavage of chloroplasts (see below), but were cautious about whether these organelles could be causative agents in the process.

Figures 6.65–6.69
NUCLEAR MIGRATION IN CLOSTERIUM LITTORALE

6.65.   The nuclei before migration, near the completed septum. × 1,800 approx.

6.66.   The nuclei during migration between lobes in the chloroplasts. Note their extremely attenuated form. Chloroplast cleavage is denoted by arrows. × 2,000 approx.

6.67.   The nuclei (arrows) after migration along the semicells to the cleavage in the chloroplasts. Expansion of the new semicells has also commenced. × 1,800 approx.

6.68.   Deformed, migrating nucleus. Note the microtubules *(t).* × 6,400.

6.69.   The microtubule center *(mc)* and its associated microtubules preceding the nucleus along the semicell. × 14,000.

SOURCE. Drs. J.D. Pickett-Heaps and L.C. Fowke, published in *J. Phycol.* 6, 189 (1970).

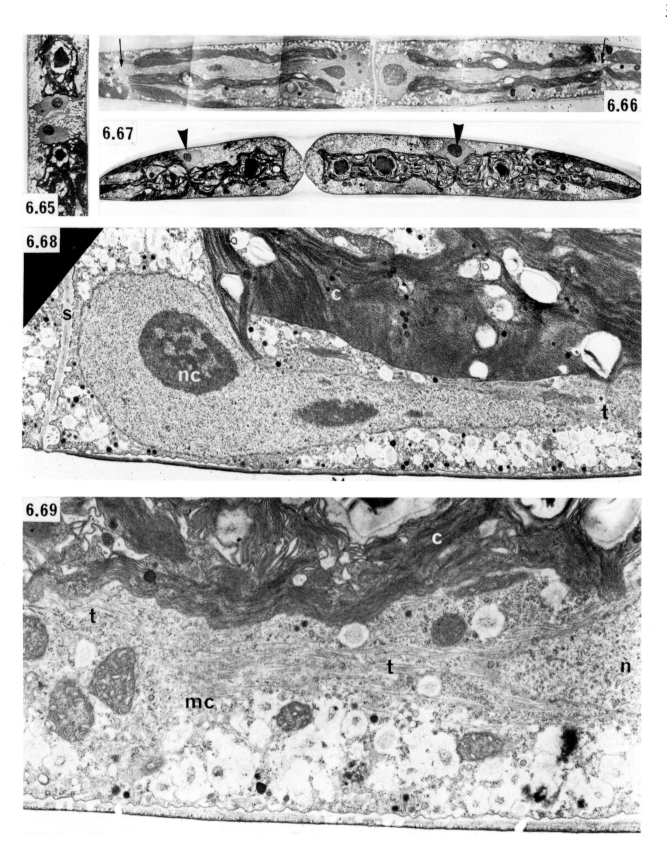

393

Division of the chloroplast appears quite autonomous, and in some species of *Closterium,* the chloroplast commences division before prophase (i.e., long before any microtubules reach the cleavage). We have stopped nuclear movement by treatment of postmitotic cells with colchicine; chloroplast division remains unaffected. Such results are in agreement with observations of other algae (e.g., *Cosmarium, Ulothrix*), where cleavage of the chloroplast again occurs during a certain distinct phase of a division cycle, either before or after mitosis and cytokinesis, and is not associated with the presence of microtubules.

There can be little doubt that microtubules are involved in movement of daughter nuclei along each semicell. After telophase, the persistent extranuclear microtubules (Figs. 6.61, 6.62) become focused upon a distinct small region of cytoplasm adjacent to the nuclear envelope. This "microtubule center" can soon be seen to contain a characteristic, if ill-defined, ball of material from which many microtubules radiate. The microtubule center moves to the side of the cell until it is opposite the cytoplasm lying in one of the grooves of the chloroplast. It then migrates along this groove (Figs. 6.63, 6.64). Microtubules run back from it toward the nucleus (Fig. 6.69); other microtubules extend forward from it, and a small number radiate irregularly in all directions. While it moves along the groove, the nucleus becomes increasingly deformed near the longitudinal microtubules emanating from the microtubule center, until a long evagination of the nucleus has become drawn out (Figs. 6.66, 6.68, 6.69), ensheathed by these tubules (cf. Figs. 6.65, 6.66). The microtubule center itself finally lodges in the constriction of the dividing chloroplast, perhaps held there by its complement of radiating microtubules which come to lie in the constriction. The nucleus, suffering remarkable deformation, moves along the same narrow groove of cytoplasm (Figs. 6.70–6.72), ensheathed by microtubules throughout this movement until it too is eventually drawn up into the chloroplast cleavage (Figs. 6.67, 6.74). Finally, the nucleus expands between the halves of the recently divided chloroplast. The narrowness of the space between the chloroplast lobes is shown in Figures 6.66, 6.70, and 6.73.

Microtubules also appear to be involved somehow in elongation and expansion of the semicell wall. Initially, there are no microtubules

Figures 6.70–6.74
NUCLEAR MIGRATION IN CLOSTERIUM LITTORALE

6.70.   Cross section of cell, showing the typically lobed chloroplast *(c)* and centrally situated pyrenoid *(py).* Nuclear migration is taking place along the groove in the chloroplast indicated by the arrow (Figs. 6.71, 6.72). × 7,000.

6.71.   Serial section of the cytoplasmic groove indicated in Figure 6.70. The group of microtubules *(t)* is engaged in nuclear migration. × 30,000.

6.72.   Serial section of the cytoplasmic groove shown in Figure 6.71. The migrating nucleus and its nucleolus *(nc),* ensheathed by microtubules, virtually fill the groove. × 24,000.

6.73.   Light micrograph. The section shaved the edge of a vegetative cell and shows two lobes of one chloroplast (*cf.* Fig. 6.66) extending the length of the semicell. The position of the nucleus is indicated by the arrow. × 900.

6.74.   The migrating nucleus just before insertion in the cleavage of the chloroplast (arrow). × 5,000.

SOURCE. Drs. J.D. Pickett-Heaps and L.C. Fowke, published in *J. Phycol. 6,* 189 (1970).

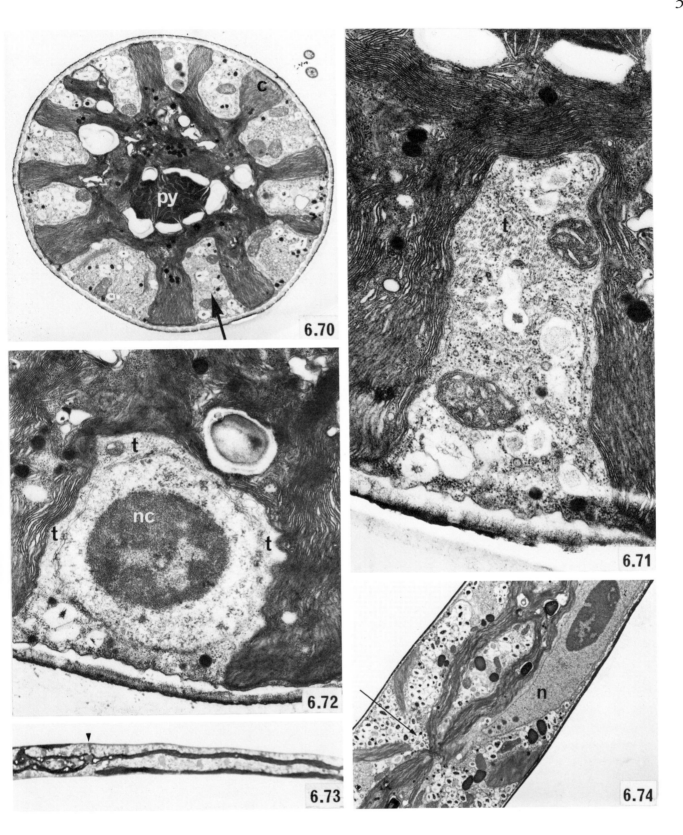

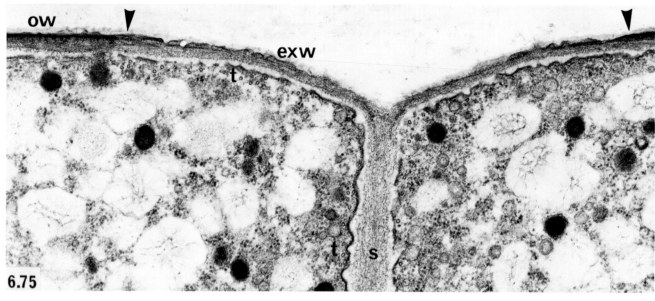

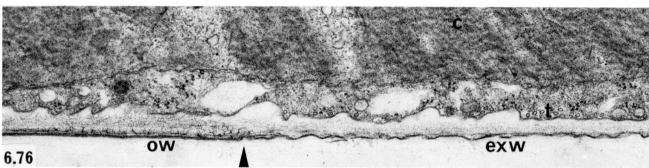

Figures 6.75, 6.76
SEMICELL EXPANSION IN CLOSTERIUM LITTORALE

6.75.   Early stage of expansion (equivalent to that in Fig. 6.59). Microtubules *(t)* proliferating near the thickening septum *(s)* have become circumferentially oriented immediately underneath the expanding wall *(exw)*. They are not present near the older wall *(ow)*. The join between older and expanding wall is indicated by arrows. × 48,000.

6.76.   Semicell expansion almost complete. The join between the two walls is again indicated by an arrow, and microtubules clearly remain only near the expanding wall, as in Figure 6.75. × 48,000.

SOURCE. Drs. J.D. Pickett-Heaps and L.C. Fowke, published in *J. Phycol. 6,* 189 (1970).

Figure 6.77
MAIN EVENTS DURING CELL DIVISION IN CLOSTERIUM LITTORALE

Diagrammatic representation. The difference in thickness of the older wall *(ow)* and the expanding wall *(exw)* is greatly exaggerated for clarity. The symmetry of the interphase cell is shown in *A*. Mitosis *(B)* and cytokinesis *(C)* destroy this symmetry. Microtubules *(t)* first appear near the septum after cytokinesis *(D)*, and during subsequent semicell expansion *(E–G)*, they line the expanding cell wall. Meanwhile the microtubule center *(mc)* forms near daughter nuclei after cytokinesis; this migrates along the cell *(D)* until it becomes embedded *(E)* in the cleavage developing in the chloroplast *(c)*. The nucleus follows the microtubule center along the cell *(E)*, eventually becoming interposed between the two halves of the divided chloroplast *(F)*. By the end of semicell expansion *(G)*, the interphase symmetry of the cell *(A)* has been restored.

SOURCE. Drs. J.D. Pickett-Heaps and L.C. Fowke, published in *J. Phycol. 6,* 189 (1970).

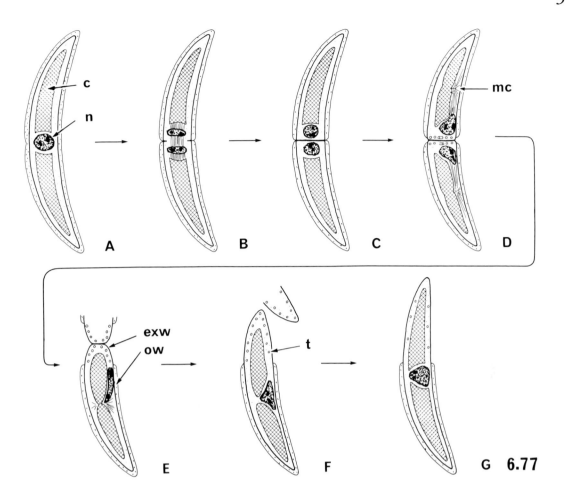

6.77

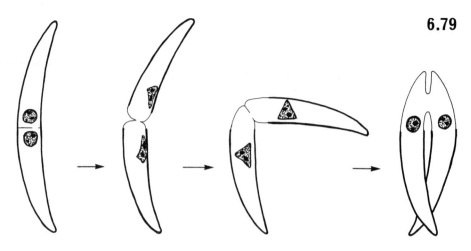

6.79

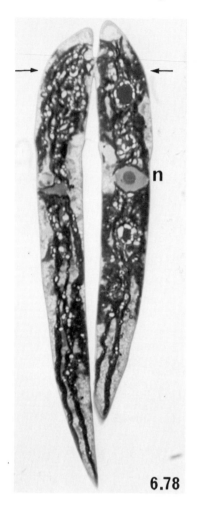

6.78

Figures 6.78, 6.79
ABNORMAL CELL DIVISION IN CLOSTERIUM LITTORALE

6.78. Abnormal, Siamese-twin pair of cells of *Cl. littorale*, joined by a neck of cytoplasm (between arrows). Two nuclei are present, one *(n)* clearly visible. × 1,400.

6.79. Diagrammatic representation of how the pair of cells in Figure 6.78 could have arisen, a result perhaps of a perforation remaining in the septum after cytokinesis.

SOURCE. Figure 6.78: Drs. J.D. Pickett-Heaps and L.C. Fowke, published in *J. Phycol. 6,* 189 (1970).

Figures 6.80–6.84
SEXUAL REPRODUCTION IN LIVE CELLS OF CLOSTERIUM MONILIFERUM AND CL. EHRENBERGII

6.80. *Cl. moniliferum.* Pairing of incompletely formed daughter cells. Papillae are just being formed. × 340.

6.81. *Cl. ehrenbergii.* Detail of the papillae. Note the hyaline gelatinous material between the tubes. × 400.

6.82. *Cl. moniliferum.* Nonsynchronous conjugation. × 330.

6.83. *Cl. moniliferum.* Typical pair of zygospores resulting from conjugation that follows cell division. × 390.

6.84. *Cl. moniliferum.* Sausage-shaped gones inside vesicular membrane, following germination of a zygospore. × 520.

SOURCE. Dr. B.E. Lippert, published in *J. Phycol. 3,* 182 (1967).

near the septum, but as the septum thickens, increasing numbers of rather randomly oriented microtubules gather adjacent to and in the plane of the septum, transversely oriented between daughter nuclei. As the daughter zyter cells start to

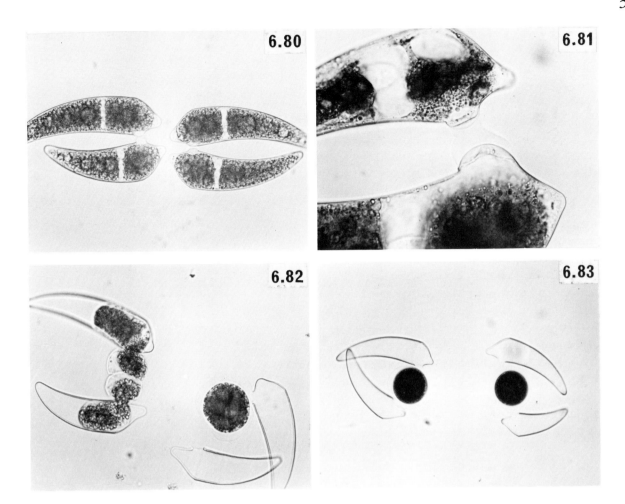

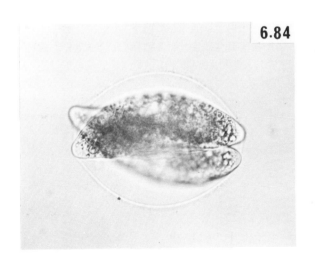

expand (i.e., by a ballooning out of the wall material of the septum in a carefully controlled fashion), these microtubules become arranged in a transverse, hoop-like configuration close to the wall, typical of many higher plant cells (Fig. 6.75). They are distributed with remarkable precision along the newly expanded wall only; none are adjacent to the older, semicell wall, and they remain thus until semicell morphogenesis is complete (Figs. 6.75, 6.76). The cytoplasm, of course, expands continuously during these events. Figure 6.77 summarizes these important morphogenetic events.

I have conducted some preliminary experiments treating dividing cells with colchicine. The action of this drug in depolymerizing microtubules in plant cells is well known (Pickett-Heaps, 1967a). When applied after cytokinesis, colchicine does not affect chloroplast division, but it prevents nuclear movement and has a variable, dislocative effect on wall synthesis and expansion. However, the experiments are still inconclusive. For example, after a longer period of exposure to the drug (12–24 hours), many nuclei seem to have moved along to their correct position at the chloroplast constriction. Expanded daughter cells are usually misshapen, but often not drastically so. Only after a much more prolonged exposure (several weeks or more) to colchicine do cells become more grossly distorted, usually developing large medial distentions. These results lead me to suspect that nuclear movement can, if necessary, be accomplished slowly without the use of microtubules, as may be the case in other cells of the Conjugales (such as *Netrium, Hyalotheca, Zygnema*), where nuclei do not become distorted during their more rapid postmitotic movement. If so, perhaps this microtubule involvement in *Closterium* represents a relatively newly evolved cytoplasmic system facilitating the rapid passage of the nucleus down a constricted and narrow passage, a problem not encountered in the other, highly vacuolate organisms mentioned above.

One further point needs to be added. As will soon become evident, constricted desmids do not utilize microtubules in any obvious way for the attainment of the beautiful symmetry of their expanding semicells after cytokinesis (Section 6.4a). Why then does *Closterium* generate large numbers of microtubules near its own expanding wall? Again I feel that the most reasonable expla-

Figures 6.85–6.88
SEXUAL REPRODUCTION IN CLOSTERIUM LITTORALE

6.85. Initiation of the conjugation tube, as a girdle of new wall material intercalated into the older wall, extending around each cell (arrows). × 4,500.

6.86. As for Figure 6.85, but in this case, the site of deposition of the new wall is highly asymmetrical in each cell. Note how a lobe of the chloroplast (c) appressed to the wall has moved away from the region of papilla formation. × 3,000.

6.87. Detail of Figure 6.85, showing the new wall material of the future papillae, between the arrows. × 1,500.

6.88. Cells with papillae half formed (see Fig. 6.89). The cells have become bent, and the cytoplasm is beginning to shrink from the tips of the semicells. × 800.

SOURCE. Drs. J.D. Pickett-Heaps and L.C. Fowke, published in *J. Phycol. 7*, 37 (1971).

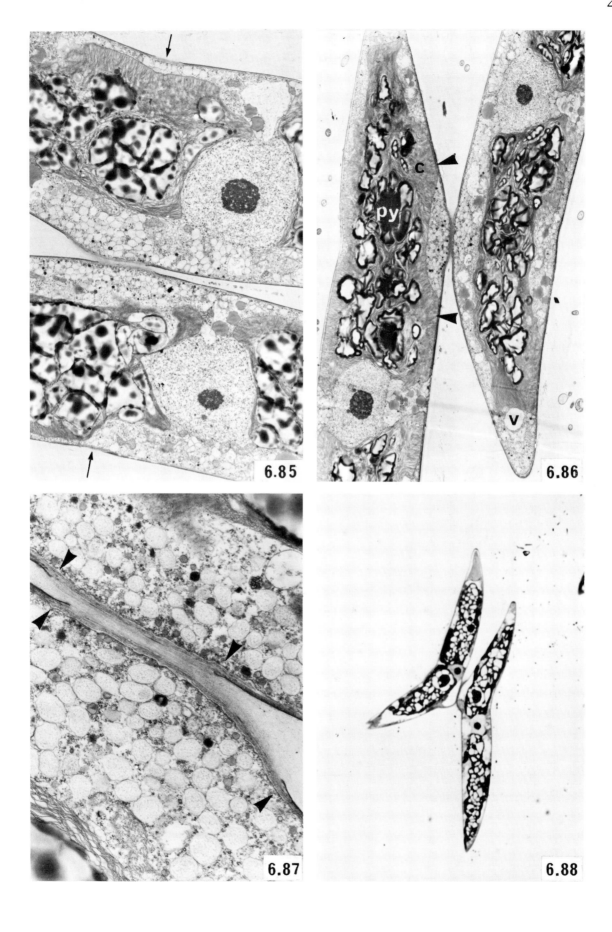

6.85

6.86

6.87

6.88

nation (in our present lack of knowledge concerning the ultrastructure of desmids generally) is that this microtubular system may also be a relatively recent cytoplasmic innovation which assists in the reestablishment of cell shape. The disposition of these tubules in *Closterium,* where the cell is circular in cross section and expands longitudinally, is similar to those in various other cylindrical algae and cells of higher plants. If colchicine is applied to *Chara* and higher plants, actively elongating cells often become spherical instead of remaining cylindrical as they grow; thus undoubtedly, wall microtubules are importantly involved in morphogenesis. Colchicine treatment of expanding cells of *Closterium* does not have such immediate and drastic effects, and so I should guess that *Closterium* has begun to utilize its wall microtubules in a fashion similar to that of these other organisms, but is not yet dependent upon them. (Needless to say, it is difficult to suggest what these microtubules might actually do!)

If the septum is incompletely formed in constricted desmids, the connected semicells remain joined by a neck of cytoplasm, and the amount of expansion they undergo is related to the extent of the septum present at attempted cytokinesis. This observation ties in with the concept that the septum's membrane is in some way preprogrammed to form the new semicell's shape (Section 6.4a) and that this preprogramming is upset if the septum is perforated. This phenomenon is sometimes encountered in *Closterium,* where the extent of wall expansion undergone by incompletely separated cells may depend upon the size and location of the perforation in the septum. For example, the incompletely separated but almost fully formed "Siamese-twin" cells illustrated in Figure 6.78 were presumably a result of expansion in daughter cells which remained joined by a small perforation on one side of the septum (as in Fig. 6.79). If true, such results argue that much of the information required to make the new semicell in *Closterium* (as is also the case in constricted desmids) resides in the cell membrane and that, therefore, the wall microtubules may be of secondary importance during morphogenesis.

The microtubule center seems to be a highly significant morphogenetic organelle, whose importance presumably resides in its ability to organize and control deployment of microtubules. I find it most significant that this structure appears

Figure 6.89
SEXUAL REPRODUCTION IN CLOSTERIUM LITTORALE

The conjugation papillae, fused between the cells, are bulging out toward one another, presumably under the influence of the enlarging papilla vacuoles *(pv).* The distinction between the older wall and the thicker, less compact wall of the papillae *(plw)* encircling the cells is clear. × 9,000.

SOURCE. Drs. J.D. Pickett-Heaps and L.C. Fowke, published in *J. Phycol.* 7, 37 (1971).

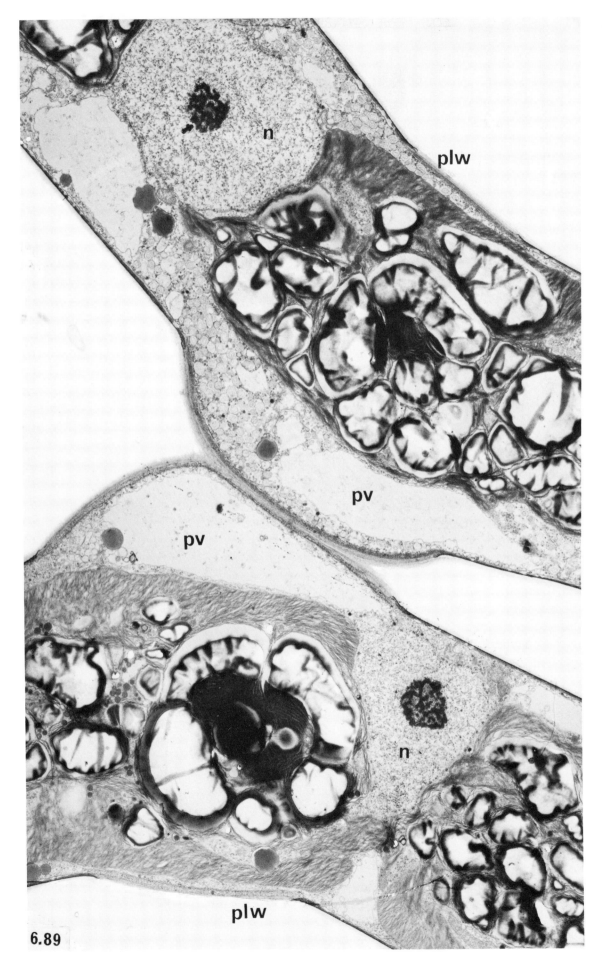

6.89

to form in the cytoplasm of the poles of the spindle, where presumably the entities responsible for organization of spindle microtubules are earlier located in an undetectable (more disperse?) state. I feel that this structure may be equivalent to a mitotic "centrosome" or microtubule-organizing center (MTOC), but one that forms after division for purposes not related directly to mitosis.

SEXUAL REPRODUCTION (CONJUGATION)

Sexual reproduction in *Closterium* has been studied by a number of workers using light microscopy. Recently, Ichimura (1971) showed that when heterothallic ("plus" and "minus") strains of *Cl. strigosum* were mixed under appropriate inductive conditions (e.g., nitrogen deficiency in the medium), sexual reproduction was preceded by one semisynchronous cycle of cell division, 5–8 hours after induction, before pairing of conjugants occurred. Such cell divisions were not observed in either strain separately subject to the same inductive conditions. Thus, as in so many other Green Algae, gamete formation is probably preceded by a cell division that is determinative in some subtle way. Furthermore, cell-free medium obtained from induced mixed cultures was able to induce this cell division in either strain separately, thus indicating the existence of some inducer secreted by sexually active cells. Such behavior is reminiscent of the induction of sexuality in *Volvox* (Section 2.2a), where the inducer, known to be first produced by male colonies, is able to induce sexuality in both male and female colonies thereafter.

Lippert (1967) has reviewed earlier work on this genus in his exhaustive documentation of conjugation in *Cl. moniliferum* and *Cl. ehrenbergii.* These two species are slightly unusual since conjugation takes place between incompletely formed daughter cells derived from an earlier (determinative) cell division (Figs. 6.80–6.82). In some strains of *Cl. moniliferum,* vegetative cells pair off before dividing and then divide, proceed with conjugation, and as a result, pairs of zygospores are formed (Fig. 6.83). Homothallic and heterothallic strains of both species have been isolated.

Sexual reproduction in *Cl. littorale* has been investigated at the ultrastructural level (Pickett-Heaps and Fowke, 1971). Conjugation takes place between fully formed cells.

Figures 6.90–6.92
SEXUAL REPRODUCTION IN CLOSTERIUM LITTORALE

6.90.   Papilla enlargement. The nuclei *(n)* are situated in the papillae. Cytoplasmic shrinkage has left the ends of the semicells filled with mucilage (arrows). × 500.

6.91.   Further cytoplasmic shrinkage. Note the emptying semicells (arrows). × 1,100.

6.92.   Detail of Figure 6.91. The papilla wall *(plw)* has become extremely thin between the cells (arrow), prior to fertilization (*i.e.,* fusion of the protoplasts). × 17,000.

SOURCE. Drs. J.D. Pickett-Heaps and L.C. Fowke, published in *J. Phycol.* 7, 37 (1971).

405

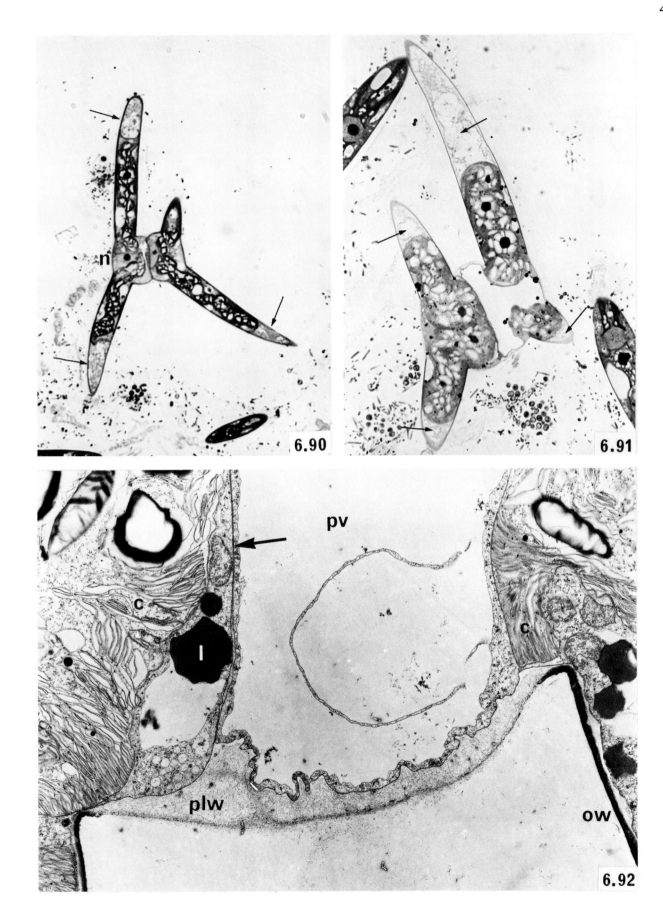

6.90

6.91

6.92

Cells about to conjugate display increasing cytoplasmic denseness and numerous lipid droplets before they pair up and become closely appressed to one another. Formation of papillae is then initiated, usually simultaneously in both cells, by the appearance around each cell of a circumferential girdle of rather loosely organized wall material inserted into and distinctly different from the mature cell wall (Figs. 6.85–6.87). This girdle is usually, but not necessarily, centrally situated (cf. Lippert, 1967). Incidentally, the conjugation tube is always formed at the isthmus of constricted desmids; the ability of *Closterium* to form the tube elsewhere along the cell is more characteristic of saccoderm desmids (Wiese, 1969). The girdles in the two cells always touch at one spot, the wall fibers of each mingling (Fig. 6.87). This new wall increases in size, but expansion is greatest at the area of mutual contact, so that the old semicells of each individual begin to kink outward and apart, and the lobes of the chloroplast may be pushed back from this region of the wall (Fig. 6.86). Soon, two distinct papillae or ballooning outgrowths of this wall appressed to one another are formed between the conjugants (Figs. 6.88, 6.89). Within each papilla there is usually (and perhaps always) a large vacuole which grows as the papilla expands. Meanwhile the cytoplasm shrinks away from the walls, and mucilaginous material accumulates in the emptying semicell (Figs. 6.90, 6.91). The terminal vacuoles soon collapse and disappear.

As far as we can tell, the papilla wall arises as a result of a rapid and localized phase of wall deposition, rather than because some change occurs in the preexisting wall (cf. *Spirogyra;* Section 6.2a). Vesicles appear to be discharged into this wall, and some microtubules are always close to it during early stages of its formation, but these disappear once the papillae have enlarged. Increasing turgor pressure in the papilla vacuole could be responsible for the expansion undergone by the papillae, even while other vacuoles are disappearing and the cytoplasm is contracting. The accumulation of mucilage in the semicells is just as likely to be a result, instead of a cause, of cytoplasmic contraction since this mucilage might not push through the cell wall if the cytoplasmic turgor pressure were not sufficiently high (indicated by the collapse of the terminal vacuoles).

Figures 6.93–6.97
ZYGOTE FORMATION IN CLOSTERIUM LITTORALE

6.93. Zygote rounding up inside the papillae, after fertilization. × 1,100.

6.94. Zygote contracting and secreting its resistant wall. The empty semicells are all interconnected by the papilla walls. One nucleus can be seen (arrow). × 1,100.

6.95. Zygote surrounded by thick wall (Fig. 6.97). × 1,100.

6.96. Cytoplasm of the zygote immediately after fertilization. The convoluted, paired membrane (arrow) contrasts with that observed between cells prior to fertilization (Fig. 6.92). The cytoplasm at this stage becomes devoid of many larger organelles and filled with ribosomes, etc. × 11,000.

6.97. Thick wall of maturing zygote, whose cytoplasm is typically becoming very dense. The wall contains seven distinguishable layers; the inner component is thick and stratified. × 21,000.

SOURCE. Drs. J.D. Pickett-Heaps and L.C. Fowke, published in *J. Phycol. 7,* 37 (1971).

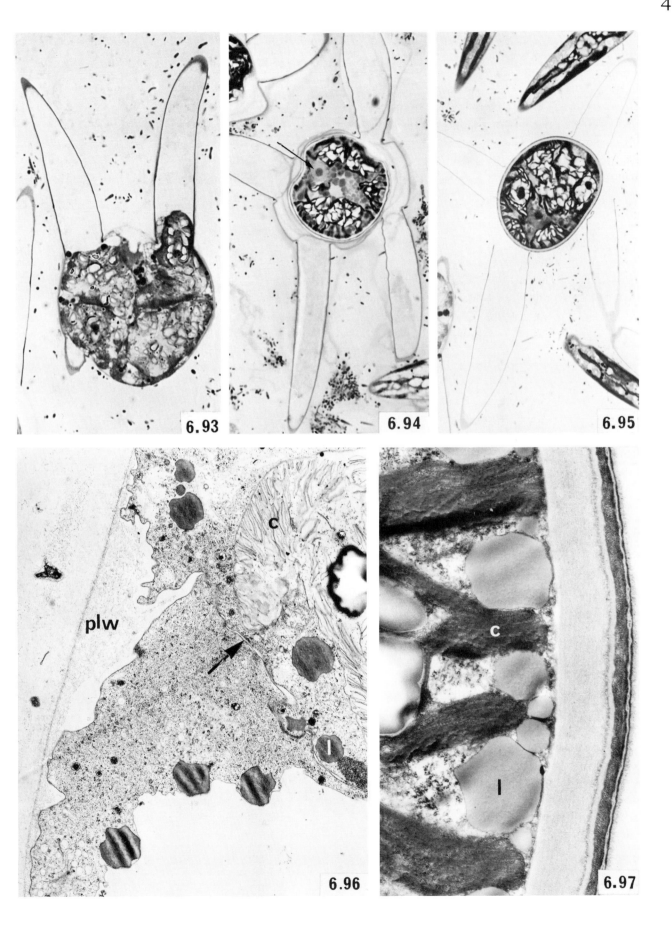

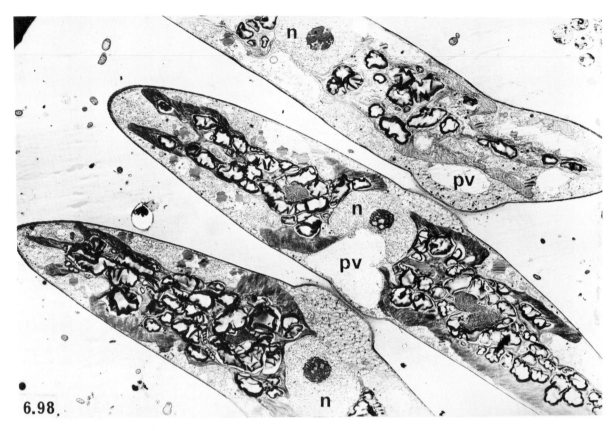

6.98.

As fertilization approaches, the protoplasts of the two gametes have now moved almost entirely into each of the considerably distended papillae. The large area of papilla wall separating the two cells becomes thin (Fig. 6.92) and finally disappears, whereupon the plasmalemmas of the protoplasts fuse. During these events, the cytoplasm and its organelles always become markedly pale in comparison with others nearby. At the ultrastructural level, this general loss of stainability is accompanied by an apparent breakdown of many of the smaller cytoplasmic inclusions. Quite large volumes of cytoplasm appear filled with a homogeneous "soup" of ribosomes and ill-defined material, but largely devoid of endoplasmic reticulum, Golgi bodies, etc. (Fig. 6.96). We have speculated that these changes indicate a breakdown of those cytoplasmic organelles used in vegetative existence but no longer needed for the zygote's changing requirements of metabolism and synthesis.

Following fertilization (Figs. 6.93, 6.96), the binucleate zygote rounds up completely inside the fused papillae, which can more correctly be thought of now as a conjugation tube. The zygote then undergoes considerable contraction while it secretes a complicated wall (Figs. 6.94,

Figure 6.98
CONJUGATION IN CLOSTERIUM LITTORALE

Attempt at triple conjugation in *Cl. littorale.* Note that the papillae are contiguous in all three cells. $\times$ 3,000.

Figures 6.99–6.101
THE CONSTRICTED DESMID ARTHRODESMUS SP.

6.99.   Live cell. The nucleolus of the centrally situated nucleus is clearly visible, flanked on either side by a chloroplast which fills most of each semicell and contains a single large pyrenoid. The cell is enclosed by a diffuse envelope of mucilage whose outer surface is revealed by the numerous bacteria on it. $\times$ 1,000.

6.100.   These cells were extracted by warm 0.1 M NaOH after fixation and before preparation for scanning electron microscopy. Even these harsh extractive conditions failed to remove all the mucilage around the cells. $\times$ 2,800.

6.101.   As for Figure 6.100. In this rare example, all the mucilage has been removed, revealing the outer wall surface. The frayed edge of the lip of the older semicell is visible. $\times$ 3,100.

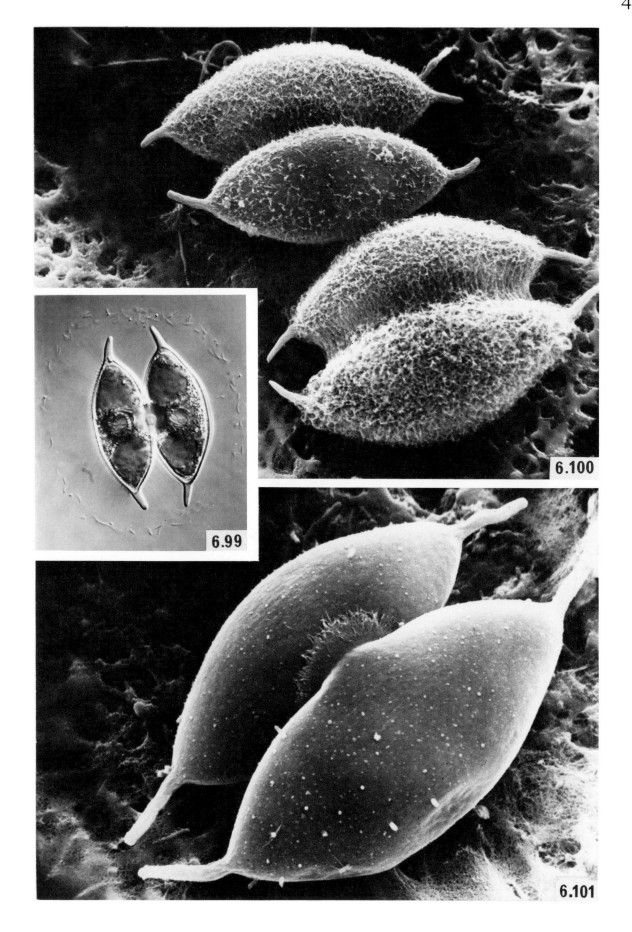

6.100

6.99

6.101

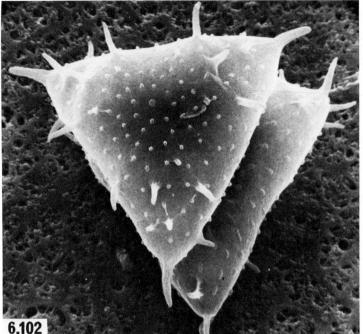

6.102

6.103

Figures 6.102, 6.103
TWO TRIRADIATE SPECIES OF STAURASTRUM

6.102. *S. cristatum.* × 2,000.

6.103. *S. gladiosum.* × 1,500.

SOURCE. Author's micrographs, published in *Trans. Am. Microsc. Soc.* 93, 1 (1974).

6.95). Electron microscopy shows that this wall contains at least seven layers, some of which are perhaps derived from the contents of Golgi vesicles. The innermost layer is massively thickened and is evenly stratified (Fig. 6.97). This complex wall enables the zoospore to withstand extremes of environmental stress. As usual, the cytoplasm of the zygospore, initially very pale, soon becomes so dense as to render ultrastructural study impossible (Fig. 6.95). As with *Spirogyra,* all the zygospores we examined appeared to be binucleate, and caryogamy probably occurs just before germination (Fritsch, 1935; Lippert, 1967). Two of the four chloroplasts in the zygote reportedly degenerate in *Closterium* during maturation (Fritsch, 1935).

Occasionally, attempts at triple conjugation can be found (Fig. 6.98). In such examples, the conjugation tubes are all in contact.

Germination of zygospores has been investigated by earlier workers (Fritsch, 1935, p. 353) and more recently by Lippert (1967). A period of dormancy (1–6 months or even more) and drying of the zygospores appears to be prerequisite for germination, the onset of which is marked by greening and enlargement of the two persistent plastids. The contents of the zygospore are then forcibly expelled within a vesicle when the wall of the spore splits. The first meiotic division follows, the protoplast cleaves, and then the second meiotic division occurs; one of the resultant

Figures 6.104, 6.105
TWO MORE SPECIES OF STAURASTRUM

6.104. *S. manfeldtii.* × 1,500.

6.105. *S. pingue.* Note that the cells in this field vary somewhat in morphology and one is a tetraradiate variant. × 650.

SOURCE. Author's micrographs, published (as stereo-pairs) in *J. Microsc.* 99, 109 (1973).

6.104

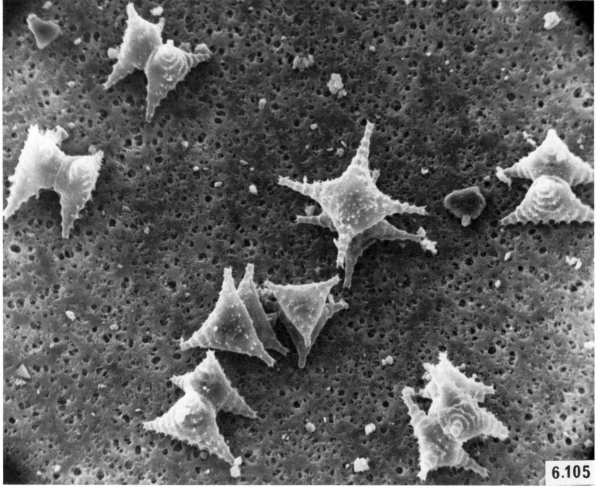

6.105

two nuclei in each cell later degenerates. This form of germination, whereby the protoplast is released from the zygospore wall before it has cleaved, is unusual in the Green Algae, but is characteristic of the desmids (Lippert, 1967; see also Sections 6.1a, 6.4c). After the cleavage, the two protoplasts or gones become sausage-shaped (Fig. 6.84), and their single chloroplast divides; finally they are released by rupture of the vesicle.

Conjugation has also been recently described at the light microscope level for two species of *Pleurotaenium* by Ling and Tyler (1972). This genus could probably be considered an unconstricted desmid. The long, truncated cells may possess small terminal spines, a series of small, circumferential ridges, and a slight central constriction. Strict characterization of such organisms into constricted and unconstricted forms is not particularly important, and the process of conjugation described in these organism closely resembles that in *Closterium*. Paired cells are held together by mucilage, and a "conjugation vesicle," apparently a small mass of polysaccharide, is secreted between and at the middle of the semicells (cf. Fig. 6.81). A papilla grows from each conjugant into this vesicle, one always earlier than the other (possible tendency toward anisogamy?). The cell contents move into the expanding papillae, and fusion of the protoplasts soon follows. Later, the thick, layered and ornamented zygospore wall is slowly secreted. Ling and Tyler also mention the confusion in terminology that has arisen concerning conjugation in desmids. I believe this confusion will probably be clarified only following ultrastructural studies on this group of algae. Thus, while mucilage is secreted between early conjugants in many species of desmids (Lippert, 1967; Ling and Tyler, 1972), several authors state that there is no wall or conjugation tube around emerging gametes, i.e., the protoplasts are naked (see references in Ling and Tyler, 1972). The very few ultrastructural results we have so far seen show the presence of a thin wall around some gametes; I see no reason why this wall should not be considered a typical conjugation tube.

6.4. Constricted Placoderm Desmids

Constricted desmids are often remarkably beautiful (e.g., Carter, 1919a,b, 1920a,b) because of their symmetry and extraordinarily com-

Figure 6.106
STAURASTRUM GRACILE

One incompletely formed daughter cell (lacking full ornamentation) just after division is indicated by the arrow. × 1,050.

SOURCE. Author's micrograph, published (as a stereo-pair) in *J. Microsc.* 99, 109 (1973).

6.106

plex and varied cell outlines. These cells can be conveniently regarded as consisting of two identical (or nearly identical) semicells joined at a median constriction, often very deep. The nucleus lies in the central "isthmus" defined by this constriction (Fig. 6.99). Each semicell is partly filled by a large and usually ornate chloroplast containing a variable number of pyrenoids. The cells of many common species may be flattened (e.g., *Cosmarium;* Fig. 6.149) or disciform (e.g., *Micrasterias;* Figs. 6.107, 6.117), in which case they have three main axes of symmetry at right angles to one another. The three resultant aspects of the cells have been termed the "front view," the "end view," and the "side view" (Fritsch, 1935; p. 338). Other genera, exemplified by the elegant *Staurastrum* (Figs. 6.102–6.106), may have a multiradiate character, being (from the end view) bi-, tri-, tetra-, penta-, or even hexaradiate (Carter, 1920b). The corners of these cells are often drawn out into long processes, and the walls may be richly ornamented with patterns of spines and smaller surface ridges and spikelets.

One semicell wall is always older than the other, but the two fit so closely together at the isthmus that they are difficult to separate, visually and physically. Many species secrete copious quantities of mucilage through a complex system of pores in the wall (e.g., Fig. 6.99) and, like other members of the Conjugales, can move actively about (Fig. 6.147). This mucilage is often very difficult to remove completely (Figs. 6.100, 6.101) and appears as small surface deposits under the scanning electron microscope (Figs. 6.109, 6.126).

The constricted desmids, like the unconstricted form *Closterium,* face considerable logistical problems each time they undergo cell division. Following mitosis, a septum divides the semicells at the isthmus (Fig. 6.118). Restoration of the beautiful interphase symmetry is a particularly fascinating, but poorly understood process, differing in several important respects from what has already been described for *Closterium* (Section 6.3a). First, the wall of the septum expands as in *Closterium,* eventually becoming the mirror image of the older semicell, which may be extremely ornate and complex (Figs. 6.112, 6.119–6.122). Second, the nucleus moves into the forming semicell, where it remains throughout expansion. Third, the chloroplast moves through the isthmus (still present as the new wall ex-

Figures 6.107, 6.108
TWO SPECIES OF MICRASTERIAS

6.107.   *M. angulosa.* Two daughter cells. The newly formed semicells enclosed by the characteristically smooth primary wall are both slightly larger than the parental semicells, which are covered by small deposits of mucilage from pores in the wall. × 960.

6.108.   *M. sol* var. *extensa.* × 1,000.

SOURCE. Author's micrographs. Figure 6.107 published in *Trans. Am. Microsc. Soc.* 93, 1 (1974). Figure 6.108 published (as a stereo-pair) in *J. Microsc.* 99, 109 (1973).

6.107

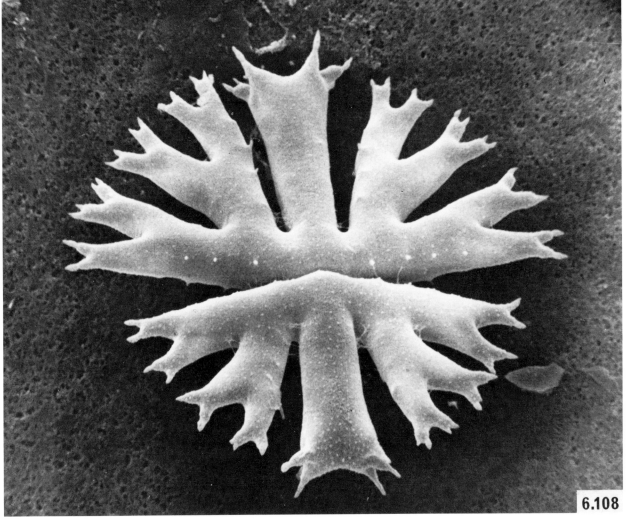

6.108

pands) and extends into the forming semicell. Finally, by the time cell expansion is essentially complete, the chloroplast cleaves at the isthmus as the nucleus moves back to this region to take up its interphase position (Fig. 6.151).

Some desmids are filamentous, the individual cells being held together often by the interlocking of ornate projections on their surface; cell division in these forms proceeds as would be expected of desmids (Fig. 6.113). We may note in passing an interesting variation in semicell wall expansion that has been developed by genera such as *Gymnozyga* (Fritsch, 1935, p. 358). In these filamentous desmids, most of the primary cell wall has been secreted *before* cell expansion takes place. The sequence of events is outlined in Figure 6.114. After mitosis, a septum bisects the cell. This septum thickens and then develops a cylindrical ingrowth of wall material on each side. When the daughter cells subsequently expand, this quite complicated wall unfolds or evaginates to give rise to the new semicell. Several species of *Spirogyra* are also known to form similar "replicate" end walls which can undergo an analogous unfolding when the filament fragments (Smith, 1950, p. 287).

Regrettably little ultrastructural work has been published on constricted desmids, although a fairly complete description of cell division has been attempted for *Cosmarium* (Pickett-Heaps, 1972g). However, some interesting work has been done on the morphogenesis of the new semicell, particularly in *Micrasterias,* which I shall review briefly along with some relevant but rather limited ultrastructural work. I shall then describe the ultrastructure of dividing *Cosmarium* cells. Before proceeding, however, I shall illustrate the diversity of form in different desmids and comment briefly upon it. The scanning electron microscope has proved valuable in this regard (Pickett-Heaps, 1973a, 1974a), although most of the constricted desmids can easily be described with the light microscope.

The symmetry of the cells is immediately apparent in these micrographs (Figs. 6.99–6.118), and while the form and pattern of ornamentation are obviously fairly characteristic of the species, variations in gross cell morphology are not unusual. These variations raise problems in the taxonomy of the organisms (see Bicudo and Sormus, 1972). First, the symmetry of the cell is not absolutely predetermined. As will be-

Figure 6.109
MICRASTERIAS RADIATA

This is a recently divided pair of cells. Note the smooth wall covering the new semicells. The morphology of the wings is somewhat variable. The wing next to the polar lobe is deeply split two out of four times in the older semicells, and one out of four in the new semicells. × 1,000.

SOURCE. Author's micrograph, published in *Trans. Am. Microsc. Soc. 93,* 1 (1974).

come apparent, forms that are normally biradiate can give rise to triradiate, uniradiate, and even aradiate cells (e.g., *Micrasterias* and *Cosmarium;* Figs. 6.126–6.131, 6.150). Likewise, triradiate species may form bi- and tetraradiate variants. Whereas one form of a given species is normally more stable and more commonly encountered than the others, in certain cases (e.g., in the aptly named *Staurastrum paradoxum*) the degree of radiation is easily varied and is perhaps affected by environmental conditions (Reynolds, 1940). In my own work on desmids (Pickett-Heaps, 1973a), cultures of the normally triradiate *Staurastrum pingue* contained a higher percentage of tetraradiate cells (Fig. 6.105) than those of other species examined *(S. manfeldii, S. gracile,* and *S. gladiosum).* Since alterations in the degree of symmetry occur in one step at cell division (Figs. 6.128, 6.130, 6.131, 6.136*B*, 6.137*B*), "Janus" cells are occasionally encountered in which one semicell has a different number of wings than the other (Figs. 6.116, 6.128, 6.130, 6.131).

Second, the size and shape of the cells vary appreciably within certain limits. In some species of *Cosmarium,* strains can be isolated that are consistently larger than normal, a change probably brought about by the development of polyploidy in the strains (Brandham, 1965). However, *Cosmarium botrytis* may become triradiate without change in ploidy (Tews, 1969). Such changes are often apparent in cultured desmids. In other cases, semicells may be appreciably fatter or slimmer than normal. It is probably safe to generalize and say that usually the volume of a cell of given ploidy is quite constant. When one or more wings are "lost" from the cell, the remaining parts of the cell such as the polar lobe (and wing if present) are larger to compensate and bring the cell volume up to normal (Figs. 6.126*A,* 6.126*B,* 6.127–6.131).

Third, the degree of ornamentation may be quite variable. While sometimes this variability can be attributed to differences between strains of the same species (again, for example, following induction of polyploidy; cf. Figs. 6.126*B,* 6.126*D*), in other cases, environmental factors can strongly inhibit the full expression of morphogenetic potential. Stunted forms of *Micrasterias radiata* predominate in cultures heavily contaminated with bacteria, and these are quite different superficially from those in normal cultures (Figs. 6.109, 6.110). However, such changes are

Figures 6.110–6.112
MICRASTERIAS RADIATA

6.110. Pair of daughter cells from a culture heavily infected with bacteria. Morphogenesis of the semicells has been severely stunted (cf. Fig. 6.109). × 830.

6.111. Live cells. Note the slightly variable morphology of the wings. × 310.

6.112. Live cells. A recently divided pair of cells is undergoing semicell morphogenesis (*cf.* Figs. 6.119–6.122). × 600.

SOURCE. Author's micrographs. Figure 6.110 published in *Trans. Am. Microsc. Soc. 93,* 1 (1974).

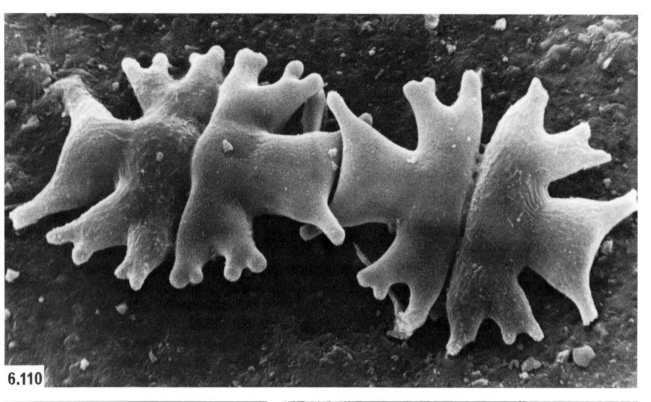

6.110

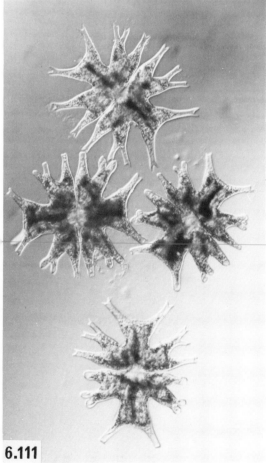

6.111

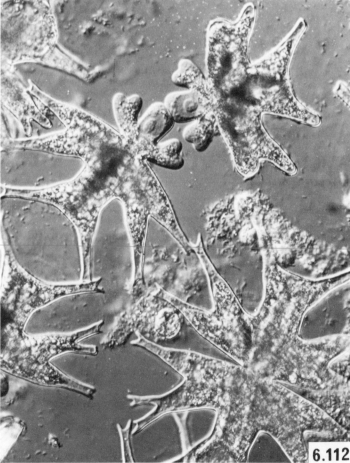

6.112

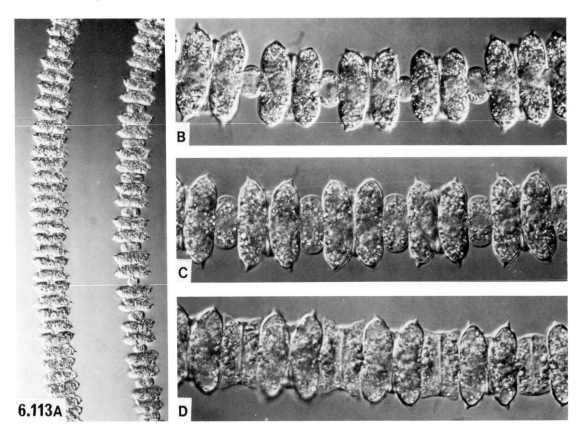

6.113A

Figure 6.113
THE FILAMENTOUS DESMID ONCHYNEMA SP.

Interphase and dividing cells are shown in *A,* and various stages of semicell formation in *B–D. A,* × 180 approx; *B–D,* × 480 approx.

usually only transitory, and stunted cells have by no means permanently lost the ability to form normal semicells during subsequent divisions. Figure 6.115 shows the result of cell division in a similarly stunted specimen of the triradiate, diploid variant of *Micrasterias thomasiana.* The older semicell, covered with mucilage pores, is very simple in outline, whereas the semicell just formed by the most recent cell division has developed the complexity of form characteristic of this particular variant (Fig. 6.126*C*).

6.4a. Micrasterias

The various species of *Micrasterias* differ considerably in the complexity of form attained by the mature vegetative cell (cf. Figs. 6.107–6.112, 6.117, 6.126). The two species mainly used in the experiments described below, *M. denticulata* and *M. thomasiana,* are fairly representative, discoidal organisms whose symmetrical semicells are separated by a deep constriction. Each semicell has a distinct central lobe, around which the wings are arrayed. The periphery of each semicell is indented by a pattern of radial invaginations. One series of these is very deep, the next series less so, and numerous minor in-

Figure 6.114
SEMICELL EXPANSION IN GYMNOZYGA

Diagrammatic representation. After mitosis, the septum forms as normal *(A).* However, cell expansion is delayed until *after* the septum has become further elaborated by a cylinder of wall material projecting into each daughter cell *(B).* Then as the cells expand *(C–E),* this cylinder evaginates or unfolds outward. Unlike other desmids, therefore, these cells deposit most if not all of their primary wall *before* undergoing semicell expansion.

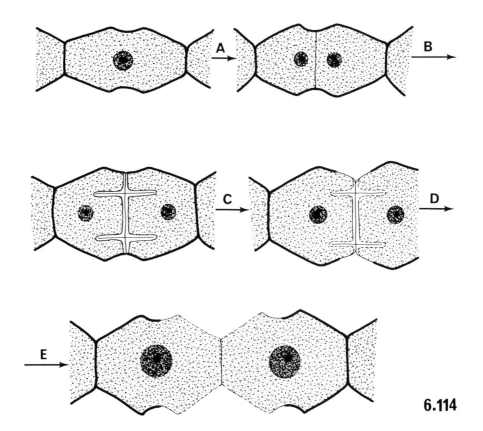

6.114

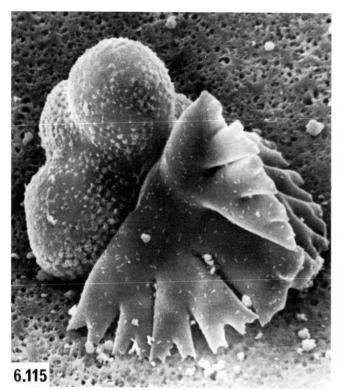

6.115

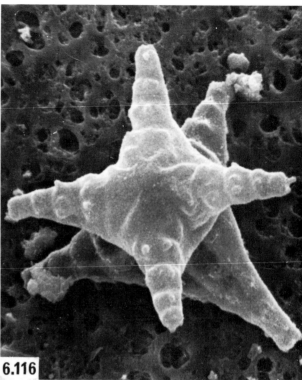

6.116

dentations ornament the outer edge of the cell. The central lobe too is ornamented by spikelets.

CELL DIVISION

Each time the cells divide, the septum balloons out and in the process attains with almost mirror-like perfection, the form of the older semicell (Figs. 6.119–6.122). This morphogenesis of a new wall is an extraordinary phenomenon when visualized by time-lapse photomicrography, as Kiermayer's (1965a) superb cinematography has shown. The expanding wall initially has a smooth profile, slightly flattened where it abuts the adjacent wall of the other daughter cell (Fig. 6.119). As expansion continues, the deepest invaginations appear first (Fig. 6.120). They apparently result from precisely localized changes in the nature of the wall that prevent its further radial expansion at these sites, which thereafter remain approximately fixed in position relative to the isthmus; meanwhile the remainder of the wall balloons out past them (Fig. 6.121). Secondary invaginations develop similarly a further precise distance from the isthmus (Fig. 6.122), and finally, the peripheral ornamentation is added as the new semicell reaches its final size.

The bulging wall derived from the septum is described by Kiermayer (1970a) as a "delicate primary wall"; the rigid secondary wall is se-

Figures 6.115, 6.116
MICRASTERIAS THOMASIANA AND STAURASTRUM PINGUE

6.115.   Triradiate variant of *M. thomasiana*. The older semicell is severely stunted in morphology owing to adverse cultural conditions. However, such a defect in morphology is not necessarily maintained during cell division. Upon subculturing, the original cell divided, and the newly formed semicell (*i.e.,* the smoothed-surfaced half) has the normal morphology (cf. Fig. 126C). × 630.

6.116.   *S. pingue.* Janus cell, one half triradiate and the other tetraradiate (*cf.* Fig. 6.105). The ornamentation of the cell is somewhat stunted owing to unfavorable cultural conditions. × 1,500.

SOURCE. Author's micrographs, published in *Trans. Am. Microsc. Soc. 93,* 1 (1974).

Figures 6.117, 6.118
MICRASTERIAS DENTICULATA

6.117.   Vegetative cell, soon after division. × 600.

6.118.   Dividing cell at telophase, with cleavage half completed. × 440.

SOURCE. Author's micrographs. Figure 6.117 published in *Trans. Am. Microsc. Soc. 93,* 1 (1974).

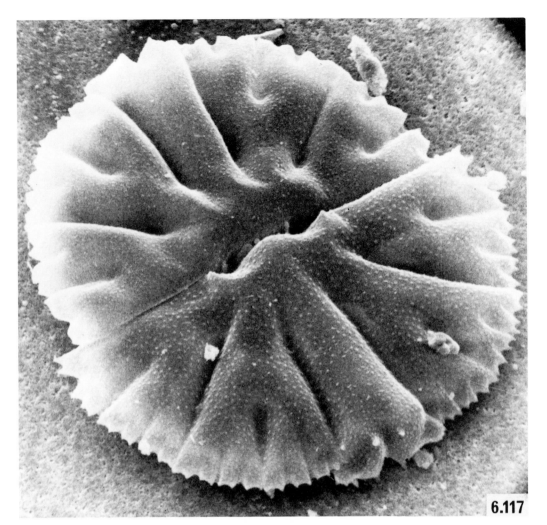

6.117

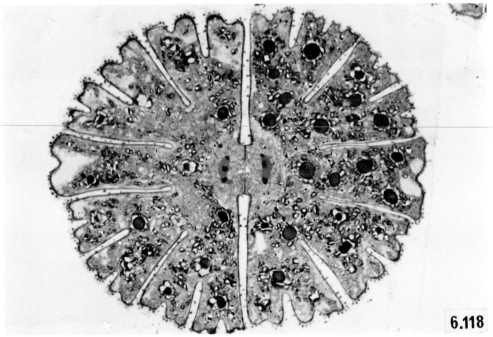

6.118

creted only after the new semicell has been completely formed, and later the primary wall is shed (Fig. 6.161). Thus, primary and secondary wall deposition are distinct processes, both temporally and functionally. In *Cosmarium,* secondary wall deposition commences when the forming semicell is about three quarters of full size (Section 6.4b). Kiermayer has demonstrated that turgor pressure is instrumental in causing the bulging wall to expand, since the wall ceases growth if cells are placed in hypertonic (0.12–0.22 M glucose) solutions. Using such treated cells, Kiermayer demonstrated that wall deposition continues in the absence or interruption of cell expansion and, most significantly, that the new wall material is deposited in a predetermined pattern (Figs. 6.124, 6.125). Thus, the accumulation of wall material at any region is related to whether that particular region of wall would have continued to expand under normal circumstances. Furthermore, if cells are prevented from undergoing any expansion at all after cytokinesis, the septum itself soon develops a reproducible pattern of uneven wall thickening, again related to the potential form of the expanding semicell (Fig. 6.123). Kiermayer (1965b) has also produced elegant time-lapse cinematographic records of these experiments.

Such results suggest that whatever determines the developmental pattern of semicell morphogenesis, it is perhaps localized in or adjacent to the plasma membrane. This suggestion is reinforced by observations of cells in which septum formation is partially inhibited so that a perforation remains in the incomplete cross wall. The subsequent expansion undergone by such permanently joined cells is dependent upon the preexisting extent of completion of the septum (as in Figs. 6.133, 6.134). Such a sequence of events explains how the Siamese-twin cells of *Closterium* (Fig. 6.78) and the asymmetrical complex cell of *Cosmarium* (Fig. 6.162) could have arisen, namely, following expansion of the wall derived from a septum that contained a small, eccentrically situated perforation.

In another fascinating paper, Waris and Kallio (1964) summarize many of their earlier experimental investigations into semicell morphogenesis. They were able to produce either polyploid, enucleate, aneuploid, or "complex" cells by centrifugation and other techniques. I shall attempt here to outline some of their results.

Figures 6.119–6.122
SEMICELL MORPHOGENESIS AFTER CELL DIVISION IN LIVE MICRASTERIAS DENTICULATA

6.119. Early bulge stage. × 370.

6.120. Initiation of marginal differentiation into five lobes. The nuclei are typically situated in the forming semicell. × 340.

6.121. Further expansion of the five lobes. The chloroplast has started moving into the forming semicell. × 370.

6.122. The five lobes have undergone two more series of marginal bifurcations to give the 17-lobe stage. Note that the polar lobe does not undergo the same sequence of bifurcations (see Fig. 6.117). × 290.

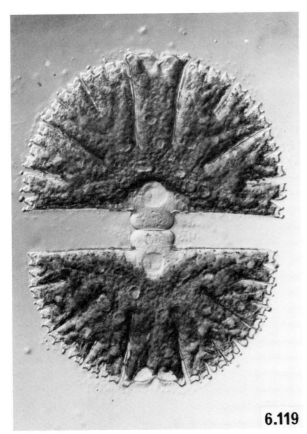

6.119

6.120

6.121

6.122

First, they demonstrated that when cells normally haploid become diploid, both the size and the complexity of peripheral ornamentation increase (Fig. 6.126*B, D*). Second, they isolated clones, some derived from spontaneous "mutants" in their cultures, which were uniradiate (i.e., considering the normal biradiate cell as consisting of the central lobe and two symmetrical wings, uniradiate cells have one of the two side wings missing; Fig. 6.126*A*). They found that this defect is fairly stable, being passed on during cell division to progeny, although spontaneous reversion to the biradiate form (subsequently stable) does occur. The uniradiate cells are both less prolific and less viable than normal cells and so tend to die out in mixed cultures. The two wings they possess are more ornate than usual, being more typical of diploid cells, presumably because the morphogenetic agents responsible for cell form and derived from the haploid nucleus now affect only one, instead of two, wings during semicell morphogenesis. Third, Waris and Kallio were able to obtain an "aradiate" form of *M. americana* whose semicells consisted only of the central lobe (as in Figs. 6.127, 6.129). This variant of the normal cell appears considerably more stable than the uniradiate form over long periods in culture; indeed, no reversion of aradiate forms (of *M. americana*) to normal was detected. A triradiate variant of *M. thomasiana* was also discovered and isolated (Fig. 6.126*C*). The appearance and inheritance during sexual reproduction of the triradiate characteristic in the biradiate *Cosmarium* will be mentioned later; Section 6.4c).

An aradiate cell of *M. torreyi*, arising from a normal cell of a culture irradiated with ultraviolet light, was also viable and successfully cloned (Figs. 6.127, 6.129; Kallio and Heikkilä, 1969). The polar lobes making up the whole cell were far larger than those of normal cells and were circular in cross section (Figs. 6.127, 6.131). The origin of this clone is interesting and demonstrates the flexibility of behavior of the organisms. The original mother cell was a Janus type, one semicell being biradiate and the other uniradiate. Upon division (as in Fig. 6.131), both newly formed semicells were aradiate. The daughter cell that was uniradiate/aradiate died, but the other biradiate/aradiate cell continued to divide, producing aradiate semicells and thereby the aradiate cell line (Fig. 6.129). This particular aradiate variant, in contrast to that isolated from

Figures 6.123–6.125
PATTERNED DEPOSITION OF PRIMARY WALL IN MICRASTERIAS DENTICULATA

6.123.   Pattern of wall deposition at the septum of a cell that was prevented from undergoing semicell expansion after cytokinesis by treatment with 0.22 M glucose for 130 minutes. The most thickened regions correspond with the areas that would have undergone most expansion in untreated cells. $\times$ 2,600 approx.

6.124.   As for Figure 6.123, but here the cell was allowed to undergo some expansion before the glucose was applied. After 130 minutes of continued wall secretion at reduced turgor, the pattern of thickening is not even and, as in Figure 6.123, is related to the degree of expansion the various regions would have undergone in untreated cells. $\times$ 1,500 approx.

6.125.   This cell was treated in similar fashion to that in Figure 6.124 at a slightly later stage in expansion. After the period of wall secretion under reduced turgor, the cell was plasmolyzed in a medium of higher osmotic strength. The cytoplasm has withdrawn completely from the forming semicell, again revealing that the pattern of wall deposition is complex and highly controlled spatially. $\times$ 1,100 approx.

SOURCE. Dr. O. Kiermayer, published in *Protoplasma* 64, 481 (1967).

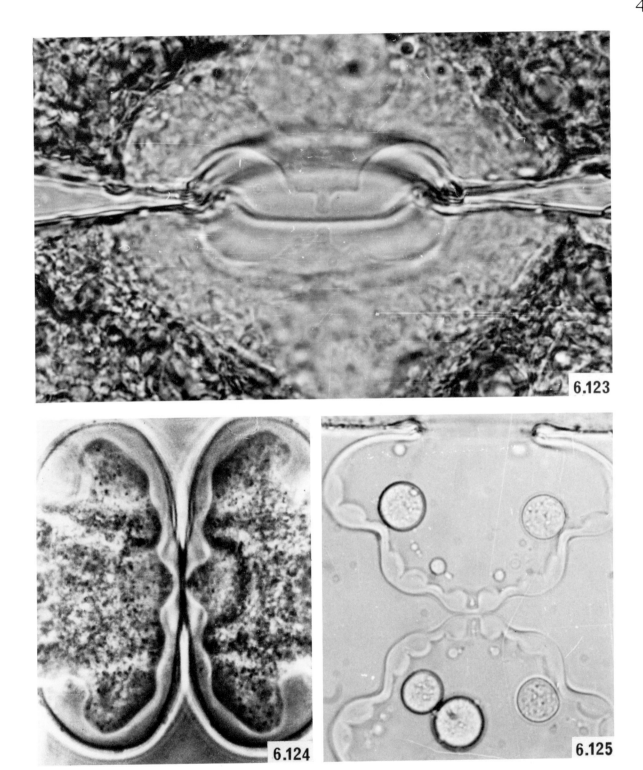

6.123

6.124

6.125

M. americana, showed a strong tendency to become uniradiate upon further cell divisions (Fig. 6.130). Believe it or not, the uniradiates thus formed had no tendency to continue reversion to the original biradiate form! Rarely, a biradiate arose directly from an aradiate cell. Kallio and Heikkilä have some beautiful micrographs showing the course of such reversions (Figs. 6.128–6.131).

Waris and Kallio (1964) have followed what happens when various dividing cells undergo cell expansion without a nucleus, which was earlier removed by centrifuging it through the isthmus prior to septum completion (forming a binucleate and enucleate set of daughter cells). Enucleate cells do not survive long, partly a result of the fragility of their newly formed wall. In all enucleate cells, the shape of the new semicell is extremely simplified, but the number of lobes in it corresponds to the number of lobes in the older semicell, whether these are aradiate or uni-, bi-, or triradiate (Fig. 6.132). The authors also review a number of experiments with complex "double" cells (Fig. 6.134) which arose when cytokinesis was interrupted experimentally, leaving a perforation in the septum (Kallio, 1959). This effect can be achieved by centrifugation, for example, when both nuclei come to lie at one isthmus (Fig. 6.133), or by treatment with chemicals such as ethanol (Kiermayer, 1966). Not infrequently, such cells arise spontaneously. Such incompletely separated cells undergo expansion to form a binucleate double cell with two isthmuses (Fig. 6.133). On subsequent division of a double cell, two normal cells are formed, while the double cell itself is maintained; once the secondary wall is formed, it is immutable (Fig. 6.135). Waris and Kallio developed the experiments still further; some double cells were centrifuged (if necessary) before division so that one nucleus was moved from its position to the other isthmus, adjacent to the second nucleus. The results of subsequent divisions of these particular cells are revealing and consistent with previous results (Fig. 6.136A). When the two adjacent nuclei underwent mitosis, the chromosome complements merged, and after mitosis, two diploid (rather than four haploid) daughter nuclei were formed. When cell expansion followed, the form of the four new semicells derived from the two septa is quite intriguing. Considering first the semicells arising from the septum separating the

Figure 6.126
STRUCTURAL VARIANTS OF MICRASTERIAS THOMASIANA

6.126A. Uniradiate, haploid form. The single wing of each semicell is larger and more ornamented than those of the biradiate, haploid cell. × 350.

6.126B. Biradiate, haploid form, the normal one found in nature. × 390.

6.126C. Triradiate, diploid form. × 550.

6.126D. Biradiate, diploid form. The semicells are obviously much larger than those of the haploid equivalent (Fig. 6.126B). × 390.

SOURCE. Author's micrographs, published in *Trans. Am. Microsc. Soc.* 93, 1 (1974).

6.126

newly formed daughter diploid nuclei, one was more ornate than normal, as might be expected from the presence of the diploid nucleus in that cell. However, the complementary semicell formed from this shared septum was normal; the equivalent effect of the complex cell's diploid nucleus on this wall seemed to have been diluted. In a somewhat analogous fashion, the other septum also formed two different semicells. One was the simplified form characteristic of enucleate cells, hardly surprising since this newly formed daughter cell was enucleate. However, its complementary semicell was normal in the complex cell, since apparently the lack of an immediately adjacent nucleus was overcome by the proximity of the diploid nucleus at the other isthmus whose morphogenetic effect was "shared." The next division in this double cell (Fig. 6.136B) was even more interesting (Kallio, 1959, 1968), since a wing was lost from the enucleate isthmus, a loss which showed up in both newly formed semicells. Such a result suggests that the absence of a nucleus from this isthmus has had a longer-term effect on the symmetry of the organism.

Kallio (1959) also describes what happens if a double cell containing only one haploid nucleus at one isthmus continues to divide (Fig. 6.137A). In this situation, there is no cytokinesis or cell expansion at the other isthmus, suggesting that at least a haploid nucleus is needed nearby to get cytokinesis. Apart from this, division proceeds as expected; both forming semicells at the nucleated isthmus are of the haploid form (Fig. 6.137A). However, the next division of the double cell is interesting; one wing may be lost from the nucleated isthmus (Fig. 6.137B). One conclusion is that some morphogenetic influences derived from the haploid nucleus have become diluted sufficiently for one of the wings to be lost. Some experiments indicate that this aspect of morphogenesis might be mediated via ultraviolet-labile intermediates. If cultures of M. sol are strongly irradiated during the interphase period between several cycles of divisions, loss of one or both wings eventually becomes common in many cells (Kallio, 1968).

Many more ingenious manipulations are described in this and other papers by these authors, which cannot be further reviewed here. They also investigated ultraviolet irradiation of the whole or part of dividing complex cells. The re-

Figures 6.127, 6.128
STRUCTURAL VARIANTS OF MICRASTERIAS TORREYI

6.127. Normal biradiate cell of M. torreyi alongside an aradiate variant. Note the much larger volume of the polar lobes in the aradiate form.

6.128. Sequence showing semicell expansion following division in a Janus (uniradiate/aradiate) variant of M. torreyi. Both new semicells have become aradiate in these specimens. The polar lobes in the aradiate semicells are larger than the polar lobe in the uniradiate semicell.

SOURCE. Drs. P. Kallio and H. Heikkilä, published in Österr. Bot. Z. 116, 226 (1969).

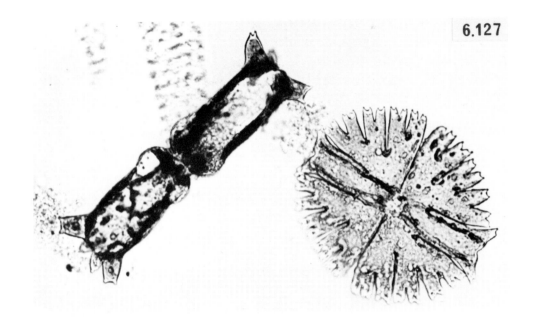

6.127

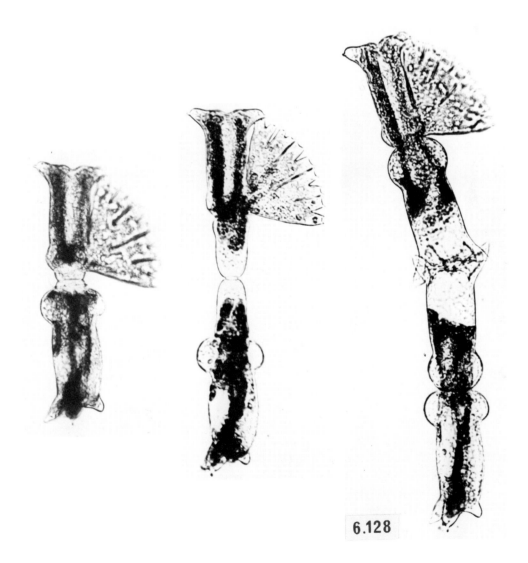

6.128

sults are again complicated and can be summarized, at the risk of oversimplification, by saying that the consequences of nuclear ultraviolet irradiation on expanding semicells were much the same as the consequences of mechanical removal of the nucleus. They summarize their results by saying "the basis of the bilateral symmetry is independent of the nucleus, whereas the degree of differentiation (i.e., ornamentation in the cell outline) depends on the nucleus." While some control of semicell morphogenesis is obviously effected by the nucleus, the symmetry (and in particular, the number of lobes formed) may only be "cytoplasmically" determined to the extent that morphogenetic material derived from the nucleus, such as certain species of RNA, could be stable and long-lived in the cytoplasm, as is the case with *Acetabularia*. This is my conclusion from Kallio's observations on double cells (Figs. 6.136*B*, 6.137*B*). Hence the caution throughout at using the term "mutant" in describing alterations in form occurring in these organisms. Waris and Kallio postulate that the three main sections of the cell contain three "axes," and they guess that these axes consist of fibrils. They then explain their results in terms of the behavior of these axes and their properties. For example, during division, these axes could be severed and then grow into the new semicells. These axes, too, may be related to strands of cytoplasm that appear to remain in characteristic positions attached to the semicell wall when expanding cells are severely plasmolyzed. However, electron microscopic study of these and similar cells has not yet revealed any cytoplasmic structure that could be equivalent to the postulated axes.

Selman (1966) has further investigated the nuclear control of semicell differentiation in *M. thomasiana* by ultraviolet microbeam irradiation of growing lobes or nuclei or by treating the cells with various inhibitory chemical agents such as dinitrophenol and actinomycin D. Again I shall attempt to summarize these results without getting too involved in details. In all cases the degree of abnormality induced experimentally depended upon the severity of the treatment.

When the expanding lobes of semicells were irradiated with highly localized ultraviolet microbeams, they ceased growing, while adjacent, untreated lobes continued to expand as normal. This defect was not transmitted to offspring during subsequent cell divisions. Similar treatment

Figures 6.129–6.131
VARIANTS OF MICRASTERIAS TORREYI INDUCED WITH ULTRAVIOLET IRRADIATION

6.129.   Division in an aradiate cell.

6.130*A–E*.   Division in aradiate cells, during which one of the forming semicells acquires a wing. *A–D*. Sequence of one cell. *E*. Different cell, showing the final result.

6.131.   Division in a Janus cell (originally a biradiate/uniradiate) during which both forming semicells have become aradiate. Note again how the newly formed polar lobes are much larger than the older ones.

SOURCE. Drs. P. Kallio and H. Heikkilä, published in *Österr. Bot. Z. 116*, 226 (1969).

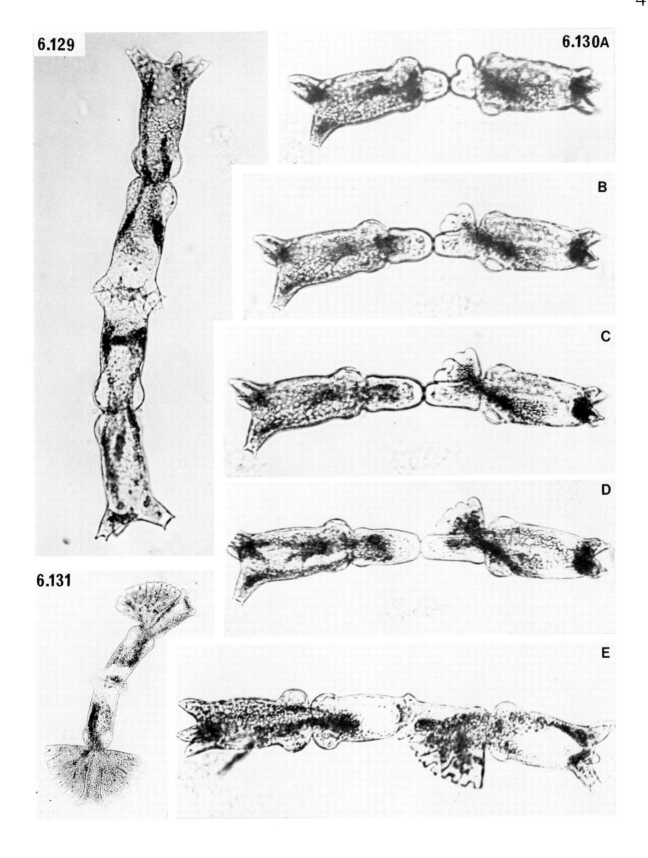

6.129

6.130A

B

C

D

E

6.131

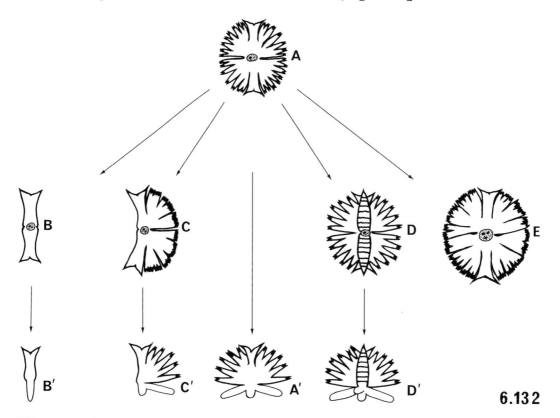

6.132

Figure 6.132
VARIANTS OF
MICRASTERIAS SP.

Diagrammatic representation. The normal biradiate haploid cell found in nature *(A)* can also give rise to aradiate, haploid *(B);* uniradiate, haploid *(C);* triradiate, diploid *(D);* and biradiate, diploid *(E)* cells. Increase of nuclear amount versus number of wings leads to increase in ornamentation of the wings *(C, E).* If these cells are allowed to undergo nuclear division and then centrifuged so that after cytokinesis one of the new semicells is enucleate, the latter undergoes some semicell expansion during which the lobes, although of very simple morphology, correspond in number and disposition to those of the parental cell *(A'–D').*

SOURCE. Redrawn after Waris and Kallio (1964).

Figure 6.133
FORMATION OF DOUBLE
CELLS OF MICRASTERIAS BY
CENTRIFUGATION

After mitosis *(A),* a cell undergoing cytokinesis is centrifuged so that one daughter nucleus passes through the incomplete septum *(B).* If cytokinesis remains incomplete, the cell undergoes partial semicell expansion from the wall of the septum *(C).* A double cell is formed with two nuclei at one isthmus *(D).*

of the central "polar" lobe usually had a different effect; treated cells did not divide again, although initially they appeared viable. Irradiation of the nucleus alone early in expansion caused eventual death of the cells; expansion continued beyond its normal extent and the cells then burst. The cell periphery was not as ornamented as normal. If

Figure 6.134
ABNORMAL CELL OF
MICRASTERIAS FIMBRIATA

Giant, abnormal double cell of *M. fimbriata,* a consequence of partial interruption of cytokinesis, followed by semicell expansion. × 330.

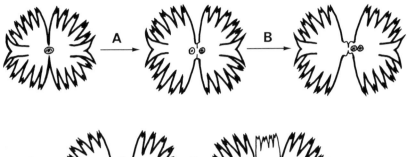

6.133

6.134

the irradiation was delayed until the wall was more developed, the lobes were bigger and rounded, but the wall did not burst and the cell survived. Subsequent division in these cells resulted in normal development. Treatment with dinitrophenol affected lobe development, but the polar and wing lobes were sometimes affected in slightly different ways. Since actinomycin D inhibits RNA synthesis, this drug was used to investigate whether RNA could be involved in semicell morphogenesis. Treatment with fairly high concentrations (ea. $100\mu g/ml$) was necessary to produce an effect—the reversible cessation of lobe expansion about half an hour after contact with the drug. Longer exposures (more than 2.5 hours) of expanding cells resulted in formation of an anucleate type of semicell which eventually ruptured and died. These and other results suggest that irradiation of the nucleus and treatment with actinomycin D have similar deleterious effects on semicell morphogenesis, both perhaps involving some interference with RNA metabolism. There is also evidence that the nuclear determination of the various phases in differentiation of lobes occurred between 30 and 155 minutes before these events were actually detectable with a light microscope.

We (Tippit and Pickett-Heaps, 1974) have repeated and extended many of these experiments with metabolic inhibitors. We have found that RNAase and puromycin treatment are unreliable for inducing defects in morphogenesis. However, cycloheximide reproducibly inhibits lobe differentiation a short while after its application to forming semicells, at whatever stage in differentiation treatment starts (Figs. 6.138–6.141); the lobes, however continue to expand. Washing out the cycloheximide after a brief period allows the cells to resume marginal differentiation even though middle stages of morphogenesis have been interrupted (Figs. 6.142, 6.143). Actinomycin D has similar effects on the cells, with a slightly longer pause between application of the drugs and visualization of its effect. Caffeine treatment, which prevents fusion of vesicles in the cell plate of higher plants, apparently also prevents discharge of vesicles into the wall of *Micrasterias*; the cells soon burst. If placed in osmotic solutions of D-mannitol as well as caffeine, the semicells do not expand and burst, but the pattern of wall thickening characteristic of reduced turgor (Figs. 6.123–6.125) never devel-

Figures 6.135–6.137
DIVISION OF DOUBLE CELL

6.135.   Result of division of a double cell that has a haploid nucleus at each isthmus. The septum formed at each isthmus gives rise to normal new semicells *(A)*.

6.136.   Results of division of a double cell that has two haploid nuclei at one isthmus (*cf.* Fig. 6.133). During mitosis, the two spindles fuse and a single *diploid* daughter nucleus arises on each side of the septum at this isthmus. During subsequent semicell expansion *(A)*, one new semicell arising from this isthmus is the larger, diploid form; the other, the normal haploid form (*n.b., two* new semicells are formed in the central double cell). At the other, enucleate isthmus, one semicell is normal, presumably because of the proximity of the diploid nucleus; the other is of the simplified, enucleate form. During subsequent divisions of this double cell *(B)*, a wing may be lost at the enucleate isthmus, a change in symmetry reflected in each of the semicells arising at this site.

6.137.   Result of division in a double cell that has only one *haploid* nucleus at one isthmus. No septum is formed at the enucleate isthmus, but otherwise the first division cycle *(A)* has the expected result. However, during subsequent divisions of the double cell *(B)*, a wing may be lost at the nucleate isthmus.

SOURCE. Figure 6.135: Redrawn after Waris and Kallio (1964). Figures 6.136 and 6.137: Redrawn after Kallio (1959).

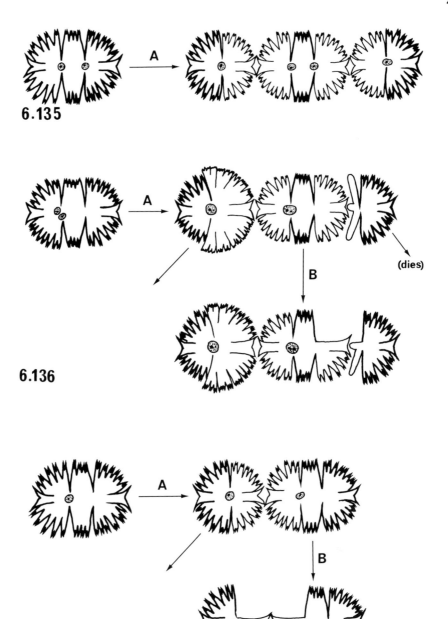

6.135

6.136

(dies)

6.137

ops. Ultrastructural examination of caffeine-treated cells reveals concentrations of vesicles in the cytoplasm.

Treatment with the drug cytochalasin B has intriguing effects. It stops cytoplasmic streaming, and a short while afterward, both semicell expansion and wall deposition cease. If the drug is removed after half an hour, a little further semicell expansion may follow, but essentially this treatment causes irreversible cessation of semicell morphogenesis, and it enables formation of defective cells consisting of one normal and one stunted semicell (Fig. 6.144*A*). Subsequent division in these cells maintained in normal conditions is interesting. A normal semicell arises from the original normal semicell (Fig. 6.144*B*); however, a miniature semicell forms from the stunted older semicell. If this latter asymmetrical cell divides again (Fig. 6.144*C*), the stunted half forms another miniature semicell, and the miniature semicell from the previous division gives rise to another miniature which is appreciably larger than itself. The net result of further divisions is an increasing number of cells of normal or near normal morphology. It appears as if the size of a semicell may have a direct effect upon the size of a new semicell it is capable of forming (perhaps a simple metabolic effect) and that a certain cell size is the most stable for a given strain or species of desmid.

A distinction has been made throughout the preceding paragraphs between "symmetry" and "ornamentation" in morphogenesis of semicells. Symmetry itself arises as a result of localized regions of wall differentiation (nonexpansion) that may be analogous to those which give rise to cleft formation. Is, therefore, a distinction between these two processes really valid? I suspect that it may be the timing of the release of information via RNA/protein synthesis that is important. Symmetry may be determined either cytoplasmically, whatever that may mean, by information released from the nucleus before mitosis and cytokinesis (leading to the appearance of the first clefts which form the lobes even in the absence of a nucleus; Fig. 6.132), and ornamentation may be determined by information released after cytokinesis.

Many interesting conclusions can be drawn from the work already done; the more obvious ones are as follows:

1. Changes in symmetry occur in discrete steps,

Figures 6.138–6.143
MICRASTERIAS DENTICULATA TREATED WITH CYCLOHEXIMIDE AFTER CYTOKINESIS

6.138.   Treatment started soon after cytokinesis. Five undifferentiated lobes of rather unequal size have formed.

6.139.   Treatment started a little later than that for the cell shown in Figure 6.138. Two of the lobes have developed an additional cleft.

6.140.   Treatment started when the daughter cells had formed five lobes (Fig. 6.121). Differentiation has proceeded as far as the nine-lobe stage before ceasing.

6.141.   Treatment started when the semicells had developed nine lobes (a little before the stage in Fig. 6.122). These lobes have undergone at least one (and in some cases two) additional series of cleft formation before differentiation ceased.

6.142.   Cell treated with cycloheximide for 90 minutes soon after cytokinesis and then incubated in normal medium thereafter. Morphogenesis was inhibited briefly after nine lobes had been formed (*cf.* Fig. 6.139) and then resumed, so that the cell margin now exhibits the finer splitting characteristic of the later stages of differentiation.

6.143.   As for Figure 6.142, showing an equivalent cell under the scanning electron microscope.

SOURCE. Drs. D. Tippit and J.D. Pickett-Heaps, published in *Protoplasma 81*, 271 (1974).

6.138

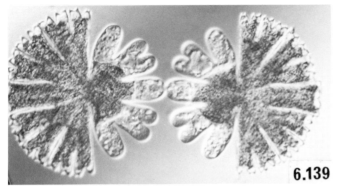

6.139

6.140

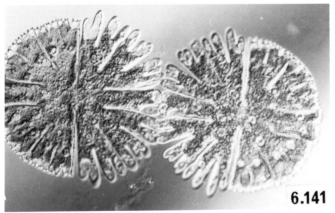

6.141

6.142

6.143

gain or loss of one or more wings taking place after division. Intermediate degrees of radiateness are not normally encountered (Figs. 6.128–6.131).

2. Symmetry is probably predetermined by the time of cytokinesis and cannot thereafter be affected by subsequent "inactivation" or removal of the nucleus (Fig. 6.132).

3. Ornamentation can be continuously variable between the extremes of none to normal morphology.

4. Ornamentation is directly influenced by the postmitotic nucleus, and control of ornamentation is probably mediated via RNA and protein synthesis. The degree of cleft formation in *Micrasterias* may be directly proportional to the quantity or concentration of the morphogenetic material(s) in the cytoplasm.

5. Changes in symmetry and ornamentations are not necessarily attributable to genetic changes.

6. Cell volume is proportional (approximately) to the ploidy of the cell. Loss of a wing, for example, will result in an increased size of the remaining wing and/or polar lobe (see particularly Figs. 6.127–6.131).

By now it will be obvious that development of form in expanding semicells of dividing desmids is a very subtle process. Waddington (1963) was unable to detect any ultrastructural evidence of cytoplasmic axes; he also says that the nuclear envelope is not "well developed" after division and that nuclear material may have been passing from the nucleus to the cytoplasm. His electron micrographs, however, are not convincing, and this latter observation has not been confirmed in my own work (e.g., on *Cosmarium*; Section 6.4b). Electron microscopy of *M. denticulata* (Kiermayer, 1968, 1970a,b) has similarly not been helpful so far in illuminating these events. Material seems to be added to the cell wall from either or both of two populations of vesicles derived concurrently from Golgi bodies. These organelles are conspicuously concentrated near the expanding cell wall (as in *Cosmarium*). There is no evidence yet of any cytoplasmic features that can be associated with differentiation of the expanding wall. Microtubules are common in the cells; they persist near the posttelophase nucleus, and many are scattered in the cytoplasm and particularly lining the wall near the isthmus (cf. *Cosmarium*). However, no specific association

Figure 6.144
MICRASTERIAS DENTICULATA TREATED WITH CYTOCHALASIN B AFTER CYTOKINESIS

The results in successive division cycles of treating differentiating daughter cells of *M. denticulata* with the drug cytochalasin B for 1 hour soon after cytokinesis, followed by transfer to normal medium. *A*. The semicells have undergone a little expansion before further growth and differentiation ceased altogether. *B*. Division, a few days later, of the daughter cell indicated by numbers *(1, 2)* in *A*. The normal-sized semicell *(2)* has formed another semicell of normal size, whereas the stunted semicell *(1)* has given rise to a miniature semicell *(3)* of relatively normal morphology. *C*. The next division in the cell indicated by numbers *(1–3)* in *B*. The original stunted semicell *(1)* again gives rise to a miniature semicell as in *B*, while the miniature semicell *(3)* in *B* now forms a semicell larger than itself and approaching normal size.

SOURCE. Drs. D. Tippit and J.D. Pickett-Heaps, published in *Protoplasma 81*, 271 (1974).

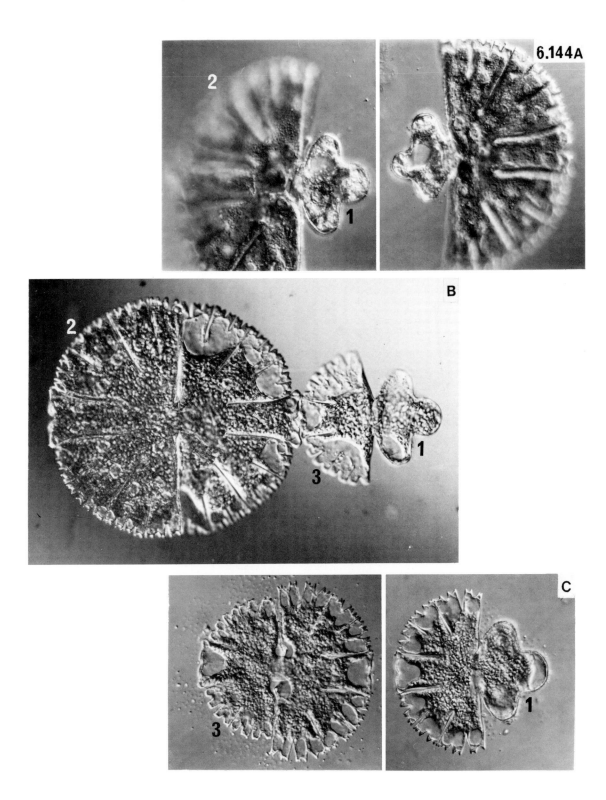

6.144A

of these organelles was found with the expanding wall. Kiermayer and his associates confirmed these results by treating expanding cells with drugs known to affect microtubules in different ways (colchicine, vinblastine, IPC). These drugs had no effect on morphogenesis of the cell wall. However, movement of the nucleus and chloroplast into the expanding semicell was usually severely affected by high (0.4%–0.7%) and perhaps rather poisonous concentrations of colchicine, suggesting that microtubules could be involved in these phenomena. Kiermayer (1970a,b) speculates that the plasmalemma itself or the cell cortex somehow inherently possesses the pattern and informational content required to form the shape of the new wall, perhaps by directing the secretion of Golgi vesicles; the nature of such patterning is totally obscure. My own investigations on cell division in *Cosmarium* (Section 6.4b) are likewise not helpful. All rather frustrating!

Kiermayer and Staehelin (1972) have used freeze-etching techniques to show massive, strap-like secondary wall fibrils which are composed of microfibrils. The cleaved face of plasmalemma is covered by particles and contains numerous small holes around the mucilage pores. They (Staehelin and Kiermayer, 1970) also showed membranous differentiation in the Golgi bodies; the distribution of particles on the cisternae changes across the stack from a minimum at the forming face to a maximum at the mature face, an observation that may be related to the way Golgi function generally (Fig. 6.146). The particles decrease in density where fenestration of the cisternae commences (Fig. 6.145).

6.4b. Cosmarium

Cosmarium botrytis is easy to maintain in culture and well known to botany students. Like other desmids, it can move about (Fig. 6.147). Normally biradiate (Fig. 6.148, 6.149), triradiate variants (Fig. 6.150) arise in small numbers under certain conditions. The two semicells are covered by a rather variable pattern of warts or ridges and a network of mucilage pores (Figs. 6.148, 6.149). Each cell contains two large chloroplasts, one in each semicell, flanking the central, flattened nucleus (Figs. 6.151*A*, 6.152). Each chloroplast normally contains two large pyrenoids (Pickett-Heaps, 1972g). The thick cell

Figures 6.145, 6.146
DIFFERENTIATION OF MEMBRANE IN THE GOLGI BODY OF MICRASTERIAS DENTICULATA REVEALED BY FREEZE-ETCHING

6.145.   The fracture plane has passed almost vertically through most of the cisternae except the last one. The association of endoplasmic reticulum (*e*) with the proximal or forming face of the Golgi body is clear. At the other face, numerous granules can be seen on the surface of the membrane, and these particles decrease sharply in density where fenestration and vesicle formation begin (between arrows). × 60,000.

6.146.   This fracture plane passes obliquely through the stack of cisternae. Two different sets of membrane faces are revealed, and in one series of these faces (*1–4*), the number of surface granules increases across the stack, reaching a maximum at the maturing face (*4*) of the Golgi body. × 60,000.

SOURCE. Drs. L.A. Staehlin and O. Kiermayer, published in *J. Cell Sci. 7,* 787 (1970).

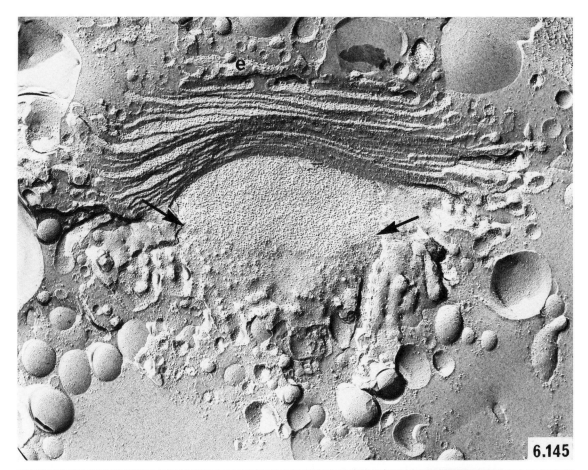

6.145

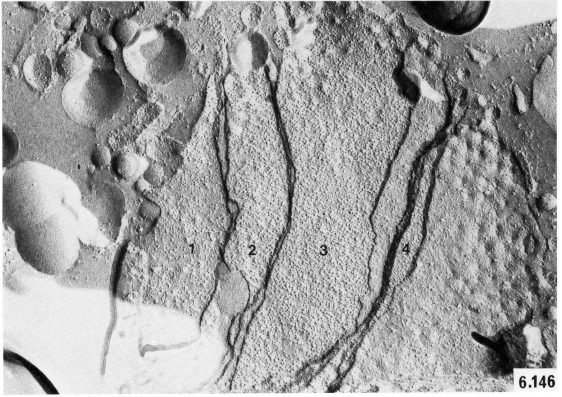

6.146

wall is traversed by numerous pores which usually have remnants of mucilage adhering to them externally in fixed and sectioned cells (Figs. 6.152, 6.159). The wall contains thick, strap-like arrays of parallel microfibrils, like those mentioned earlier in *Micrasterias* (Lott, Harris, and Turner, 1972; see also Mix, 1966). At the isthmus, the wall of the older semicell possesses a lip (Fig. 6.152) formed at a previous division (see Fig. 6.153). Numerous microtubules encircle the interphase cell inside the isthmus (Fig. 6.160), just as in *Micrasterias* (Kiermayer, 1968, 1970a). The ultrastructural appearance of cell division in *Cosmarium* has been described by Pickett-Heaps (1972g).

CELL DIVISION

Mitosis. Premitotic cells (Fig. 6.151*A*) are very dense compared with cells following division (Fig. 6.151*J*). During early stages of prophase, a girdle of newly deposited wall material is laid around the inside of the isthmus, and the semicells concurrently separate slightly, a sure sign of imminent division (Fig. 6.151*B*). The microtubules that were previously concentrated around the isthmus now disappear, and during prophase the nucleus becomes ensheathed by a typical extranuclear spindle with broad poles located adjacent to the chloroplasts. The nucleolus disintegrates as the chromosomes condense. By metaphase (Fig. 6.151*C*), the nuclear envelope has dispersed, although membranes frequently coat some of the numerous small double chromatids aligned on the metaphase plate (Fig. 6.154). The future cross wall or septum has by this stage normally grown inward a little from the girdle of new wall material around the isthmus. The spindle is typical, and the numerous chromosomal and interzonal microtubules seem to be oriented toward several regions in the broad poles. Remnants of the nucleolar material can usually still be discerned at this stage. Anaphase (Figs. 6.151*D*, 6.152) and telophase (Fig. 6.151*E*) follow as usual. The telophase nuclei remain widely separated until cytokinesis is finished (see below). The spindle microtubules do not all disappear but instead become increasingly focused toward one or more regions in the pole (cf. *Closterium*) and remain there intact as the daughter nuclei enlarge and return to the interphase condition.

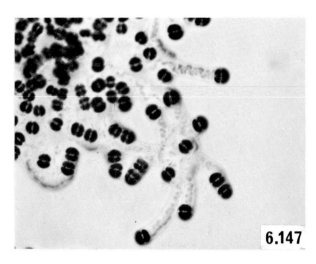

6.147

Figure 6.147
COSMARIUM BOTRYTIS GROWING ON AGAR

These cells, like most desmids, can move actively, if slowly about. The trails left on the agar surface by the moving cells are clear. The paired cells have recently undergone division.

Figures 6.148–6.150
SCANNING ELECTRON MICROGRAPHS OF COSMARIUM BOTRYTIS

6.148.   Pair of daughter cells after division. The new semicells are smooth-surfaced and covered by the primary wall, which is devoid of mucilage pores characteristic of the secondary wall. × 1,600.

6.149.   Another cell from the same culture as that shown in Figure 6.148. Note the subtly different surface morphology of the two examples. Such variation is common in desmids. × 1,500 approx.

6.150.   Triradiate variant of *C. botrytis*. × 1,100 approx.

SOURCE. Author's micrographs. Figure 6.148 published in *J. Phycol.* 8, 343 (1972). Figure 6.149 published in *Trans. Am. Microsc. Soc.* 93, 1 (1974).

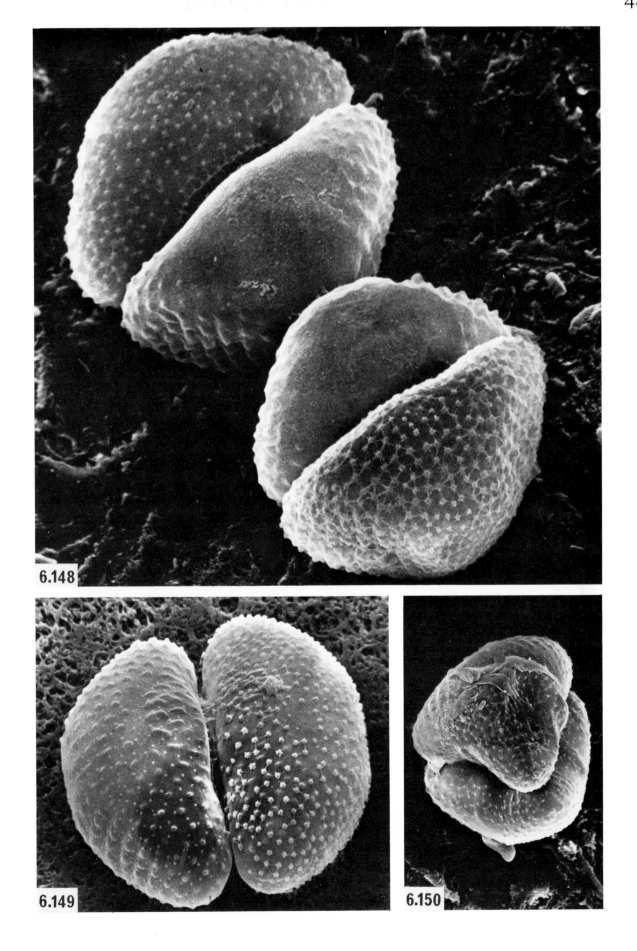

6.148

6.149

6.150

Cytokinesis and semicell expansion. During mitosis, the septum slowly grows inward, becoming about half-formed by telophase (Fig. 6.152) and completed sometime thereafter (Fig. 6.151 *E*). How the septum grows is not clear. Its diffuse fibrous wall material (Fig. 6.154) resembles that inside nearby vesicles, but direct evidence for incorporation of these vesicles into the wall has not been obtained. At no stage were microtubules in any way significantly associated with growth of the septum.

Cell expansion begins soon after the septum is formed. The cells move apart, and the septum stretches and begins to split (Figs. 6.151*E–H*, 6.153; cf. *Closterium*). Since the wall does get thinner during expansion, material obviously must be added to it continuously. The expanding wall consists of loosely aggregated fibrils (Fig. 6.155), somewhat oriented at the outside but aggregated into clumps near the plasmalemma. Its appearance clearly suggests that wall material is secreted as small masses of tangled fibrous material (contents of vesicles?) and that the fibrils become increasingly oriented parallel to the plasmalemma as the wall expands and the fibrils, secreted earlier, move to the outside of the wall. Golgi bodies collect preferentially in the forming semicell in large numbers (Fig. 6.151*G*), and they probably form the vesicles which may be engaged in wall secretion.

Expansion of the soft, extensible primary cell wall (Fig. 6.153) has to be carefully controlled spatially, since it usually achieves an almost perfect mirror image of the older semicell. Just how this control is exercised by the cell cannot yet be ascertained. No microtubules are ever present near this wall. Microtubules return soon after telophase to the isthmus (Fig. 6.160), but they invariably are confined to the nonexpanding region of the wall, and become more symmetrically arrayed around the isthmus only when the new secondary wall is thickening. (This sequence of events differs markedly from what has been described in *Closterium*.) The cytoplasm of these expanding semicells is delicate and difficult to fix, which adds to the difficulties of trying to interpret electron micrographs. Fairly late in expansion, the primary wall develops a pattern of ridges (Fig. 6.156), the forerunners of the future wall ornamentation (Figs. 6.148, 6.149). How the wall develops its pattern is also quite obscure, but

Figure 6.151
CELL DIVISION IN COSMARIUM BOTRYTIS

A. Interphase. The single chloroplast containing two large pyrenoids fills much of each semicell. The interphase nucleus, as always, is centrally situated in the isthmus. *B.* Prophase. The semicells have moved very slightly apart as a tiny girdle of new wall material, just visible in this micrograph, is deposited around the inside of the isthmus. The nucleolus is dispersing as the chromosomes condense. *C.* Metaphase. Note the ingrowth of the future septum just visible at the isthmus. *D.* Early anaphase. *E.* Telophase. The daughter nuclei are re-formed, and cytokinesis is almost, but not quite, complete. *F.* Early semicell expansion. The nuclei always lie in the forming semicell, and the chloroplasts are just beginning to move into it too. *G.* Later stage in semicell formation. The tiny dense objects in the peripheral cytoplasm of the forming semicell (arrows) are Golgi bodies. Note the pyrenoids beginning to move through the isthmus. *H.* Semicell formation essentially complete. *J.* As for *H*, but the chloroplasts are now dividing at the isthmus, and the nuclei (one arrowed) are moving back to the isthmus. All × 1,000.

SOURCE. Author's micrographs, published in *J. Phycol.* 8, 343 (1972).

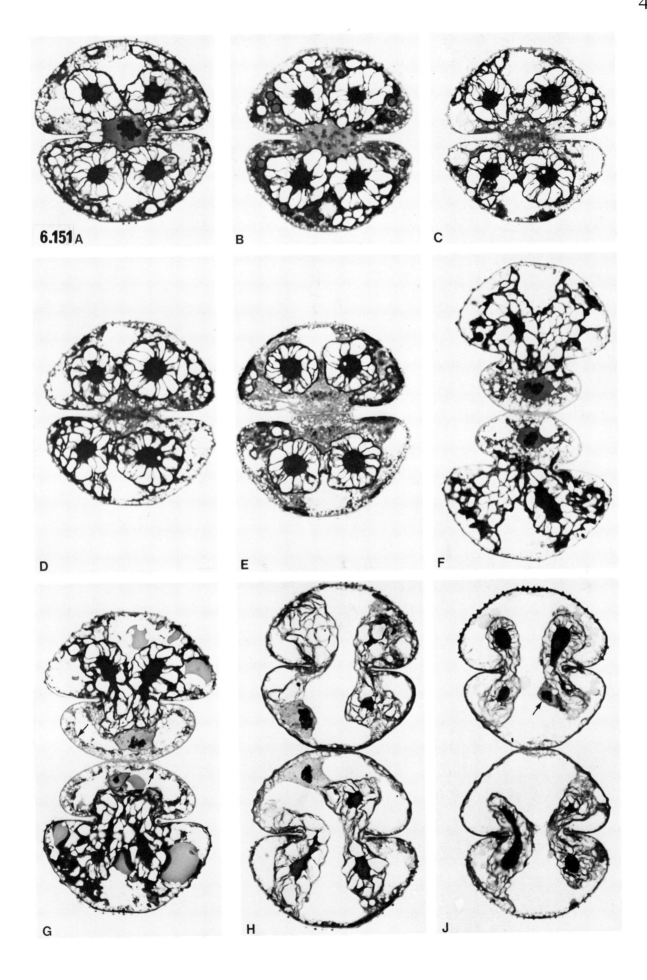

6.151A

the plasmalemma is obviously more closely attached to the wall between the ridges than to the ridges themselves and is, therefore, probably differentiated spatially in some way related to the pattern of wall development. This has been further supported by examining cells in which slight plasmolysis has occurred. Some micrographs suggest that secretion of the primary wall virtually ceases before expansion is complete. Between the ridges, the wall fibers appear to have become more interwoven, whereas on the ridges, the thinner wall appears more highly oriented and stretched (Fig. 6.156). The wall between the ridges could have become less extensible because of its more interwoven matrix of wall fibers, but equally likely, this appearance could be ascribed to a localized inhibition of cell expansion mediated by some other agent, and manifested by the alternation of thicker and thinner wall regions. Without further careful observation and experimentation, it is still impossible to separate cause and effect.

During expansion, the nucleus always moves into the forming semicell (Figs. 6.151*F–H*), and invariably, numerous microtubules are found around the nucleus, but their three-dimensional arrangement is complex and difficult to ascertain from thin sections. One is inevitably reminded of the extranuclear sheath of microtubules in *Closterium* that appears to be involved in postmitotic nuclear movement (Section 6.3a). However, the analogy may be misleading in some respects; for example, the nuclei move in different directions in the two genera (i.e., in *Closterium* the nucleus moves back into the older semicell and the movement appears to be a more precisely controlled event). In *Cosmarium,* the single chloroplast in each daughter cell begins to move through the isthmus (Figs. 6.151*F,* 6.153) and then around the edge of the expanding semicell (Fig. 6.151*G, H*). In thin sections, this migrating region of the chloroplast usually appears to consist of two distinct lobes. The pyrenoids elongate considerably, and roughly half of their material then passes into the new semicell with the chloroplast (Fig. 6.151 *J*). Finally, during the latter stages of secondary wall deposition, a constriction appears in the chloroplast precisely in the plane of the isthmus, and then the chloroplast divides in two. A little while later, the nucleus migrates back into the isthmus between the chloroplast halves (Fig.

Figure 6.152
LATE ANAPHASE IN COSMARIUM BOTRYTIS

This micrograph shows many features of the interphase cell, including the two chloroplasts each containing two pyrenoids *(py),* mucilage pores *(mp)* in the thick, warty wall, and the way the two semicells are joined. The daughter nuclei are re-forming. The semicells always move slightly apart during mitosis (cf. Fig. 6.151*A, B*), and the early ingrowth of the septum at the isthmus *(is)* can be seen. × 3,800.

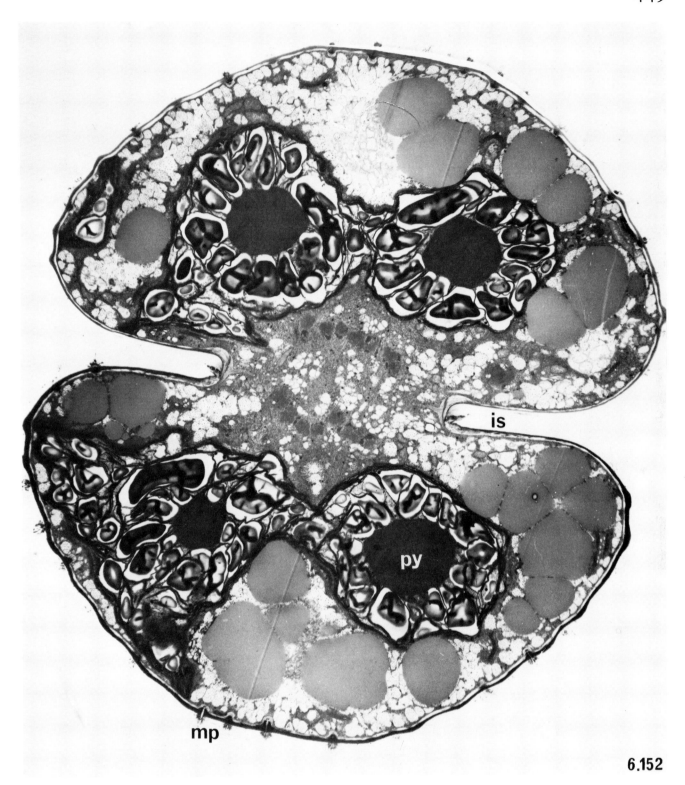

6.152

6.151*J*). I did not find any microtubules associated with it during this movement. The two daughter cells become highly vacuolated during cell expansion, and as might be expected, their cytoplasm is much less dense than that in the single premitotic cell.

Secretion of the secondary wall is initiated a little before the cells are fully expanded. This wall is totally different in appearance from the primary wall (Figs. 6.157, 6.158, 6.160), being even and dense; when heavily stained it reveals evidence of fibrillar substructure. How it is secreted is not clear. During formation, it matches and then accentuates the pattern of ornamentation already established by the primary wall. The sites of future mucilage pores are marked from the inception of secondary wall deposition by homogeneous plugs of material, quite distinct from the rest of the wall and surmounted by the plasmalemma, whose cytoplasmic surface at this site is covered by particles or bristles (Fig. 6.158). The plug traverses the entire secondary wall, which is often thickened around it; presumably the plug is eroded away to form the functional pore (Fig. 6.159), a process I have not investigated. This formation of pores raises another interesting problem concerning morphogenesis of the semicell. Most, if not all, of the form of the future semicell has been determined during expansion of the primary wall (as is particularly obvious in the more ornate constricted desmids such as *Micrasterias*). However, the primary wall is devoid of pores and smooth-surfaced, apart from its gross ornamentation (Fig. 6.148; see also Figs. 6.107, 6.109, 6.115). Therefore, at the commencement of secondary wall deposition, another feature of morphogenesis, the pattern of pores, becomes expressed. Presumably the pores themselves are templated by the localized secretion (cf. the specialized plasmalemma) of special material capable of being removed later. The plasmalemma may be differentiated for future pore formation during deposition of the primary wall, so that secretion of the plug material is initiated only when the secondary wall has been started. It has been impossible to tell whether establishment of the pattern of pores (and of spines, for example, in species of *Staurastrum*) is a separate morphogenetic event from that which determines the overall form of the semicell. The primary wall is shed later (Fig. 6.161).

Figure 6.153
SEMICELL EXPANSION IN COSMARIUM BOTRYTIS

The shape of the expanding primary wall is already beginning to duplicate that of the older semicells. The daughter nucleus is always situated in the forming semicell, and in this example, the chloroplasts are starting their migration into it. × 3,400.

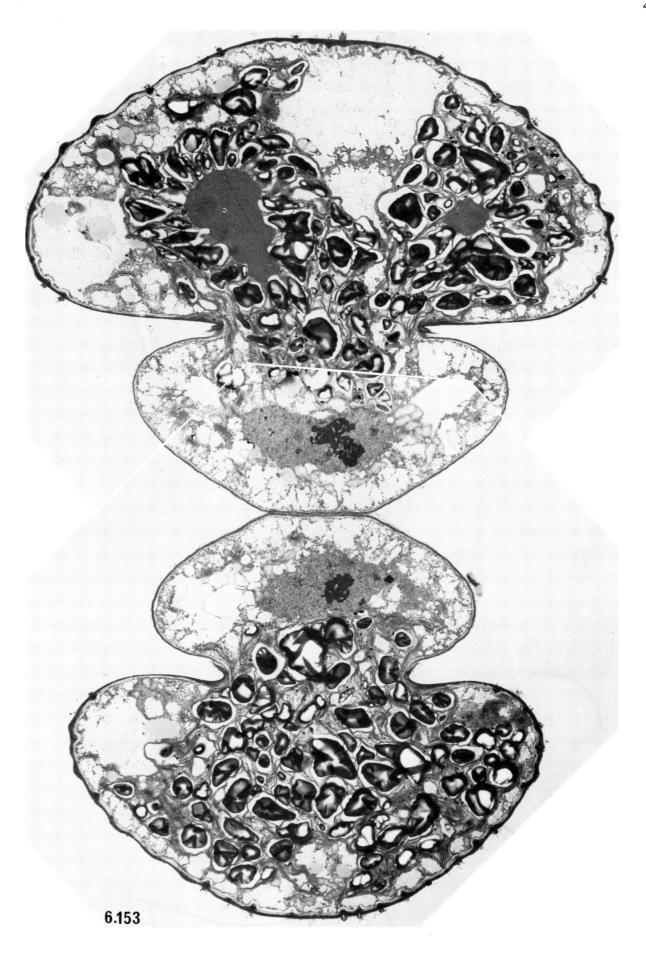

6.153

These electron microscopic findings do not really clarify the enigmatic nature of semicell morphogenesis. Indeed, they raise another problem already discussed in Section 6.3a: why should microtubules be intimately associated with the expanding semicell of *Closterium* and not of *Cosmarium, Micrasterias,* and presumably other constricted desmids?

Abnormal cells which have resulted from incomplete cleavage are not uncommon in cultures of *Cosmarium.* An example is shown in Figure 6.162, and Figure 6.163 explains how I think such a cell arose, a result of an eccentrically situated hole remaining in the septum after cytokinesis. Such a result is analogous to others already described for *Micrasterias* and *Closterium* (see earlier).

6.4c. Sexual Reproduction in Constricted Placoderm Desmids

As emphasized earlier, gametes in all members of the Conjugales are ameboid and not flagellate. The process of conjugation in the constricted desmids quite closely resembles what has already been described for *Closterium* (Section 6.3a). Fritsch (1935) supplies some earlier references concerning this process; perhaps the earliest comprehensive account of it is given by De Bary (1858). Conjugation, as usual, may be heterothallic or homothallic. In many species it is rare (Starr, 1959), perhaps because we do not know how to induce it, while in other species it is quite common. For example, the homothallic strain of *Cosmarium botrytis* maintained in our laboratory (Cat. No. LB 953 from the Indiana Culture Collection; Starr, 1964) has been reliably induced to conjugate (Fig. 6.164) by merely transferring actively growing cells on to the surface of 1% agar made up in the culture medium. Sexual reproduction in some species can be induced by exposing cultures to an atmosphere rich in carbon dioxide (e.g., Starr, 1955b, 1964; Winter and Biebel, 1967) or subjecting them to nitrogen deficiency.

Conjugation has been studied in detail for *C. botrytis* by Starr (1954a) and Brandham (1966, 1967), and this organism is probably representative of many other desmids. Cells begin to pair off within a day of mixing compatible strains of heterothallic species. After pairing, they move about until they lie flat against one another with

Figures 6.154–6.157
MITOSIS AND WALL FORMATION IN COSMARIUM BOTRYTIS

6.154.   Metaphase spindle. Note how the septum is just starting to grow inward from the girdle of new wall material laid down inside the isthmus *(is)* as the semicells separated for mitosis. × 9,000.

6.155.   Secretion of the primary wall. Just outside the plasmalemma, the wall material has apparently been secreted as discrete lumps (contents of vesicles?). Further out, stretching of this material has apparently given it a more oriented form. × 20,000.

6.156.   Appearance of ridges in the wall. These arise late in expansion, and the wall is alternately thickened and thinned out. No cytoplasmic feature can be associated with this differentiation. × 54,000.

6.157.   Secretion of the secondary wall underneath the primary wall after cell expansion is essentially complete. Note the pores (arrows; see Fig. 6.158) confined to the secondary wall. × 36,000.

SOURCE. Author's micrographs, published in *J. Phycol.* 8, 343 (1972).

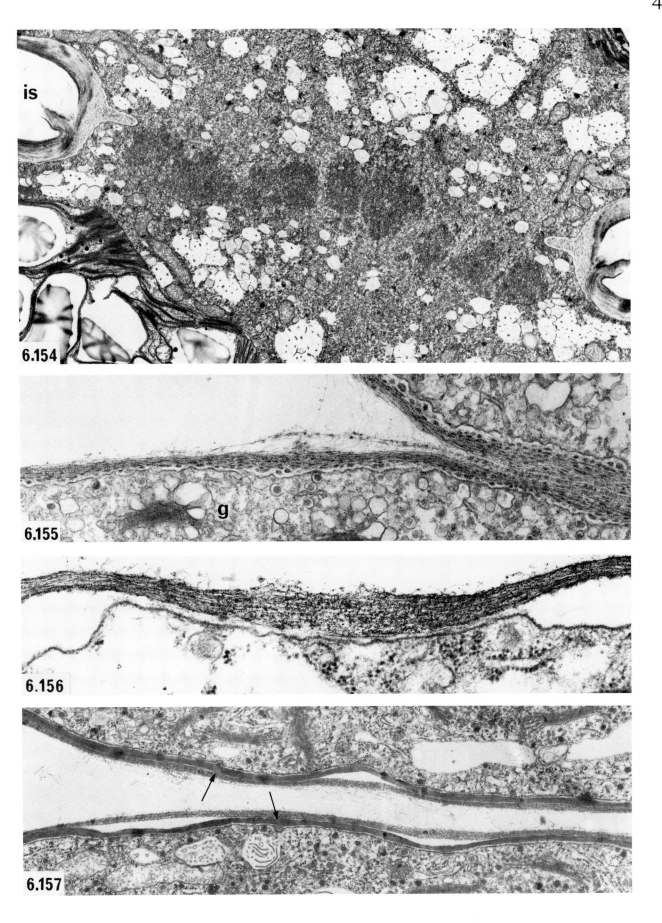

6.154

6.155

6.156

6.157

one isthmus at right angles to the other. This "reorientation" stage lasts about 4 hours, and fusion of gametes follows around 17 hours later. The paired cells cease moving and secrete mucilage around themselves which causes them to separate slightly. Then the isthmus splits apart at one side (Fig. 6.166) and a protuberance of cytoplasm enclosed by a hyaline vesicle is extended toward a similar one from the other cell. The protoplasts migrate quite rapidly out of the semicells (i.e., in about 30 minutes) and then they fuse. After fertilization, the zygote remains enclosed by the thin hyaline cell wall, probably analogous to the conjugation tube of *Closterium* (Fig. 6.164). Contractile vacuoles now appear on the surface briefly (Starr, 1954a), actively discharging their contents into the external medium. The zygote then slowly contracts while a wrinkled wall is secreted around it. Spines then appear on this wall (Fig. 6.165) and the zygote now enlarges to roughly its original size. Eventually its cell wall is thick and resistant; the ornamentation of spikes on zygospores (Fig. 6.167) is often characteristic of the species. The whole process from pairing to the appearance of fully developed spines lasts about 100 hours in *C. botrytis*.

Chemotaxis seems to be important in attracting compatible cells together. Brandham records anisogamous behavior in these supposedly isogamous cells, distinguishing one strain of gametes from the other by using morphological variants of cells. One of the conjugants, previously quite active, becomes essentially passive at the pairing stage, while that from the other strain moves actively toward it. One of his isolates was able to react in both these ways if mixed with appropriate strains. He, therefore, extends a definition of "anisogamous" to cover such subtle differences between gametes. However, I believe the term should be retained in its present usage, indicating that the gametes are of the same size and morphology, since it now appears likely that other small but significant differences may exist between isogametes in many other algae (see Sections 2.1b, 3.7b).

I have attempted some ultrastructural work on conjugating cells of *Cosmarium,* but this has been largely unsuccessful, since the gametes are extremely fragile. Some scanning electron micrographs are included here which show the split at the side of the isthmus (Fig. 6.166) that allows

Figures 6.158–6.161
WALL DIFFERENTIATION IN COSMARIUM BOTRYTIS

6.158.   Immature pore in the secondary wall. It does not extend through the primary wall. The pore is first seen as an amorphous plug under a differentiated region of the plasmalemma. × 88,000.

6.159.   The functioning pore in a vegetative cell. Apparently it arises from the dissolution of the plug seen in Figure 6.158. × 45,000.

6.160.   The isthmus during secondary wall secretion. The primary wall layer is indicated by the large arrows. Microtubules always line the plasmalemma around the isthmus (*is*) of interphase cells. Here, these microtubules (*t*) are reappearing, but they are always asymmetrically distributed most near the older semicell (*i.e.,* as here between the small arrows) until the secondary wall deposition is complete. × 34,000.

6.161.   Fully formed daughter cells shedding their primary wall. × 800.

SOURCE. Author's micrographs. Figures 6.158–6.160 published in *J. Phycol. 8,* 343 (1972).

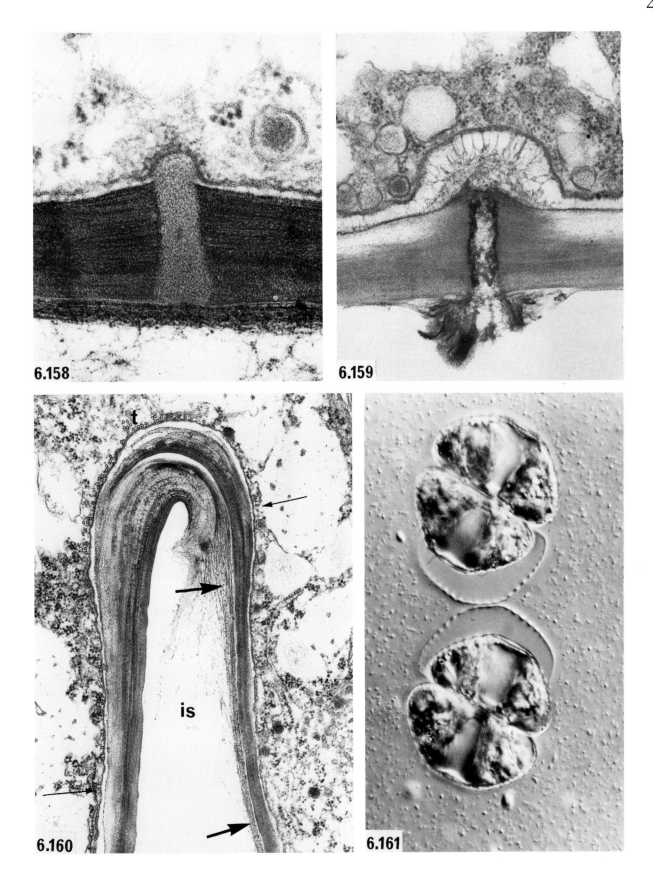

6.158

6.159

6.160

t

is

6.161

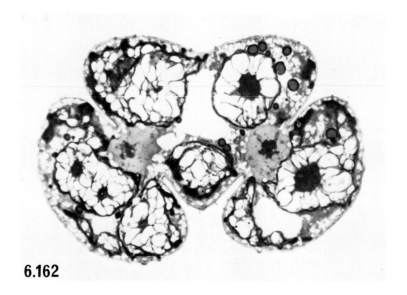

6.162

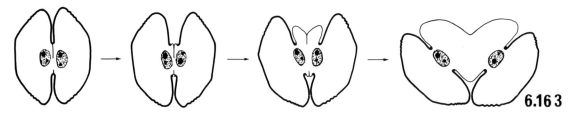

6.163

Figures 6.162, 6.163
ABNORMAL COSMARIUM BOTRYTIS CELL

6.162.   Abnormal, binucleate *C. botrytis* cell, probably a result of incomplete cytokinesis as shown in Figure 6.163. Similar double cells are shown in Figures 6.78 and 6.134. × 1,000.

6.163.   Diagrammatic representation of how the abnormal cell of *Cosmarium* in Figure 6.162 might have arisen, following incomplete cytokinesis in which a small, asymmetrical perforation remained in the septum (cf. Fig. 6.133).

SOURCE. Author's micrograph and diagram, published in *J. Phycol.* 8, 343 (1972).

the gamete to escape, and the cells' orientation with respect to one another, maintained even when the zygote is interposed between them (Fig. 6.165). The large and spiky zygote is covered with a deceptively thin conjugation tube, which has resisted quite strenuous attempts at removing it (e.g., treating the zygotes with cellulases, sodium hydroxide, etc.; Pickett-Heaps, 1973d).

Figures 6.164–6.167
CONJUGATION IN COSMARIUM BOTRYTIS

6.164.   Live cells. The zygote on the left is seen soon after fertilization, inside the conjugation tube. The other is at a later stage of maturation with zygote wall being secreted. × 450 approx.

6.165.   Scanning electron micrograph of a mature zygote still enclosed by the filmy but tough conjugation tube. Note the orientation of the now empty parental cell walls, at right angles to one another. These cells incidentally are slightly different in morphology from those in Figures 6.148 and 6.149. × 2,000 approx.

6.166.   Empty cell vacated by a gamete. The cell has split open at the isthmus on one side of the cell (CF. Fig. 6.165). × 2,400 approx.

6.167.   Spines of a mature zygote. × 6,000 approx.

SOURCE. Author's micrographs. Figures 6.165–6.167 published in *Trans. Am. Microsc. Soc.* 93, 1 (1974).

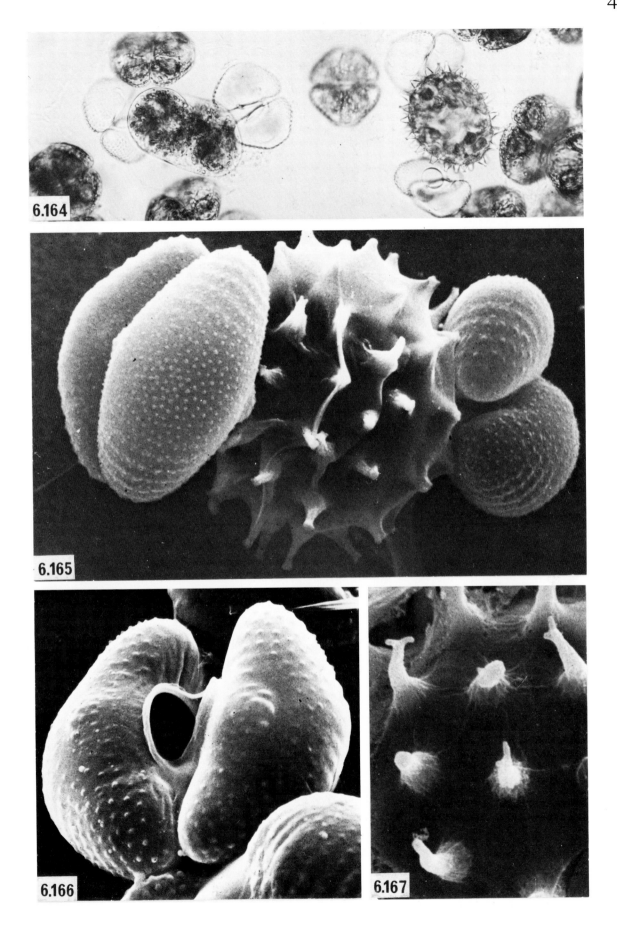

6.164

6.165

6.166

6.167

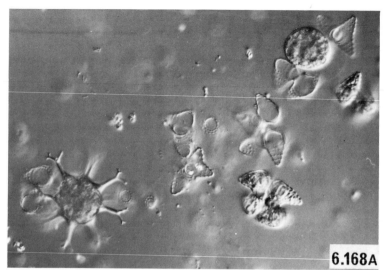
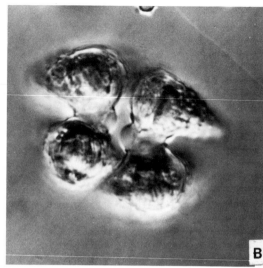

Kies has followed conjugation in *Micrasterias papillifera* using time-lapse cinematography (this film was not available at the time of writing, so I cannot comment on it; however, the superb technique of the Institut für den Wissenschaftlichen Film, Gottingen, will doubtless be evident again). Included here are some light micrographs, kindly supplied by Dr. R. Lenzenweger, of conjugation in *Micrasterias rotata* (Fig. 6.173), and the life cycle of this organism has been set out in Figure 6.172.

Extrusion and formation of spines in the maturing zygospore of *M. papillifera* are dependent upon turgor pressure, as is cell expansion of the semicell after division (Kies, 1968, 1970a,b). The spines grow apically; their thin outer wall stains with ruthenium red, and once formed, they become filled with the thick secondary wall as the protoplast retracts from within them. This secondary wall contains much cellulose, and Kies notes the similarity between primary and secondary wall deposition in dividing cells and this sequential deposition of the two wall layers (primary and secondary exospore) enclosing the zygote. The primary exospore is later shed from the zygote, which secretes further thick layers of wall material inside the secondary exospore. These layers undoubtedly contribute to the cell's ability to withstand desiccation, cold, etc. (cf. the zygote of *Closterium;* Section 6.3a; Fig. 6.97).

Conjugation has also recently been described in *Staurastrum gladiosum* (Winter and Biebel, 1967). The cells pair at right angles to one another, as described earlier. The semicells stretch apart slightly when the cytoplasmic protuberance is formed. Gametes never emerge

Figure 6.168
CONJUGATION IN AN UNIDENTIFIED SPECIES OF STAURASTRUM

In *A,* three different stages can be seen in this species isolated by the author. The central pair of cells is just forming conjugation papillae, the young zygote at the top right has rounded up, and the older zygote at the bottom left has formed its intricate spines. *B* shows the pair of cells in *A* at higher magnification. *A,* × 600 approx.; *B* × 1,600 approx.

Figures 6.169–6.171
CONJUGATION IN ANOTHER LARGE, BIZARRE SPECIES OF STAURASTRUM

This desmid has now been identified by Dr. Paul Biebel as *S. furcigerum* Breb., var. *armigerum* Nordst. It was isolated, like those in Figure 6.168, by the author from a lake in the Colorado Rockies.

6.169. Maturing zygote with fully formed spines between the empty parental cell walls. Much debris has collected around the cell. × 580.

6.170. Vegetative cell, showing the beautiful symmetry of the species. × 1,400.

6.171. Zygote of this same species, partially cleaned up before being prepared for scanning electron microscopy. × 800.

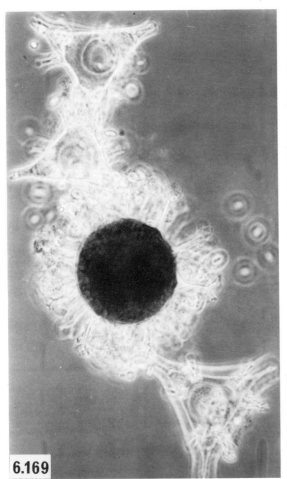

6.169

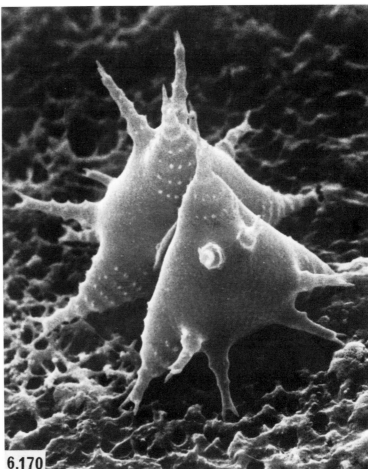

6.170

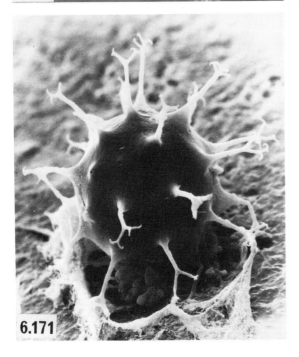

6.171

quite synchronously, and once released they tend to enlarge. Fertilization follows the erosion of the wall of the conjugation tube between them. Spiky zygotes are formed within 24 hours of fertilization. Winter and Biebel state that no conjugation tube is formed, but this statement probably reflects a difference in definition of the term. In my opinion, the wall enclosing an escaping gamete deserves this name, as in *Closterium* (see earlier comments). I have illustrated conjugation in this book using two species of *Staurastrum* I have isolated from mountain lakes, which undergo sexual reproduction readily in culture (Figs. 6.168–6.171); one of these species is particularly ornate and quite large (Fig. 6.170). The zygotes are beautiful (Fig. 6.168A, 6.169); that in Figure 6.171 has been slightly damaged by the rather vigorous treatment it was subjected to for cleaning the debris, conjugation tube, etc. off it (cf. Fig. 6.169).

Germination of zygotes is interesting for several reasons. The number of cells formed from the zygotes is normally two, but sometimes only one (Starr, 1959). Germination, as usual, is difficult to induce at will, and Starr (1949) outlines a method using heat treatment that often works with many zygospores. In *Cosmarium botrytis* (Starr, 1955a) the zygospore wall ruptures at germination, apparently by an intake of water, since the released protoplast is quite large. Starr says that this release of an undivided protoplast (as in *Closterium*) is characteristic of placoderm desmids but does not occur in other Green Algae. The number of chloroplasts in the cell is difficult to determine from Starr's account; he mentions eight chromatophores containing one pyrenoid each, per zygote after fertilization. I suspect there are really only four chloroplasts (= chromatophores), since vegetative cells contain two chloroplasts, each containing two pyrenoids; each of these two chloroplasts is deeply lobed and often appears as two (Pickett-Heaps, 1972g). However, the protoplast released from the zygote probably does contain two chloroplasts, as Starr (1955a) suggests, which move to opposite sides of the protoplast (now shrinking considerably) as the first meiotic division takes place. After the second meiotic division, the protoplast cleaves and one of the two nuclei in each daughter cell or gone degenerates (see also Fritsch, 1935; p. 353). The gones are then released from

Figure 6.172
LIFE CYCLE OF MICRASTERIAS ROTATA

Diagrammatic representation. The typical vegetative cell *(A)* undergoes cell division as shown in *B–D*. Sexual reproduction commences with pairing of cells *(E)*, which form conjugation tubes *(F–I)* and undergo fertilization *(J)*. The mature ornamented zygote *(K)*, upon subsequent germination, releases its contents into a large vesicle *(L)*, in which are formed the two gones *(M)*. The gones are constricted, but otherwise simple in morphology, and are soon released from the vesicle *(M, N)*. The gones now undergo cell division, during which the newly forming semicells acquire the morphology of the species *(O–S)*. The products of this division are asymmetrical *(S)*; during the next cycle of division, they produce normal individuals *(T–W)*. Note, however, that the simple semicells derived from the gone can never attain normal morphology, and so some such asymmetrical cells *(S)* persist in cultures derived from germinating zygotes. See Fig. 6.173.

SOURCE. Dr. R. Lenzenweger, published in *Mikrokosmos 57*, 270 (1968).

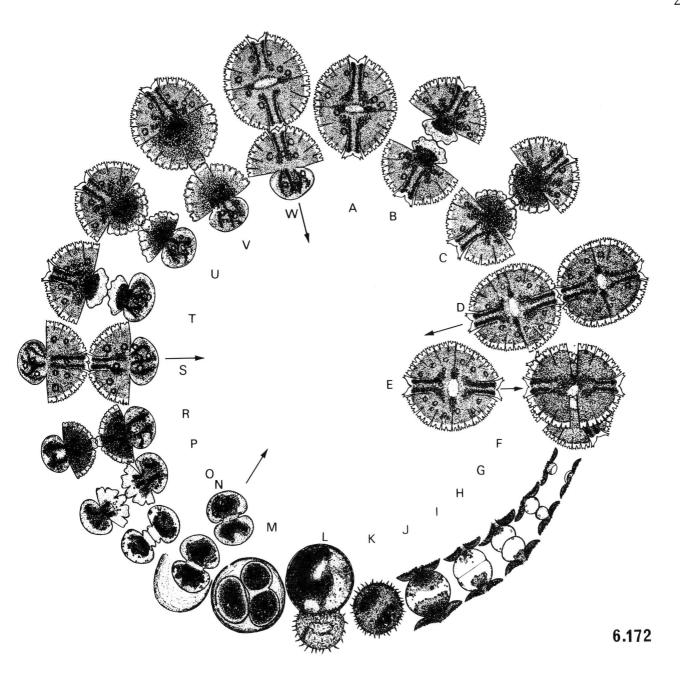

6.172

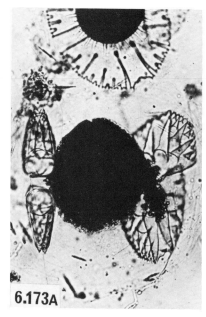

6.173A B C

the thin wall that originally enclosed the liberated protoplast.

The gones at this point are simple constricted cells. It is only after the first true vegetative division that the newly forming semicells take on the shape characteristic of the species of desmid. This sequence of events is even more dramatic in more ornate desmids such as *Micrasterias rotata* (Figs. 6.172; 6.174–6.177; Lenzenweger, 1968a,b). The simple form of the semicells from the gones is, of course, retained subsequently unaltered through cycles of cell division by one of the daughter cells (Figs. 6.178–6.181). This delay in the expression of vegetative morphogenetic form following meiosis is likely significant, but I cannot suggest why.

Korn (1969) has investigated chloroplast inheritance during both sexual and asexual reproduction in *Cosmarium turpenii*. Unfortunately this account may be confusing, since the author also says there are four chloroplasts (each containing one pyrenoid) per cell. While it may be that different species contain different numbers of chloroplasts, I rather suspect (see previous comments) that light microscopy is misleading and that there really are only two per cell as in *Cosmarium botrytis* (Pickett-Heaps, 1972g).

The appearance of triradiate cells (Fig. 6.150) of *Cosmarium* is often associated with sexual reproduction. For example, triradiate cells are not infrequently encountered among cells conjugating on agar. Starr (1958) and Brandham and Godward (1964) have investigated the ap-

Figure 6.173
CONJUGATION AND ZYGOTE FORMATION IN MICRASTERIAS ROTATA

Three stages in conjugation and zygote formation in *M. rotata*. See Figure 6.172 *E–K*.

SOURCE. Dr. R. Lenzenweger, unpublished micrographs.

Figures 6.174–6.177
GERMINATION IN LIVE CELLS OF MICRASTERIAS ROTATA

6.174. Ornamented zygote *(z)* rupturing and releasing its contents into a vesicle. See Figure 6.172 *L.* × 530 approx.

6.175. The two gones inside the vesicle released from the zygote. See Figure 6.172*M.* × 870 approx.

6.176. The simple form of the gone. See Figure 6.172*N.* × 1,200 approx.

6.177. The gone at cytokinesis. See Figure 6.172 *O–S.* × 1,100 approx.

SOURCE. Dr. R. Lenzenweger, published in *Mikrokosmos 57,* 270 (1968).

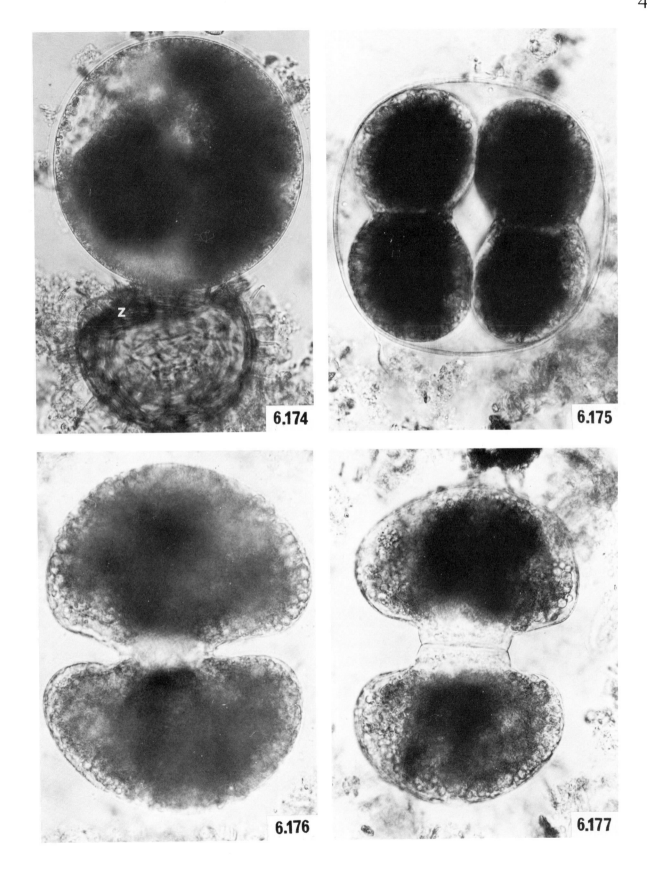

6.174

6.175

6.176

6.177

pearance and inheritance of this trait during asexual and sexual reproduction. Starr, working with *C. turpenii,* found that larger than normal cells appeared spontaneously in cultures, probably a result of polyploidy. In clonal cultures of these biradiate cells, tri- and tetraradiate variants appeared and could be isolated from the original cultures on rare occasions. The tri- and tetraradiate variants of many species of *Cosmarium* resemble some species of *Staurastrum.* During vegetative reproduction, the larger cells normally produced daughter semicells similar to themselves. However, the tetraradiate form was highly unstable and tended to revert to the triradiate form. The latter could also give rise to biradiate cells. The larger cells contained more pyrenoids than normal; their cellular volume was, however, relatively constant whether they were bi-, tri-, or tetraradiate. Triradiate cells in this heterothallic species behaved normally during sexual reproduction, and could conjugate either with other triradiate cells or normal biradiate cells of the appropriate complementary strain (Fig. 6.-182). However, germination of the hybrid zygotes did not give viable progeny, and cytological examination of the meiosis revealed chromosomal abnormalities and polyploidy. Germination of zygospores derived from normal haploid conjugants is also interesting. Zygospores occasionally released only a single cell (instead of the normal two), and this could be a small triradiate or the larger diploid triradiate variant. During culture, the former tends (during cell division) to revert back to the normal biradiate form.

Brandham (1965) confirmed that diploidy accompanied persistent cell enlargement in *C. botrytis,* and in these larger cells the triradiate form was the more stable. Cultures of biradiate cells eventually came to contain 97% triradiate cells. Furthermore, the ratio of cell volume between triploid and diploid zygospores and the rare haploid parthenospores was close to 3:2:1. He also described another fascinating facet of sexual reproductive behavior: each diploid cell usually pairs off with two haploids but conjugates with one. Polyploids probably occur in nature and give rise to high chromosome counts. Chromosome numbers in *Netrium,* for example, can vary up to 592 (Godward, 1966).

Brandham and Godward (1964) investigated the haploid triradiate variant in heterothal-

Figures 6.178–6.181
CELL DIVISION IN GONES OF MICRASTERIAS ROTATA

6.178.   Early stage of semicell expansion following the first division in a gone. See Figure 6.172*P.* × 940 approx.

6.179.   Later stage in cell expansion. × 780 approx.

6.180.   The asymmetrical cell resulting from the first division in a gone. See Figure 6.172*S.* × 920 approx.

6.181.   Result of the next cell division. A normal cell is formed, but the other daughter cell necessarily remains asymmetrical. See Figure 6.172*T–W.* × 650 approx.

SOURCE. Dr. R. Lenzenweger. Figures 6.179, 6.181 published in *Mikrokosmos 57,* 270 (1968).

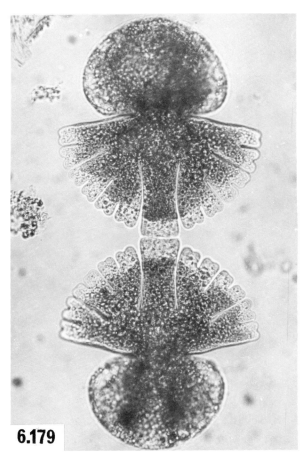

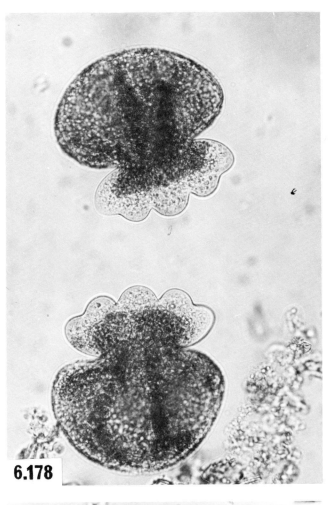

6.178

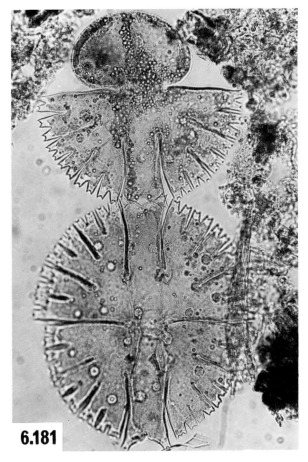

6.179

6.180

6.181

lic and homothallic strains of *C. botrytis.* These variants were never found in vegetative cultures, but quite often in those derived from germinating zygospores. In contrast to Starr's results, Brandham and Godward found that either or both of the pair of cells arising from a zygote could have this trait. The small triradiates had the same cell volume as normal cells but grew more slowly; they tended to revert to the biradiate form quite easily. "Reduced" triradiate cells were encountered which the authors feel were in the process of reverting back to the normal form. Once reversion had occurred, it was apparently permanent. Most of the triradiate clones behaved normally during conjugation, and the authors investigated the progeny of some resultant zygotes to determine how the triradiate trait was inherited through sexual reproduction. They discovered that around 12% of the progeny were triradiate, regardless of whether one or neither of the conjugants was triradiate. They concluded that the triradiate character is not under genetic control. It should be stressed again that the immediate products of zygote germination, the gones, are very simple cells (see earlier comments) and that their future form does not become apparent until after the first vegetative cell division. Brandham and Godward conclude that the gones are highly plastic and devoid of the cytoplasmic symmetry present in the normal vegetative individual.

6.182

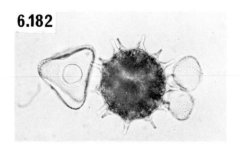

Figure 6.182
COSMARIUM TURPINII

Conjugation between bi- and triradiate variants. × 385.

SOURCE. Dr. R.C. Starr, published in *Am. J. Bot. 45,* 243 (1958).

7

THE CHARALES

The Charales pose for the systematist and tax-
onomist considerable problems (Chapter 8).
They have many features in common with other
Green Algae. Fritsch (1935, p. 466) considers
them "merely a very highly specialized side line
of the Chlorophyceae." Wood (1952) and Desikachary and Sundaralingam (1962) suggest an
affinity with the Chaetophorales. However, many
authorities maintain that they are probably closer
to the bryophytes than to algae (Scagel et al.,
1965; p. 293); Bold (1967) and Round (1971),
for example, prefer to place them in a division or
subphylum of their own (Charophyta), separate

from other Green Algae. My own limited work inclines me toward the latter view, since these organisms possess several features not normally encountered elsewhere in the Chlorophyceae (e.g., the method of cytokinesis and the complexity of reproductive structures and cells). Since this book is primarily concerned with cell structure and function in the Green Algae, I have included the Charales partly to draw attention to these unique features and also because the organisms themselves are so interesting.

The main reasons why the Charales stand out from other Green Algae can be summarized as follows:

1. Vegetative structure. The plants are highly complex in form, although the structural complexity is still achieved by various semifilamentous arrangements of single cells. The "stem" consists of a regular alternation of tiny, discoidal cells (nodes) separated by large —often very large indeed—highly vacuolate, coenocytic internodes (Figs. 7.1–7.3). In some species of *Chara* these latter cells are ensheathed by a layer of small, beautifully patterned cortical cells (Fig. 7.17). From each node arises a regular whorled pattern of laterals ("leaves") whose own structure again consists of a regular alternation of nodal and internodal cells. Tiny leaflet cells usually emerge from the nodes of the laterals too, and the terminal cell of each lateral soon stops dividing and elongates before growth ceases. The vegetative thallus, which may in extreme cases be 1 meter or more in length, is attached to some substrate by multicellular, branched rhizoids ("roots"; Fig. 7.1). Growth is initiated by cell division in the dome-shaped apical cell (Figs. 7.3, 7.66). The lower of the resulting daughter cells divides again. The lower of the two cells formed by this second division does not divide further, but undergoes enormous enlargement to become in due course the next internodal cell (Figs. 7.2, 7.3). The other, upper cell undergoes a complex series of divisions, often highly asymmetrical, and always strictly controlled spatially, to form the complex of cells at each node (in particular, the cells that will give rise to the laterals and the cortical cells). This region of tissue, therefore, resembles a meristem; furthermore, a miniature version of the apical complex of cells, an "axillary shoot," also is

Figures 7.1–7.3
CHARA FIBROSA

7.1. Female plants alongside a centimeter rule. The internodal cells, between branches and the laterals of limited growth, may be over 2 cm long. Rhizoids, invisible in this micrograph, are situated at the base of the main filament (small arrows). The apical growing regions (Fig. 7.2) are at the other ends of each plant (large arrows). × 0.37.

7.2. Apical region of female plant. Oogonia are situated at the nodes of the laterals (one is arrowed; see Figs. 7.63 *et seq.*). × 10 approx.

7.3. Section of apical meristem. The subapical cell has recently divided. The upper daughter cell would have become a node *(nd)* and undergone many further divisions to form the laterals. The other (basal) daughter cell would never have divided again, and instead would have formed a giant internodal cell *(in)*. The older laterals consist of alternating nodal and internodal cells too. The upper laterals had not yet differentiated into this node/internode pattern, and one contains a prophase nucleus (double arrow). A telophase (triple arrow) at the node is initiating the formation of the next lateral. The internodes characteristically have their chloroplasts peripherally arranged (arrowheads). The internode in Figure 7.6 is a bit younger than that designated *(in*)*. Some of the cells at the internode were giving rise to oogonia; see the metaphase (single arrow). The paired arrows show the cells that would have undergone division (see Figs. 7.17, 7.18) to form the complex sheath of cortical cells that envelops each internode (Fig. 7.17; see also cortical cells in Fig. 7.66). × 500 approx.

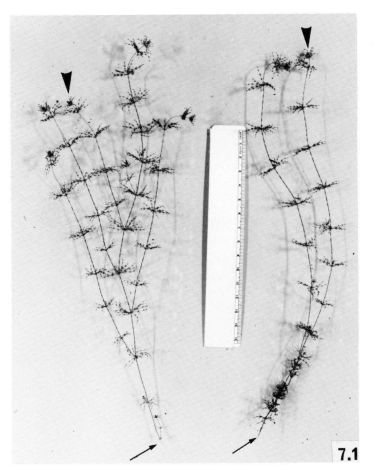

7.1

7.2

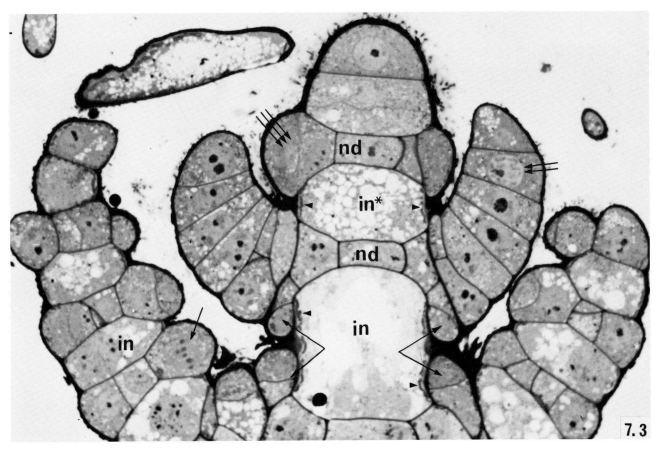

7.3

formed at internodes (Fig. 7.26). This shoot can give rise to well-developed branches on the main stem (Fig. 7.1) or may become the growing point of the plant if the original apical complex is damaged or destroyed. The tendency of daughter cells to differentiate into a regular sequence of cells of different sizes (i.e,. nodes and internodes) is reflected also in the pattern of cortical cells ensheathing the internodes of *Chara* (see later).

2. Sexual reproduction. Sexual reproduction is always oogamous, and the reproductive organs have acquired a degree of complexity unmatched elsewhere in the Green Algae. The structure of the antheridia and oogonia will be described in Sections 7.4–7.6. In particular, the reproductive cells are enclosed by a specialized sheath of sterile cells, a considerable advance in evolutionary development over virtually all other Green Algae.

3. Protonemal stage. Fritsch (1935, p. 447) lists as another advanced characteristic of the Charales the occurrence of a protonemal stage during germination of the zygote (Chapter 8). This protonema is a lengthening filament that emerges from the zygote along with rhizoids and undergoes a series of divisions forming large and small cells (not really nodes and internodes). The new stem arises initially as an offshoot from the protonema and soon becomes the dominant, erect portion of the thallus.

Some other interesting points can be mentioned here. In the greatly enlarged internodes, the nucleus gradually multiplies by amitosis. The numerous small chloroplasts are arranged along the wall in longitudinal rows (Figs. 7.63–7.65) which usually exhibit a slight spiral in larger cells. Green (1964), using time-lapse cinematography of living cells, has unequivocally shown that these chloroplasts replicate by fission. Cytoplasmic streaming in the large internodes of these organisms is most spectacular, with the layer of cytoplasm internal to the stationary chloroplasts swirling around the cell in a carefully defined path. A slightly wider gap than normal between the rows of chloroplasts (visible in several places in Figs. 7.63–7.65) marks the interface between the upstreaming and downstreaming flow of cytoplasm. Streaming in the Characeae has been especially studied by Jarosch (1958), Kuroda (1964), Kamiya (1968), and Kamitsubo (1966a,b, 1972).

Figures 7.4–7.8
VEGETATIVE CELLS OF CHARA FIBROSA

7.4.   Detail of figure 7.6, showing an element of endoplasmic reticulum *(e)*, apparently dilating to form a small vacuole *(v)*. $\times$ 14,000.

7.5.   Golgi body sectioned in the plane of its cisternae. Note the fenestrated form of the cisternae and the parallel, fibrous components in the stack (arrow). $\times$ 48,000 approx.

7.6.   A rapidly vacuolating nodal cell (see *in** in Fig. 7.3). In several regions (arrows) vacuoles appear to arise from endoplasmic reticulum (Fig. 7.4). $\times$ 4,6000 approx.

7.7.   Plasmodesmata traversing a wall, in longitudinal section. $\times$ 75,000.

7.8.   Field of the numerous plasmodesmata in a wall, here in cross section. $\times$ 57,000.

SOURCE. Author's micrographs, published in *Aust. J. Biol. Sci. 20*, 539 (1967).

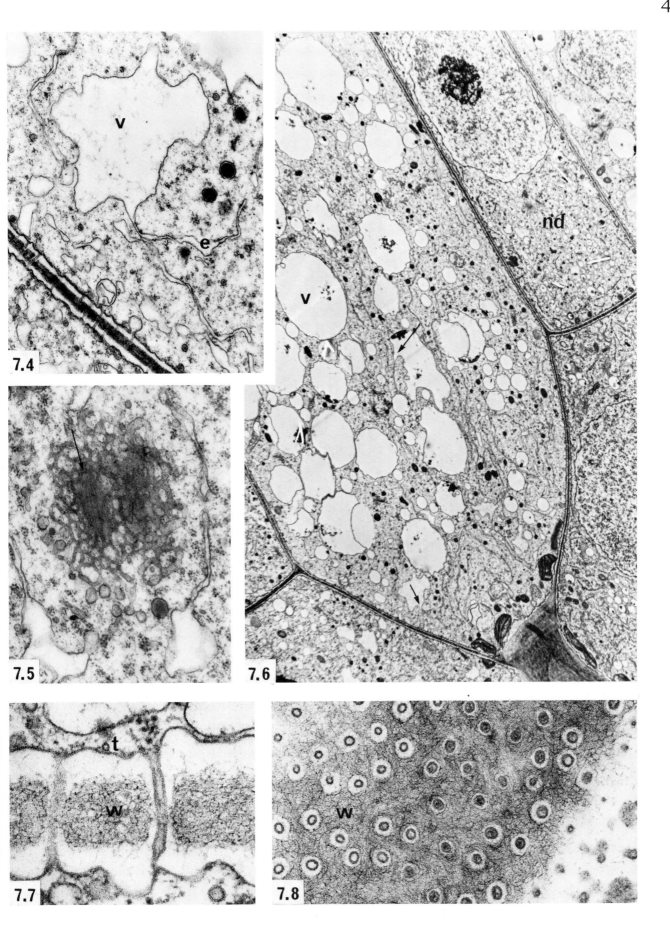

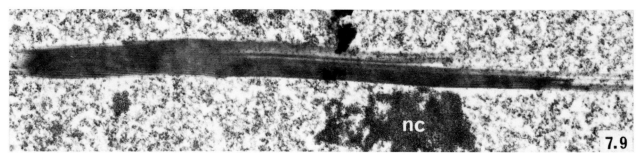

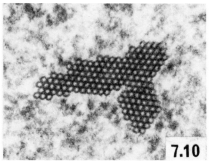

Figures 7.9, 7.10
TUBULAR COMPONENTS OF CHARA NUCLEI

Transverse and longitudinal sections through the characteristic crystalline array of tubular components present in most large nuclei of *Chara* (and *Nitella*). Figure 7.9, × 15,000 approx.; Figure 7.10, × 40,000 approx.

These workers have managed to isolate in vitro motile cytoplasmic fibrils from drops of cytoplasm. Similar fibrils can be discerned just internal to the chloroplasts, at the interface between the stationary and moving cytoplasm (Fig. 7.11). Kamitsubo centrifuged chloroplasts away from regions of the wall in *Nitella,* so as to be able to visualize these fibrils in vivo; linear fibrillar structures, over 100μ in length, were visible attached to the cortical gel layer along which moved cytoplasmic granules in the direction of streaming. Furthermore, these linear fibrils were occasionally observed to develop kinks, which eventually overlapped; touching portions of the fibril fused and then separated from the main fibril, thereby giving rise to polygonal loops (Fig. 7.14), which underwent rotation for a short while before disappearing. The polygonal fibrillar loops were motile in two different ways, either by rotating vigorously as a relatively inflexible unit, or by the concurrent propagation of bending waves around their periphery (Kamitsubo, 1966a). These microfibrils observed in vivo are almost certainly equivalent to bundles of fine microfilaments detected with the electron microscope (see below). Kamitsubo (1972) concludes that they are indeed the cytoplasmic entities involved in generating cytoplasmic streaming.

The internodal cells are so large that they constitute an excellent and widely used experi-

Figures 7. 11–7.14
FILAMENTS ASSOCIATED WITH STREAMING IN INTERNODAL CELLS OF CHARA

7.11.   Linear fibrils in the intact internode of *C. australis.* They lie just above the layer of chloroplasts arranged in rows and probably represent the morphological entity responsible for generating cytoplasmic streaming. × 2,400.

7.12.   Internode of *C. fibrosa,* showing a long, thin bundle of microfilaments just above the peripheral layer of chloroplasts *(c).* × 10,000.

7.13.   Detail of Figure 7.12, showing the microfilaments more clearly. Some microtubules *(t)* are near the wall. × 22,000.

7.14.   Motile, polygonal loop of microfilaments in living endoplasm. The bent corners of this polygon were transmitted as waves around the loop. × 1,800.

SOURCE. Figures 7.11, 7.14: Dr. E. Kamitsubo, published in *Protoplasma 14,* 53 (1972). Figures 7.12, 7.13: Author's micrographs, published in *Aust. J. Biol. Sci. 20,* 539 (1967).

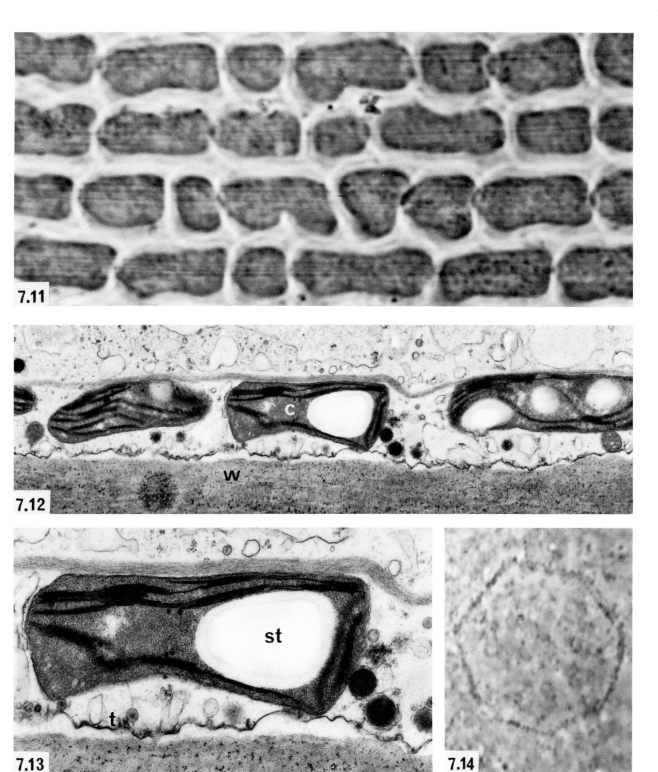

mental material for studying various biological phenomena, for example, ion fluxes in living cells. Microelectrodes can easily be inserted through the wall into the cytoplasm and on into the vacuole to measure potentials. Green and his coworkers in a number of publications (e.g., Green, 1965; Green and King, 1966; see review by Green, Erickson, and Richmond, 1970) have used *Nitella* for investigating the anisotropic nature of cell wall expansion, which they relate to the anisotropy of the wall itself (in particular, the highly oriented disposition of the structural cellulose microfibrils in it). Green (1968) has designed a micromanometer to measure and monitor directly the turgor pressure inside a living *Nitella* cell (see also Green and Stanton, 1967). These fascinating experiments are somewhat outside the scope of this book, and the interested reader is particularly referred to the review of Green, Erickson, and Richmond (1970).

Older cells become somewhat calcified, particularly the mature oogonia which have been recovered as fossils from as early as the Silurian period. These organisms then obviously belong to an ancient family, and Fritsch (1935) suggests that they may be regarded as a remnant of one of the many evolutionary lines that attempted to colonize the land.

7.1. Vegetative Cells

Only comparatively young cells of *Chara* have been examined in detail ultrastructurally (Pickett-Heaps, 1967b), since older, very large internodes are difficult to fix. (The species described in much of this work is *Chara fibrosa,* which may perhaps be better described as *C. preissii;* Proctor, personal communication). All cells are interconnected by extremely numerous plasmodesmata, lined by the plasmalemma, which penetrate through the typical fibrous cell wall (Figs. 7.7, 7.8). The nucleus of a young cell is large and contains a most conspicuous nucleolus (Fig. 7.3) ; the chromatin is finely divided and dispersed (Fig. 7.4). Older nuclei in internodal cells become large and pleomorphic. Many nuclei contain closely packed arrays of densely staining microtubular structures (Figs. 7.9, 7.10) whose significance is not understood. Golgi bodies are numerous, often with highly fenestrated cisternae; fibrous components are sometimes visible between the cisternae (Fig. 7.5). The chloro-

Figures 7.15, 7.16
MITOSIS IN CHARA FIBROSA

7.15.   Prophase, metaphase in forming antheridium (see Fig. 7.25). Spindles are acentric and open. Unusual is the adherence to the chromosomes of dense material derived from the nucleolus. × 2,800.

7.16.   Late anaphase, tip of forming lateral. These particular cells always had a concentration of mitochondria *(m)* at the apical pole. Nucleolar material (arrows) is still adhering to chromosomes. The spindle is barrel-shaped like that of higher plants. Note the aggregation of spindle fibers at the interzone *(i)*, preparatory to formation of the phragmoplast. × 8,700.

SOURCE. Author's micrographs, published in *Aust. J. Biol. Sci. 20,* 883 (1967) and *21,* 255 (1968).

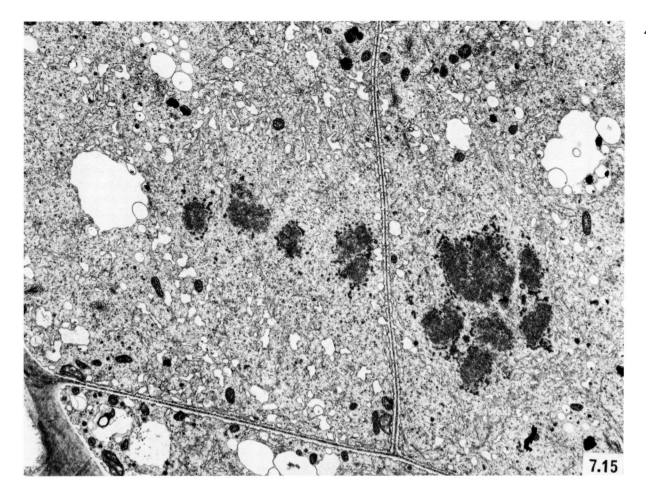

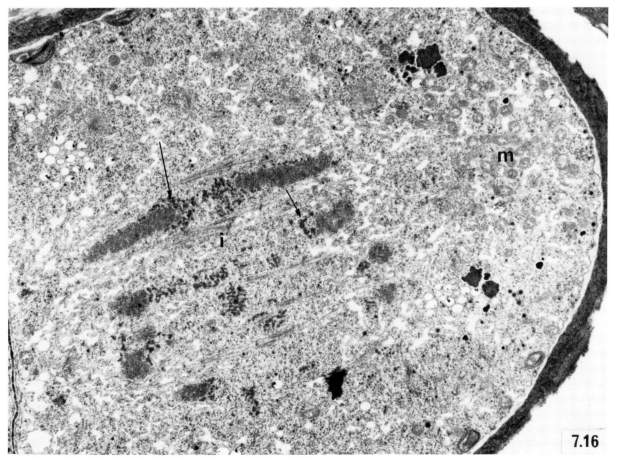

plasts, always situated adjacent to the wall, are quite small (Fig. 7.13) and never contain pyrenoids. Just internal to them, at the interface between stationary and streaming cytoplasm of large nodal cells, are often seen bundles of 50–100 filaments (Nagai and Rebhun, 1966; Pickett-Heaps, 1967b), each about 50 Å in diameter and many microns long (Figs. 7.12, 7.13). They are oriented parallel to the direction of cytoplasmic streaming in live cells and seem almost certain to be involved in it (see earlier comments). For example, colchicine treatment of internodes, which destroys microtubules, does not affect streaming (Pickett-Heaps, 1967b). Conversely, cytochalasin B reportedly stops streaming in *Nitella* (Wessells et al., 1971) and this drug is known to disrupt microfilamentous systems without affecting microtubules (Schroeder, 1970). Bradley (1973) confirmed these results in *Nitella,* but found that the microfilaments remained structurally unaltered by cytochalasin, even though streaming had ceased.

The plasmalemma of cells is underlain by typical transversely oriented microtubules (Fig. 7.13) which, when removed by colchicine (Pickett-Heaps, 1967b), cause actively growing cells in vivo to lose their cylindrical shape and round up as they enlarge. Green has demonstrated that the organized deposition of cellulose microfibrils in the wall is disrupted in these circumstances. Furthermore, by constraining such treated cells within jackets, he showed that randomized deposition of microfibrils continued even if the cells were prevented from bulging outward. Thus, the bulging itself does not cause the random microfibril deposition (see Green, Erickson, and Richmond, 1970, p. 722). It follows that cell shape is probably controlled (in part) by the disposition of microfibrils in the wall, which in turn is affected in some subtle way by the microtubules adjacent to the microfibrils but on the other side of the plasmalemma (as also appears to be the case in higher plants). Since oriented deposition of microfibrils is resumed upon colchicine removal, the disposition of microtubules cannot be controlled by the preexisting wall microfibrils (Green, Erickson, and Richmond, 1970).

Following division of the subapical cell, daughter cells about to differentiate into pairs of nodes and internodes (Fig. 7.3) were compared carefully (Pickett-Heaps, 1967b). No significant

Figures 7.17–7.19
CELL DIVISION IN CHARA FIBROSA

7.17. Layer of cortical cells growing via cell division, apically and basally, to cover the internodes *(in)*. The layer is cylindrical, and so this section cuts only part of it. Even these cortical cells have the node/internode cell pattern so typical of *Chara*. The three cells at each "node" are formed by two asymmetrical cell divisions. Two such asymmetrical prophases (arrows) are clearly visible in this micrograph. × 900 approx.

7.18. Asymmetrical telophase, following a prophase such as those visible in Figure 7.17. × 4,500.

7.19. Cell plate, very early stage of cross-wall formation. The new wall is initiated among the mass of convoluted membranes and vesicles, possibly derived from the Golgi *(g),* gathering in one plane in the longitudinally oriented microtubules *(t)* of the phragmoplast. Even mitochondria *(m)* tend to collect in this region. × 18,000.

SOURCE. Author's micrographs. Figures 7.17, 7.18 published in *Aust. J. Biol. Sci. 20,* 883 (1967).

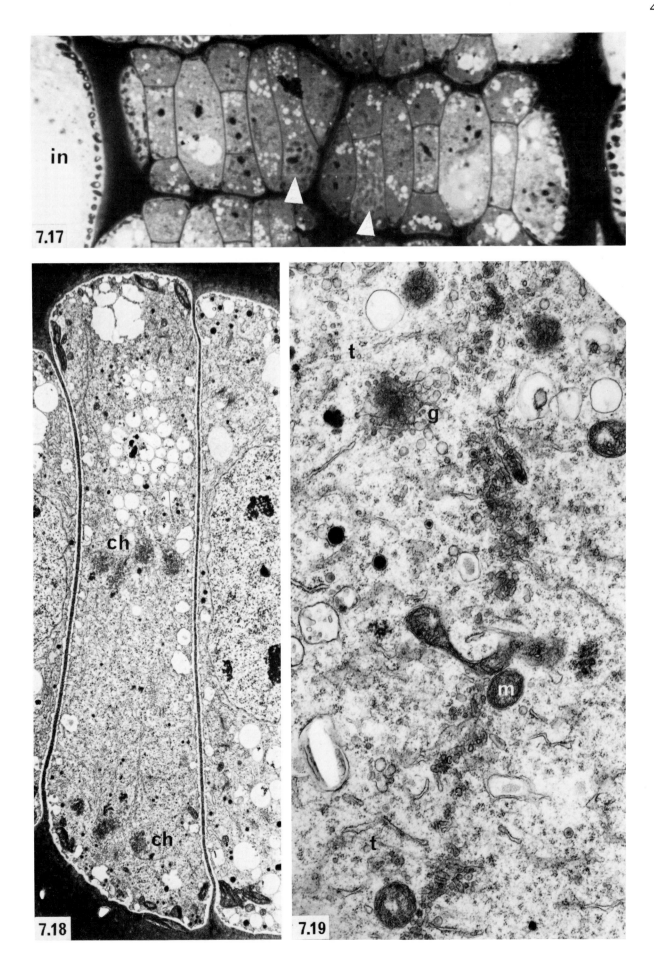

difference in ultrastructure could be detected between them. Initially, they appear identical and even later when a considerable difference in size has arisen, there is little to indicate how this differentiation is accomplished structurally. However, vacuolation in a future internodal cell (Fig. 7.6) is probably initiated by extensive dilations of elements of endoplasmic reticulum, forming numerous small vacuoles (Fig. 7.7), which in turn fuse to give rise to the large central vacuole. Young cells of *Chara* were also remarkable for the number and variety of odd-shaped inclusions contained within membrane-bound, vacuole-like, bodies. Their nature and significance are obscure. The streaming cytoplasm of older cells also contains a variety of fascinating inclusions.

7.2. Cell Division

Preprophase and prophase are particularly interesting in *Chara* for two reasons. First, some divisions are asymmetrical and predictable, particularly those that give rise to the distinctive pattern of cortical cells surrounding the internodes and those that form the antheridia (Fig. 7.25) and oogonia (Fig. 7.62). The nucleus is highly polarized before division in forming cortical cells (Fig. 7.17); the spindle is asymmetrically sited (Fig. 7.18), and cytokinesis subsequently divides the cell unequally. In the epidermis of leaves of higher plants, similar asymmetrical divisions establish the pattern of cells in the stomatal complex, for example (Pickett-Heaps and Northcote, 1966b; Pickett-Heaps, 1969b–d). One fascinating phenomenon, invariably encountered in such preprophase cells of higher plants, whether they are polarized or not, consists of a fairly massive grouping of microtubules near the cell wall in a position that predicts with great accuracy the eventual site where the cell plate would have joined the older cell walls, and thereby predicts the plane of cell division. This phenomenon is probably quite general in higher plants (Pickett-Heaps, 1973c). Consequently, preprophase nuclei, particularly when polarized (e.g., Fig. 7.17), were carefully scrutinized in *Chara* for evidence of an analogous "preprophase band" of microtubules. No trace of them could be found (Pickett-Heaps, 1967c). This observation indicates that preprophase bands are not essential for accomplishing cell divisions that are precisely controlled spatially, and so argues

Figure 7.20
TELOPHASE IN CHARA FIBROSA

Cytokinesis in tip of a lateral (cf. Fig. 7.16). Note concentration of mitochondria *(m)* remaining at apical pole. Re-forming (highly lobed) daughter nuclei *(n)* are far apart as the cell plate *(cp)* forms in the longitudinally oriented microtubules of the phragmoplast (invisible at this magnification). × 8,000.

SOURCE. Author's micrograph, published in *Aust. J. Biol. Sci.* 20, 883 (1967).

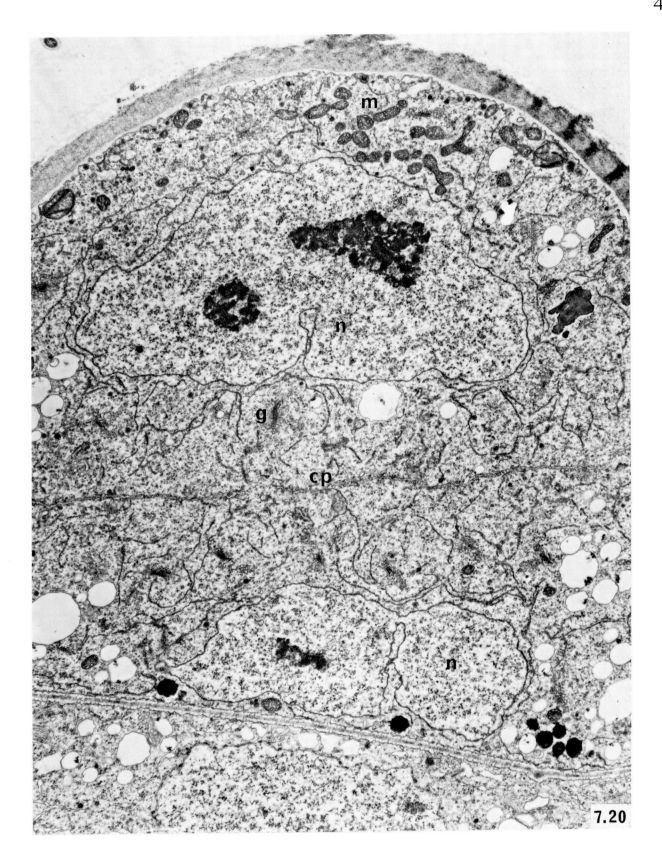

7.20

against a suggestion that the band is instrumental in achieving such asymmetrical divisions in higher plants (see Pickett-Heaps, 1974d).

Second, condensing prophase chromosomes are ill-defined and rather poorly stained, but the dense nucleolar material disperses somewhat during prophase and coats these chromosomes (Fig. 7.15). This particular phenomenon is relatively unusual (but is found, for example, in *Spirogyra;* Section 6.2a) and caused some confusion in the early literature, since in many preparations the chromosomes appeared to arise from the nucleolus. During subsequent phases of mitosis, this nucleolar material is transported on the chromosomes (Fig. 7.16) and appears to be incorporated into nucleoli re-forming in daughter nuclei. Thus, the nucleolar material is "persistent," which may be a relatively primitive feature of mitotic systems (Pickett-Heaps, 1970c). In higher plants, the nucleolus disperses during division. So, in these two particular respects (the absence of the preprophase band and the persistence of nucleolar material), the mitotic apparatus in *Chara* differs from that in higher plants (e.g., Pickett-Heaps and Northcote, 1966a). In other respects it is quite similar.

Light microscopists studying division in vegetative cells of *Chara* were unable to decide whether the spindles contained centrioles (see Fritsch, 1935; Karling, 1926). In my own study of numerous mitotic and interphase vegetative cells, I have never encountered any centriole or analogous structure. During prophase (Fig. 7.15), a typical, if rather poorly developed, extranuclear spindle of microtubules is formed, and by metaphase (Fig. 7.15) the nuclear envelope disperses and the cells have broadly defined poles with both continous and kinetochore microtubule systems. Anaphase (Fig. 7.16) and telophase (Fig. 7.18) follow rather as in higher plants.

Cytokinesis (Fig. 7.20) occurs by cell plate formation within a phragmoplast (Fig. 7.19). The nuclei remain far apart at telophase (see also Fig. 7.37), in contrast to the many other algae described earlier, and clusters of longitudinally oriented microtubules collect and proliferate by anaphase (Fig. 7.16) in the interzone region between the two cells. The tubules, however, are never as numerous as they are in the phragmoplasts of higher plants. The cell plate (Fig. 7.19) contains numerous vesicles and masses of

Figures 7.21–7.23
RHIZOIDS OF CHARA FOETIDA

7.21.   Group of live rhizoids, showing stratified organization of larger cell organelles. At the tip are the statoliths *(sl),* further back the nucleus *(n),* and then the vacuole *(v).* $\times$ 128.

7.22.   Cross section through a rhizoid, slightly back from the tip. The rhizoid had been rotated through 90°. The statoliths collect at the lower surface, and the Golgi vesicles at the upper. $\times$ 5,000.

7.23.   Diagrammatic representation of the cytoplasmic organization at the tip of a rhizoid. Golgi vesicles *(vs),* some being incorporated into the wall, collect at the very tip, with the statoliths close behind them. Further back are the Golgi bodies *(g),* mitochondria *(m),* plastids *(pd),* etc.

SOURCE. Dr. A. Sievers, published in *Z. Pflanzenphysiol. 53,* 193 (1965) and *51,* 462 (1967) and *Protoplasma 64,* 225 (1967).

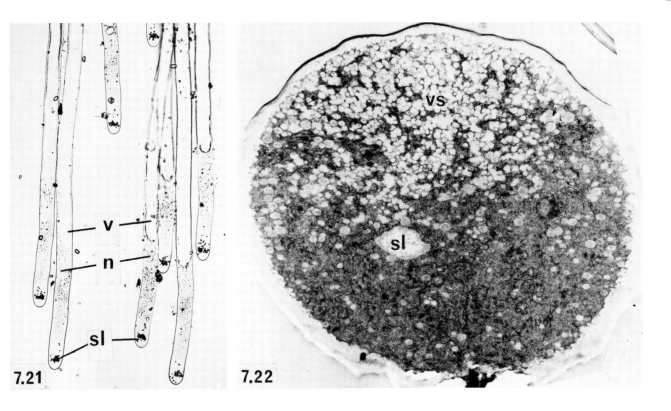

7.21

7.22

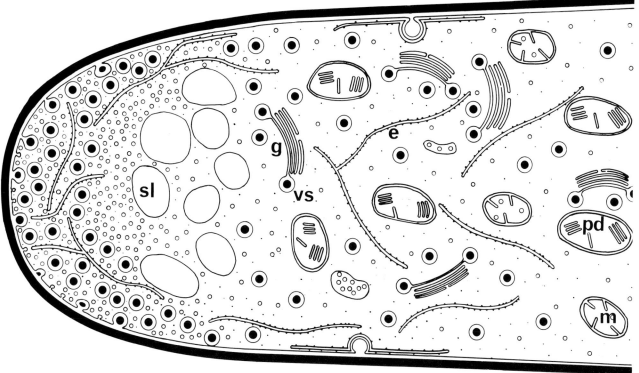

7.23

smooth, tubular membrane profiles; some or all of these appear to fuse to form the cross wall. Thus, *Chara* (and *Nitella;* Turner, 1968) is the only alga, apart from *Coleochaete,* shown to possess a cytokinetic mechanism similar to that of higher plants (although *Spirogyra* may be evolving in this direction; Section 6.2a.). Claims that other algae do resemble higher plants in this regard have not yet been substantiated (Pickett-Heaps, 1972a). Furthermore, the cell plate formed after certain mitoses in *Chara* (e.g., during formation of the oogonia [Fig. 7.67] and axillary shoots) may have a complex curved but predictable shape, as is also the case in certain cell divisions in higher plants; this complexity has not to my knowledge been found in other Green Algae. I regard these similarities between *Chara* and higher plants as highly important but their full significance is not yet assessable.

7.3. Rhizoids

The numerous, colorless rhizoids arising from the base of the vegetative thallus serve to hold the plant body firmly to some substrate; they may also function in vegetative propagation, forming new shoots. The rhizoids are positively geotropic, and Sievers (1965, 1967a,b) has investigated their ultrastructure in *Chara foetida,* with emphasis on the mechanism of the geotropic response.

Most of the elongation of these narrow, cylindrical cells (Fig. 7.21) takes place at their tip, and consequently, extension in the rhizoids resembles in many respects that of root hairs, pollen tubes, and some fungal hyphae. Sievers (1967a) distinguishes five zones at the end of the rhizoid. The zone at the extreme tip where wall growth is greatest (Fig. 7.23) contains numerous Golgi vesicles and microvesicles, endoplasmic reticulum, and ribosomes. The contents of the Golgi vesicles resemble the material of the wall, into which they almost certainly are discharged. Slightly further back, these vesicular constituents are arranged mostly around the periphery of the cell, the central region containing a high concentration of microvesicles and ribosomes. (This organization quite strongly resembles that of the spitzenkorper in the hyphae of certain fungi; Grove and Bracker, 1970.) The plasmalemma in this region is smoother, with less evidence of the

Figure 7.24
RHIZOID FROM CHARA FOETIDA

Geotropic response of a rhizoid from *C. foetida,* after having been turned through 90°. *A,* zero time; *B,* 3 minutes later; *C,* 6 minutes; *D,* 10 minutes; *E,* 15 minutes; *F,* 25 minutes. Note downward movement of statoliths, followed by extension and downward growth of the rhizoid. × 670.

SOURCE. Drs. A. Sievers and K. Schröter, published in *Planta* 96, 339 (1971).

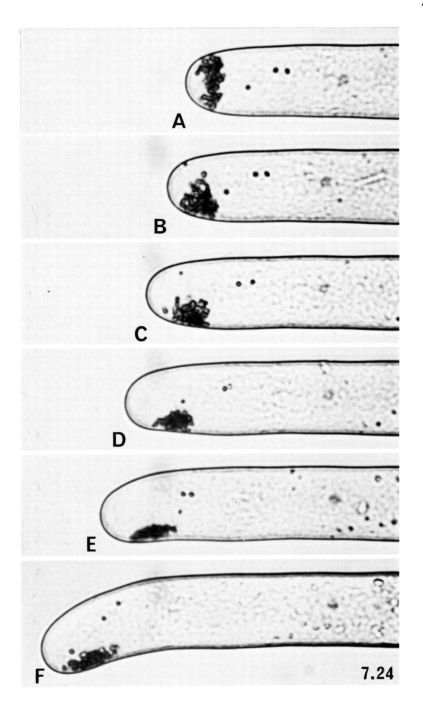

7.24

7.25

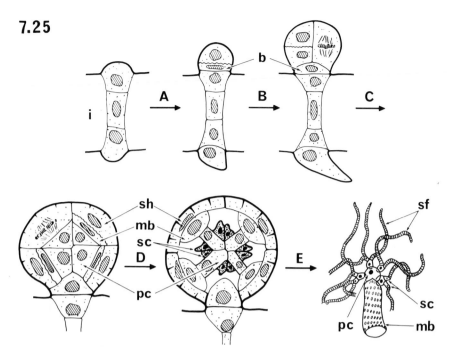

Figure 7.25
FORMATION OF THE
ANTHERIDIUM IN CHARA

Diagrammatic representation; stages are not all at the same scale. An inner cell at the node enlarges and divides twice *(A)*, forming a basal cell *(b)*. After further enlargement, the upper cell divides into octants *(B)* (i.e., the original cell is cut up by three planes of divisions all at right angles to one another). Each octant then undergoes two periclinal divisions *(C)*. The eight inner cells thus formed are the primary capitula *(pc)*, the eight middle cells are the manubria *(mb)*, and the eight outer are the shield cells *(sh)*. Neither manubria nor shield cells divide again. During further considerable enlargement *(D)*, the shield cells become bright orange; a most complex pattern of wall ingrowth turns each into a labyrinth of cytoplasmic channels. The manubria enlarge radially, projecting into the lumen of the antheridium. The primary capitula divide several times to form secondary capitula *(sc)*, and from the latter *(E)* arise the long spermatogenous filaments *(sf)* by repeated divisions.

SOURCE. Author's diagram, published in *Aust. J. Biol. Sci. 21*, 255 (1968).

secretion of Golgi vesicles at the tip. The third zone back from the tip contains additionally about 30–60 statoliths (Figs. 7.21, 7.22), rather large, membrane-bound inclusions containing a characteristic material which may be somewhat

Figures 7.26–7.29
DIFFERENTIATION OF
ANTHERIDIA IN CHARA
FIBROSA

7.26.   Three forming antheridia at the nodes of a lateral (cf. Fig. 7.25). Note the prophase nucleus (arrow) about to undergo the second periclinal division in one octant. The outer wall of this young complex already has started forming the wall ingrowths illustrated in Figure 7.28. Also visible is an axillary shoot *(ax)* which can form a meristematic region similar to that in Figure 7.3 and then give rise to a branch in the filament (Fig. 7.1). × 510.

7.27.   Whole antheridium, fixed and mounted. The outer layer of shield cells is clearly visible. × 35 approx.

7.28.   Differentiation of the outer wall of a shield cell. These wall ingrowths eventually partition the cell into a labyrinth (Fig. 7.29). × 4,800.

7.29.   Later stage of differentiation of a shield cell, which is now highly vacuolated and complex in morphology. The chloroplasts *(c)* always move to the inner surface. × 4,400.

SOURCE. Author's micrographs. Figures 7.26, 7.27, 7.29 published in *Aust. J. Biol. Sci. 21*, 255 (1968).

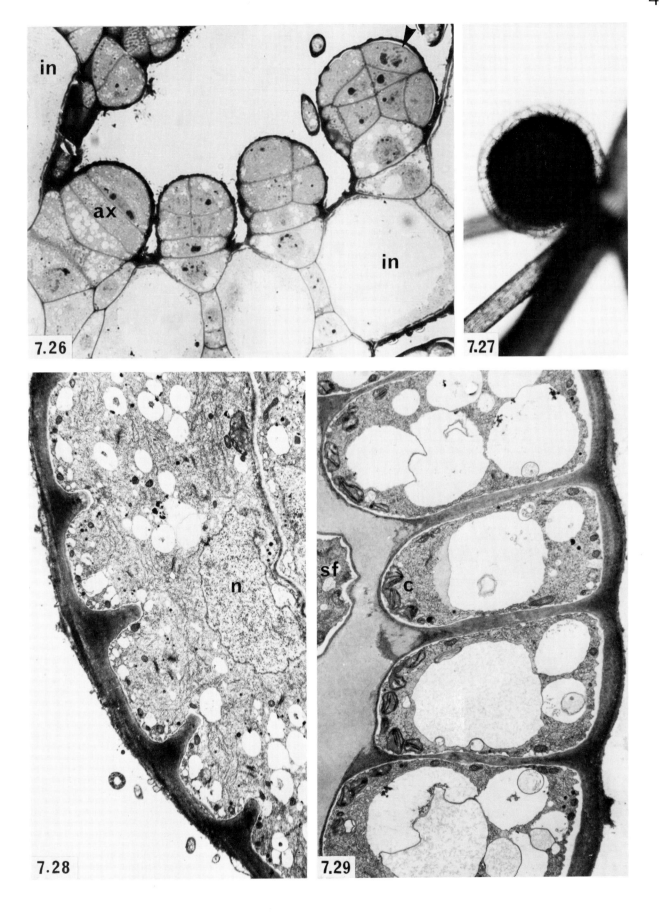

7.26

7.27

7.28

7.29

variable in texture and appearance. The fourth zone back, behind the statoliths, contains the larger cell organelles (plastids, mitochondria, etc.) and the numerous (ca. 25,000 per rhizoid) Golgi bodies. These latter organelles are described as being "polarly differentiated," forming vesicles on only one side of the cisternal stack. Zones five and six contain the nucleus and large central vacuole, respectively (Fig. 7.21). The microtubules in the rhizoid bear no obvious relation to the polarized structure and growth of the rhizoid itself.

To investigate the geotropic response of rhizoids, Sievers (1967b) rotated the rhizoids until they were horizontal and then fixed them for electron microscopy after having maintained them in this orientation for various periods of time. Within 10 minutes of such treatment, curvature cannot be detected, and much of the top organization described above remains relatively undisturbed. The statoliths of the third zone, however, settle in the lower half of the cell, in contrast to their random distribution in the central region of zone three in vertically growing rhizoids (Figs. 7.24). Furthermore, Golgi vesicles accumulate in the upper region of these cells (Fig. 7.22), and they are discharged into the cell wall in greater numbers than is normal for this region. This latter phenomenon can obviously be related to the greater extension of the upper region of the wall which results in curvature downward of the rhizoid (Fig. 7.24E). Sievers infers that the redistribution of statoliths in horizontal rhizoids caused by gravity results in the displacement of masses of Golgi vesicles upward, an explanation that goes some way in explaining the mechanism of the geotropic response.

These results were later confirmed (Sievers and Schröter, 1971) when it was clearly shown that if a rhizoid was moved from the vertical to the horizontal, the upper surface of the rhizoid did indeed grow faster than the lower surface. Statoliths, accumulating in a particular region under the influence of gravity, inhibited wall growth adjacent to themselves. During normal downward growth, they inhibit wall deposition a short distance behind the tip, thus promoting even deposition at the tip itself. If rhizoids were placed in a medium of high osmotic pressure, the statoliths then accumulated abnormally in the extreme tip of the rhizoid, whereupon wall expansion took place immediately behind them, lead-

Figures 7.30–7.34.
DIFFERENTIATION OF ANTHERIDIA IN CHARA FIBROSA

7.30. Chloroplasts in a very young shield cell. They are relatively normal in morphology and contain few granules. The antheridium at this stage is essentially colorless. × 10,000.

7.31. Shield cell from an antheridium that has opened to release the sperm. The shield cells split apart and flatten out. Their labyrinthine form is clear from this micrograph. They are bright orange in color, owing to their numerous pigmented chromoplasts (differentiated chloroplasts). × 180 approx.

7.32. Chromoplasts (pigmented chloroplasts) in the orange, mature shield cell. Note the accumulation of granules in each. The granules likely contain the pigment. × 13,000.

7.33. Fairly mature manubrium projecting into the center of the antheridium (cf. Fig. 7.25). The nucleus (n) is large, and the cytoplasm contains numerous large vesicles (vs). × 3,700.

7.34. Detail of cell equivalent to that in Figure 7.33. It appears as if the vesicles are being discharged into the wall (w), perhaps contributing to the mucilage (ml) filling the extracellular space inside the antheridium. × 17,000.

SOURCE. Author's micrographs, published in *Aust. J. Biol. Sci. 21*, 255 (1968).

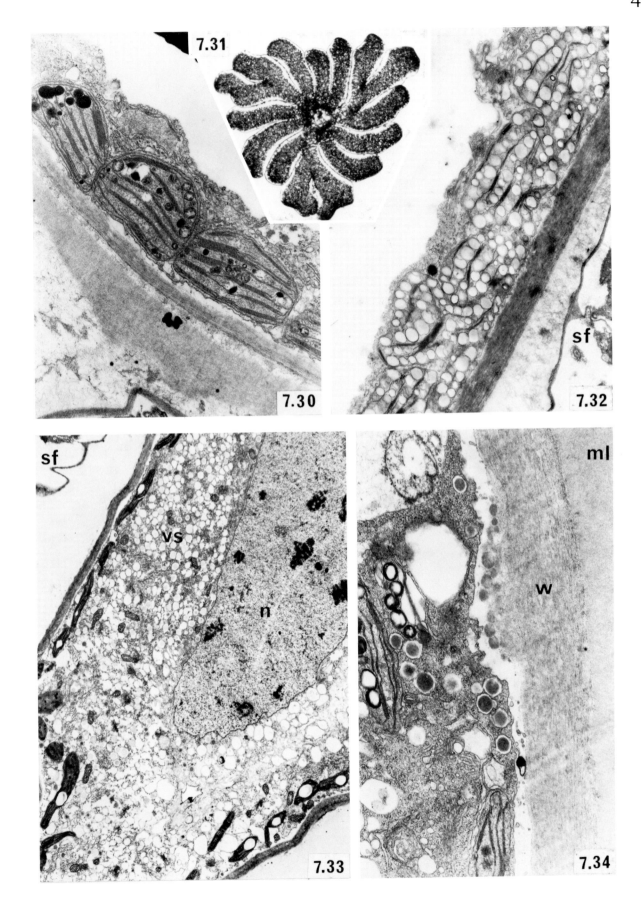

7.31

7.30

7.32

sf

sf

vs

n

7.33

ml

w

7.34

ing to a distention of the subapical portion of the rhizoid. The system is obviously self-regulating under normal conditions. Hejnowicz and Sievers (1971) have proposed a mathematical model of this system.

7.4. Formation of the Antheridia

The term antheridium is perhaps incorrect when used to describe the male reproductive structure in *Chara,* since that structure is highly complex and contains a number of sterile cells. Consequently, some authors prefer the term "globule" (and similarly, "nucule" for the oogonium). However, I shall use the popular term which should not cause any confusion.

The antheridia are bright orange, spherical structures, easily visible with the naked eye. They are borne usually singly at the nodes of the laterals (branches). They are striking objects, since the eight outer sterile shield cells (see below) contain an intricate, regular, and interwoven pattern of wall thickenings. In monecious species, one or sometimes two oogonia are situated at this same site. In the diecious species described here, *Chara fibrosa,* male plants are far outnumbered by females, and each carries a large number of antheridia at different stages of development. Each lateral bears three antheridia, the most mature being that at the outermost node; and the older the lateral, the more mature the set of antheridia on it. When development of the spermatozoids is complete, the eight outer shield cells separate almost entirely and fall away, appearing as eight beautiful flat orange rosettes (Fig. 7.31). This exposes the spermatogenous threads (see below) to the medium.

The formation of the antheridium demonstrates (Pickett-Heaps, 1968a) how a precisely controlled series of cell divisions, followed by divergent forms of intracellular differentiation in the resultant daughter cells, can give rise to highly complex structures (Fig. 7.25). Differentiation commences with the swelling of the innermost (ventral) small cell at the node on the lateral. This cell then undergoes two divisions, which cut off two discoidal basal cells (Fig. 7.26). It then enlarges and three distinct series of mitoses follow, the first two being longitudinal and at right angles to one another and the third transverse. Thus the original cell has been cut up into octants of closely similar size. Next, there are

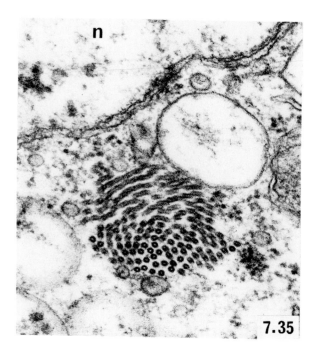

n

7.35

Figure 7.35
MICROTUBULE BUNDLE IN CHARA

Cross section of a bundle of microtubules which is often found wandering somewhat haphazardly through the cytoplasm of primary capitulum cells. The significance of these tubules is unknown. × 45,000 approx.

SOURCE. Author's micrograph, published in *Aust. J. Biol. Sci. 21,* 255 (1968).

Figures 7.36, 7.37
CELL DIVISION IN SPERMATOGENOUS FILAMENTS OF CHARA AND NITELLA

7.36.   *C. fibrosa,* metaphase. The spindle, skewed diagonally in the cell, is centric. Two centioles *(cn),* oriented at right angles to one another, define the position of the pole toward which the spindle microtubles *(t)* converge. × 27,000.

7.37.   *Nitella* sp., telophase. The daughter nuclei *(n)* typically remain widely separated. The phragmoplast is forming, and the plane of the cell plate (arrows) is marked by an accumulation of mitochondria as well as small vesicles. One centriole is visible (double arrow). × 10,000 approx.

SOURCE. Author's micrographs. Figure 7.36 published in *Aust. J. Biol. Sci. 21,* 655 (1968).

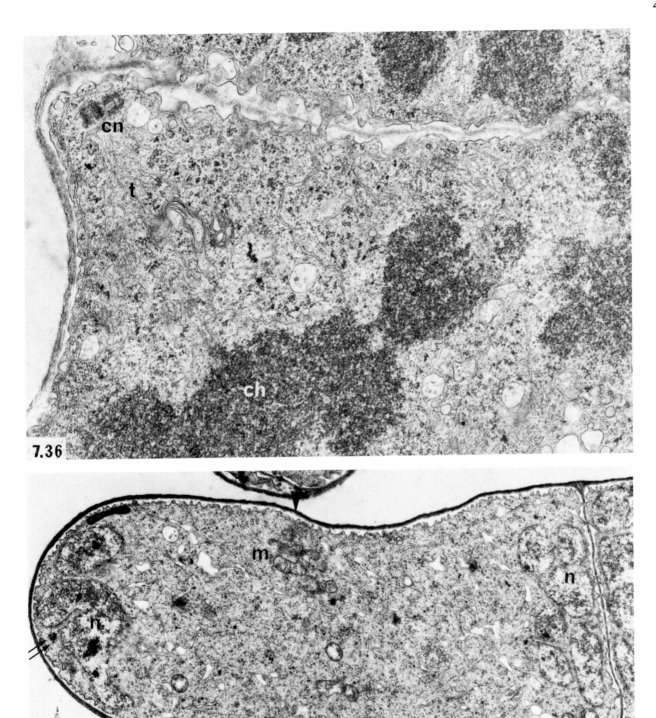

7.36

7.37

two periclinal divisions in each octant, so that the antheridium now contains a regular array of 24 cells, arranged in three concentric layers of 8. The outermost layer does not divide further, but differentiates into the bright orange shield cells. The middle layer also does not divide subsequently, but each cell gives rise to an enlarged "manubrium." The innermost group of cells, called the "primary capitula," however, continues to divide, forming first a small number of secondary capitula. Each secondary capitulum then gives rise to one or two long fine filaments of cells by repeated cell division. Once division has ceased in these "spermatogenous threads," each cell in it differentiates into a motile sperm cell (Section 7.5). Throughout all these events the whole structure enlarges considerably (Fig. 7.27).

At the ultrastructural level, mitosis and cytokinesis in the forming antheridium (Fig. 7.15) are typical of those in vegetative cells. In particular, centrioles are absent, and there is nothing to indicate how the precise planes of cell division are controlled. The first periclinal division in each octant cuts off the primary capitula, and the second division in the outer of these two daughter cells gives rise to the manubria and shield cells.

Soon after their formation, the shield cells begin to differentiate. They enlarge, becoming vacuolate, and all the chloroplasts move to the innermost wall surface. The bright orange color of the antheridium is probably associated with a marked accumulation of lipoidal droplets within these chloroplasts (cf. Figs. 7.30, 7.32). When the shield cells split apart to release the spermatozoids, the chloroplasts (or chromoplasts, as perhaps they should now be called) are bright orange under the light microscope; the rest of the contents of the antheridium is colorless. The outer wall of the shield cells also begins to differentiate as invaginations grow across from the outer to the inner wall (Fig. 7.28). The pattern formed by these ingrowths is highly complex so that the mature shield cells have a labyrinth-like structure (Fig. 7.29), but as far as I can tell, they still remain single cells (Fig. 7.31). Wall microtubules are particularly numerous around the wall invaginations, an association that reminds one strongly of a similar relation between wall microtubules and growing thickenings in xylem cells of higher plants. Furthermore, differentia-

Figures 7.38, 7.39
EARLY STAGES OF SPERMATOGENESIS IN CHARA FIBROSA

7.38.   Undifferentiated cells in the spermatogenous filament after cell division has ceased. The cytoplasm fills the walls. The paired centrioles *(cn)* have elongated (cf. Fig. 7.36) but are still in the central region of the cell. The simple plastids *(pd)* are scattered in the cytoplasm. Endoplasmic reticulum interconnects cells through pores (plasmodesmata?) in the wall (arrows). × 15,000.

7.39.   Early differentiation. The cytoplasm has begun to shrink away from the walls *(w)*. The centrioles (now basal bodies) have moved to the plasmalemma and have just begun to extrude the two flagella *(f)*. A dense structure close to the basal bodies (arrow) contains a flat row of microtubules, the manchette, embedded in it. These microtubules elongate as differentiation proceeds. In this cell, the nucleus has already become appressed to the manchette (see Fig. 7.40), which runs between the arrows. The plastid may also be moving to its future position adjacent to the nucleus (see Fig. 7.44). × 17,000.

SOURCE. Author's micrographs, published in *Aust. J. Biol. Sci. 21,* 655 (1968).

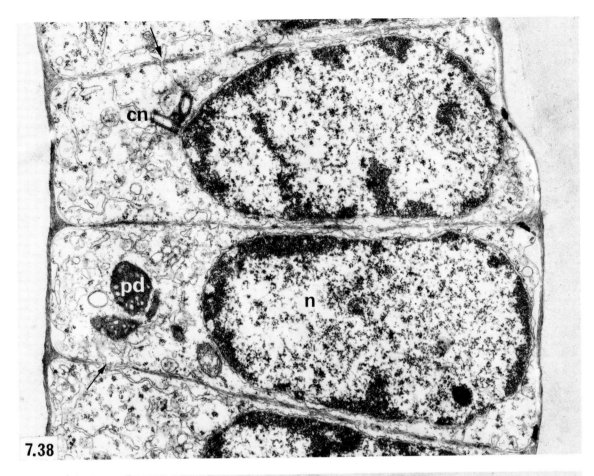

7.38

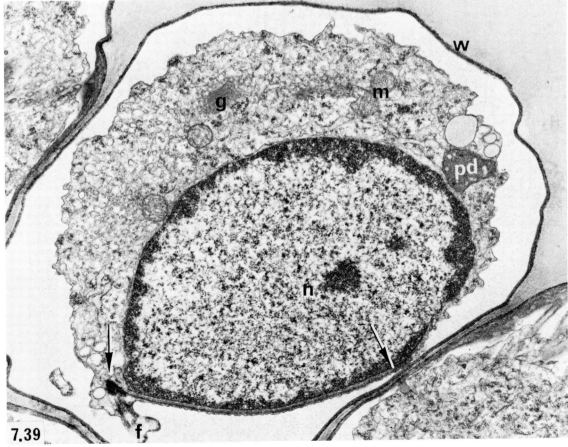

7.39

tion of shield cells is completely disrupted by treatment with colchicine, which also prevents normal xylem wall development (Pickett-Heaps, 1967a).

The manubria are initially thin discoidal cells sandwiched between the shield cells and the primary capitula (Fig. 7.25). As the antheridium grows, these cells enlarge considerably and elongate into its center (Fig. 7.33). The extracellular space inside the mature antheridium is filled with a homogeneous, lightly stained material. I suspect this is mucilage secreted by the manubria, an activity initiated soon after they are formed. Their cytoplasm contains numerous large vesicles, probably derived from Golgi bodies, and these vesicles appear to discharge their contents through the plasmalemma (Fig. 7.34).

The primary capitula and their derivatives, the secondary capitula, are often hard to tell apart, both remaining quite small and containing dense cytoplasm. Their mitoses are normal and lack centrioles. Many of these cells contain an intriguing and quite characteristic bundle(s) of varying size composed of microtubules that wind through the cytoplasm (Fig. 7.35). The significance of these bundles is not understood, but they are also encountered in forming oogonia (Section 7.6). Cells also always contain an unusual aggregation of vesicles and ill-defined dense material enveloped by membranes. Elsewhere in the same cells, spherosomes and irregularly shaped lipid droplets are often ensheathed by elements of endoplasmic reticulum (Pickett-Heaps, 1968a).

The spermatogenous threads are created as a result of a succession of transverse cell divisions in cells derived from the secondary capitula (Fig. 7.25). The first few cells thus formed are quite large, but soon, as the filaments elongate, the cells all become progressively smaller, remaining always densely packed with cytoplasm. Their chloroplasts are really plastids (i.e., they have dedifferentiated, being small and having simple internal structure). Sometime during the growth of these threads, centrioles appear de novo in the constituent cells. I have expended considerable effort in trying to elucidate their development without reaching any really definitive conclusions. I could never find anything resembling a centriole in cells of short (i.e., containing ca. 10–20 cells) spermatogenous threads. Cells of older threads contain two centrioles, and during

Figures 7.40–7.43
SPERMIOGENESIS IN CHARA FIBROSA

7.40.   Detail of Figure 7.39, showing one of the manchette microtubules (large arrows) running along the edge of the cell, with the nucleus pressed to it. Note also the dense material (small arrow) into which the manchette is embedded (Fig. 7.43). × 40,000.

7.41.   Early flagellar extrusion. A spindle-shaped fibrous structure interconnects the basal bodies, which are still coplanar. × 53,000.

7.42.   As for Figure 7.41, but possibly a stage in the movement of the basal bodies into the almost parallel, slightly staggered arrangement characteristic of the mature sperm cell. × 68,000.

7.43.   Manchette microtubules at an early stage in spermiogenesis. The number of tubules increases later. The three other microtubules near the basal body (bb), usually present in such cells, are of unknown significance. × 50,000.

SOURCE. Author's micrographs, published in *Aust. J. Biol. Sci.* 21, 655 (1968).

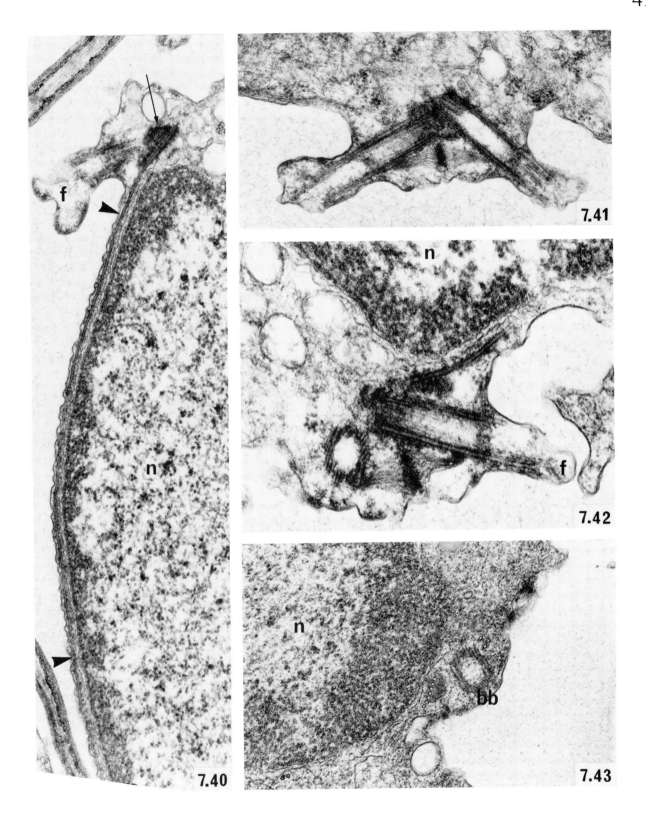

7.40

7.41

7.42

7.43

mitosis, two are present at each pole of the spindle (Fig. 7.36) where they act as the foci of the spindle microtubules (see also Turner, 1968). In other words, the spindles become typically centric. Thus, somewhere between these two developmental stages, the centrioles must be assembled de novo. The search for some possible formative stages was considerably hampered by the density of the cell's cytoplasm, but some centrioles found (Pickett-Heaps, 1968b) were certainly less well developed structurally (e.g., in length, complexity, and stainability) than those observed a little later in differentiation and therefore had probably just been formed.

I believe this example demonstrates clearly the unimportance of the centriole in the formation and function of the spindle in cells generally (Pickett-Heaps, 1969a). In essence, *Chara* has a perfectly functional acentric spindle which, during a certain stage of differentiation in highly specialized spermatogenous tissue, is transformed into a centric spindle, in many ways typical of the classic spindles illustrated in all textbooks. It is difficult to see why the centriole should be so important in centric mitoses if it is not necessary prior to this. Incidentally, cytokinesis in these spermatogenous cells is always achieved by the phragmoplast/cell plate system (Fig. 7.37), whether or not the spindles are centric (Turner, 1968; Pickett-Heaps, 1968b).

7.5. Spermatogenesis

Differentiation of spermatozoids from spermatogenous cells involves drastic changes in the morphology and distribution of many cell organelles and the elimination of others. The mature spermatozoid is long, thin, and coiled, at least soon after its release (Fig. 7.57, 7.58). The prominent nucleus, greatly condensed and elongated, has a posterior appendage containing several plastids; at the other end is a row of mitochondria. Two posteriorly directed flagella diverge slightly and are inserted parallel to the row of mitochondria, a short distance from the anterior end of the organism. All extraneous cytoplasm is eliminated from around the spermatozoid. The formation of such an extremely specialized cell (Fig. 7.59) obviously provides an excellent example of cytoplasmic differentiation. Spermatogenesis has been studied in both *Chara* and *Nitella* by early workers such as Belajeff

Figures 7.44, 7.45
MIDSTAGES OF SPERMIOGENESIS IN CHARA FIBROSA

7.44. Considerable shrinkage of the cytoplasm while the Golgi bodies *(g)* are apparently very active. The manchette now runs from near the basal body *(bb)*, along the somewhat convoluted profile of the nucleus *(n)*, to the plastids *(pd)*, all neatly lined up and appressed also to the microtubules (Fig. 7.46). × 23,000.

7.45. Later stage than that in Figure 7.44. The manchette is now so long that it has become coiled in the cell and consequently is only partly in the plane of the section. The plastids are now elongated and have accumulated starch. Small lipid droplets are collecting near them. Between the nucleus and the flagellar base *(fb)*, mitochondria (arrows) have also begun to line up on the manchette. × 30,000.

SOURCE. Author's micrographs, published in *Aust. J. Biol. Sci. 21*, 655 (1968).

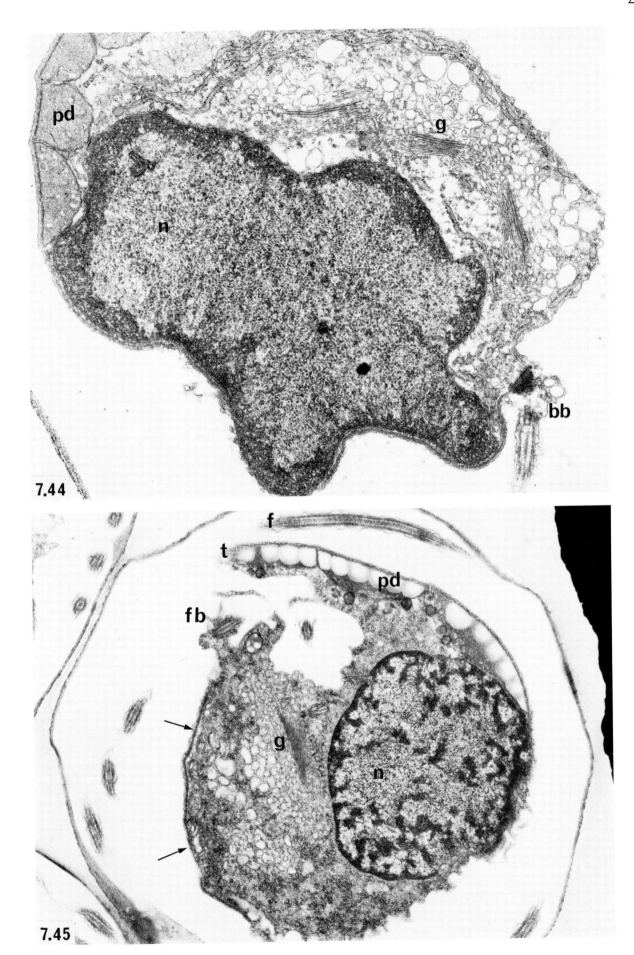

7.44

7.45

(1894) and Mottier (1904) and more recently by Turner (1968) and Pickett-Heaps (1968b). This present account is a combined summary particularly of the last two papers concerning *N. missouriensis* and *C. fibrosa*, respectively.

Spermatogenous cells in any one filament are all interconnected by numerous plasmodesmata, through which often run elements of endoplasmic reticulum (Fig. 7.38). Thus, the fairly close synchrony of cell divisions and subsequent cytoplasmic differentiation is not surprising. An early sign of incipient spermatogenesis after the cessation of cell division is the elongation of the centrioles, which remain near the nuclear envelope, situated fairly centrally within the cell (Fig. 7.38). Throughout differentiation, the protoplast of each spermatid slowly and steadily withdraws from the parental cell wall (Fig. 7.39 et seq.). Protoplasmic connections are maintained for a considerable period (Fig. 7.48), and so the spermatid displays an increasingly irregular outline until eventually all these connections are ruptured. This cytoplasmic contraction is soon initiated as the nucleus moves to one side of the cell (Fig. 7.39). Chromatin collects around the periphery of the nucleus while the nucleolus disperses. The paired, elongating centrioles come to lie near the plasmalemma; interconnecting the centrioles there now appears a fibrous, spindle-shaped structure which later disappears (Figs. 7.41, 7.42). The centrioles are initially oriented at about 120° to one another (Figs. 7.38, 7.41), but as differentiation proceeds and the spermatids elongate, they appear to move slowly toward one another until in the mature spermatozoid they are almost parallel but staggered slightly along the cell, a conformation also found in the sperm of some bryophytes. Although the flagella are eventually somewhat heterokont (i.e., unequal), this could be a characteristic that arises from equivalent basal bodies. Manton (1965) argues that such a heterokont condition rules out a derivation of the bryophytes (and presumably the Charales) from an ancestral *Chlamydomonas*-like cell. This view is further discussed in Chapter 8. Early in differentiation, extension of typical flagella commences and continues steadily during differentiation, so that eventually the two long flagella which have been formed coil around several times inside the confines of the cell wall. Both flagella and basal bodies (i.e., centrioles that have formed flagella)

Figures 7.46–7.48
MIDSTAGES OF SPERMIOGENESIS IN CHARA FIBROSA

7.46. Detail of Figure 7.44, showing the plastids *(pd)* next to the nucleus *(n)* lined up on the microtubules of the manchette (arrows). × 30,000.

7.47. Tangential section, showing the surface of flagella covered with tiny diamond-shaped scales. × 70,000.

7.48. Cross section of cells probably equivalent to that in Figure 7.45. Microtubules of the manchette (arrows) are appressed to the plastids and mitochondria *(m)*. The cytoplasm between cells is probably still interconnected, although cytoplasmic shrinkage is well advanced. × 25,000.

SOURCE. Author's micrographs, published in *Aust. J. Biol. Sci. 21*, 655 (1968).

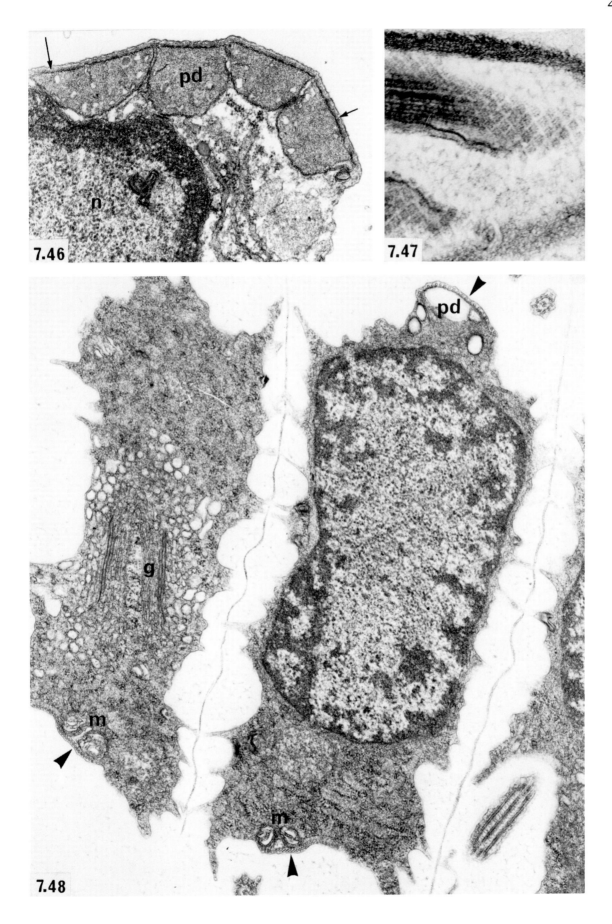

7.46

7.47

7.48

are quite typical, except that rather surprisingly the basal bodies degenerate in the mature spermatozoid. They appear to lose their distinguishing characteristics (Pickett-Heaps, 1968b; Moestrup, 1970), and most, if not all, of the triplet tubules revert to doublets, while the cartwheel structure at the end of the basal body is lost (Turner, 1968).

When the nuclei of very young spermatids have shifted to one side of the cell in *Chara,* a dense mass of material appears near the centrioles (Figs. 7.39, 7.40) which I have termed the "manchette adjunct." Inside this manchette adjunct is formed a planar row of densely stained microtubular structures (Fig. 7.43) which increase in number to around 20. In view of its apparent function in the cell, the manchette adjunct seems to represent a conspicuous form of microtubule-organizing center (Pickett-Heaps, 1968b). Turner (1968) does not draw attention to a similar structure in *Nitella,* but some of his micrographs indicate a somewhat analogous structure is present in that organism (Turner, 1968; Figs. 21, 22, 26). A most important morphological structure, the flat band of up to 26 manchette microtubules, arises from this manchette adjunct and grows steadily in length. Like the elongating flagella, it is obliged to adopt a spiral configuration within the confines of the cell wall. A profound and most significant rearrangement of cell organelles accompanies its elongation. The following sequence of events has been firmly established for *Chara* (and may happen in *Nitella* too). First, the nucleus becomes appressed to the elongating manchette (Figs. 7.39, 7.40). As the latter grows past the nucleus, the plastids next line up on the manchette, closely appressed to one another (Fig.s 7.44, 7.46). Then later, the mitochondria also line up on the manchette, but in this case· between the nucleus and the manchette adjunct/basal body complex (Figs. 7.45, 7.48). Next, the nucleus continues its condensation and elongation along the manchette (Fig. 7.50). Much of the ground nucleoplasm appears to be lost as the rather granular chromatin increasingly differentiates into dense filaments or lamellae (Figs. 7.49, 7.51), which in *Nitella* are reported to consist of helically arranged subfibrils. The lamellae in the nucleus (now considerably elongated; Fig. 7.50) further condense (Fig. 7.53) until the nucleus of the mature sperm is very dense and quite homogeneous in texture

Figures 7.49, 7.50
LATER STAGES OF SPERMIOGENESIS IN CHARA AND NITELLA

7.49.   *N. missouriensis,* cytoplasmic shrinkage well advanced. The nucleus now is also condensing into densely staining lamellae (fig. 7.51). The two flagella *(f)* are long and make several turns in the cell. The manchette now also runs around the cell several times. Visible in this micrograph are cross sections of the manchette near the flagellar insertion *(fi),* further along near the mitochondria *(m)* and nucleus, and finally the plastids *(pd).* Note also the conspicuous differentiated region in the wall (arrow), which will open to form the pore through which the sperm escapes. × 15,000.

7.50.   *C. fibrosa,* nuclear condensation well advanced as the dense lamellae are beginning to fuse (Fig. 7.53). Because the sperm is now so long and coiled in the cell, each section can only show part of its structure. Mitochondria are just visible on the manchette (large arrow). Condensation of the cytoplasm is now almost complete. Note the scales on the flagella (small arrows). × 17,000.

SOURCE. Figure 7.49: Dr. F. R. Turner, published in *J. Cell Biol. 37,* 370 (1968). Figure 7.50: Author's micrograph, published in *Aust. J. Biol. Sci. 21,* 655 (1968).

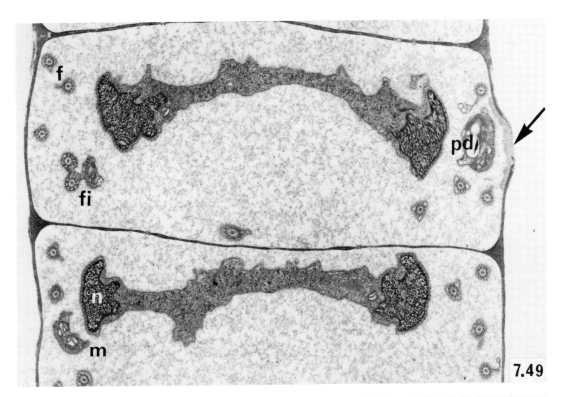

7.49

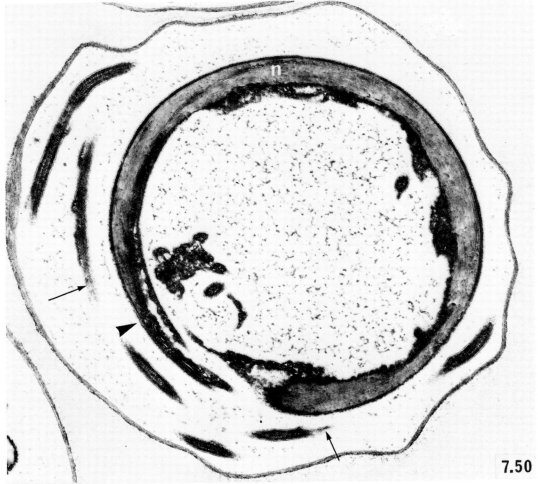

7.50

(Fig. 7.54). During elongation of the manchette, another characteristic structure called the "head piece" (Turner, 1968) or "vesicular adjunct" (Pickett-Heaps, 1968b) appears near the two basal bodies. Its significance is unknown. It is a dense ring-like layered cap surrounded by a layer of uniform vesicles, and eventually it becomes situated in the anterior part of the spermatozoid. During spermiogenesis, other cell constituents also undergo various changes. The Golgi bodies appear particularly active during early stages of differentiation (e.g., Fig. 7.44), and the cisternae of the endoplasmic reticulum accumulate dense material (Fig. 7.46). However, later in spermatogenesis these organelles, along with free ribosomes, etc., slowly break down and eventually disappear. The surface of the spermatozoid and its flagella are covered with tiny, close-patterned, diamond-shaped scales (Figs. 7.47, 7.50, 7.56; Pickett-Heaps, 1968b; Turner, 1968). Moestrup (1970) has shown that the scales covering the body are slightly smaller than those on the flagella (0.06μm versus 0.075μm). By analogy with the work of Manton and her coworkers (e.g., Manton et al., 1965), the scales could be expected to be formed within the vesicles arising from the Golgi apparatus. However, this has not been clearly demonstrated, although Turner (1968; p. 388) says that his (as yet) unpublished results indicate this to be so. The presence of scales on the sperm cells is unusual in Green Algae and further points to the rather isolated phylogenetic situation of this group (Round, 1971). This matter is raised again in Chapter 8. During later stages of differentiation, the simple plastids, initially devoid of starch (Figs. 7.39, 7.46), accumulate considerable quantities of this storage material, and small lipoid droplets gather near the plastids (Figs. 7.45, 7.48).

The mature sperm cell remains partly ensheathed along its length by the microtubules (Fig. 7.59), which run right up to the flagellar end of the sperm (Fig. 7.56). Moestrup (1970) points out a difference, possibly taxonomically significant, in the arrangement of these microtubules between *Chara* and *Nitella,* and it is likely that these tubules are not all equivalent. One particular group of four individual tubules close to, but separate from, the manchette, can almost always be discerned at the flagellar end of the spermatozoids of *Chara* (Fig. 7.51). The microanatomy of the tubular systems in these cells

Figures 7.51–7.56
TERMINAL STAGES OF SPERMIOGENESIS IN CHARA AND NITELLA

7.51. *C. fibrosa.* Stage equivalent to that in Figure 7.49. The nucleus now contains a mass of densely stained lamellae. The structure of the manchette at the flagellar insertion *(fi)* is visible. About 25 evenly spaced tubules partially enclose the row of mitochondria, and usually another group of about 4 tubules is situated nearby (arrow). As the manchette runs along the body of the sperm, the number of microtubules decreases; 12 are seen here next to the condensing nucleus (double arrow) and further along still, about 9 lie along the plastids (Fig. 7.55). × 43,000.

7.52. *C. fibrosa.* Tangential section of the manchette near the plastids (the central pale bodies are starch grains). The evenly spaced arrangement of the constituent microtubules *(t)* is obvious. × 43,000.

7.53. The dense lamellae of condensing nuclei (Figs. 7.49, 7.51) fuse, as shown in this micrograph, until the nucleus is very dense and homogeneous (Fig. 7.54). × 73,000.

7.54. *C. fibrosa.* Fully condensed nucleus in a mature sperm. Manchette microtubules can be seen. Note also the thin, even layer enclosing the body of the sperm, surmounted by scales (Fig. 7.47). × 76,000.

7.55. *C. fibrosa.* Plastids, full of starch, in a mature sperm. × 57,000.

7.56. *N. missouriensis.* Flagellar end of a mature sperm. Mitochondria *(m)* are lined up on the manchette microtubules, adjacent to the flagellar insertion. Scales are visible on the flagella (arrow). × 46,000.

SOURCE. Figures 7.51–7.55: Author's micrographs, published in *Aust. J. Biol. Sci. 21,* 655 (1968). Figure 7.56: Dr. F.R. Turner, published in J. Cell Biol. 37, 370 (1968).

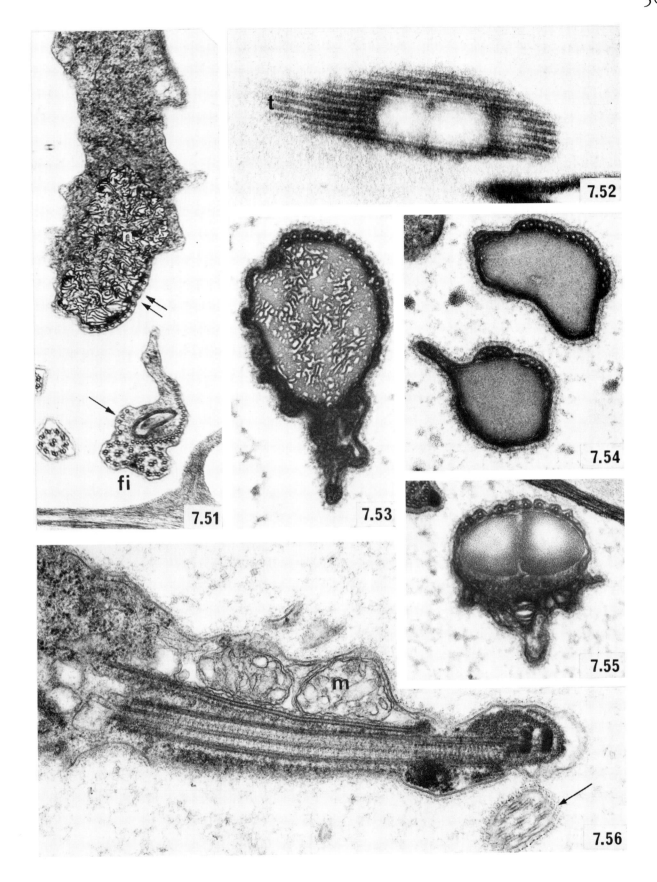

has not yet been definitely worked out. In *Chara,* it appears as if the number of microtubules in the manchette decreases along the length of the sperm. There may be up to 28 tubules near the mitochondria and flagellar insertion (Figs. 7.48, 7.51); these decrease at the nucleus (Fig. 7.51) and perhaps drop below 9 in number at the plastids (Fig. 7.55). The microtubules are virtually touching one another near the mitochondria (Fig. 7.51), but have become fairly evenly separated near the nucleus (Fig. 7.51) and plastids (Figs. 7.52, 7.55).

The spermatozoids escape through pores in the wall, with each gamete emerging "backward" (Moestrup, 1970); the flagella always emerge last, even when they have already started beating inside the mother cell wall. Turner (1968) has clearly shown that the site of the pore is predetermined by an accumulation of granular material between the spermatid membrane and the wall (Fig. 7.49). The wall itself is later broken down at this site, and the perforation is plugged by the pore material, which itself later disperses to release the spermatozoid. Strangely enough, I have never seen an equivalent structure in *Chara.*

It should be quite clear from the foregoing account that the manchette microtubules (whose assembly is perhaps initiated and controlled by the manchette adjunct) are extremely important agents involved in the redistribution of cell constituents and the enormous elongation undergone by the spermatid during differentiation. This appears to be true for spermatogenesis in a wide range of organisms, although some workers have suggested that control of nuclear shape may reside within the condensing nucleus itself (Fawcett, Anderson, and Phillips, 1971). Turner (1970) has further demonstrated the importance of these microtubules by treating antheridia with colchicine. Mitosis and cytokinesis in such spermatogenous filaments were, of course, completely disrupted. Short-term treatment of filaments and spermatids early in differentiation produced spermatozoids with branched and/or abnormal numbers of flagella, the latter presumably reflecting abnormal distributions of centrioles following disrupted mitosis; the size of the spermatozoids also varied. The singlet and doublet tubules inside these flagella were also abnormal in distribution and number. If the spermatids were treated for prolonged periods (e.g., over 1 week) they had no flagella or manchette, and the

Figures 7.57–7.59
MATURE SPERM OF NITELLA MISSOURIENSIS

7.57.   Fixed preparation of whole spermatogenous filaments and released sperm. × 670.

7.58.   Fixed sperm under the light microscope. The mitochondria *(m),* flagella *(f),* nucleus *(n),* and plastids *(pd)* are labeled. × 2,700.

7.59.   Diagrammatic Reconstruction. The flagella *(f)* are inserted at the anterior end near the mitochondria *(m).* Next in line along the manchette microtubules is the highly condensed and greatly elongated nucleus *(n).* Last is the row of plastids *(pd).* See Figure 7.58.

SOURCE. Dr. F.R. Turner, published in *J. Cell Biol. 37,* 370 (1968).

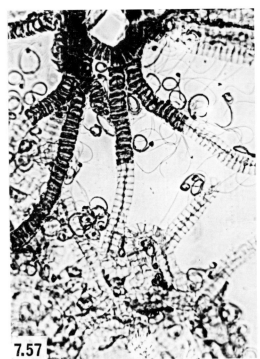

7.57

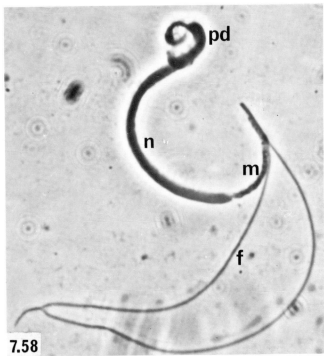

7.58

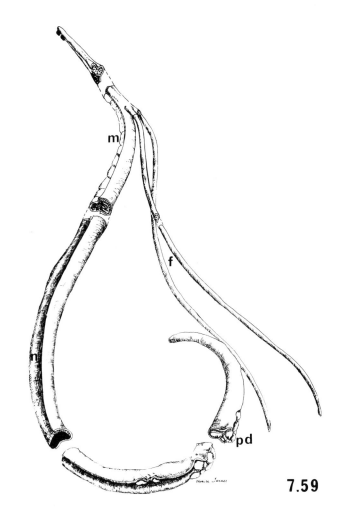

7.59

nucleus, while undergoing an apparently normal condensation, did not elongate (Fig. 7.60). The cell contents remained centrally situated as irregularly shaped masses with no order evident in the disposition of either the mitochondria or the plastids. When such cells were allowed to recover from the treatment, various large and small aggregations or clusters of microtubules appeared in the malformed spermatids (Fig. 7.61). Some of these were triplet as well as doublet and singlet microtubules, the latter apparently forming from the accumulation of tubule precursors that build up in the presence of colchicine. It was also clear that once formed, the tubules of the flagella and manchette were more resistant to dissolution by colchicine than other cytoplasmic and spindle microtubules.

Another, rather speculative suggestion can be made here. Earlier (Section 2.1) I indicated that perhaps "rootlet templates" were important structures in motile algal cells, being associated with basal bodies and perhaps involved in the formation of the cytoskeletal framework of microtubules that both affects the shape of the cell and transmits the forces of flagellar beating over the entire cytoplasm. It is possible to interpret the manchette adjunct as a modified form of rootlet template, and the manchette as a highly evolved form of rootlet microtubular system, an interpretation that goes a little way toward explaining how this complex form of gamete differentiation originally arose from primitive ancestral motile cells.

We may also note that while motile cells in *Chara* differ significantly from those of most other Green Algae (with the notable exception of *Klebsormidium* and *Coleochaete*), they have several features similar to spermatozoids of some bryophytes (Chapter 8).

7.6. Oogenesis

The oogonia of the Charales are most beautiful. Like the antheridia, they are large, being visible to the naked eye, and are borne at the nodes of the laterals (Fig. 7.2). In monecious species, both reproductive structures are thus situated with the oogonium sited on the tip side of the node; some species (e.g., *Nitella flexilis;* Fritsch, 1935; p. 454) may have two oogonia in this position. In the diecious species described here, *Chara fibrosa,* the immature oogonia are

Figures 7.60, 7.61
COLCHICINE TREATMENT OF DEVELOPING SPERM OF NITELLA MISSOURIENSIS

7.60. Sperm midway through differentiation, having been in colchicine for 11 days. The nucleus *(n)* has begun to condense, and the plastids *(pd)* are accumulating starch, but the cells are totally disorganized morphologically, compared with normal cells (e.g., Fig. 7.49). Flagella are absent. × 21,500.

7.61. Part of developing sperm treated with colchicine for about a week and then allowed to grow in the absence of colchicine. Microtubules *(t),* here associated with membrane-like structures, have reappeared, but the sperm still suffers gross morphological deformities. × 55,000.

SOURCE. Dr. F.R. Turner, published in *J. Cell Biol. 46,* 220 (1970).

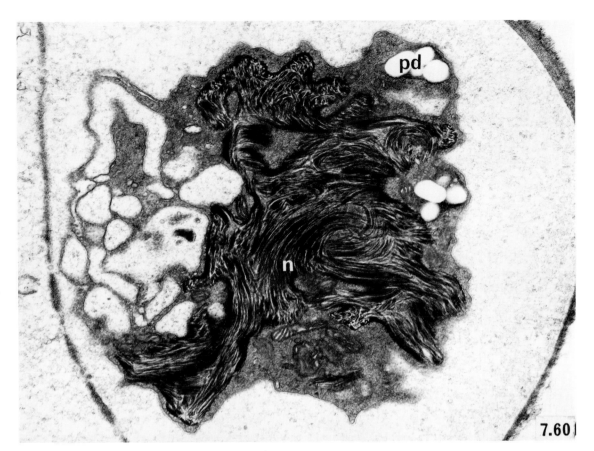

7.60

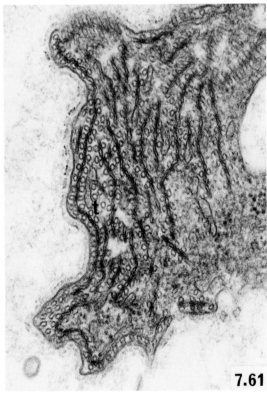

7.61

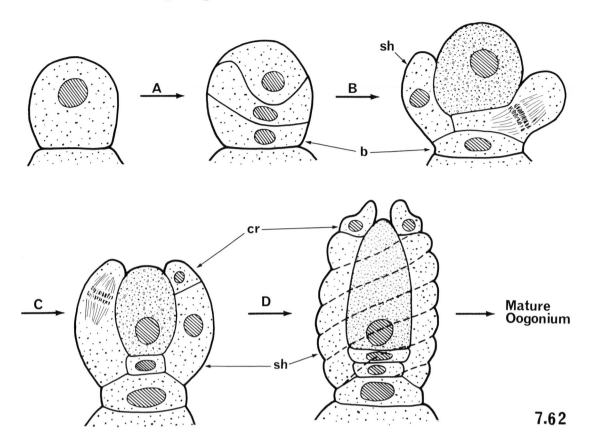

7.62

Figure 7.62
FORMATION OF OOGONIAL COMPLES OF CELLS IN CHARA

Diagrammatic representation. Differentiation starts (as with the antheridium; Fig. 7.25) when an inner cell at the node of a lateral enlarges and cuts off two cells *(A)*. The basal cell does not divide further and becomes the stalk or basal cell *(b)*. The middle cell divides five times to cut off five sterile sheath cells *(B)*. The sheath cells *(sh)* divide once each *(c)* to form the corona *(cr)*, and as they elongate, they assume a spiral course around the enlarging oogonium *(D)*. The oogonium itself cuts off one more basal cell *(D)* and enlarges enormously, accumulating massive quantities of oil, starch, and other reserve material.

pale yellow-brown and they become dense and black after fertilization.

The formation of the oogonial complex (Fig. 7.62) makes an interesting comparison with the formation of the antheridium (Fig. 7.25), since in both cases, a highly organized series of cell divisions forms the basic architecture of the complex, and the products of division differenti-

Figures 7.63–7.67
OOGENESIS IN CHARA FIBROSA

7.63–7.65. Different stages in the formation of the oogonial complex; live cells. The earliest stage shown is indicated by the arrow in Figure 7.64. An almost mature oogonium is shown in Figure 7.68. All × 54.

7.66. Apical meristem of female plant (cf. Fig. 7.3). This specimen was fixed in permanganate. Oogonial complexes (large arrows) are forming at the nodes of the laterals. A number of branchlet cells surrounds the oogonial complex (see Fig. 7.63 et seg.), and these also arise from the base of the oogonium, which complicates the geometry of the complex (as shown in Fig. 7.62). Note how the internode *(in)* is becoming ensheathed by cortical cells (paired arrows), growing apically and basally from the nodes (cf. Figs. 7.3, 7.17). × 200 approx.

7.67. Very young oogonial complex (see Fig. 7.62), *b* indicates the basal cell. Note the curvature of both cell walls (one is a cell plate, *cp*). × 1,800 approx.

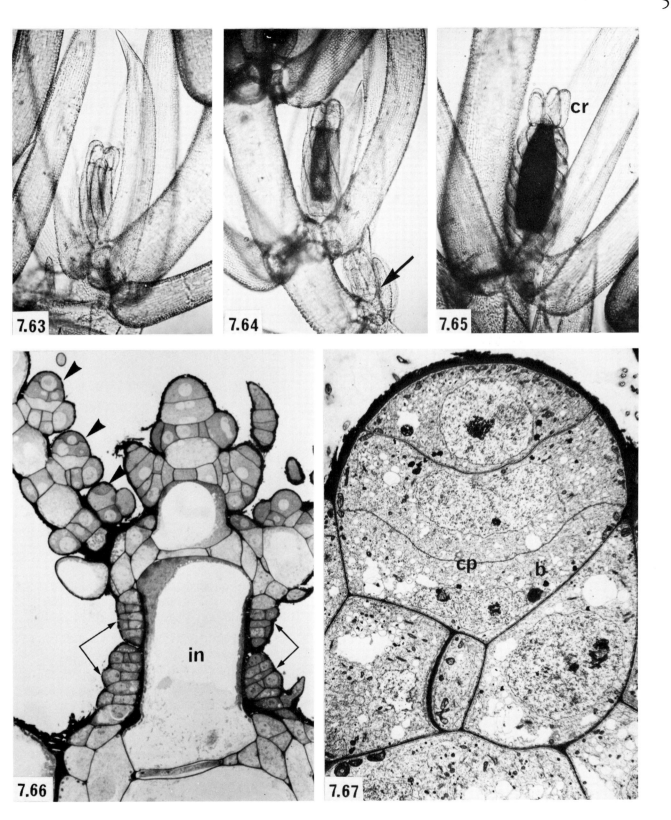

7.63

7.64

7.65 cr

7.66 in

7.67 cp b

ate in different ways. As with the antheridium, differentiation commences with enlargement of the nodal cell, followed by two divisions which cut the original cell in three (Fig. 7.62*A*). The cell plate of these divisions is always characteristically curved (Figs. 7.66, 7.67). Very young complexes also always have extra cells arranged at their bases which later develop into the leaflet cells surrounding the complex (Figs. 7.63, 7.64, 7.66, 7.68, 7.69). The basal cell of the three does not divide again and remains small and inconspicuous at the base of the oogonium (Fig. 7.70). The middle cell divides five times (Fig. 7.62*B*) forming the sterile sheath cells. As the oogonial complex enlarges, each sheath cell divides once more (Figs. 7.62*C*, 7.69), thereby forming the characteristic "coronet" of cells that surmounts the complex (Figs. 7.63–7.65, 7.68). Characteristic of the genera *Nitella* and *Tolypella* is the creation of two such coronal cells per sheath cell. As the complex grows further, these sheath cells assume an increasingly spiral course around the central oogonial cell (Figs. 7.63–7.65, 7.68), always clockwise (Fritsch, 1935), while the coronal cells remain erect. Meanwhile the oogonium itself enlarges considerably, initially mostly by elongation (Figs. 7.63–7.65), but later by becoming more rotund (Fig. 7.68). The oogonium also cuts off one more small basal cell (Fig. 7.62*D*); in *Nitella,* three such basal cells are formed. The oogonium soon becomes very dense indeed (Figs. 7.64, 7.65, 7.68) as it accumulates vast quantities of storage material, such as lipid and starch. When the oogonium is mature, slits appear between the coronet and shield cells through which the sperm enter to achieve fertilization (Fritsch, 1935; p. 460).

After fertilization, the oogonium secretes a thick, resistant wall; the inner walls of the sheath cells also thicken. Calcium carbonate and silica may impregnate these walls to varying extents. As mentioned earlier, the female reproductive structure has been clearly identified in fossil form (Fritsch, 1935—see Figs. 8.49–52). Germination is apparently preceded by meiotic divisions, and so the life cycle is probably haploid. As also emphasized earlier in this chapter, the way the new vegetative thallus arises is unusual for Green Algae, involving the initial appearance of the protonemal thread of cells from which the new plant body emerges as an erect branch.

Figures 7.68–7.71
OOGENESIS IN CHARA FIBROSA

7.68.   Almost mature, live oogonial complex (cf. Figs. 7.63–7.65). Note the corona (arrow). × 54.

7.69.   Young oogonial complex. Note the prophase nucleus (arrow) in one of the young sheath cells. This division would have cut off one of the cells of the corona (see *C* in Fig. 7.62). × 350 approx.

7.70.   Older oogonial complex (see also Fig. 7.72), probably equivalent to the upper complex in Figure 7.64. × 210 approx.

7.71.   Cytoplasm of a fairly young oogonium. It is already becoming difficult to fix properly (cf. the younger cell in Fig. 7.72). Many membrane systems (e.g., of the chloroplasts) appear to be disintegrating, and lipid *(l)* is accumulating along with starch in the cell. Note the membrane complex (arrow), a common feature of these cells. Soon after this stage, the cells become very dense and quite impossible to handle for microscopy with present-day methods. × 9,000 approx. *Inset.* Lipid droplet from slightly younger oogonium, enveloped by close-fitting elements of endoplasmic reticulum.

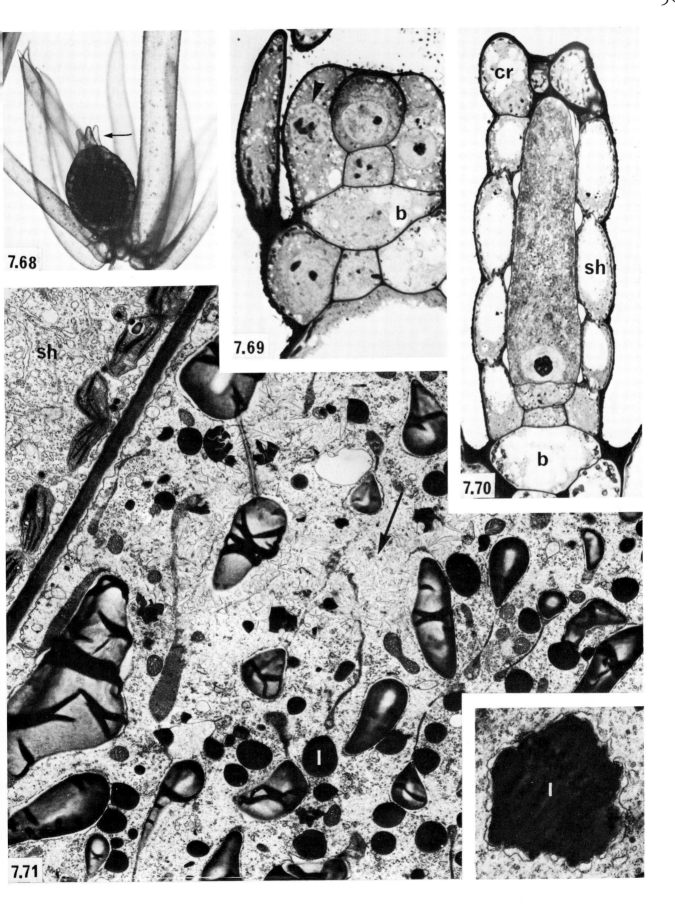

Unfortunately, the differentiation of the oogonial complex cannot yet be fully studied at the ultrastructural level. The quantity of reserve material in the older oogonia renders them totally impossible to fix and embed using present-day methods. The contents of the oogonium in such cases can be scraped out of a hardened plastic block as a sticky mess! Nevertheless, we can examine some of the earlier stages of oogenesis. The intricate cross walls of the forming complex (Figs. 7.62A, 7.67) are formed in a typical phragmoplast. There is nothing unusual in any of the divisions that form the complex (centrioles have never been encountered, which is hardly surprising). The oogonium always has a large, basally situated nucleus (Figs. 7.64, 7.69, 7.70, 7.72) with a most prominent nucleolus; although small vacuoles are present, it never undergoes vacuolation to the extent of the surrounding cells (Figs. 7.70, 7.72). The cytoplasm of the young oogonium fills with a variety of organelles and inclusions. Most conspicuous is the differentiation of the chloroplasts (Fig. 7.73), which proliferate and become greatly elongated or cup-shaped (Fig. 7.72). They soon lose most of their photosynthetic membranes and form large starch grains (becoming amyloplasts; Fig. 7.71). When young, these chloroplasts contain the conspicuous microtubule-like structures (Figs. 7.74–7.77; Pickett-Heaps, 1968c) found in many other Green Algae (Hoffman, 1967). These tubules, often in bundles (Fig. 7.74), are appreciably larger than cytoplasmic microtubules (Fig. 7.75) and exhibit a helical cross-banded substructure at high magnification (Figs. 7.76, 7.77; Hoffman, 1967). They should not be confused with cytoplasmic microtubules, being dissimilar in many ways (they are, for example, insensitive to colchicine; Pickett-Heaps, unpublished observations). Apparently, the chloroplast membranes disintegrate during maturation (Fig. 7.71); however, the increasing difficulties of fixation may mean that this observation can be ascribed to an artifact. In the forming oogonium, lipid inclusions proliferate (cf. Figs. 7.72, 7.73) and these may often be found in close association with elements of endoplasmic reticulum (inset, Fig. 7.71), as in the capitular cells of the antheridium. Also often encountered is the bundle of microtubules winding through the cytoplasm (Fig. 7.75), analogous to those in the capitula (Fig. 7.35). I shall not attempt to describe all the other more unusual

Figures 7.72, 7.73
OOGENESIS IN CHARA FIBROSA

7.72. Young oogonium (equivalent to that in Fig. 7.64). The basally situated nucleus is always large. Cell organelles are becoming differentiated, particularly the plastids, which are elongating and accumulating starch. Compare this cell with the adjacent sheath (sh) and supportive cells. × 2,400 approx.

7.73. Intersection of oogonium (o), sheath (sh), and supportive cells. Note how the chloroplasts (c) have differentiated, being quite different in each of these three cell types. × 5,000 approx.

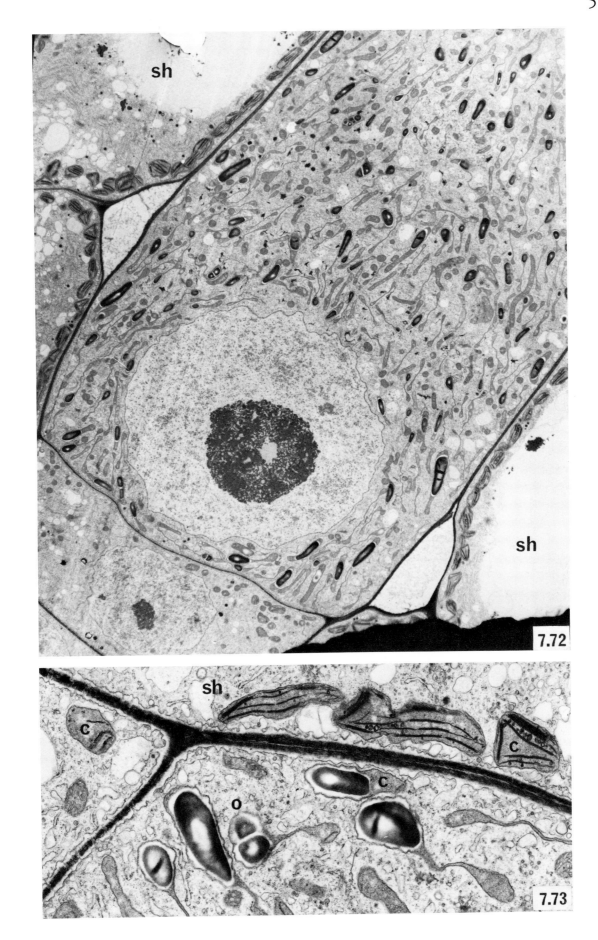

7.72

7.73

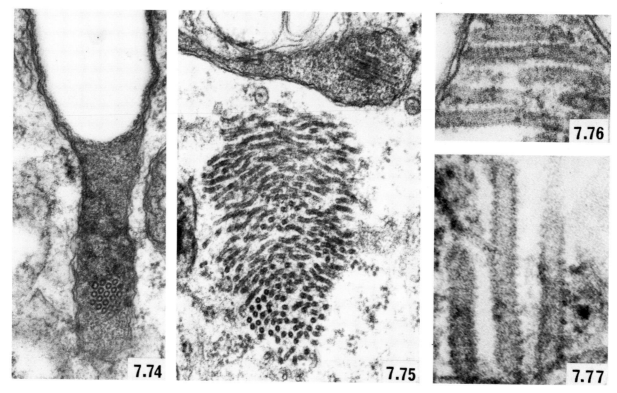

7.74   7.75   7.76   7.77

inclusions in the oogonia, since their significance is obscure. Notable even in younger cells is a relatively clear region of cytoplasm at the apex of cells from which all large organelles are excluded. This region probably represents the receptive spot of the mature oogonium (Fritsch, 1935; p. 460). Many of the cell organelles seem to break down during maturation; however, as repeatedly stressed, fixation difficulties render artifacts increasingly likely. Even in Figure 7.71, taken from a comparatively immature oogonium, probably younger than that in Figure 7.65, the cytoplasm shows evidence of such breakdown. Notable in these cells are the large aggregations of membrane-like material associated with lipid. Mitochondria remain normal in appearance up to this stage. Investigation of the structure of the mature oogonium, fertilization, meiosis, and germination will have to await the development of improved methods of tissue preparation.

Figures 7.74–7.77
MICROTUBULE STRUCTURES IN OOGONIAL PLASTIDS OF CHARA FIBROSA

7.74.   Group of tubules in cross section. The plastid was accumulating starch (as in Fig. 7.72). × 49,000.

7.75.   The tubules in longitudinal section. In the cytoplasm nearby is a cross section of the microtubular bundle often found in young oogonia and primary capitula (Fig. 7.35). Note the difference in diameter of these two types of tubules. × 58,800.

7.76, 7.77.   The tubules at higher magnification in longitudinal section, showing their helical, banded substructure. Figure 7.76, × 164,000; Figure 7.77, × 267,000.

SOURCE. Author's micrographs, published in *Planta 81*, 193 (1968).

8

EVOLUTION AND CELL MORPHOLOGY

". . . God forbid that we should lay the mummifying hand of orthodoxy on these iconoclastic outpourings!"
—D. J. Carr

The micromorphology of the interphase and dividing cells described in the preceding chapters shows both extensive diversity in detail and elements of consistency. This chapter represents a preliminary attempt to discuss possible phyletic affinities revealed by comparison of the cells' ultrastructure. Reasons for adopting such an initially restricted approach to phylogeny are given in Chapter 1. Unfortunately, many important and relevant organisms have not yet been examined in any detail, and so for many interesting Green Algae (particularly in the siphonous lines of evolution), there are virtually no pertinent recent ultrastructural data; these gaps in our knowledge suggest specific organisms that could be profitably investigated. A comparison will be made with various other more classically derived phylogenies at the end of this chapter.

I shall start with a list of suppositions I have derived from data on the ultrastructure of cells generally, and particularly algal cells, which have (in my opinion) significant phylogenetic implications. The reader can assess their importance for himself.

8.1. Cytoplasmic Systems: General Conclusions

1. *The unicellular condition is primitive, and the colonial tendency demonstrates evolutionary advancement.* This statement probably needs little justification and is an implicit assumption in most phylogenetic discussions. The old risky adage that ontogeny recapitulates phylogeny may well be applied here, since even the most complex forms of Green Algae usually retain their ability to return to the uninucleate, flagellate condition for sexual and asexual reproduction, and, as will be seen later, the morphology of such motile stages may be of particular phylogenetic importance. During this transformation, the gametes or zoospores acquire many features and morphology of their primitive ancestors (see Section 8.2) and often—but not always—bear obvious structural resemblances to simpler extant organisms such as the biflagellate *Chlamydomonas* or the quadriflagellate *Carteria*.

2. *Centrioles, basal bodies, and flagella are primitive organelles.* Although variability does exist (Klein and Cronquist, 1967), flagella, basal bodies, and centrioles in eucaryotes show remarkable structural consistency. There are differences, for example, in the structure of the transitional zone between the flagellum and basal body when plants are compared with various animal cells, and a few functional flagella are known that lack one or both central tubules (Section 2.1). Apart from such discrepancies, the basic architecture of the centriole and its relation to the spatial arrangement of the flagellar tubules are maintained almost universally in eucaryotic cells possessing either or both of these organelles. It follows that if an archetypal eucaryotic cell ever did exist, it probably acquired a centriole before it began to

Figures 8.1–8.3
ZOOSPORES OF MICROTHAMNION SP.

8.1. Cross section of cell near the flagellar bases. The symmetrically arranged two- and six-membered rootlets are obvious. × 63,000.

8.2. Longitudinal section through the slightly asymmetrical cell. Note particularly the rootlet microtubules *(r)* extending up the cell from the basal bodies *(bb)*. × 26,000.

8.3. The elongated, biflagellate zoospores in vivo. × 900.

SOURCE. Drs. M.W. Watson and H.J. Arnott, published in *J. Phycol.* 9, 15–27 (1973).

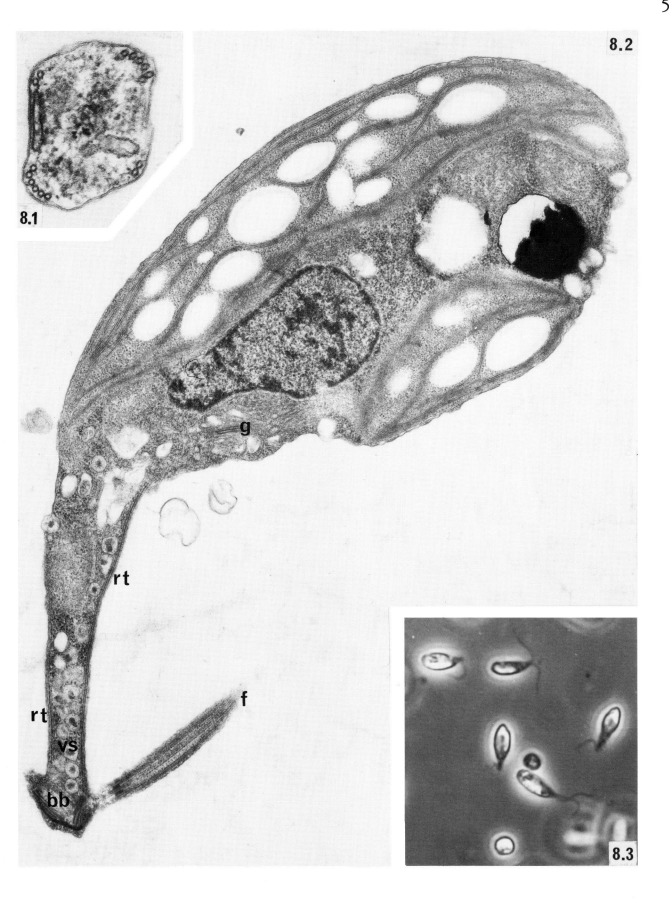

8.1

8.2

g

rt

rt
vs
f
bb

8.3

diversify very far. However, some arguments can be made against a monophyletic origin of this organelle (e.g., Klein and Cronquist, 1967; p. 158). Many highly evolved extant organisms lacking centrioles probably lost them during evolution as they became unnecessary for reproduction (paragraph 11, below). These organelles are conspicuously absent in one other large and diverse division of the eucaryotic plant kingdom, the Red Algae, and the significance of their absence there is probably profound but difficult to assess (see later).

3. *Closed spindles are primitive.* If the definition of "closed" spindles is taken widely and includes those enclosed by a persistent nuclear envelope throughout mitosis, with or without small openings such as polar fenestrae, then closed spindles are common in the Green Algae except for a few, usually advanced genera. Closed spindles are also a feature of some primitive fungi (see Pickett-Heaps, 1969a, 1972a), the one Red Alga so far investigated (*Membranoptera;* McDonald, 1972), many protozoa that undergo both mitosis and amitosis (e.g., Stevenson and Lloyd, 1971a,b), and several primitive algal groups whose mechanisms of mitosis and spindle structure are quite different from those of the majority of cells so far documented (dinoflagellates, Kubai and Ris, 1969; euglenoids; Leedale, 1970). The complementary statement and logical deduction —namely, that open spindles are generally advanced features—also appears valid, since such spindles are usually present in more highly evolved organisms; examples include the basidiomycetes in the fungi, higher plants and some complex algae, and a variety of animal cells (Pickett-Heaps, 1972a). *Klebsormidium* and *Stichococcus* appear to be exceptions.

Closed spindles have been retained by a number of organisms, and it is not clear why this should be so. Ross (1968) has suggested that retention of the nuclear envelope during mitosis might be important because it would prevent interaction (e.g., fusion) of spindles in coenocytic cells. This idea goes a little way toward explaining, for example, why slime molds have closed spindles in plasmodia and open spindles in the uninucleate amebae (Aldrich, 1969).

Likewise, it is not always obvious why some organisms should use open spindles; once the nuclear envelope breaks down, the nucleoplasm

Figures 8.4–8.8
ZOOSPORES OF SCHIZOMERIS LEIBLEINII

8.4–8.6. Live zoospores displaying a variable number of flagella, from three to seven.

8.7, 8.8. Two sections through a quadriflagellate base. The basal bodies are symmetrically arrayed and equivalent, as are the rootlet microtubules (*rt*) and striated fibers radiating out from them. Both × 50,000 approx.

SOURCE. Drs. T.R. Birbeck, K.D. Stewart, and K.R. Mattox, published in *Phycologia 13,* 71 (1974).

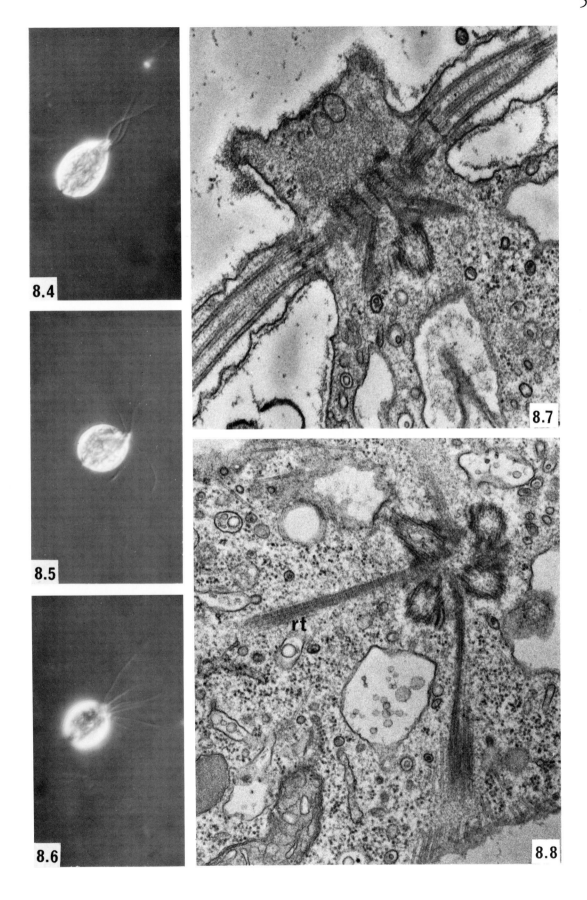

8.4

8.5

8.6

8.7

rt

8.8

is invaded by organelles such as mitochondria, endoplasmic reticulum, and Golgi bodies, and these are all removed once the daughter nuclei re-form. I have argued (Pickett-Heaps, 1974f) that perhaps open spindles become more efficient when (1) the spindle becomes larger and contains increasing numbers of microtubules, and (2) these microtubules are used as well during interphase for a variety of other functions. Numerous studies (some examples have been given earlier in this book) show that extranuclear microtubules enter the spindle of many cells at prometaphase and that the cytoplasm is normally temporarily depleted of tubules during mitosis. This entrance of tubules is not however apparent in those primitive cell types in which the spindle is completely closed (i.e., has no polar fenestrae), in which event cytoplasmic (extranuclear) microtubules may remain conspicuous and numerous during mitosis as in ascomycetes (Zickler, 1970) and *Vaucheria* (Ott and Brown, 1972). Breakdown of the nuclear envelope could simply allow large numbers of microtubules rapidly to permeate large prometaphase nuclei.

4. *Closed, centric spindles are primitive.* This logical deduction from paragraphs 2 and 3, above, is supported by the observation that most cells possessing persistent centrioles, whether the cells are primitive or highly evolved, position the organelles at the poles of the spindle (e.g., aquatic phycomycetes, algae generally, and most animal cells). If the centriole makes up part of the cell's microtubule-organizing centers (MTOCs) involved in organizing the spindle tubules (Pickett-Heaps, 1971a), their situation at the poles is logical. In euglenoids and dinoflagellates, centrioles or basal bodies reportedly remain dissociated from the spindle; we cannot yet tell if these constitute exceptions to the principle.

5. *Furrowing or cortical constriction is a primitive mechanism of cytokinesis.* Annular ingrowth of the cell membrane divides procaryotic cells in two. Similar ingrowth of the plasmalemma apparently also partitions the cell in diverse algae. Other cells divide by an ingrowing constriction of the whole cell cortex (e.g., many algae, protozoa, and animal cells). Whether this latter mechanism is analogous or related to the former, or represents a development of it, is debatable and will

Figures 8.9, 8.10
SPERMATOZOID DIFFERENTIATION IN MARCHANTIA AND SPHAEROCARPOS

8.9.   Differentiating spermatozoids of *Marchantia*. In one cell, the large arrow marks the single band of manchette tubules near the multilayered structure, next to which is a mitochondrion *(m)*. In the adjacent cell, the manchette sectioned at a different level (small arrow) contains fewer tubules. Note the stellate structure in the transition zone between basal body and flagellum (double arrow), typical of plant flagella.

8.10.   Differentiating spermatozoids of *Sphaerocarpos*. In each cell, the large arrow marks the multilayered structure, the smaller arrow the manchette tubules. The laminations of the multilayered structure are oriented skew to both the orientation of the basal bodies *(bb)* and the manchette tubules.

SOURCE. Figure 8.10: Dr. S.E. Frederick, unpublished micrograph.

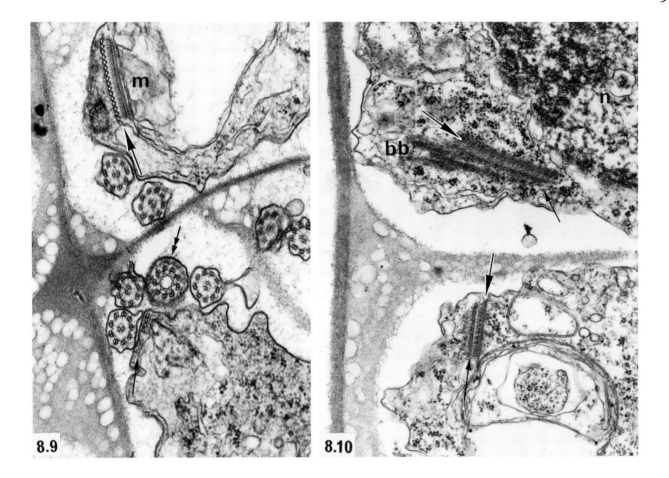

8.9

8.10

not be discussed further. Many larger animal cells have filamentous systems associated with the ingrowing furrow. There is increasing evidence that such filamentous systems are contractile (Schroeder, 1970), indicating that perhaps the filaments have been evolved to assist growth of a cleavage furrow through larger cells. So far as I know, filaments have not yet been recorded near the furrows of cleaving algal cells, which when large, are often vacuolated and therefore possibly offer less resistance to such membrane ingrowth.

6. *Cell plate formation is a comparatively advanced form of cytokinesis.* The collection of vesicles into a localized region of the cell (defined by microtubular arrays; see paragraphs 7 and 8, below), and their subsequent fusion to form a wall dividing daughter cells, are characteristic of various more advanced Green Algae and all higher plants. This method of cross-wall formation has not yet been clearly documented in any other group of organisms to my knowledge. It seems a logical development once a plant cell had evolved mechanisms of wall deposition utilizing precursors initially synthesized in the cytoplasm and packaged within membrane-bounded vesicles which are then secreted into the cell wall (e.g., during mucilage secretion). The advantage of cell plate formation over furrowing (paragraph 5) is not understood at present; nevertheless, cells utilizing either phycoplasts or phragmoplasts (paragraphs 7 and 8) have consistently favored this method of cross-wall initiation during evolutionary advancement (Pickett-Heaps, 1972a). Two organisms, *Spirogyra* and *Microspora,* may even now be evolving from furrowing toward utilization of a cell plate, since each appears to use both furrowing and vesicles for cross-wall formation. However, the microtubular systems concurrently engaged in cytokinesis are quite different, *Spirogyra* having a rudimentary phragmoplast and *Microspora* a phycoplast.

7. *The phycoplast is a fairly primitive cytokinetic structure in the Green Algae.* This proposition may be more controversial. However, the phycoplast seems to be present in a wide variety of Green Algae (and no where else yet), whether the cells are fairly advanced (e.g., *Oedogonium, Hydrodictyon, Stigeoclonium,*) or are primitive (e.g., *Chlamydomonas, Chlorella*). The phycoplast is used in conjunction with cytokinesis by cortical or

Figure 8.11
PORTION OF A MATURE SPERM OF EQUISETUM

This section through the anterior end shows the multilayered structure (arrows) sectioned four times (i.e., through two gyres of the helical band of flagella). There is the usual single flat band of manchette microtubules and mitochondria (*m,* cf. Figs. 8.9, 8.10) associated with the multilayered structure and the numerous basal bodies (*bb*). The microtubules band increases in width as it passes through the gyres sectioned here. × 40,000 approx.

SOURCE. Dr. G. Duckett, unpublished micrograph.

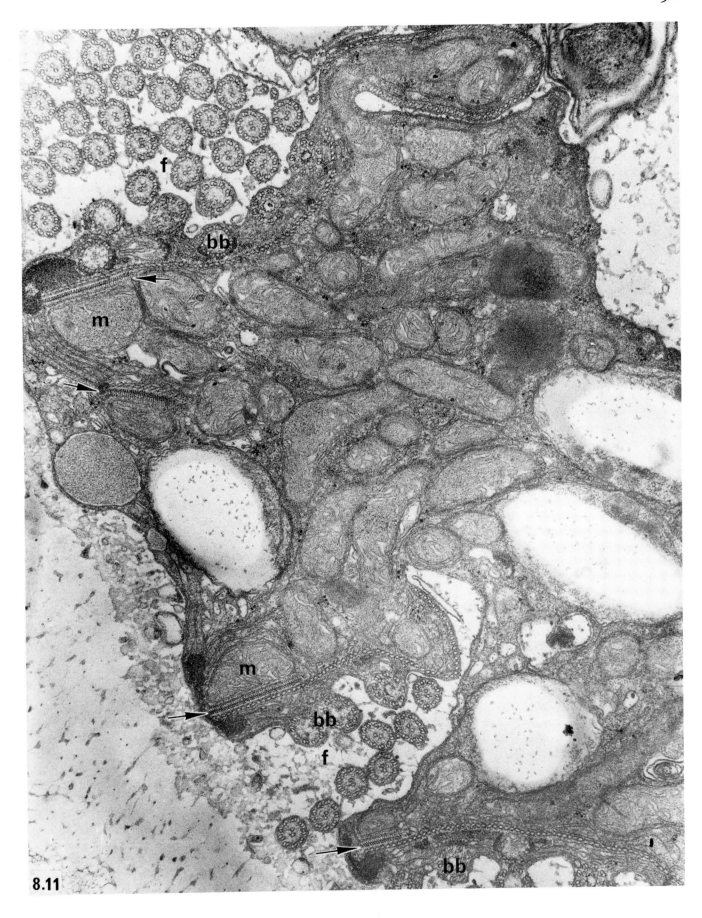

8.11

membrane furrowing (paragraph 5) or cell plate formation (paragraph 6). Just how primitive the phycoplast is I find difficult to assess. As mentioned later (Section 8.3c), the persistent interzonal spindle may well be more primitive (and this latter feature is present in several Green Algae), so that the phycoplast may be a derived condition. In any event, phycoplast-containing cells have been around long enough for considerable evolutionary diversification to have occurred in them.

8. *The phragmoplast is an advanced cytokinetic system.* Phragmoplasts appear to be universally present in higher plants undergoing cytokinesis. However, they are probably rare in algae despite the widespread assumption in the literature that many Green Algae undergo cell division in the same way that higher plants do. Phragmoplasts have been recorded so far only in *Coleochaete*, the very advanced and complex Charales, and perhaps in rudimentary form in *Spirogyra*, which may be in the process of evolving a phragmoplast at present. (I should guess that a few more isolated examples of phragmoplast-containing Green Algae may well turn up in the Ulotrichales.) Since I cannot see how the phycoplast could have given rise to the phragmoplast, I deduce that the latter appeared perhaps in isolation (as in *Spirogyra*) and only in certain groups of Green Algae, a proposition to be discussed later.

9. *Centrioles per se are not necessary for mitosis.* A few years ago this statement might have been considered heretical. Today, there are so many examples of acentric spindles or spindles that function with or without centrioles (e.g., in *Chara,* bryophytes) that it is difficult to maintain the formerly accepted theory that centrioles have any role in spindle formation or function—even in some mammalian cells (Szollosi, Colarco, and Donahue, 1972)

10. *Absence of centrioles from vegetative cells does not mean that these cells have lost the ability to form the organelles.* A number of organisms are known not to contain centrioles for much of their life cycle, and yet these organelles have been shown to appear before differentiation into certain, usually flagellate forms of the cell. Examples include some of the protozoa (Fulton and Dingle, 1971), slime molds (Aldrich, 1969), diatoms (Manton,

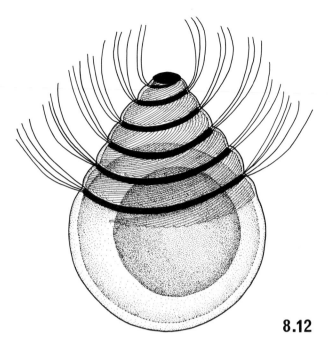

8.12

Figure 8.12
SPERMATOZOID OF ZAMIA.

Diagrammatic representation. The massive flagellar apparatus, containing about 10,000–12,000 flagella, winds spirally around the cell forming about six gyres. (The cell measures ca. 280 by 180μm and is largely filled by the nucleus.) From the flagellar apparatus, about 60,000 manchette microtubules extend diagonally rearward just beneath the cell membrane. Only a few of these flagellar components are shown in the diagram.

SOURCE. Redrawn after Norstog (1968).

Figure 8.13
FLAGELLAR BASE OF SPERMATOZOID OF ZAMIA INTEGRIFOLIA

This section through the tip of the spermatozoid shows part of the descending spiral of the flagellar bases and massed flagella from the bases further into the cell. The bases insert into the *vierergruppe (vg).* × 4,500.

SOURCE. Dr. K. Norstog, published in *Am. J. Bot. 61,* 449 (1974).

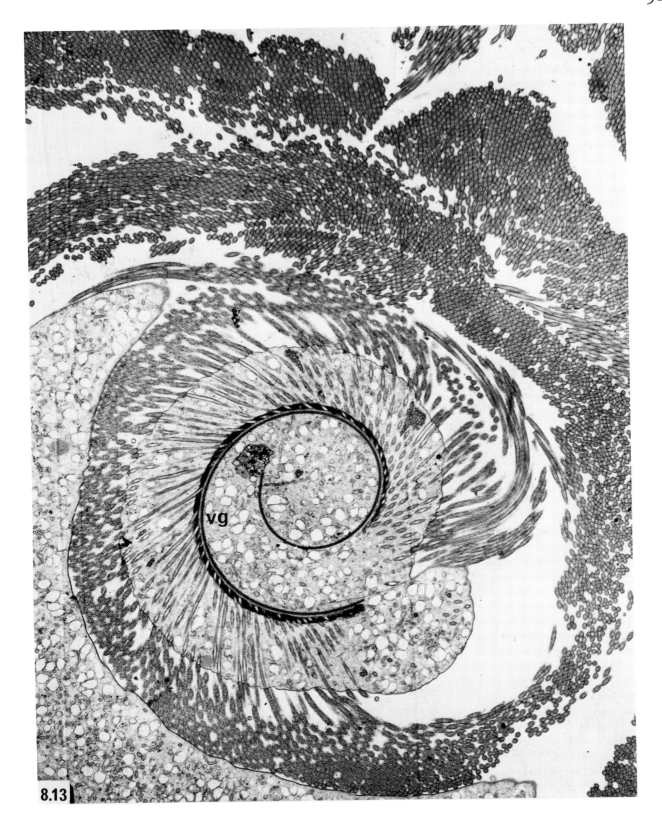

8.13

Kowallik, and von Stosch, 1970), bryophytes (Moser and Kreitner, 1970), and some Green Algae (e.g., *Oedogonium, Chara*).

11. *Absence of centrioles in certain cases may be a sign of evolutionary advancement.* This proposition can be justified in two circumstances; first, when an organism has evolved to the extent that it is no longer dependent upon an aqueous environment for reproduction (e.g., angiosperms, ascomycetes, and basidiomycetes), rendering a flagellar apparatus redundant; and second, when an organism has circumvented the need to keep centrioles permanently formed and therefore replicating during the cell cycle (paragraph 10). Presumably in the latter situation, the organism replicates and then, during division, partitions some form of informational material into daughter cells, whose potential can become expressed structurally as centrioles only when these latter organelles are needed for flagella formation (see examples in paragraph 10). The formation of new centrioles in each mitotic cycle in centric cells probably represents a small but significant investment in synthetic, energy-requiring processes. This investment is probably profitable when it allows the organism to form zoospores quickly. The zoospore of *Oedogonium* contains about 100 basal bodies; imagine the complications if this number of centrioles had to be replicated and then partitioned to daughter cells every time *Oedogonium* divided! Suppression of centriole formation until the centrioles were needed was probably necessary if *Oedogonium* was to compete successfully over millions of years of evolution. Similar suppression of centriole formation might also have constituted a small but perceptible advantage to those organisms in which the vast majority of vegetative cells in a plant body would never become flagellate (e.g., the Charales and bryophytes).

8.2. Phylogenetic Significance of Motile Cell Morphology

It is generally accepted that the present variety of Green Algae arose originally from tiny uninucleate, flagellate forms (see paragraph 1 above), and many highly evolved genera retain the ability to re-form small motile cells for asexual and sexual reproduction. The Conjugales constitute the only order that appears to have

Figures 8.14, 8.15
SPERMATOZOID OF ZAMIA INTEGRIFOLIA

8.14. Longitudinal section through the spiral flagellar apparatus. The successive profiles of the flagellar bases and the vierergruppe are labeled *1–7*. As in Figure 8.13, the first section of the vierergruppe (about one half turn) is devoid of basal bodies *(1, 2)*, which appear in transverse section on one side *(3, 5, 7)* and skew on the other *(4, 6)*. × 2,900.

8.15. Detail of the vierergruppe, showing basal bodies *(bb)*, manchette or spline microtubules *(t)*, and the multilayered structure. × 95,000.

SOURCE. Dr. K. Norstog, published in *Am. J. Bot. 61*, 449 (1974).

8.14

8.15

entirely dispensed with a flagellate phase. Some primitive land plants such as the bryophytes and ferns still utilize motile gametes for sexual reproduction, and although these plants have evolved so that they no longer require an aqueous environment for vegetative existence, they still need a film of water present on the plant body (gametophyte), through which the flagellate sperm swim to the oogonium. Of course, the most advanced land plants have entirely dispensed with the need for water when carrying out sexual reproduction, and the pollen grain represents an extraordinarily successful reproductive cell, most resistant to desiccation, whose motility has been assured by the employment of a variety of vectors such as wind, animals, insects, birds. Particularly intriguing are the larger land plants, *Ginkgo* and the cycads, which use pollen in the normal way but which just before fertilization differentiate a most complex, multiflagellate sperm cell in the pollen tube. (In fact, a "haustorium," or complex of pollen tubes growing from the pollen grain, pushes the latter toward the receptive ovule, and the sperm forms in the old pollen cell wall.) This sperm cell is released deep inside the ovule and it consummates the union of the male and female cells, needing to move only a tiny distance to do so. The primitive cycads and *Ginkgo,* therefore, provide beautiful examples of extant intermediates in the evolution of sexual reproduction of the angiosperms and more advanced gymnosperms, having evolved the pollen grain but still utilizing a motile sperm.

When Green Algae differentiate motile cells, they may need to re-form primitive cellular structures that are not needed in their sessile, thick-walled vegetative condition. A flagellar apparatus, complete with contractile vacuoles, rootlet microtubules, and sometimes even the basal bodies, is formed de novo. In some cases, cells that now reproduce by nonmotile autospores still reveal their motile ancestry at the ultrastructural level—*Kirchneriella,* for example, by re-forming rudimentary centrioles de novo apparently only during mitosis and cytokinesis (Section 3.3a), and *Tetraedron,* by the transient rootlet templates appearing near the centrioles during cleavage of the cytoplasm (Section 3.5a; Fig. 3.35). *Microspora* can easily form flagellate cells, but even during vegetative cell division, a transient band interconnects the centrioles (Fig. 4.55), a structure that could be interpreted as resulting from a

Figures 8.16, 8.17
ZOOSPORES OF LEPTOSIRA

In longitudinal section (Fig. 8.16) the rootlet microtubules (arrows) extend up the periphery of the cell from the basal bodies *(bb).* A transverse section (Fig. 8.17) reveals, however, that there are only two sets (arrows) of these tubules (cf. Fig. 8.1), numbering about nine in this example. Figure 8.16, × 23,000 approx.; Figure 8.17, × 29,000.

Figure 8.18
SOME VOLVOCINE LINES OF EVOLUTION

Diagrammatic representation showing the development of various colonial forms from a simple chlamydomonad type of cell. The constituent cells of the colonial forms remain uninucleate and essentially separate from one another, although cytoplasmic strands interconnect cells in some genera for at least part of their life cycle.

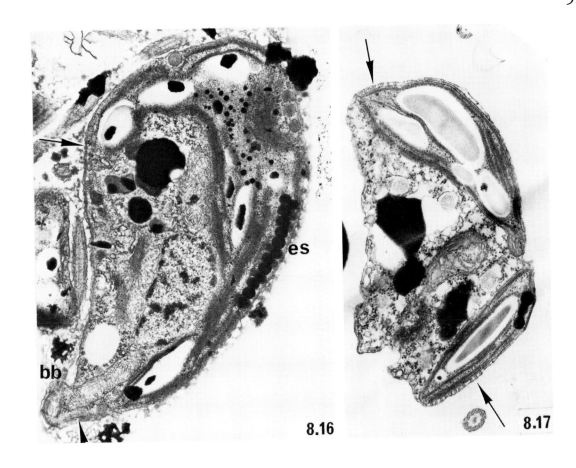

es

bb

8.16

8.17

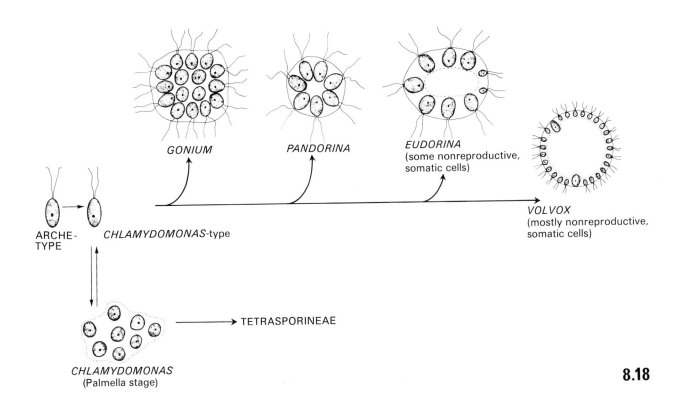

GONIUM

PANDORINA

EUDORINA
(some nonreproductive,
somatic cells)

VOLVOX
(mostly nonreproductive,
somatic cells)

ARCHE-
TYPE

CHLAMYDOMONAS-type

TETRASPORINEAE

CHLAMYDOMONAS
(Palmella stage)

8.18

preliminary phase of re-forming a flagellar base.

If the motile unicell represents the primitive condition, we could expect that the structure of such a motile cell might tell us something about the phylogeny of the organism that differentiates it. This conclusion has been most strongly supported by the ultrastructural work described previously in this book and summarized below.

Zoospores in the Green Algae can already be divided into two distinct classes based on their micromorphology (more such classes can be anticipated when more algae have been studied). The first class I have termed "chlamydomonad" (Pickett-Heaps, 1974c), for the obvious reason that the structure of the motile cell can be related to the basic structure of *Chlamydomonas,* particularly with respect to its flagellar and microtubular apparatus. Birbeck, Stewart, and Mattox (1974) use the term "chlorophycean" in this context. Cells of this type are usually reasonably symmetrical about one plane at least. When biflagellate, the rootlet microtubules radiate from between the two basal bodies in an X or cruciate (Moestrup, 1972) pattern (Section 2.1; Fig. 2.24). The precise number and disposition of these microtubules are somewhat variable (see Moestrup, 1972), and frequently two sets of two different rootlet complexes are encountered. The number and disposition of such rootlet microtubules may have a more subtle phylogenetic significance, as suggested by Manton (1965), than is evident from our present limited information. Many volvocalean genera and the zoospores of various higher Green Algae are quadriflagellate. Since *Chlamydomonas* and other biflagellate forms may have one or two extra, nonfunctional basal bodies near the flagellar apparatus (Section 2.1), quadriflagellate forms could be regarded as those in which all four basal bodies have become (or else have remained) functional. The zoospore of *Stigeoclonium* (Manton, 1964b; Section 4.6b) has been shown to contain two pairs of basal bodies in the cruciate pattern, with two pairs of slightly different rootlets radiating from them. Also of importance is a striated fiber associated with one pair of these rootlets (see below).

In many such cells, additional microtubules (which I have called "secondary cytoskeletal microtubules"; Pickett-Heaps, 1974c) are also found lying near the cell's periphery; the sperm of *Volvox* and the zoospores of *Hydrodictyon, Pediastrum,* and *Sorastrum* indicate how this second-

Figure 8.19
MITOSIS IN CHLAMYDOMONAS MOEWUSII

The spindle of this species seems to be entirely closed. The basal bodies *(bb)* replicate before division and separate near the poles, as indeed they have now been reported to do in *C. reinhardii* as well (contrary to the description given in Chapter 2; see text).

SOURCE. Drs. R.E. Triemer and R.M. Brown, paper in preparation.

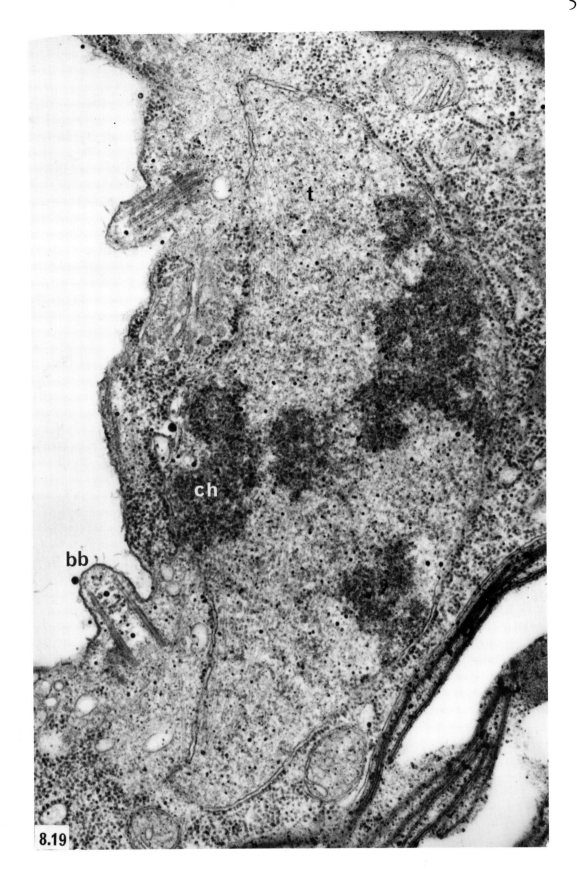

8.19

ary system may have become increasingly important, for example, in the evolution of oogamy and colony formation.

More recently, the zoospores of *Microthamnion* and *Shizomeris* have been described. In *Microthamnion* (Watson and Arnott, 1973), the zoospore is elongated (Fig. 8.3), quite highly differentiated, and somewhat asymmetrical (Fig. 8.2), but nevertheless it conforms, in my opinion, with the basic chlamydomonad morphology; its rootlet microtubules are in four groups, two containing about six and two containing two tubules, alternating around the cell periphery (Fig. 8.1). There is also a fine striated rhizoplast lying between the basal bodies in the nucleus. The zoospore of *Shizomeris* (Birbeck, Stewart, and Mattox, 1974) is usually quadriflagellate (Fig. 8.5), but now the four flagellar bases are equivalent, as are the four rootlet microtubule systems, each with an associated striated fiber. These components are symmetrically arranged within the cell (Figs. 8.7, 8.8); the secondary cytoskeletal system is extensive.

The zoospore and sperm of the Oedogoniales at first sight seem to pose a special problem in classification. However, although the flagellar apparatus is enormously complex, I consider it as repetitive units whose fundamental components (Pickett-Heaps, 1974c) reveal a chlamydomonad ancestry. Between each of the basal bodies is the typical rootlet template/rootlet microtubule system plus some secondary microtubules, and these are intimately associated with a striated fiber (Fig. 5.61) similar to that found in *Stigeoclonium*. The impressive fibrous ring (Fig. 5.59) could have been originally derived from a fibrous band such as that interconnecting the basal bodies of *Chlamydomonas*. Manton (1964b) concluded that such a multiflagellate base could not have arisen from an organism like the zoospore of *Stigeoclonium*, but the flagellar base in zoospores of *Shizomeris* can be considered as a ring of four basal bodies (Birbeck, Stewart, and Mattox, 1974). Furthermore, the zoospore of *Shizomeris* may also have between two and eight flagella (Figs. 8.4, 8.6), instead of four (see references in Birbeck, Stewart, and Mattox, 1974). Although Birbeck and coworkers have not yet been able to examine zoospores with eight flagella, a ring-like array of basal bodies seems likely to be present. As Manton has suggested, the multiflagellate condition probably

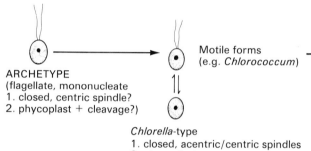

ARCHETYPE
(flagellate, mononucleate
1. closed, centric spindle?
2. phycoplast + cleavage?)

Motile forms
(e.g. *Chlorococcum*)

Chlorella-type
1. closed, acentric/centric spindles
2. phycoplasts

8.20

Figure 8.20
SOME SIPHONOUS LINES OF EVOLUTION

Diagrammatic representation. These are characterized by the tendency for organisms to become increasingly multinucleate. Nuclear division and cell division are usually entirely separate events. Only some genera in the chlorococcalean lines of evolution have been investigated in sufficient detail ultrastructurally to give taxonomically significant data (e.g., all chlorococcalean organisms so far described have phycoplasts in some form).

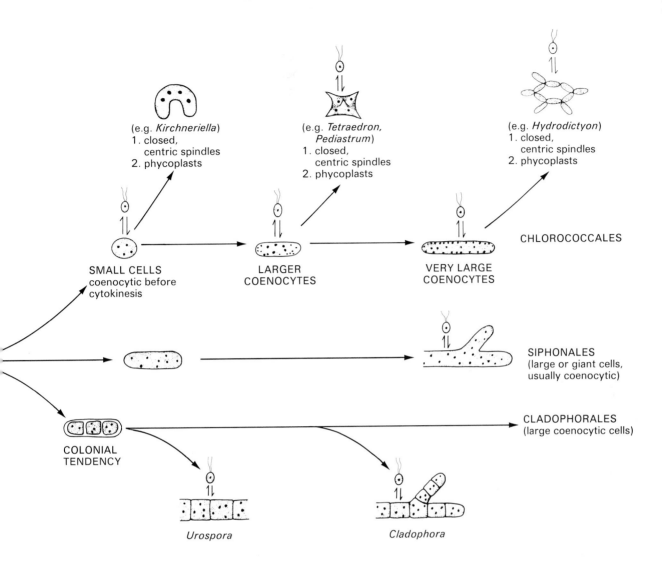

(e.g. *Kirchneriella*)
1. closed,
   centric spindles
2. phycoplasts

(e.g. *Tetraedron, Pediastrum*)
1. closed,
   centric spindles
2. phycoplasts

(e.g. *Hydrodictyon*)
1. closed,
   centric spindles
2. phycoplasts

SMALL CELLS
coenocytic before
cytokinesis

LARGER
COENOCYTES

VERY LARGE
COENOCYTES

CHLOROCOCCALES

SIPHONALES
(large or giant cells,
usually coenocytic)

CLADOPHORALES
(large coenocytic cells)

COLONIAL
TENDENCY

Urospora

Cladophora

arose a number of times independently, even in this assemblage of organisms.

I hope it has not escaped the attention of the reader that all the organisms described above are characterized also by having closed mitotic spindles and, more importantly, the phycoplast during cytokinesis. The zoospores of all Chlorococcales so far described, many members of the Ulotrichales, and vegetative cells of the Tetrasporales and Volvocales fall into this category.

The second class of motile cells I have called the "bryophytan" type (the "characean" type of Birbeck, Stewart, and Mattox, 1974). These motile cells display considerable or extreme cellular asymmetry. The flagella are probably not exactly equivalent to one another, being inserted into the cell asymmetrically. The microtubular system consists of a single layer of microtubules arranged in a band that is often flattened and sometimes associated with some secondary cytoskeletal microtubules. This band or manchette runs the length of the cell, and at one end is intimately associated with the basal bodies. The four algae so far described in this category are *Chara, Nitella, Coleochaete,* and *Klebsormidium.* Of great significance is the multilayered structure that is also associated with the end of the microtubule band in *Coleochaete* and *Klebsormidium* (Sections 4.2b, 4.3c). The multilayered structure is present in all the motile gametes so far described of bryophytes and other higher land plants that form sperm (e.g., Figs. 8.9–8.11, 8.15). A great elaboration of the presumably original biflagellate apparatus has taken place in ferns and cycads, for example—an evolutionary development equivalent perhaps to the elaboration of the multiflagellate apparatus in *Oedogonium* from simpler forms. However, the intrinsic asymmetry of the bryophytan type of motile cell seems to have prevented the formation of flagellar rings as in *Oedogonium,* and instead, spiral bands of flagella have resulted (Figs. 8.11–8.14). In terrestrial plants, the multilayered structure is usually associated with one or more mitochondria, and the manchette has become enormous in width as the number of the flagella in the system increases. The structure of these motile cells is extremely complex, and the few descriptions we have can be only briefly summarized here.

In the biflagellate gamete of *Marchantia,* the two basal bodies are inserted, one somewhat be-

Figure 8.21
SOME ULOTRICHALEAN LINES OF EVOLUTION

Diagrammatic representation. The cells of the organisms are almost always uninucleate. The colonial tendencies give diverse forms to the plant bodies. The basically filamentous form can become branched, discoidal, parenchymatous, etc., following the development of the ability to control cell division in a plane away from that of the original filament axis. The lines of evolutionary advance have been split in this diagram on the basis of whether the organism possesses either a phycoplast or the persistent spindle/phragmoplast during cytokinesis.

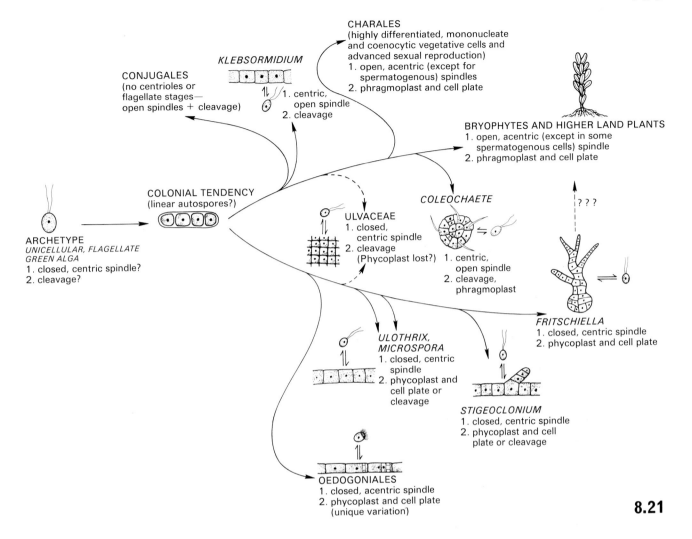

CONJUGALES
(no centrioles or
flagellate stages—
open spindles + cleavage)

KLEBSORMIDIUM
1. centric,
open spindle
2. cleavage

CHARALES
(highly differentiated, mononucleate
and coenocytic vegetative cells and
advanced sexual reproduction)
1. open, acentric (except for
spermatogenous) spindles
2. phragmoplast and cell plate

BRYOPHYTES AND HIGHER LAND PLANTS
1. open, acentric (except in some
spermatogenous cells) spindle
2. phragmoplast and cell plate

COLONIAL TENDENCY
(linear autospores?)

ULVACEAE
1. closed,
centric spindle
2. cleavage
(Phycoplast lost?)

COLEOCHAETE
1. centric,
open spindle
2. cleavage,
phragmoplast

ARCHETYPE
*UNICELLULAR, FLAGELLATE
GREEN ALGA*
1. closed, centric spindle?
2. cleavage?

???

FRITSCHIELLA
1. closed, centric spindle
2. phycoplast and cell plate

*ULOTHRIX,
MICROSPORA*
1. closed, centric
spindle
2. phycoplast and
cell plate or
cleavage

STIGEOCLONIUM
1. closed, centric spindle
2. phycoplast and cell
plate or cleavage

OEDOGONIALES
1. closed, acentric spindle
2. phycoplast and cell plate
(unique variation)

8.21

hind the other, over the multilayered structure and on either side of the axis of the manchette microtubules, and slightly skew to them. The microtubules number about 20 in this region (Fig. 8.9) but soon decrease to about 6 where they run around the body of the gamete (Kreitner, 1970; Carothers and Kreitner, 1967, 1968). The multilayered structure and nearby row of tubules are often called the "vierergruppe" (e.g., Carothers and Kreitner, 1967). The sperm of *Sphaerocarpos* (Fig. 8.10), *Blasia,* and *Phaeoceros* are similar (Carothers, 1973; Moser, 1970; Frederick and Pickett-Heaps, unpublished data), as is that of *Polytrichum* (Paolillo, Kreitner, and Reighard, 1968a,b). In the multiflagellate sperm of *Marsilea* (Rice and Laetsch, 1967; Myles, personal communication), the flagella are arrayed around the spirally organized manchette, which now contains about 25 microtubules. In *Equisetum* (Fig. 8.11; Duckett, 1973), the manchette of the sperm contains up to about 300 microtubules, and in the extraordinary sperm cell of *Zamia,* a cycad (Fig. 8.12), there are about 10,000–12,000 flagella associated with a spiral manchette of about 60,000 microtubules (Figs. 8.13, 8.14; Norstog, 1967, 1968) and the multilayered structure (Fig. 8.15). The microtubules presumably serve to attach the flagellar apparatus to the large body of the sperm (Fig. 8.12). In all these organisms, as far as is known, elongation and/or differentiation of the sperm cell appear to be intimately associated with elaboration of the manchette microtubules, as in spermatogenesis in the Charales (Section 7.5). In broadest terms, the complexity of such sperm can be visualized as having arisen from predecessors having the simpler morphology of the sperm of the bryophytes. I feel that the ultrastructure of these sperm and in particular their possession of the multilayered structure, argue for an essentially monophyletic origin of higher plants. A cautionary note should be injected here. I shall frequently talk about the origin of the bryophytes and higher forms of green plants as possibly monophyletic. This does not imply that the modern bryophytes represent intermediates in the evolution of the higher forms. Rather, many of the bryophytes appear to be reduced forms and perhaps in evolutionary cul-de-sacs.

All the organisms in this second group (i.e., both algae and higher plants), which have the bryophytan motile cell, also have open spindles

Figures 8.22–8.27
CELL DIVISION IN TRICHOSARCINA AND PSEUDENDOCLONIUM

8.22–8.24. *Trichosarcina polymorpha.*

8.25–8.27. *Pseudendoclonium basiliense.*

8.22, 8.25. Preprophase. Precocious development of the cleavage furrow *(cf)* before mitosis. The centrioles (*cn,* arrow) are always sited near the furrow at this stage.

8.23, 8.26. Metaphase. The cleavage furrow is pronounced. The closed, fenestrated spindle has centrioles (arrow) placed to one side of the pole (cf. Figs. 4.64, 4.65).

8.24. Telophase, cytokinesis. The cleavage furrow is passing between the daughter nuclei, still widely separated at this stage.

8.27. Serial section of the upper pole of the spindle in Figure 8.26, showing the pair of centrioles to one side of the polar fenestra.

SOURCE. Drs. K.R. Mattox and K.D. Stewart, paper in preparation.

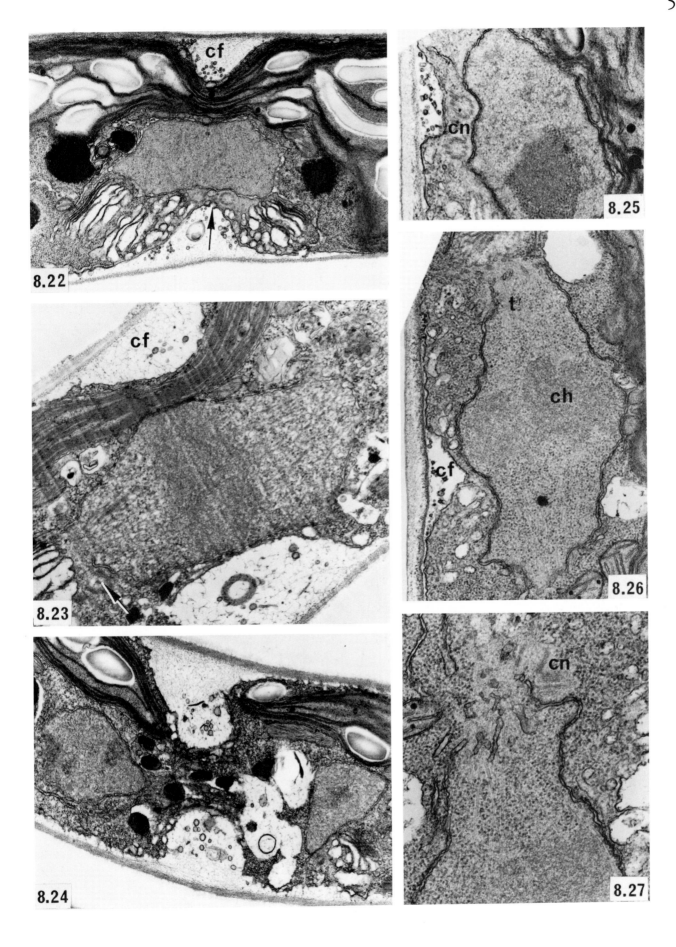

and either the persistent telophase spindle or the phragmoplast during cytokinesis. Such results are bound to be of phylogenetic significance, a matter taken up later in this chapter.

There are several types of zoospores which cannot be so easily characterized. For example, the zoospore of *Leptosira* (Pickett-Heaps, unpublished observations), a genus placed in the Ulotrichales, has only two flat bands of microtubules running along each side of an elongated zoospore (Figs. 8.16, 8.17), perhaps because two of the four cruciate groups have disappeared entirely. The gamete of the siphonous Green Alga *Acetabularia mediterranea* also possesses some unusual features. Woodcock and Miller (1973) say that only one band of 15–20 tubules runs the length of the cell; the flagella emerge at 180° to one another, and the basal bodies actually overlap each other (i.e., they do not lie end to end). The significance of these observations will emerge only when further detailed ultrastructural work has been carried out on the organisms concerned and on their relatives.

8.3. Cytomorphologic Indications of Phylogenetic Affinities

8.3a. The Volvocine Lines of Evolution

The volvocine lines of evolution (Scagel et al., 1965) are reasonably easy to follow and are the least controversial. As discussed in Chapter 2, evolutionary advance in the Volvocales (and to some extent, the Tetrasporales) shows several coincident and expected trends. The size of the colonies increases, and there arise both specialization and division of labor in colonial cells (Fig. 8.18) and a progression from isogamy to oogamy in sexual reproduction. It is emphasized that the organisms specifically mentioned on diagrams by name should not be considered directly related; they rather illustrate evolutionary trends. It is not difficult to visualize how the flat colony characteristic of *Gonium* could have evolved into a ball-like cluster of cells such as in *Pandorina*. Many smaller colonial forms which are spherical when fully developed are derived after inversion from colonies that were previously plakeal in form (including *Pandorina*, for example, which undergoes inversion; Smith, 1950; p. 97). The rather exceptional flat colony of *Platydorina* (Fig. 2.5) is also derived

Figures 8.28–8.30
CELL DIVISION IN CYLINDROCAPSA BREBISSONII

8.28. Interphase cells. Each has secreted its own wall, while remaining in a sheath derived from older cell walls. Strands of mucilage project from the filament. The centrioles (arrow) are visible in one cell. × 4,400 approx.

8.29. Metaphase. The closed, fenestrated spindle has broad poles, with centrioles sited to one side of the fenestrae (cf. Figs. 8.23, 8.27). The cleavage furrow (*cf*) is well developed before prophase. × 10,000 approx.

8.30. Early telophase. The ingrowing cleavage furrow is already passing between daughter nuclei (cf. Fig. 8.24). × 10,000 approx.

SOURCE. Drs. J.D. Pickett-Heaps and K. McDonald, published in *New Phytol.* (in press).

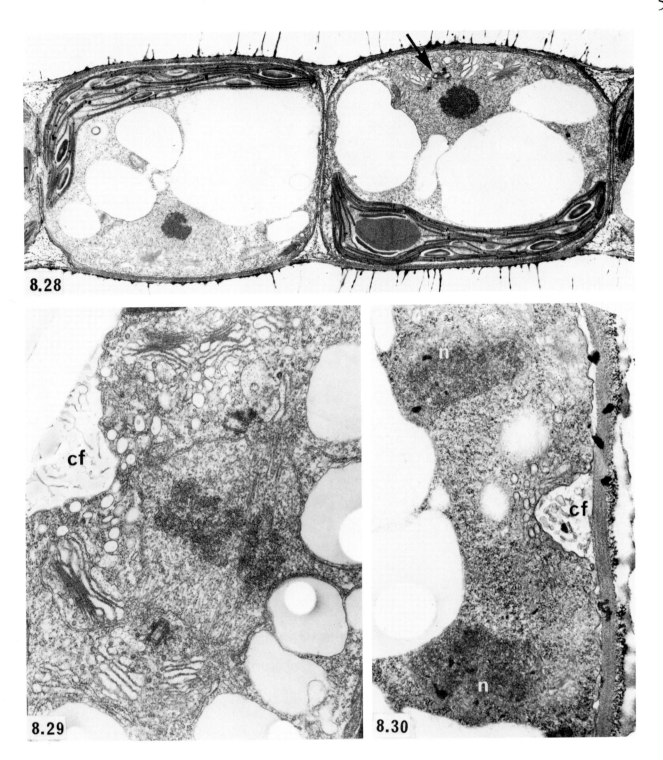

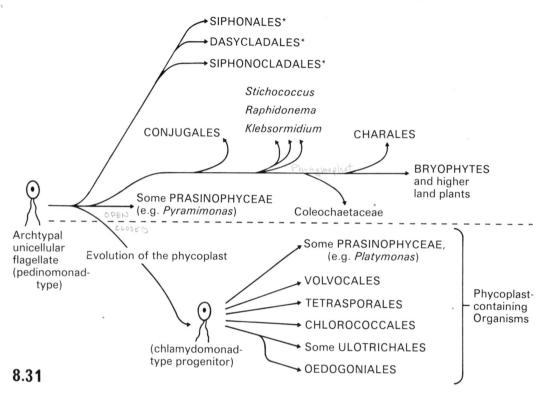

8.31

Figure 8.31
ORIGIN OF PHYCOPLAST-CONTAINING ORDERS

A phylogenetic scheme that shows the phycoplast-containing orders (and genera such as *Platymonas*) as having an essentially monophyletic origin from a chlamydomonad progenitor. These are separated from those orders and some specific ulotrichalean and prasinophycean genera that have persistent telophase spindles or phragmoplasts during cytokinesis. This latter group is considered to have provided the progenitors of higher land plants. The truly ancient cell type may be exemplified in some respects by present-day pedinomonad cells. Those orders marked by an asterisk (e.g., Siphonales*) are included in a most tentative manner, since there is little relevant ultrastructural information available. Hopefully, their taxonomic affinities may be revealed by appropriate research in the near future.

SOURCE. Drs. J.D. Pickett-Heaps and D. Ott, published in *Cytobios* (in press).

Figures 8.32–8.34
SCALES ON THE SURFACE OF REPRODUCTIVE CELLS OF PSEUDENDOCLONIUM BASILIENSE

8.32, 8.33. Scales in surface and cross section on the body of the zoospores. The flagella *(f)* apparently lack them. Figure 8.32, × 41,000; Figure 8.33, × 44,000.

8.34. Similar scales (arrows) on the surface of *nonmotile* autospores of this same alga.

SOURCE. Drs. K.R. Mattox and K.D. Stewart. Figures 8.32, 8.33 reproduced by permission of the National Research Council of Canada from *Can. J. Bot. 51*, 1425 (1973).

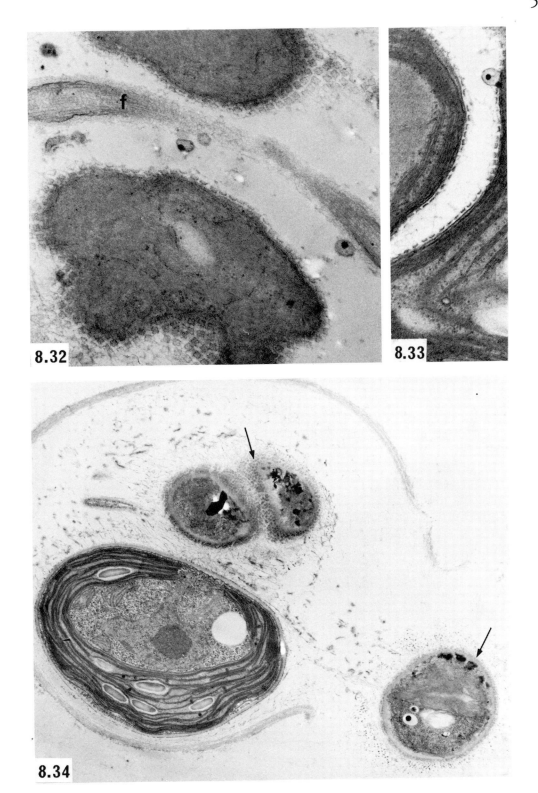

8.32

8.33

8.34

from a plakea that undergoes inversion, but the young spherical colony thus formed flattens immediately afterward (Smith, 1950; p. 102; see Chapter 2). It appears as if cellular elongation mediated via microtubules may be of great importance in achieving inversion; the small male sperm cells are also elongated, and perhaps the evolution of the male gamete in these forms is related to the evolution of inversion.

The other volvocine line of evolution leading to the Tetrasporineae (Tetrasporales) appears to constitute only a slightly different, less organized variation on the colonial tendency, with cells being randomly dispersed within a mucilaginous matrix. Motility in these cells usually appears only after reproduction (Fritsch, 1935) and may be of relatively short duration. Similar suppression of motility has also occurred in most other lines of evolution and can be considered to have been important in the evolution of the Chlorococcales and filamentous forms of Green Algae. This trend toward loss of motility is easily understandable, since once situated in a suitable environment, plant cells often have no great need to move around.

At this stage, it looks as if we shall find considerable ultrastructural conservatism in all these lines of evolution. The basic morphology of the interphase cell probably has been only superficially elaborated upon, and mitosis and cytokinesis are likely to be similar in all genera. Johnson and Porter (1968) report that the basal bodies of *Chlamydomonas reinhardii* are not associated with the spindle pole during mitosis. My doubts about the validity of this conclusion (Pickett-Heaps, 1973c; Chapter 2) have since proved justified. Coss (1974) and Triemer and Brown (personal communication) confirm the presence of basal bodies at or near the poles of *C. reinhardii* and *C. moewusii* (Fig. 8.19), which brings this genus into line with the observations on *Volvox* and *Tetraspora*. *C. moewusii* is reported (Triemer and Brown, in preparation) to have no polar fenestrae, which might indicate that it is slightly more primitive in this respect than *C. reinhardii*. As will be seen later, such completely closed spindles are found in various rather primitive organisms, including the Green Alga *Pedinomonas*.

Another line of volvocine evolution may be represented by the few members of the family Spondylomoraceae (sensu Smith, 1950), for example, in which the colonies do not undergo

Figures 8.35–8.38
CELL DIVISION IN THE PRASINOPHYTE PLATYMONAS SUBCORDIFORMIS

8.35. Metaphase. Note the position of the flagellum (*f*) adjacent to the metaphase plate of chromosomes (*ch*), inside the closed spindle.

8.36. Metaphase. Basal bodies (arrows) are clearly situated on either side of the cleavage furrow, which is just forming. The vesicles derived from the Golgi bodies (*g*) contain densely stained material (scales?) destined for secretion. × 22,000.

8.37. Telophase, cleavage. The cleavage furrow (*cf*) is growing through a phycoplast between the flagellar bases (cf. Fig. 8.36). × 12,000.

8.38. Later cleavage. Tubules (*t*) of the phycoplast are visible. The massive rhizoplast (*rh*), always conspicuous in interphase cells, is re-forming and connecting the basal bodies (*bb*) to the nucleus (*n*). × 22,000.

SOURCE. Drs. K.D. Stewart, K.R. Mattox, and C.D. Chandler, published in *J. Phycol. 10.* 65 (1974).

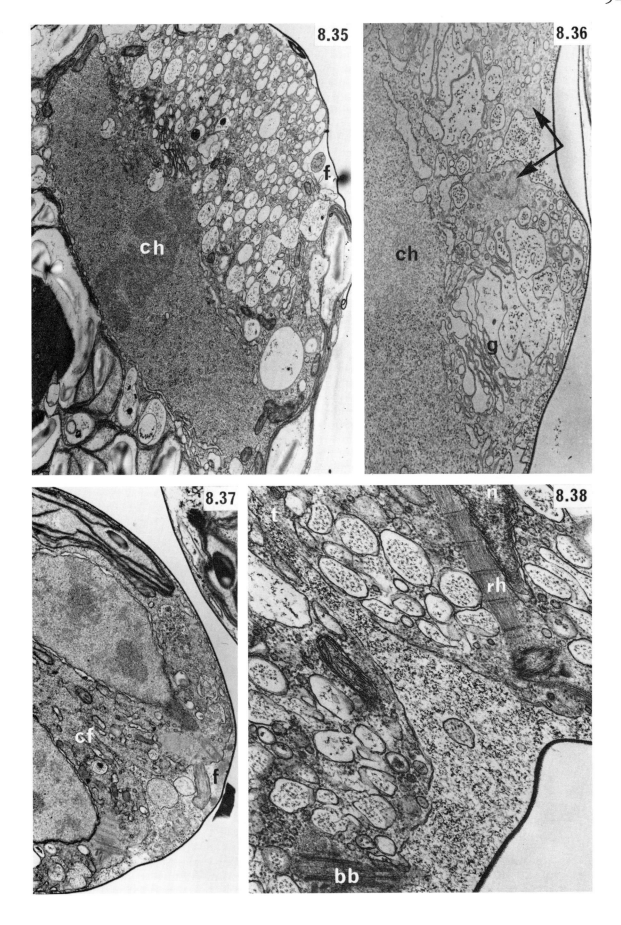

Figure 8.39
PEDINOMONAS MINOR

These tiny cells have their single long flagellum inserted into a slight groove. Some flagella are covered with fine hairs. × 2,000.

inversion since the constitutive cells are all oriented parallel to the long axis of the colony. So far as I am aware, no ultrastructural work has been done on this family.

8.3b. The Siphonous Lines of Evolution

The siphonous lines of evolution (Fig. 8.20) are distinguished by the tendency toward formation of large coenocytic cells, in contrast to the ulotrichalean lines (the tetrasporine lines in some schemes; e.g., Scagel et al., 1965), in which the cells remain relatively small and predominantly uninucleate. In other words, siphonous organisms have partly or completely lost the capacity for vegetative division (Smith, 1950). It seems a reasonable assumption that some of these siphonous lines began from a type of unicell exemplified by *Chlorococcum / Chlorella*, "with suppression of motility [occurring] again and again along different evolutionary lines" (Fritsch, 1935; p. 147); motility often reappears briefly during reproduction. We already have evidence that *Chlorella* itself is a highly plastic and morphologically variable organism that has perhaps only recently lost the ability to form flagellate cells (Section 3.2). The few ultrastructural observations we have demonstrate that species of *Chlorella* both have and lack centrioles; other strains may possess the perinuclear envelope characteristic of many more advanced Chlorococcales. Furthermore, one species of *Chlorococcum* examined (Section 3.1) has no centrioles in vegetative cells, yet this genus can be motile briefly after cell division

Figures 8.40–8.42
CELL DIVISION IN PEDINOMONAS MINOR

8.40. Metaphase. The spindle is entirely closed (i.e., devoid of polar fenestrae), with a flagellum and additional nonfunctional basal body *(bb)* at each pole. × 25,000.

8.41. Cell undergoing cleavage, showing a single flagellum at each pole. × 8,400.

8.42. Cleavage. The daughter nuclei *(n)* are far apart as the cleavage furrow *(cf)* passes between them. × 23,000.

SOURCE. Drs. J.D. Pickett-Heaps and D. Ott, published in *Cytobios* (in press).

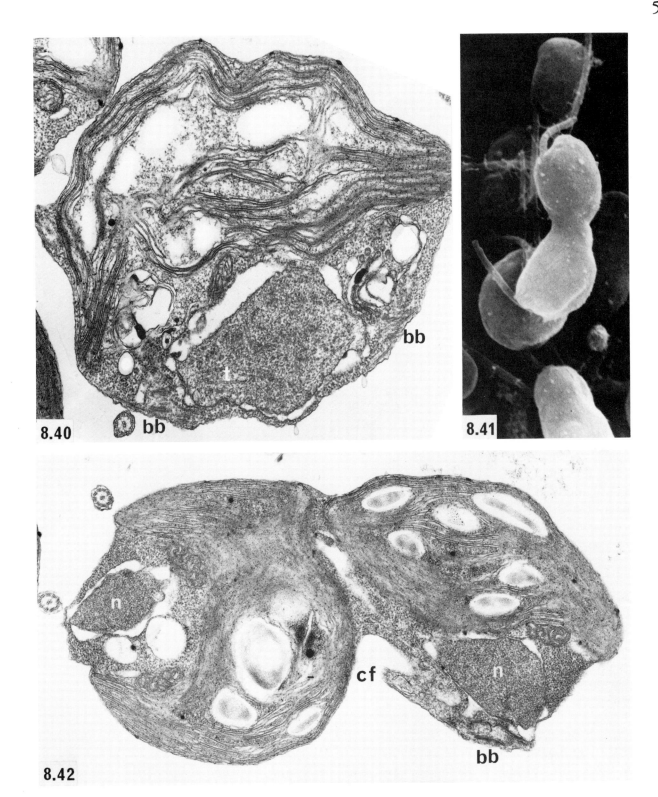

8.40

8.41

8.42

(Fritsch, 1935; p. 145), again demonstrating that the loss of the centriole from vegetative cells does not always result in loss of motility (paragraph 10 above) and may indicate evolutionary advancement. *Chlorococcum* has retained the option of forming zoospores or autospores, apparently in contrast to *Chlorella.*

The chlorococcalean branch of the siphonous lines of evolution (Fig. 8.20) shows some consistently characteristic features among a heterogeneous collection of organisms. Again, it is emphasized that the organisms named on the diagram may not be directly related and are used only as examples. Many different branches must have occurred in the siphonous lines, and the Chlorococcales are almost certainly polyphyletic (Fritsch, 1935; p. 147). There is far too little information available to allow finer gradations in phylogeny using the limited morphological criteria I have chosen. However, certain features common to many of the organisms in this group support the contention that they are indeed related to one another. As evolution progressed in this line, there has been a tendency for cytokinesis to become increasingly divorced from mitosis, giving rise to the coenocytic condition. In *Chlorella,* cleavage appears complete after each of a set of mitoses, as in *Chlamydomonas.* In other small cells (e.g., *Kirchneriella*), cleavage does follow each cycle of mitosis, but the first cleavage is only partial. As the cells become larger, cytokinesis does not follow mitosis; instead it tends to be delayed and then occur simultaneously throughout the multinucleate cytoplasm. In some organisms, motility has apparently been totally suppressed (e.g., *Chlorella, Kirchneriella*) and autospores are formed; centrioles, however, may still be present. In other genera, zoospores or gametes appear only under exceptional circumstances while autospores constitute the normal reproductive cells (e.g., *Tetraedron, Scenedesmus*). In the most advanced members, motile and sessile forms may alternate, and the organisms have more complex variations in behavior during sexual reproduction (e.g., polyhedra formation in *Pediastrum* and *Hydrodictyon*).

All members of the Chlorococcales so far investigated use phycoplasts for cytokinesis, as do the Volvocales. During evolutionary advancement, the phycoplast appears to become an increasingly autonomous structure. In simpler cells, it always arises between daughter nuclei. In

Figures 8.43–8.45
PYRAMIMONAS SP.

8.43. Interphase cell with its four lobes. The four flagella are inserted in a deep depression. × 3,900.

8.44. Cell with eight flagella, presumed to be about to undergo division. × 2,500.

8.45. Late division stage. × 3,000.

SOURCE. Author's micrographs from material supplied by Dr. R.E. Norris.

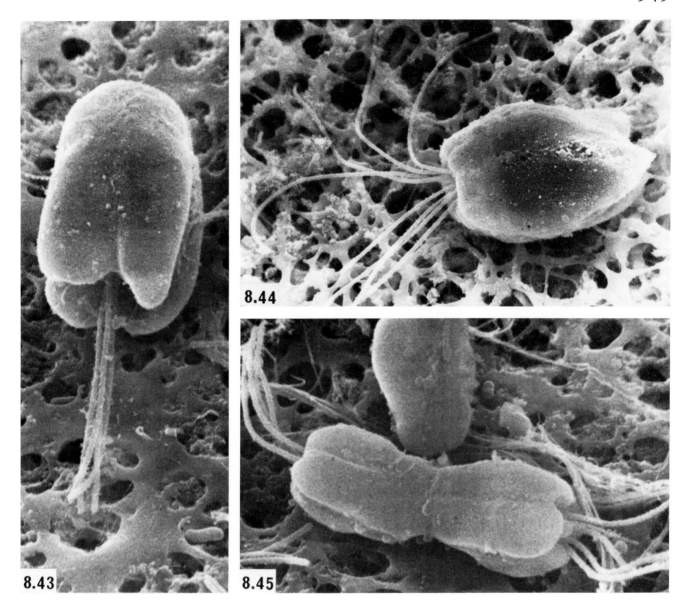

8.43

8.44

8.45

cells producing a number of autospores or motile zooids at one time, the phycoplast now appears between all nuclei. Finally, in *Hydrodictyon* (Section 3.7), it appears under the tonoplast before cytoplasmic cleavage, enabling the vacuolar envelope to be cleaved off first. An analogy can be made with the phragmoplast in certain higher plants cells, notably *Haemanthus* endosperm (Bajer, personal communication), in which phragmoplasts in a syncytium form simultaneously between daughter and nondaughter nuclei.

Almost all chlorococcalean genera examined have closed, centric spindles; the two exceptions so far discovered have closed, acentric spindles (*Chlorella* and possibly *Chlorococcum*), and the existence of a centric form of *Chlorella* demonstrates that this discrepancy is of little significance. Finally, many members of this group possess the distinctive perinuclear envelope, and it is especially informative that *Chlorella,* representative perhaps of the ancestral chlorococcalean cells that gave rise to more advanced forms, can divide either with or without this membrane around its nucleus.

As for the other two general lines of evolution (Fig. 8.20), represented by the Siphonales (very large, coenocytic, and non- or rarely septate cells) and the Cladophorales (large, coenocytic cells), there is virtually no extensive ultrastructural work that gives us any indication of their affinities. It will be especially interesting to discover what sort of spindle apparatus the various genera of Siphonales and Cladophorales possess, and even more important, what microtubular systems, if any, are involved in cytoplasmic cleavage during zoospore formation. The presence of phycoplasts (cf. *Hydrodictyon*) might be expected if these groups do have any affinity with the Chlorococcales.

8.3c. The Ulotrichalean Lines of Evolution

Next we come to the ulotrichalean lines of evolution (Fig. 8.21). Many people have used the nomenclature "tetrasporine lines" in this context, and they often suggest that these uninucleate, "colonial" cells were derived from colonial Tetrasporales (e.g., Klein and Cronquist, 1967; p. 232). Other authorities (e.g., Smith, 1950) use the term "tetrasporine" to describe only the evolutionary tendency of organisms to become increasingly immobilized while

Figure 8.46
INTERPHASE CELL OF PYRAMIMONAS PARKEAE

The cell and its flagella *(f)* are covered by beautifully intricate scales. Tiny, diamond-shaped scales in close-packed array lie next to the plasmalemma of the entire cell. The flagella are additionally covered by larger, somewhat circular scales with a central spike, whereas the body is covered also with square, basket-shaped scales. These scales are formed within vesicles (arrows) derived from the massive Golgi bodies *(g)*. A large rhizoplast *(rb)* and an associated dense organelle (peroxisome?) lie between the nucleus *(n)* and basal bodies *(bb),* which are themselves interconnected by fibrous bands. × 19,000.

SOURCE. Drs. R.E. Norris and B.R. Pearson, paper in preparation.

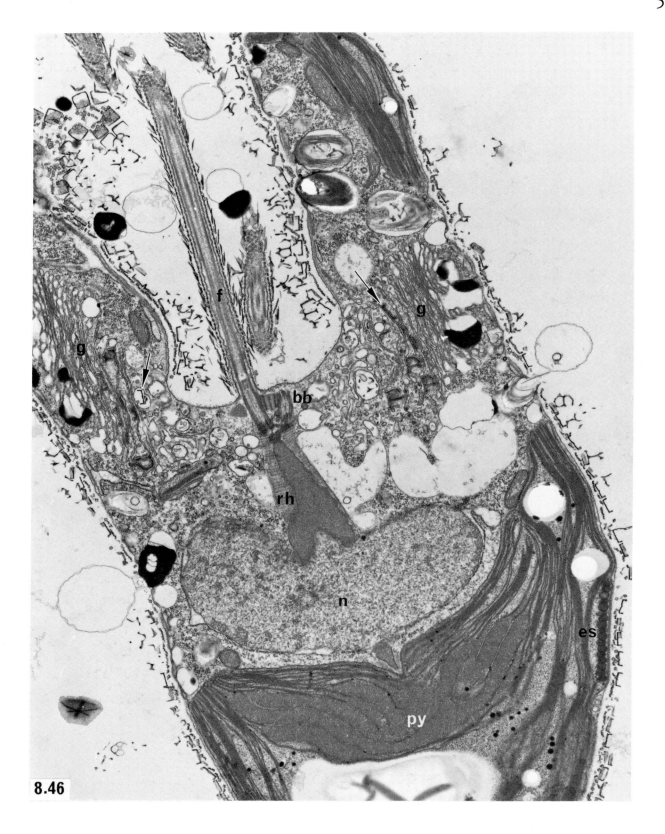

8.46

retaining the ability to divide vegetatively (as exemplified by *Tetraspora* itself). Some authors refute the possibility that the Tetrasporales gave rise to the Ulotrichales on various grounds, and so I have adopted this terminology which does not prejudice the issue regarding the affinities of these genera.

A distinctive feature of the ulotrichalean lines is their maintenance of the uninucleate condition in almost all circumstances. However, considerable morphological diversification has taken place within these organisms which, it is generally believed, have provided the progenitors of the higher plants. Ulotrichalean evolution is often illustrated with organisms such as *Ulothrix, Stigeoclonium,* and *Fritschiella,* and it is easy to follow the rationale. *Ulothrix* is representative of primitive filamentous Green Algae (Fritsch, 1935), and little imagination is needed to visualize how branched and thallus-like structures can arise when the cells in a filament become able to divide in more than one plane. The morphology of interphase cells of *Ulothrix* and *Stigeoclonium* is similar, and division is almost identical except that in the former the nuclear envelope is more vesiculated and dispersed than in the latter (Chapter 4). These algae use the phycoplast/cell plate system of cytokinesis, as apparently does *Fritschiella* (Section 4.8a). What little information we have, therefore, is consistent with the notion that such organisms have reasonably close affinities with one another.

Classic taxonomy places *Klebsormidium* (and *Stichococcus*) close to these typical ulotrichalean algae. However, the microanatomy of the former two genera suggests that such an affinity is at best distant. *Klebsormidium* and *Stichococcus* have an open spindle and no phycoplast; daughter nuclei remain far apart at telophase, and the cell divides by cleavage. These and other differences (summarized later) oblige me to place such genera on a quite separate evolutionary line from *Ulothrix* et al. There are important implications if such a separation is valid; before they are discussed, it is necessary to consider an important problem concerning the evolution of higher plant cells.

Throughout this book and in several papers (Pickett-Heaps, 1969a, 1972a), I have stressed the difference between the phycoplast and the phragmoplast, the two microtubular systems involved in cytokinesis in green plant cells. It is important to summarize now the reasons for my

Figure 8.47
TRANSVERSELY SECTIONED CELL OF PYRAMIMONAS PARKEAE

This cell is at metaphase. The spindle is open. Note the two groups of four flagella *(f)* whose basal bodies are associated with the spindle poles. Golgi bodies *(g)* lie in the lobes (cf. Fig. 8.46). Scales and also mastigonemes (arrows) are visible within their vesicles. × 19,000.

SOURCE. Drs. R.E. Norris and B.R. Pearson, paper in preparation.

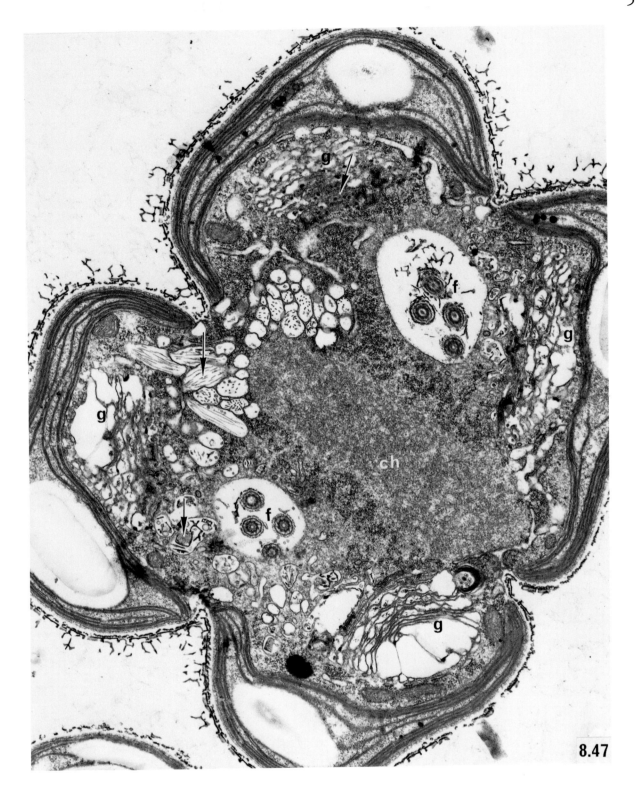

8.47

concern with these systems so that reader can judge for himself whether the cytoplasmic microtubular system involved in cytokinesis has any value as a phylogenetic indicator.

First, what is the function of the microtubules in the phycoplast? I believe they are involved in guiding the cleavage furrow or cell plate through the cytoplasm in such a way as to reliably partition the daughter nuclei, morphogenetic centers, and other organelles into daughter cells (see discussion in Section 2.1a). These tubules need not necessarily be involved in the mobility of the furrow (although they could be), since ingrowing cleavage furrows bisect many types of cells without any need for microtubules nearby (e.g., in the Conjugales). Thus, we might expect to find phycoplasts in those Green Algae in which cleavage has to be carefully controlled spatially. One such situation occurs when the spindle collapses after telophase and the daughter nuclei approach one another, apparently the usual condition in many Green Algae. Why the nuclei should come back together in such a manner is not obvious, but this movement could originally have been consequent upon (1) the collapse of the interzonal spindle before cleavage is well under way, and (2) the possibly elastic nature of the cytoplasm of small spherical or ovoid unicells lacking rigid cell walls which pushes the nuclei back together once the rigid spindle has disappeared. However, in *Oedogonium*, for example, the movement of the nuclei back together appears a precisely directed and rapid phenomenon taking place within a rigid cell wall, so here, cytoplasmic elasticity is unlikely to bring the nuclei together. Whatever the cause of this post telophase movement, if cytokinesis is not carefully directed between the nuclei, then binucleate and enucleate cells could easily result. Another situation requiring precise control of cleavage occurs during the formation of uninucleate cells from the cytoplasm of coenocytes (e.g., zoosporogenesis in *Hydrodictyon*, during which a separate, carefully controlled cleavage also gives rise to the vacuolar envelope). There are multinucleate cells, of course, which can cleave into uninucleate fragments without phycoplasts (e.g., during spore formation in fungi). That the phycoplast is indeed important during cytoplasmic cleavage in *Pediastrum* and *Hydrodictyon* is indicated by the observed effect of colchicine on zoosporogenesis; cleavage apparently

Figure 8.48
CYTOKINESIS IN
PYRAMIMONAS PARKEAE

The cleavage furrow (arrows) is growing between widely separated daughter nuclei *(n)*. There is no sign of a phycoplast. The nuclear envelopes are re-forming, although some spindle microtubules *(t)* are still present. × 18,000.

SOURCE. Drs. R.E. Norris and B.R. Pearson, paper in preparation.

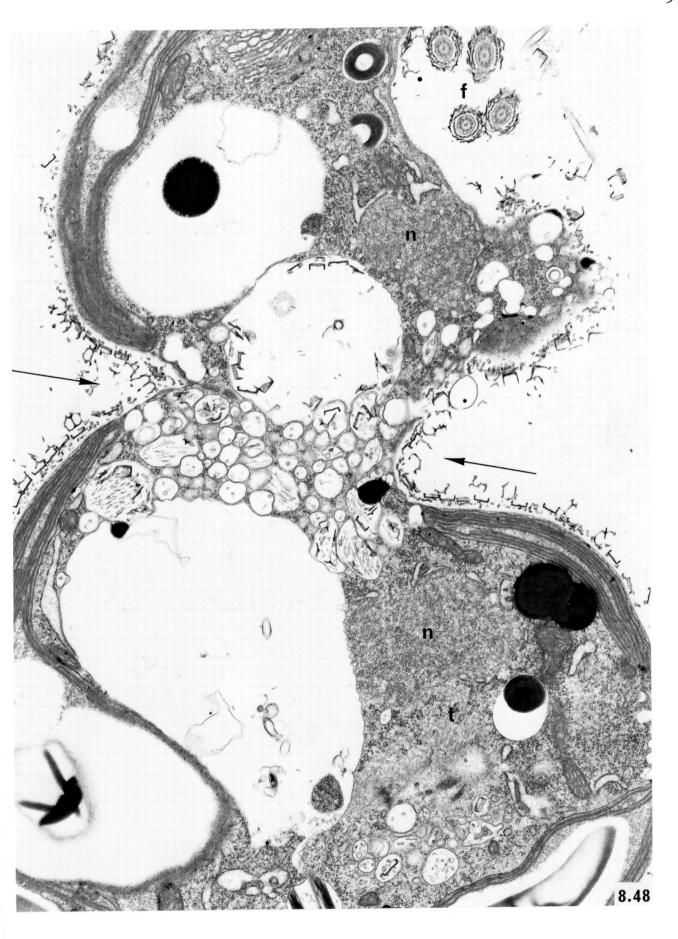

8.48

continues to some extent in the presence of this drug but entirely randomly, and in *Pediastrum,* a mass of undifferentiated cytoplasm is eventually released from treated cells into the swarmer vesicle (Section 3.8; Fig. 3.146).

Hence the phycoplast appears to be a primitive structure present in many Green Algae, whose genesis was perhaps necessitated by collapse of the interzonal spindle apparatus. In other words, because the daughter nuclei approach one another before cytokinesis is well under way, an additional structure may be vital in ensuring that cell cleavage takes place only between them.

We should now also consider how some Green Algae manage cytokinesis without phycoplasts. Again, we have to extrapolate rather dangerously on meager evidence, but I think this is justifiable if it suggests testable hypotheses. One type of cell that obviously does not need a phycoplast is a coenocytic cell undergoing simple vegetative cleavage (i.e., septation). Once nuclei become numerous in a cell, they need not be segregated precisely equally between daughter cells at cytokinesis. Thus we can guess that phycoplasts are unlikely to feature during septation of vegetative cells in the Cladophorales.

Reliable separation of daughter nuclei can be achieved in the absence of a phycoplast, even when the nuclei approach one another; if cleavage is well developed by anaphase so that the interzonal spindle is cut by this furrow, then the daughter nuclei will remain on either side of the ingrowing cross wall. This appears to happen in *Ulva* (Section 4.7a) and has recently been confirmed to take place in *Trichosarcina* and *Pseudendoclonium* (Stewart and Mattox, personal communication). The spindle in these cells is quite distinctive, the wide cleavage furrow being initiated even before prophase (Figs. 8.22, 8.25) and becoming pronounced by metaphase (Figs. 8.23, 8.26). The poles of the closed spindle are broad, and the centrioles are situated to one side of the polar fenestrae (Fig. 8.27). The cleavage furrow seems to cut the interzonal spindle at late anaphase (Fig. 8.24), and some microtubules are oriented in the plane of this furrow. However, I believe that these cells possess neither the true persistent spindle nor the phycoplast. Our own work on the interesting oogamous genus *Cylindrocapsa* (Fig. 8.28; Pickett-Heaps and McDonald, in preparation) has revealed a mitotic and cytokinetic apparatus entirely equivalent to that

Figures 8.49–8.52
FOSSIL CHAROPHYTA

8.49 *A, B.* Lateral and apical views of the gyrogonite (fossilized oogonium, minus its sterile sheath and coronal cells) of *Tectochara dutemplei* (Watelet) Grambast (in the family Characeae); Lower Eocene. × 50.

8.50. Utricle of *Perimneste micrandra* Grambast (family Clavatoraceae); Hauterivian. The utricle is a supplementary envelope of cells surrounding the oogonium in primitive Charophyta. The family Clavatoraceae ranges from the Upper Jurassic to the Uppermost Cretaceous. Fossilized antheridia (arrows) are clearly discernible in this utricle. × 60.

8.51. The rather more elaborate utricle of *Septorella ultima* Grambast (family Clavatoraceae); Maestrichtien (Uppermost Cretaceous). × 50.

8.52 *A–C.* Lateral, apical, and sectional views of the utricle of *Embergerella cruciata* Grambast (family Clavatoraceae); Upper Barremian. The longitudinal section *(c)* of this utricle reveals the enclosed gyrogonite. × 50.

SOURCE. Dr. L.J. Grambast, unpublished micrographs.

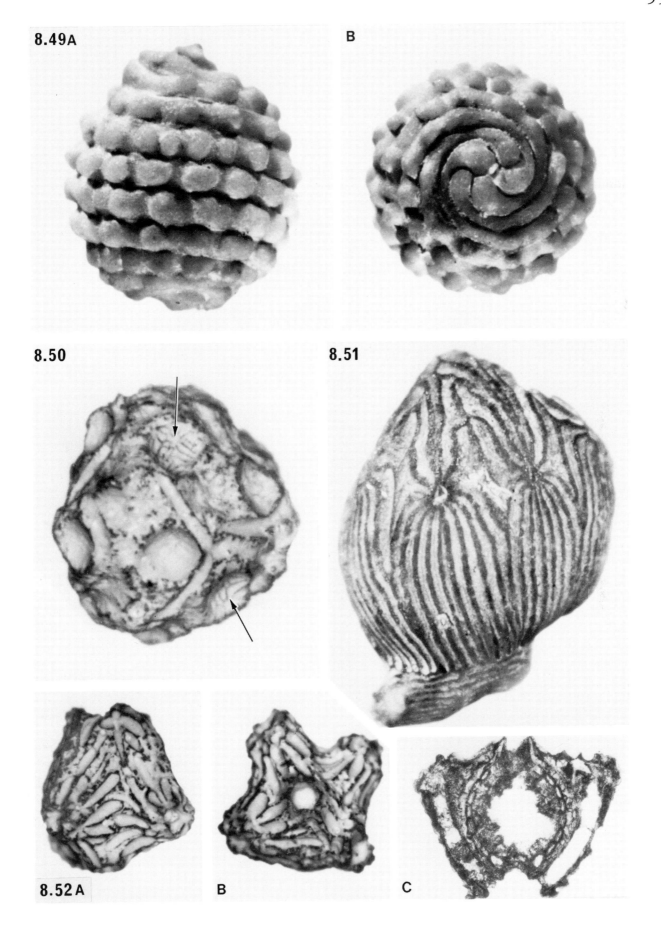

8.49A

B

8.50

8.51

8.52 A

B

C

of *Trichosarcina* and *Pseudendoclonium* (Figs. 8.29, 8.30). I suspect that this distinctive form of cell division may serve to define yet another natural group of organisms in the Ulotrichales. The justification for this step may seem meager, but I find the consistency of cell division systems impressive, and so I have indicated a rather equivocal position for such a group, loosely defined now as the "Ulvaceae" (Mattox and Stewart, personal communication), in Figure 8.21. Mattox and Stewart (1973b) suspect that *Ctenoclodus circinnatus* and *Ulothrix zonata* are also likely candidates for inclusion in this group.

Another type of cell not needing a phycoplast would be one in which the two daughter nuclei are held far apart after mitosis, usually by a persistent interzonal spindle; an ingrowing cleavage furrow is then virtually certain to segregate daughter nuclei if it starts growing at the midpoint of an elongating anaphase spindle that remains elongated at telophase (Pickett-Heaps, 1974f). This situation exists in those members of the Conjugales studied so far; no phycoplast is present, and the nuclei do not approach one another after division, at least not until the furrow is practically complete (e.g., *Closterium*). Within this order is an extant species (*Spirogyra*) that gives us a clue to how the phragmoplast could have arisen. The example of *Spirogyra* (Section 6.2a), furthermore, strongly suggests to me that persistence of the interzonal spindle apparatus was a prerequisite for evolution of the phragmoplast. I believe that these matters must be taken into account when trying to guess from what sort of algae the higher plants arose. *Spirogyra* also shows us that the phragmoplast could easily have evolved in relative isolation (the closely related *Mougeotia* does not appear to have this rudimentary phragmoplast). We cannot yet tell whether the acquisition of the phragmoplast/cell plate system confers upon its possessor some appreciable advantage over its competitors or whether this cytokinetic system assists in the formation of complex plant forms (Pickett-Heaps, 1972a).

I find it difficult to believe that many of the more complex ulotrichalean algae which have phycoplasts could be derived from the phyletic lines that gave rise to higher plants as postulated by numerous authors (see the latter part of this chapter). This so far represents very much a minority opinion. I do, however, see one interesting possibility. Let us now reconsider *Klebsor-*

Figure 8.53
AUTOSPORE FORMATION

Diagrammatic representation. After mitosis, the cytoplasm cleaves into daughter cells which secrete their own cell wall *(A)*. The parental wall later falls apart to release the growing autospores *(B)*.

SOURCE. Author's diagram, published in *New Phytol. 72,* 347 (1973).

Figure 8.54
POSSIBLE INTERMEDIATE STAGE IN EVOLUTION OF FILAMENTOUS FORM BY AUTOSPORE-FORMING CELLS.

Diagrammatic representation. (cf. Fig. 8.53). The cells are loosely held together (e.g., by mucilage). After nuclear and cell division *(A)* as in Figure 8.53, the cells remain held together. The parental wall ruptures in a rather carefully predetermined manner (i.e., equatorially) and during growth of the daughter cells *(B)*, this wall remains held around the ends of the daughter cells.

SOURCE. Author's diagram, published in *New Phytol. 72,* 347 (1973).

Figure 8.55
ANOTHER POSSIBLE INTERMEDIATE STAGE IN EVOLUTION OF FILAMENTOUS FORM

Diagrammatic representation (cf. Figs. 8.53, 8.54). Two, not necessarily concurrent, modifications to wall secretion have been made. First, during cytokinesis *(A, C),* a *single integral* wall is laid down between daughter cells, instead of two separate autospore walls (perhaps one consequence of replacement of cleavage by cell plate formation). Second, the daughter cells do not secrete end walls which always remain enclosed by parental wall. Cell expansion (i.e., elongation; *B, D)* otherwise proceeds as in Figures 8.53 and 8.54. The wall now consists of interlocked H-shaped segments.

8.53

8.54

8.55

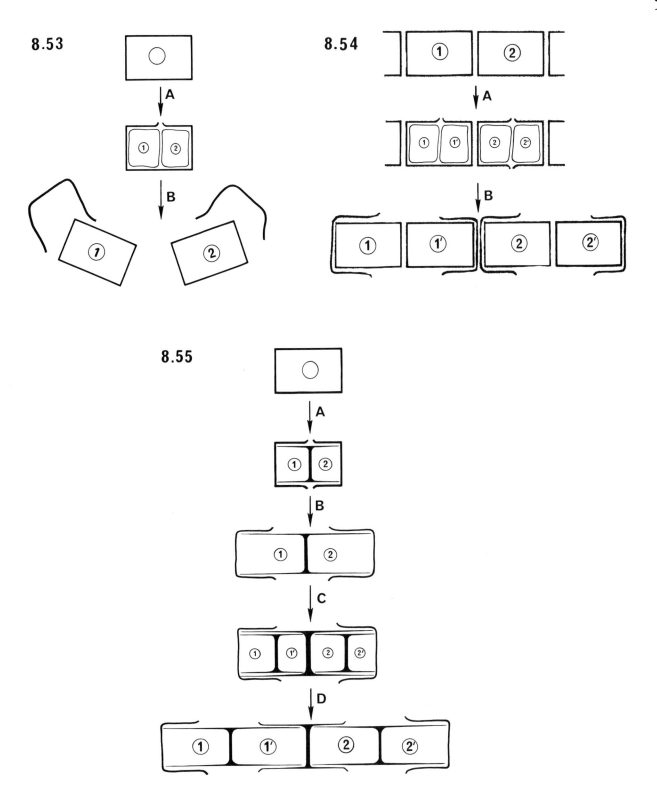

midium. It possesses a persistent interzonal spindle, and cleavage is achieved by membrane furrowing. *Spirogyra* tells us that this may be the sort of cell that could evolve a phragmoplast. We now have discovered an advanced member of the Ulotrichales, *Coleochaete,* which has a true phragmoplast. So the intriguing possibility now exists that there are at least two separate ulotrichalean lines of evolution in organisms that have always been lumped together (Fig. 8.21). One line, as previously described, can be represented by phycoplast-containing cells such as *Ulothrix, Stigeoclonium,* and *Fritschiella,* and contains many members of the Chaetophorales as delineated by Fritsch (1935). A separate line contains *Stichococcus, Raphidonema, Klebaormidium,* and *Coleochaete* (and the Charales). The latter line in my opinion is the one that contained the progenitors of higher plants. Thus a key organism in the evolution of the higher plants could conceivably have been a simple unicellular form with a persistent telophase spindle; this organism was perhaps the real progenitor of all the other Green Algae (Section 8.3d), and it may have evolved directly into filamentous forms like *Klebsormidium* and *Stichococcus.* Then these filamentous forms, having evolved the phragmoplast at some stage, continued to evolve morphologically in a manner similar to that displayed by extant phycoplast-containing members of the Ulotrichales, but separate from them. The acquisition of the phragmoplast may have enabled these hypothetical organisms to evolve further than the present ulotrichalean algae had been able to do.

Our confidence that this proposed phylogenetic separation is valid has been vastly increased by the structure of the zoospores of *Klebsormidium, Chara,* and *Coleochaete,* all of which are quite different from the zoospores of the phycoplast-containing ulotrichalean algae, being more akin to those of bryophytes (Section 8.2).

The affinities of the Conjugales are rather uncertain, whatever criteria one uses for erecting phylogenies. If we compare various taxonomic features in the mitotic and cytokinetic systems of *Klebsormidium* with those of the Conjugales and certain Ulotrichales (Table 8.1; Pickett-Heaps, 1972h), *Klebsormidium* (and *Stichococcus*) has a closer affinity with the former than the latter, even though there are two distinctly different suborders, the Euconjugatae and the Desmidioideae, in the Conjugales, which is usually re-

Figure 8.56
WALL STRUCTURE AND METHOD OF CELL ELONGATION IN MICROSPORA

Diagrammatic representation of a possible development of the situation outlined in Figure 8.55. The interphase cell is enclosed by two interlocked H-shaped wall segments (cf. Fig. 4.51). During elongation *(A,B),* these segments move apart as a *separate* cylinder of new wall is laid down around the cell (Fig. 4.52). Following the next cell division *(C, D),* the new cross wall is integrated into this cylinder, thereby converting it into another H-shaped segment.

Figure 8.57
WALL STRUCTURE AND METHOD OF CELL ELONGATION IN OEDOGONIUM

Diagrammatic representation of events when components are in the form of H-shaped segments. The interphase segments are asymmetrical and possess a clearly defined discontinuity with one another, giving rise to the rupture site (cf. Fig. 8.54). The ring (wound response) is situated at the wall weakness before and during division *(A).* Wall rupture and ring expansion *(B)* are followed by the secretion of a complete new H-shaped segment inside the material of the ring. The new cross wall, it will be noted, forms part of this new segment, just as it does in *Microspora* (Fig. 8.56D), although the method used for cell elongation is totally different.

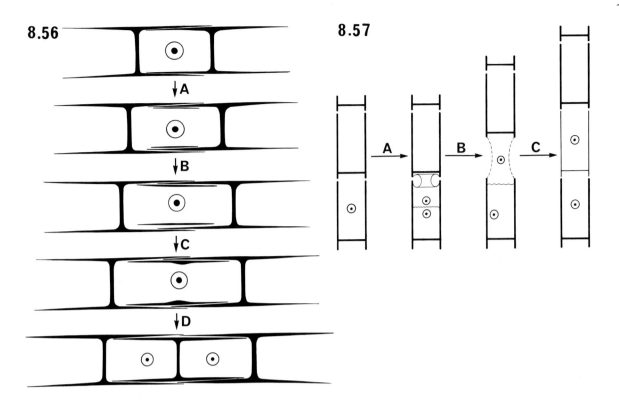

8.56

8.57

Table 8. 1.

A COMPARISON OF KLEBSORMIDIUM WITH OTHER MEMBERS OF THE ULOTRICHALES AND CONJUGALES

Characteristic	Conjugales[a]	Klebsormidium	Other Ulotrichales[b]
Spindle type	Open	Open	Closed (or semiclosed)
Centriole present	No	Yes	Yes
Pole-to-chromosome distance constant during anaphase	Probably in *Spirogyra*	Yes	Yes
Persistence of microtubules near nucleus after telophase	Yes	Yes	No
Persistence of the interzonal microtubules for some while after telophase	yes, particularly in *Spirogyra*	Yes	No
Membranes coating mitotic chromosomes	Sometimes	Sometimes	No
Spindle collapse at telophase	No	No	Yes
Phycoplast	No	No	Yes
Cleavage	Yes	Yes	No[d]

[a] Based on *Spirogyra, Closterium, Cosmarium, Micrasterias,* and some observations on *Netrium.*

[b] Based on *Ulothrix, Stigeoclonium, Microspora.*

[c] *Stichococcus* closely resembles *Klebsormidium,* but the former is acentric.

[d] Some may cleave, especially during zoosporogenesis and gametogenesis; this remains unresolved at present.

garded as being polypheletic (Chapter 6). The notable discrepancy between *Klebsormidium* and the Conjugales is the existence of centrioles in the former. However, this may not be highly significant since *Stichococcus* and *Raphidonema* are acentric, and the loss of centrioles in the Conjugales probably has accompanied their evolutionary advancement (paragraph 11, earlier) as the organisms have evolved methods of reproduction not requiring flagellate stages. I feel that *Klebsormidium* and the Conjugales are more closely related than normally suspected (Pickett-Heaps, 1972h). An affinity between the Ulotrichales and the Conjugales has been proposed on rather ill-defined grounds by several authors (e.g., Chapman, 1962).

Two other groups awkward to place are the Oedogoniales and perhaps the Charales. Since the Oedogoniales possess a phycoplast, an affinity with one main ulotrichalean line of evolution is reasonable, but the complexity of cell division—the zoospore, sexual reproduction, etc.—in the Oedogoniales suggests a divergence at an early stage. The Charales are even more difficult to place. They probably diverged from the other Green Algae a long time ago or else perhaps more recently after the second line of ulotrichalean evolution (*Klebsormidium, Coleochaete*) had developed a phragmoplast. On the basis of micromorphology alone, a primitive relative of this class could be considered as a possible progenitor of higher plants (Pickett-Heaps, 1972a). Most phycologists have discarded this proposition on a number of other grounds which I am not really qualified to examine, but some reassessment of this possibility is now timely.

This discussion leads me to the more radical and novel phylogenetic summary given in Figure 8.31, which includes the Volvocales, Chlorococcales, and other groups into the one scheme. As mentioned earlier, a satisfying way (at least to me) to split up the Green Algae is simply to divide the phycoplast-containing cells from those that do not possess this structure. This approach solves several problems, and it has already enabled us to test predictions—in particular, that the zoospore of *Klebsormidium* might be of considerable phylogenetic interest. Since the persistence of the interzonal spindle is so widespread in all classes of organisms (including fungi and animal cells), it is probably a very primitive feature. If so, it would appear that the progenitors of

Figure 8.58
POSSIBLE INTERMEDIATE STAGE IN EVOLUTION OF FILAMENT IN THE OEDOGONIALES.

Diagrammatic representation. This diagram is similar to Figure 8.54. A primitive autospore-forming cell divides *(A, B)*, and the daughter cells are loosely held together by mucilage. As for Figure 8.54, the parental wall must rupture in a carefully controlled fashion (or else a filament could not form), and segments of parental walls accumulate between cells. However, this particular cell type is presumed to have acquired polarity, one result of which is the movement of the wall's rupture site from the center of the cell (as in Fig. 8.54) toward one end.

Figure 8.59
UNUSUAL WALL STRUCTURE CHARACTERISTIC OF THE OEDOGONIALES.

Diagrammatic representation of a logical evolutionary development of the ancestral filament depicted in Figure 8.58. As in a diagrammatic sequence outlined previously (Figs. 8.53–8.55), the wall segments are presumed to have evolved into H-shaped segments; however, these segments are asymmetrical (as in Fig. 8.57), since the original siting of the rupture site was asymmetrical (Fig. 8.58). A consequence of such asymmetry is that the segments cannot overlap in a structurally sound manner (Cf. Figs. 8.55, 8.56), and thus the wall rupture site continues to figure significantly in the structural integrity of the filament (necessitating the development of an efficient wound-response mechanism). The new wall layers formed after each division are cylindrical, tapering in thickness where they are enclosed within the parental wall (exactly as in Fig. 8.55, 8.56), but again because of the asymmetry of the wall, the layers enclosing daughter cells differ markedly in thickness. The apical daughter cell secretes a thick wall, whereas the basal daughter cell forms a much thinner, tapered wall inside the older wall layers. The filament here is shown arising from a germling *(A)*, to emphasize the polarity of the cell and ultimately the filament *(B)*. The wall of *Oedogonium* and *Bulbochaete* is demonstrably layered in precisely this manner. For examples see text and the illustrations in Chapter 5.

8.58

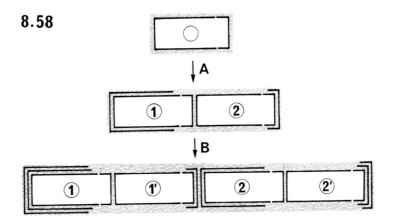

8.59

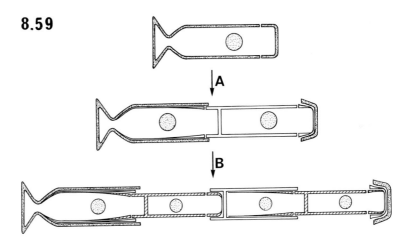

some extant phycoplast-containing Green Algae (including *Chlamydomonas*-like forms) arose from a more primitive stock, perhaps rather suddenly following some mutation or series of mutations affecting the organization of the persistent telophase spindle. The archetypal cell type thus arising acquired the phycoplast and then diversified in the usual different directions, becoming filamentous, coenocytic, etc., and thereby giving rise to the Volvocales, Tetrasporales, Chlorococcales, many members of the Ulotrichales, and the Oedogoniales. The direct line from the more primitive ancestor with a persistent spindle gave rise to primitive filamentous forms of Green Algae such as *Klebsormidium* and to more highly evolved groups such as the Conjugales, the Coleochaetaceae, the Charales, and eventually perhaps, the bryophytes. In these lines, the phragmoplast may have evolved several times in isolation (as is probably happening now in *Spirogyra*). The position in Figure 8.31 assigned to such groups as the Siphonales, the Siphonocladales, and the Dasycladales cannot be specified, and they have been most tentatively included on the diagram for interest's sake and hopefully to stimulate relevant critical and detailed ultrastructural work. The information we possess about them at present is too limited to use more confidently.

Evidence to support this scheme has been summarized elsewhere (Pickett-Heaps and Marchant, 1972) and has been already discussed in part above. We should also note that (1) *Klebsormidium* and *Coleochaete* produce only a single, biflagellate zoospore per cell, lacking an eyespot; in contrast, other ulotrichalean algae form 1–16 zoospores per cell, which are bi- or quadriflagellate and which have eyespots; (2) the phycoplast-containing ulotrichalean algae so far described undergo isogamous or anisogamous sexual reproduction (as apparently also does *Klebsormidium*); the complex oogamous reproduction of *Coleochaete* and the Charales has no counterpart yet in phycoplast-containing genera, and it is distinctly reminiscent of sexual reproduction in the bryophytes (nowhere else in the Green Algae are the oogonia protected by sterile cells); (3) all phycoplast-containing algae have closed or essentially closed spindles, whereas those genera I have placed in association with the phyletic line leading to the higher plants have open spindles.

Figures 8.60, 8.61
CELL ELONGATION IN GERMLINGS OF THE XANTHOPHYTE TRIBONEMA AEQUALE

8.60A. Detail of the wall of the germling in Figure 8.60B, showing the dislocation in it. × 24,000.

8.60B. Young germling. Its wall possesses a circular discontinuity and is composed of two overlapping segments. The region indicated by the arrow is shown in Figure 8.60A. × 8,800.

8.61A. A germling similar to that in Figure 8.60B, after having undergone some cell elongation. The two overlapping segments enclosing the germling originally have moved apart as a girdle of new wall is intercalated between them. The region indicated by the arrow is shown in more detail in Figure 8.61B. × 10,800.

8.61B. Detail of Figure 8.61A, showing the girdle of new wall between the segments of the older wall. × 24,000.

SOURCE. Dr. D. Ott, unpublished micrographs.

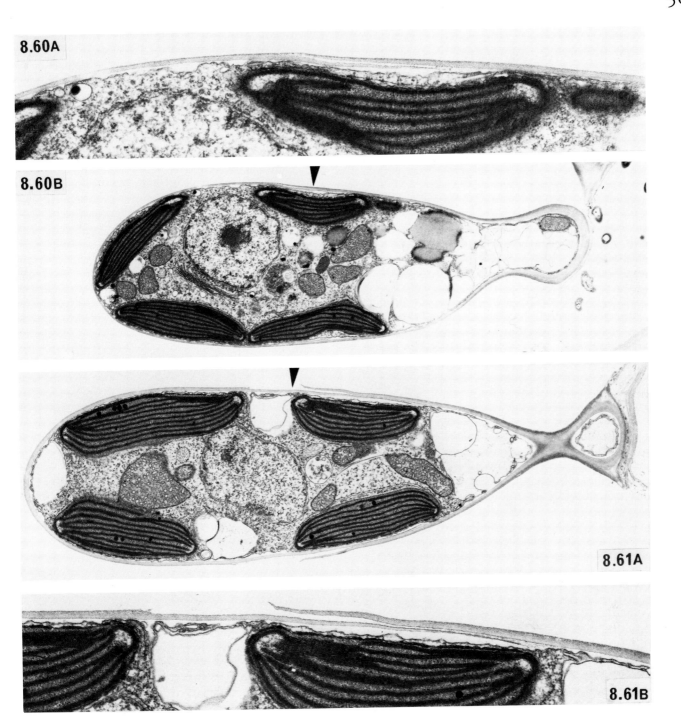

Why this should be so is unclear, but the dichotomy is at least provocative.

The taxonomic importance of scales on zoospores and gametes now seems unclear. Scales, particularly the small, diamond-shaped variety, are a common feature of the Prasinophyceae (Figs. 8.46–8.49) and characterize the motile cells of *Coleochaete* and the Charales. Some prasinophytes also have phycoplasts (*Platymonas*, Figs. 8.37, 8.38; Stewart, Mattox, and Chandler, 1974). Mattox and Stewart (1973a) report scales on zoospores and autospores of *Pseudendoclonium* and *Trichosarcina* (Figs. 8.32–8.34). Perhaps we should await more information before assigning any specific taxonomic significance to these features.

I can mention one new and strong piece of evidence that supports our scheme, and which was obtained quite independently. Frederick, Gruber, and Tolbert (1973), investigating peroxisomal systems in green plants, discovered that two functionally analogous enzymes could be distinguished in different genera. Many Green Algae, including *Chlorella*, *Dunaliella* (Volvocales), *Microspora*, and *Stigeoclonium*, contain the enzyme glycolate dehydrogenase. The other enzyme, glycolate oxidase, characteristic of the leaves of higher plants, was detected in a number of liverworts, mosses, *Psilotum*, *Selaginella*, *Equisetum*, and three aquatic angiosperms including *Elodea*. It was also found in *Coleochaete*, *Klebsormidium*, *Nitella*, *Spirogyra*, and *Netrium*! This distribution of the two distinct systems is entirely consistent with the phylogeny given in Figure 8.31, and yet it is inexplicable by any other classic phylogeny of the Green Algae I have seen.

An interesting, somewhat opposing view to this rationale has been presented but not further pursued by Stewart, Mattox, and Floyd (1973). They recognize that a basic difference exists between the phycoplast and the phragmoplast, but debate the possibility that the latter could have evolved from the former, which would support the current consensus regarding the origin of higher plants from phycoplast-containing chaetophoralean progenitors. Their paper is valuable and stimulating, particularly their attempts to reclassify the ulotrichalean algae more logically. However, I find their arguments regarding a possible origin of the phragmoplast from the phycoplast rather unpersuasive.

Stewart, Mattox, and Floyd (1973) suggest that the "phragmoplast may have evolved from the early stages of phycoplast development when the tubules are not yet transverse." As I see it, this suggestion weakens their argument in two ways. First, the immediate implication is that it is the persisting longitudinally oriented microtubules which are important in the evolution of the phragmoplast. Second, these longitudinally oriented microtubules are ephemeral in phycoplast-forming cells, almost certainly representing the remnants of the collapsing spindle. Finally, these authors do not address the most important question of all: which is more ancient, the phycoplast or the persistent spindle? In my opinion, the evidence overwhelmingly supports the notion that the persistent spindle (often in slightly different guises) is by far the more commonly encountered in dividing eucaryotic cells of all types and, therefore, must constitute the more primitive structure (Pickett-Heaps, 1974f). If so, the argument of Stewart, Mattox, and Floyd (1973) obliges us to assume that the Green Algae abandoned the persistent spindle, developed the phycoplast, and then redeveloped a persistent spindle apparatus with many of the characteristics, say, of animal cells and which they probably previously possessed. While not impossible, this seems unlikely. The other alternative is to assume that the phycoplast is the more primitive structure, and then either the Green Algae evolved quite separately from all other organisms which independently came to possess the persistent telophase spindle, or else all other organisms with persistent spindles later evolved from the phragmoplast-containing Green Algae. Other suggestions can, of course, be made, which become increasingly devious and, in my opinion, unlikely.*

8.3d. The Archetypal Green Algal Cell: Chlamydomonad or Prasinophyte?

I can find no objection to the premise that a chlamydomonad-type cell constituted a progenitor of many other Green Algal cells. The

---

*More recently, Stewart and Mattox, (1975) have reviewed the implications of the dichotomy in green algal evolution evident from the difference between the phragmoplast and phycoplast. Their paper is an excellent review for botanists and phycologists interested in new attempts to reclassify the green algae.

question can be extended: is such a cell ancestral to all the other Green Algae? A consensus of phycologists would certainly answer yes (e.g., Scagel et al., 1965; Klein and Cronquist, 1967). My answer, from a comparison of mitotic, cytokinetic, and certain morphological features, is a firm no! The main reason is simple and quite clear cut: *Chlamydomonas* and its relatives all possess the phycoplast. The overwhelming majority of living cells possess persistent telophase spindles in some form or other (Pickett-Heaps, 1974f). Since the phycoplast-containing genera are obviously Green Algae, they surely must be a group derived from other Green Algae.

The question, of course, now arises: what are or were the really primitive cell types ancestral perhaps to all the Green Algae? Our suspicion (Pickett-Heaps and Marchant, 1972) centers upon a group of Green Algae hardly mentioned earlier in this book, called the Prasinophyceae. The exact status and limits of this group have been debated and altered constantly. It contains a rather motley collection of small flagellates whose diminutive size has made classic taxonomy with the light microscope difficult. The varied taxonomic history of this group can be seen from the papers of Chadefaud (1960), Round (1963), Parke and Manton (1965), Ettl and Manton (1964), and Christensen (1966). Cell division in three members of this group has only just been described, and the observations are most relevant and intriguing.

It has already been mentioned that *Chlamydomonas reinhardii* has (as should be expected) a centric spindle (Section 8.3a). However, no sooner had this species been shown to conform to expectations than Stewart, Mattox, and Chandler (1974) illustrated that the basal bodies of the prasinophyte *Platymonas* were indeed at the metaphase plate (Figs. 8.35, 8.36).* During cleavage (Figs. 8.37, 8.38), this organism was found to have a phycoplast, with a cleavage furrow bisecting the cell as in *Chlamydomonas*. This particular prasinophyte, therefore, may be a quite highly evolved form of the phycoplast-containing, unicellular flagellate—highly evolved

because it now has separated the basal bodies from the MTOCs presumably located at the poles during mitosis. I agree with Stewart, Mattox, and Chandler (1974), who say that a wide taxonomic separation of *Platymonas* from *Chlamydomonas* seems—at the moment—unjustifiable.

We (Pickett-Heaps and Ott, 1974) were concurrently examining another prasinophyte, the uniflagellate *Pedinomonas minor* (Fig. 8.39). In both interphase morphology and during cell division, *Pedinomonas* turned out to be quite different from *Platymonas*. *Pedinomonas* possesses a subtle asymmetry in the disposition of its two main flagellar rootlets, one having two and the other three microtubules (see Ettl and Manton, 1964). The flagellar apparatus contains an extra, nonfunctional basal body. If this basal body were to become functional during the evolution of a different species, the latter would probably be heterokont. Conversely, if the flagellar apparatus were to double up, then a derived decendant might have two isokont flagella, two nonfunctional basal bodies, a cruciate flagellar rootlet system, etc., akin to what *Chlamydomonas* now possesses—and indeed, somewhat like the replicated flagellar apparatus of *Pedinomonas* itself going into prophase. During division, *Pedinomonas* reveals primitive characteristics. The spindle is entirely closed (i.e., devoid of fenestrae; Fig. 8.40), a condition probably more primitive than a fenestrated spindle. Basal bodies and the flagellum remain at each pole throughout mitosis (Fig. 8.41). Most importantly, cytokinesis is effected by cleavage between widely separated daughter nuclei (Fig. 8.42); the spindle persists and no trace of a phycoplast is apparent. We feel that these cytological features indicate that the pedinomonad cell type is indeed primitive and that it represents the most likely candidate so far described which exemplifies the really primitive progenitor of other Green Algae. Hence, its rather tentative position at the base of the phylogenetic tree in Figure 8.31.

The third prasinophyte recently described is *Pyramimonas* (Norris and Pearson, 1973), whose interphase morphology is considerably more complex than that of any other unicell described in this book. Each cell has four lobes and four flagella inserted into a deep depression in one end of the cell (Fig. 8.43). The cell (Fig. 8.46) is covered with a layer of scales of two different

*Stewart and Mattox (personal communication) have now shown that *Pleurastrum* and *Leptosira*, semifilamentous ulotrichalean genera, very closely resemble *Platymonas* during division, perhaps indicating the existence of yet another phyletic line!

sizes and morphology, produced in the large Golgi bodies. The four flagellar bases are interconnected by striated fibers, and a massive rhizoplast joins them to the nucleus. During division, the flagellar bases duplicate (Fig. 8.44), and each pole thereafter is marked by four flagella (Figs. 8.45, 8.47). During mitosis (Fig. 8.47), the spindle is open, and most importantly, during cytokinesis (Fig. 8.48) a furrow grows between widely separated daughter nuclei; there is no evidence of a phycoplast. The organism, therefore, logically does not belong with either *Pedinomonas* or *Platymonas* in Figure 8.31, and seems to fit best above the dotted line. (The reader may wonder why I have dealt so cursorily with these important examples. Regrettably, the information above, which deserves a separate chapter, became available only after most of this book was written.)

My conclusion is, therefore, that the presently defined group of prasinophytes contains a heterogeneous collection of genera, some of very ancient stock, which may well have been ancestral to all other Green Algae.

One final point can now be emphasized. I believe the phycoplast-containing Green Algae will turn out to constitute a *unique* and highly circumscribed group of organisms. Up till the time of writing, they are the *only uninucleate* (i.e., noncoenocytic) organisms known that do not keep their interzonal spindle intact during cytokinesis (*Cylindrocapsa, Pseudendoclonium,* and their relatives are not true exceptions, since they commence and partially complete cleavage while the continuous spindle is involved in mitosis). I have elsewhere discussed the possible primitiveness of the continuous spindle (Pickett-Heaps, 1974f) and have suggested that this structure may have been of great importance in the evolution of the eucaryotic condition. Put simplistically, it could have ensured the formation of two uninucleate daughter cells each time a cell divided by holding the nuclei apart as far as possible during cytokinesis so that the cleavage furrow always passed between them—a necessary function perhaps to avoid the creation of enucleate and binucleate cells. So far as we know, the continuous spindle is present in all eucaryotic cells in some form or other, including those poorly understood mitoses which apparently proceed without the utilization of kinetochoretubules (e.g., in certain dinoflagel-

lates, *Euglena,* some protozoa). Only these Green Algae have been able to dispense with it during cytokinesis, by evolving a (probably) unique set of microtubules whose ultimate function is the same as that of the persistent, continuous spindle.

8.4. A Comparison of Phylogenies

In concluding this book, it is appropriate to comment on how these phylogenetic and taxonomic ideas compare with others that are derived from more classic criteria.

There is now little dispute about what organisms should be classified as Green Algae. Inevitably, a few genera defy convenient categorization, but generally speaking a number of characteristics, such as pigmentation and chemistry of the cell wall, define this group quite well (e.g., Fritsch, 1935; p. 5; Bold, 1967; p. 11; Scagel et al., 1965; p. 146; Chapman, 1962; p. 1). (We may also note in passing that the structure of the chloroplast and the arrangement of thylakoids within it have recently been recognized as features of considerable value in taxonomy. The Green Algae all have similar thylakoid systems; Gibbs, 1970.) Yet within the Green Algae, we find tremendous diversity. They have evolved practically every type of simple plant body; they occupy a tremendous range of habitats; and they have developed an extensive range of sexual and asexual reproductive behavior (Fritsch, 1929). Specialization of reproductive cells has evolved independently of the development of vegetative morphology, so that while some simple unicells may have developed oogamy, more complex filamentous and thalloid forms remain isogamous. Some highly evolved Green Algae have quite complex reproductive structures (*Coleochaete*); yet in other supposedly related and advanced organisms, no such development has been achieved (e.g., *Stigeoclonium*).

The fossil record of algae is scanty and provides us with little information regarding algal evolution, and particularly that which gave rise to land plants (Fritsch, 1949). For a detailed summary of the fossil record, see Banks et al. (1967), Black et al. (1967), Banks (1970), Chalonar (1970), and Schopf (1970). As could be expected, calcareous algae have been best preserved, and records of the Dasycladaceae, for example, extend back to the Cambrian at least.

The Charales (or Charophyta) with their distinctive calcareous oogonia, have also an increasingly detailed taxonomic record coming to light (Grambast and Lacey, 1967). I have included in this book some micrographs of fossil Charophyta, kindly supplied by Dr. Grambast, which should be quite recognizable to the reader (Figs. 8.49–8.52).

Most of the time, we have to erect our phylogenies using extant organisms to illustrate possible evolutionary steps; this approach is reasonable because although some genera have apparently remained relatively unchanged over extremely long periods of time, related organisms have evolved considerably in various directions (e.g., in the volvocine lines of evolution; see Fott, 1965). It is also significant that the same or similar evolutionary trends have become expressed in many different algal classes (e.g., Smith, 1950, p. 9; Fritsch, 1935, p. 26); for example, nonmotile, "tetrasporine" forms are described in all classes. Studying only the "ultimate twigs of an evolutionary tree" (Prescott, 1968) generates great difficulties. The example of the Charales demonstrates all too clearly the limitations in this procedure, since that order (or class, if the reader prefers it), like many others in the algae, stands conspicuously alone, a relic of most uncertain antecedents. Fott (1965) mentions objections to the theory that the simplest ancestral Green Algae were like those of present-day unicellular flagellate Volvocales. In refuting them, he maintains that cells simpler in structure than these flagellates (for example, the coccoid forms of the Chlorococcales) have in reality lost some of the organelles obligatory for the primitive flagellate condition. His argument appears reasonable, since the cells of more advanced algae are often simplified as well (e.g., they contain neither a flagellar apparatus nor contractile vacuoles), but these cells are transformable to the primitive flagellate condition for reproduction, acquiring the necessary primitive features.

There is considerable consensus in the literature that *Chlamydomonas* represents the primitive type of cell from which the other Green Algae arose. Klein and Cronquist (1967) note that the present-day *Chlamydomonas* is probably not like an archetypal cell in all respects, although they stress that such an organism may also be representative of progenitors from which ultimately all higher organisms evolved. Manton (1965), however, argues that the heterokont condition (where the two flagella are unequal in length, appearance, etc.) is of "extreme antiquity" and she states that this view is "on the whole, not controversial," which then implies that the isokont Volvocales are, relatively speaking, not as primitive as certain other groups of algae. Thus the heterokont Chromophyta could be older than the isokont Green Algae. Since bryophytes are or were heterokont, Manton further says that "the common textbook practice of treating *Chlamydomonas* as an unequivocally primitive type is thus valid, if at all, only for certain groups of green algae, but not for the mainstream of phyletic change leading to vegetation of the land." In contrast, Klein and Cronquist (1967, p. 229) say that "scarcely a man is now alive who disagrees with the idea that the Volvocales constitute a basic group from which the other taxa of green algae have evolved." By now, it should be obvious that I strongly side with Manton in this matter. One other point can be made here. The spermatozoids of the Charales are slightly heterokont (Section 7.5), but this condition arises from basal bodies that initially are outwardly identical and arrayed at an angle to one another, as in *Chlamydomonas*. One could conclude, therefore, that the heterokont condition in this case has arisen from isokont ancestry (which I personally do not believe to be the case). Neither should we forget that the flagella in at least some Volvocales (*Volvox*; Section 2.2a) are not quite identical; one is formed much earlier than the other, and there is no reason to suppose that both always function identically. In many isokont Green Algae, one flagellum is consistently longer than the other (Moestrup, 1972). The distinction between heterokonts and isokonts may become difficult to define in such cases.

8.4a. Phylogeny of the Flagellar Apparatus

It is valuable at this point to review briefly the phylogeny of the flagellar apparatus itself in a slightly wider context than was set out in Section 8.2. Manton (1965) has summarized earlier work showing the importance of this complex apparatus in phylogeny, and many of her remarks are increasingly supported by more recent work. She stresses the differences between the transitional structure in basal bodies of plant and other cells; in particular, the stellate structure (Fig. 8.9; Chapter 2) has not been recorded outside the plant kingdom. She also provides examples of the slight but significant variation in the microtubular systems of flagellar roots between various Green Algae. A general lack of information at the moment makes the value of this latter character in phylogeny difficult to assess; Moestrup (1972) summarizes the information we presently have regarding rootlet systems.

Since the basic structure of the centriole and flagellum is so consistent throughout eucaryotic organisms, these organelles could be monophyletic and very primitive (see earlier comments), and one conclusion might be that this is the only flagellar system which is functionally effective. However, the precise structure of the flagellum, almost universally possessed by motile eucaryotes, may not be of key importance; one may also argue that its universality is due to the possibility that the first organisms to acquire this organelle gained such an advantage over their competitors that they rapidly prevailed everywhere. Several authors have tried to explain the origin of the flagellum in terms of a derivation from bacterial flagella, perhaps gathered in a group (Klein and Cronquist, 1967). Margulis, in several publications (see particularly Margulis, 1970), has suggested that the flagellum had an origin from a symbiotic bacterium which turned into a "9 + 0" homologue. Other authors refute this viewpoint for a variety of reasons (Klein and Cronquist, 1967, pp. 157 et seq.) and state that the eucaryotic flagellum has no direct affinity with the bacterial flagellum. I support this latter view for various reasons, principally because of the dissimilarity between the single tubules and the fused doublets of the peripheral fibers of the eucaryotic flagellum and the much simpler structure of the bacterial flagellum (Lowy and Spencer, 1968). To those who maintain that a derivation from bacterial flagella is the only reasonable explanation for the appearance of eucaryotic flagella, I would answer that we have virtually no information regarding the transition of procaryotic cells into eucaryotic cells and perhaps the elaboration of a centriole should be regarded as one aspect of this deeper mystery, no less puzzling than the acquisition of a nuclear envelope. My view is that the cell at some stage evolved microtubular systems, perhaps in connection

with achieving eucaryotic cell division (Pickett-Heaps, 1974f) and that the centriole evolved during increasing utilization and differentiation of the cell's mitotic MTOCs, an elaboration that conferred upon the cell such advantageous capabilities that those organisms possessing it soon became increasingly dominant. Further (Pickett-Heaps, 1974f), I should argue that the problem of accounting for the origin of mitosis itself is inseparable from the problem of explaining the acquisition of a membrane-bounded nucleus, and both may have constituted the major evolutionary advance in the transition of procaryotic to eucaryotic cells. It may be most significant that, so far as I am aware, *all* eucaryotic organisms possess microtubules, sometimes displaying them only while undergoing mitotic or meiotic processes, whereas a number do not possess centrioles, a lack that cannot always be convincingly ascribed to a loss of centrioles late in evolutionary development.

The Red Algae could well have diverged from the other lines of evolution before the centriole evolved (Christensen, 1962; Manton, 1965). Again, I favor this view since it seems to me rather unlikely that such a useful structure if originally present, would have disappeared later from all the progenitors of the Red Algae, which presumably have always been aquatic. Thus, since the Red Algae (or rather one Red Alga) possess mitotic microtubules (McDonald, 1972), it can be argued that the flagellum/centriole system evolved after some form of microtubule/ MTOC system in the cell had already evolved to achieve mitosis. The densely stained cylinder that appears briefly at the poles of prophase and metaphase spindles in the Red Alga *Membranoptera* (McDonald, 1972) could perhaps be distantly related to the centriole, but any morphological resemblance is tenuous indeed. Investigations of other Red Algae would obviously help our speculation considerably. Another possibility is that the Red Algae could have themselves been somewhat intermediate between procaryotes and other algae, and they certainly possess some striking similarities to the blue-green algae (e.g., Allsopp, 1969). In either case, I reach the same conclusion—that the mitotic apparatus evolved first, and then later some organisms developed the basal body from it.

But we are faced with yet another problem: how then do we account for the radically different mitotic structures that are possessed by dinoflagellates and euglenoids (which may also be present in some protozoa; see Leedale, 1970)? These latter organisms often have centrioles. I am forced to assume that their ancestors diverged from the other centric eucaryotes after the centriole had evolved and that they successfully experimented with the primitive spindle apparatus, if indeed there was only one basic apparatus (a risky assumption). The dinoflagellate's spindle apparatus can be interpreted as one more specialized variant of a primitive spindle with chromosomes attached to the nuclear envelope instead of the cell membrane as in procaryotic cells. A rationale to explain the origin of an extranuclear continuous spindle from an intranuclear spindle has been attempted elsewhere (Pickett-Heaps, 1974f). If the dinoflagellate spindle forms outside the nucleus between separating polar MTOCs, and then later sinks into it, then such a spindle has a certain similarity to that of diatoms (Manton, Kowallik, and von Stosch, 1969, 1970). What then was the primordial spindle like? Perhaps the study of a careful selection of algae, fungi, and protozoa will allow us to guess —a speculative venture that has the merit that whatever we decide, we shall probably never know the answer!

It may not in fact be difficult (at least in principle) to test these hypotheses. The method immediately suggesting itself could involve sequencing microtubular proteins (a tedious business) and applying phylogenetic analysis to the variance obtained in the sequences (Dickerson, 1972). Microtubular protein probably has a conservative structure, since it has most important and complex properties, both physiologically and structurally as a macromolecule. It might also be an ancient protein (Pickett-Heaps, 1974f). The method, to be useful, requires sufficient variation in the sequence between different organisms to allow some analysis. The preliminary data of Luduena and Woodward (1973) are most pertinent in this respect. Microtubular protein from a variety of sources always seems to consist of two distinct components, α— and β—tubulin. Luduena and Woodward obtained both tubulins from different sources, the doublet tubules of sea-urchin sperm and the brain of embryo chicks. Each of the four proteins was sequenced for about 25 residues from the N-terminal end. There were no differences between the two α—

tubulins and only one between the β—tubulins. This result, although preliminary in nature, suggests a powerful conservatism in the protein structure, since the organisms supplying the protein are far apart phylogenetically. Furthermore, α— and β—tubulin, although distinctly different proteins, reveal marked similarities in the sequenced portion of the molecule studied. This observation supports one's feeling that flagellar doublet and cytoplasmic microtubules are similar polymers, built from similar subunits, but it is not yet clear what alteration in the components is necessary to obtain the fused wall of the doublet. (Similar conservatism is becoming evident in the protein structure of bacterial flagellin; Delange et al., 1973.)

Assuming enough variance in the amino acid sequence will be present for valid comparisons to be made, I should predict that Red Algal tubulins may be significantly different from those of all other eucaryotes which have or had centrioles (or were derived from centric ancestors). The fused doublet tubule of the flagellum may have evolved following a mutation in the microtubule protein. This process has happened, for example, in the evolution of the various hemoglobins and could be consequent upon the duplication of the protein's gene, followed by mutations appearing in one of the copies thus arising. This procedure allows mutation and variation in an important protein to be developed and maintained without necessarily lethal consequences (Dickerson and Geis, 1969, p. 61). Such a process may originally have given rise to the differences between α— and β—tubulin, and later to a possible tubulin derivative involved in the creation of doublet tubules. Thus, sequencing tubulins in selected organisms (particularly Red Algae, dinoflagellates, euglenoids, protozoa, and fungi as well as higher organisms) may provide profound phylogenetic insights and perhaps settle once and for all the question of the origin of the centriole—a structure that I believe evolved from the MTOCs at the pole of a primitive spindle *after* the microtubule/MTOC systems had already evolved (Pickett-Heaps, 1972a, 1974f). These ideas do not suggest how the centriole came to acquire its exquisite nine-fold symmetry, an enigma that remains with any theory concerning the origin of this fascinating organelle.

The concept that the centriole is an autonomous, self-replicating organelle, possibly possessing its own DNA, and derived from a symbiotic flagellate bacterium, is being increasingly undermined (Fulton, 1971; Pickett-Heaps, 1972a), and even now in mammalian cells, this organelle probably is not necessarily continuous through generations of organisms (Szollosi, Colarco, and Donahue, 1972). I believe we should regard the centriole/flagellum merely as one differentiation of the cell's MTOC/microtubular system—as indeed the rootlet template/rootlet microtubules and the spindle apparatus also appear to be.

In summary, perhaps we should not be beguiled into attributing to the centriole the properties it seems to have acquired by popular acclaim over the years. Debating which came first, the flagellum or the centriole (discussed in several places by Klein and Cronquist, 1967), appears to me as irrelevant as debating the problem of the chicken and the egg, as is considering whether "either the flagellum or the centriole (or both) could be lost separately in the course of evolution" (Klein and Cronquist, 1967, p. 224). In this vein, I feel that Fulton (1971) in his excellent review of the centriole, misdirects his emphasis when considering its phylogeny. He says that the odd structures that serve as cell centers during mitosis (centrosomal structures in ascomycetes, basidiomycetes, diatoms, etc.; see Pickett-Heaps, 1972a) "are so unlike the pinwheel [i.e., centriole] that they probably represent independent evolution." Yet there is increasing evidence, in my opinion anyway, that the centriole is just one highly organized structure differentiated by the mysterious entity that is the "cell center." It appears to me as if the centriole has often diverted attention from the relevant and more important central issue of how the membrane-bounded eucaryotic nucleus and its mitotic apparatus arose.

8.4b. The Volvocine Lines of Evolution

Several authors (e.g., Fritsch, 1929, p. 108) stress the likelihood that there may be several parallel lines leading to the organization of the colonial Volvocales, and ultrastructural work has not clarified this possibility so far. There seems little doubt that the Tetrasporineae (or Tetrasporales of some authorities), sessile for most of their life cycle, are closely related to the Chlamydomonadineae, many of which can also form sessile palmella stages (see Chapter 2). Several authors derive the ulotrichalean lines of evolution from the palmelloid Tetrasporales (e.g., Klein and Cronquist, 1967, p. 228; Scagel et al., 1965; Dittmer, 1964) because of their colonial tendency. There are semifilamentous forms of the Volvocales in the suborder Chlorodendrineae (Fritsch, 1935, p. 130) which have not been described in this book. While some members of the Tetrasporineae are essentially unicellular, many advanced forms are colonial and thus demonstrate a colonial tendency similar to that seen in the Chlamydomonadineae (Chapter 2; Klein and Cronquist, 1967, pp. 229 et seq.). The concept that parallel evolution in different algal classes gave rise to similar types of organisms is stressed in many books (e.g., Fritsch, 1935). We may note in passing that the three genera *Tetraspora, Chlorococcus,* and *Phaeosphaera* provide an excellent example of this parallel evolution in the three classes Chlorophyceae, Xanthophyceae, and Chrysophyceae, respectively (Rhodes and Stofan, 1967).

Klein and Cronquist, in commenting on mitosis in *Chlamydomonas* and some other volvocalean organisms, state that other members of the Green Algae (including, for example, *Oedogonium* and *Spirogyra*) are probably similar and that the "pattern of mitosis" in Green Algae is of little phylogenetic significance, a view also somewhat shared by Leedale (1970). However, for reasons that I hope are now obvious, mitosis is of real phylogenetic interest. Klein and Cronquist's attempts to derive the Euglenophyta and Pyrrophyta from the Volvocales are surely incorrect at our present state of knowledge, although they admit their evidence for this decision is inadequate (Klein and Cronquist, 1967, p. 230). In particular, these organisms possess permanently condensed chromosomes during interphase, a most unusual feature. Furthermore, the spindle apparatus of both *Euglena* (Leedale, 1970) and some dinoflagellates (Leadbeater and Dodge, 1967; Kubai and Ris, 1969; Soyer, 1971) has marked and most important differences from that in the Volvocales and other Green Algae. It seems safest to regard both these groups as very ancient and taxonomically isolated (Round, 1971; Leedale, 1970).

The status of the Prasinophyceae is contentious (see Section 8.3d). Many authors include these organisms in the Volvocales. However,

Round (1971) makes a good case for putting them into their own subphylum (the Prasinophytina) of equivalent status to the Euglenophytina, the Charophytina, and the Chlorophytina (the rest of the Green Algae).

8.4c. The Siphonous Lines of Evolution

The siphonous lines of evolution have one dominating theme, the tendency of the cells to become larger and larger coenocytes. The Chlorococcales are a heterogeneous group, but their derivation from simple unicellular forms of the Volvocales is not disputed by any authority I have read. In the Cladophorales, septation of the coenocyte is reasonably frequent, and large, often branched filamentous types of cells have arisen. Fritsch (1935) believes this group has only superficial resemblances with the Siphonales, in which septation is rare or nonexistent. The Cladophorales have been included in the Ulotrichales by some (e.g., West and Fritsch, 1926), but while an extremely early derivation from a ulotrichalean ancestor is often considered likely, this line of evolution seems distinct from the present-day Ulotrichales (Fritsch, 1935). The Siphonales could have evolved from forms like the Cladophorales, but the consensus apparently favors a derivation from the Chlorococcales. The Vaucheriaceae do not fit well into the Siphonales, and Manton (1965) emphasizes that, had the value of flagellar apparatus as a phylogenetic character been realized earlier, these organisms' recent transfer to the Xanthophyceae would have taken place much earlier. We may note two fascinating and provocative observations in passing. *Vaucheria* is often somewhat vaguely associated with the water molds (e.g., see Manton, 1965). Fritsch (1929) lists some of the characteristics so difficult to account for in *Vaucheria* and mentions a possible common origin with the phycomycetes too. *Vaucheria* has a closed, centric spindle devoid of polar fenestrae, with the centriole situated outside the nuclear envelope at the poles (Ott and Brown, 1972). This type of spindle is not yet recorded in the Green Algae, but—tantalizingly—is characteristic of the phycomycetes *Catenaria* (Ichida and Fuller, 1968), *Saprolegnia* (Heath and Greenwood, 1968; Howard and Moore, 1970), and *Blastocladiella* (Lessie and Lovett, 1968). Even more intriguingly, in *Vaucheria* and the phycomycete *Saprolegnia,* each Golgi body is appressed to a mitochondrion with an element of endoplasmic reticulum sandwiched between the two organelles, an unusual arrangement not yet recorded elsewhere (Ott and Brown, 1974).

8.4d. The Ulotrichalean Lines of Evolution and the Origin of Higher Land Plants

The origin of the ulotrichalean lines of evolution is the subject of some argument. Fritsch (1935, p. 198) strongly comes out in favor of a direct descent from a motile unicell, stating that the "germination of every zoospore recapitulates the evolution of the filament" and denying the possibility of the derivation from palmelloid organisms (e.g., the Tetrasporineae) as is proposed by Klein and Cronquist (1967), Scagel et al. (1965), and others. The example of zoospore germination above is well known in Green Algae, being clearly demonstrated, for example, in the growth of *Ulva* (Section 4.7) and *Shizomeris* (Mattox, Stewart, and Floyd, 1974), whose plant body arises from a single-celled germling via a filamentous stage. Even the gametophytes of liverworts and ferns and the protonemata of mosses commence growth by forming a filament of cells (Bold, 1967), and Bower (1935, p. 496) emphasizes how the structure of certain ulotrichalean algae resembles the haploid phase of archegoniate plants. In the Volvocales, cell division is closely linked to reproduction in that the daughter cells cast off their old wall and separate. In contrast, in sessile forms the division furrow in primitive filamentous forms of Green Algae probably cut the cell in half within the parental wall and the tendency for successive cleavages to occur at right angles to one another (as in *Chlamydomonas*) had to be suppressed. Cytokinesis thereby also cut the single spherical type of chloroplast, and if the daughter cells subsequently elongated, then the filament of ulotrichalean-type cells, each containing a girdle-shaped chloroplast, likely resulted (Fritsch, 1935, p. 198).

Forest (1956, p. 142) emphasizes how difficult it is using extant organisms to demonstrate the beginnings of a filamentous tendency, and the complexity of the processes by which unicellular algae developed into filaments should not be underestimated (Fott, 1965). Neither should the evolutionary significance of this step be overlooked. It was *fundamentally important for the further development of complex plant forms,* and indeed, for the eventual appearance of higher land plants. An explanation of how the filament might have evolved has been proposed by Pascher, quoted in Fott (1965). According to his theory, vegetative reproduction gave rise to rows of autospores, at first perhaps held together by mucilage, and the parental cell wall enclosed little of the daughter cells' protoplast. Presumably, the daughter cells later became increasingly cylindrical (e.g., as in the extant genus *Geminella;* Fritsch, 1935, p. 207). Recent observations on the wall structure of *Microspora* (Section 4.5; Pickett-Heaps, 1973b) and other algae have led me to a more detailed elaboration of this general theory, summarized as follows. Figure 8.53 illustrates the fate of the wall in primitive, autospore-forming cells. After nuclear and cell division, each autospore secretes its own wall, and the parental wall is sloughed off (as is the case with *Kirchneriella, Tetraedron,* and many other genera). In Figure 8.54, the cells adhere to one another by mucilage, and after cell division the parental walls are held between the daughter cells or "autospores"; examples of extant organisms that may illustrate this condition include *Binuclearia* and *Radiophyllum* (Fritsch, 1935, p. 207). Filaments of this form can obviously fragment easily. To some extent, *Cylindrocapsa* (Fig. 8.28) and related genera also fit this category, with individual cells, each with its own wall, being held together by a sheath derived from older wall material. Even this primitive stage of evolutionary development imposed constraints of considerable significance upon the cells: (1) each cell had to lose the primitive tendency, presuming it previously had it, of forming multiple daughter cells; (2) the plane of division had to be carefully controlled; and (3) the parental wall probably had to be preprogrammed in some way so as to split in the right place—if splitting was random, then a ball of cells would result. The need for the wall to contain a carefully positioned potential rupture site used in later cell elongation is important too in considering the evolution of the Oedogoniales (see later) and indeed many other algae.

Two significant developments could lead to the filamentous form illustrated next in Figure 8.55. First, daughter cells secreted a single, integral cross wall between themselves at cytokinesis instead of two autospore walls, a change that may have been related to the evolution of a cell plate during cytokinesis. (So far as I know, all plants that have true cell plates are filamentous or more complex in form; the cross wall between daughter cells does not usually split apart easily.) Second, they did not secrete end walls (or else these end walls were thin), since the ends now remained enclosed by the older parental cell wall. Cell growth, as in the autospore-forming cell, could then be accomplished by longitudinal extension concurrent with continued wall deposition. The wall did not need to split at cell division, except at the first division (e.g., of a zoospore germinating into a filament), and it obviously now consisted of interlocking H-shaped segments. It had acquired one most important attribute: it was considerably stronger and had far less tendency to fragment into segments, the latter a characteristic of simpler ulotrichlaean algae. One possible further step is the condition observed in present-day *Microspora* (Section 4.5; Fig. 8.56). The only new development now is that during interphase cell expansion, the H-shaped segments do not elongate, but merely slide apart to accommodate extension; new wall deposition gives rise to a separate cylinder, tapering in thickness toward each end, inside the interlocking H-shaped pieces. This cylinder itself is converted to the next H-shaped segment by subsequent cross-wall formation at cytokinesis. During evolutionary advancement, the wall segments could have become ever more closely integrated with one another and consequently increasingly difficult to separate into components. (Perhaps those algae which now possess discrete, segmented walls lost the ability to secrete a wall that could expand uniformly during cell enlargement.) These ideas help explain why many simpler ulotrichalean filamentous algae break up and release individual cells from the filament (i.e., just as autospores are released from parental cells) and also the origin of the segmented wall characteristic, for example, of *Microspora,* where distinct phases of wall deposition, associated with sequences of cell division, form a slightly discontinuous wall structure that can be broken up into H-shaped segments (also reported to exist in *Klebsormidium* and *Ulothrix;* Jane and Woodhead, 1941). Likewise, in *Cylindrocapsa,* newly formed "autospores" secrete their own wall inside the layered parental wall, a new layer being added to the end wall per division (Fig. 8.28). I am not yet

571

convinced that such rationales do entirely explain the ulotrichalean filaments' origin, but they serve to emphasize a rather neglected phylogenetic problem.

Cell division in some filamentous ulotrichalean algae clearly reveals their unicellular (chlamydomonad) ancestry. As Mattox and Stewart (1974) point out, the disposition of cell organelles during interphase, mitosis, and cytokinesis often resembles that in *Chlamydomonas* itself, if we assume that the cell in a filament is oriented so that its ancestral flagellar apparatus is located on one side wall, a region devoid of the peripheral chloroplast. This orientation is indeed indicated by the situation of the chloroplast and interphase centrioles of many such cells. Thus, during mitosis, the centrioles replicate and then move to the poles of the spindle; after telophase, they move back to their premitotic position near the edge of the cell (as happens, of course, also in the many chlorococcalean genera). Cleavage progresses fastest from this side of the cell, either early in mitosis (e.g., as in *Trichosarcina, Pseudendoclonium,* and *Cylindrocapsa*) or later (e.g., in *Microspora*). We can picture such filaments as stacks of unicells adhering laterally, with chlamydomonad-type cleavages proceeding most rapidly from the "flagellar" end of the cytoplasm. A little reflection will show that this is just about the only way such chlamydomonad unicells could form a truly filamentous colony. If cleavage proceeded in any other plane, or if the cells adhered in any other fashion and continued to utilize this form of cytokinesis, an ordered, single-stranded filament could not arise. The plane of cleavage must be perpendicular to the axis of the filament. This rationale, by the way, helps explain a problem that has long perplexed phycologists: why the flagellar apparatus of the Oedogoniales should arise on one side of the filament during zoosporogenesis, apparently thereby effecting a change in the polarity of the cell. In fact, zoosporogenesis in this class may only represent a more extreme illustration of what must happen in other ulotrichalean cells of chlamydomonad ancestry—more extreme simply because the flagellar apparatus is so large and conspicuous. Yet again, cell division and zoosporogenesis turn out to be important keys to understanding algal evolution.

To pursue this line of logic further, it is not difficult to perform the similar mental gymnastics necessary to transform the pedinomonad type of cell division into that displayed by *Klebsormidium* —in other words, to form a filament from unicells in the upper line of algal evolution postulated in Figure 8.31. The analogy does not give quite such a clear-cut correspondence, but it is still illuminating. In *Pedinomonas* (and *Pyramimonas*), the flagella and basal bodies stay at the poles of the spindle even during growth of the cleavage furrow between the widely separated daughter nuclei. In *Klebsormidium,* likewise, the centrioles remain at the poles during cytokinesis, a reflection of the fact that this genus does not form a phycoplast. In *Klebsormidium,* the positioning of the flagellar apparatus within the parental wall changes during zoosporogenesis (Marchant, Pickett-Heaps, and Jacobs, 1973) and the germinating zoospore does not form a holdfast, so it is difficult to guess from the zoospore what orientation of an ancestral flagellate might be displayed in the organization of the vegetative cell in the filament.

One puzzle remains in these speculations that we should not ignore, even if it cannot yet be explained: why should the centrioles remain at the poles of certain phycoplast-forming ulotrichalean genera such as *Ulothrix?*

I have spent considerable time pondering the subtleties of wall morphology and the method of cell division employed by the Oedogoniales, and I have found my conclusions helpful in understanding this rather extraordinary group of organisms. I hope the reader also finds further theoretical speculation helpful; if too much, skip the next paragraphs.

The Oedogoniales can be considered to have walls made up from highly asymmetrical H-shaped segments (Fig. 8.57) which do not overlap. New intercalary H-shaped segments are formed essentially during one short period at the end of cell elongation; the cross wall, it will be remembered, coalesces and thickens as the new inner wall layer is deposited around the apical daughter cell, inside the stretched-out material of the ring (Section 5.1b). Thus, the newly formed H-shaped segment *abuts* and does not significantly overlap the older wall segments, and it is not structurally integrated with them. Consequently, the discontinuity arising in the wall obviously represents a weakened region. If we now follow some of the rationale proposed earlier, we can postulate what is (in my opinion) a reasonable sequence of events that could have led to the wall morphology of the present-day Oedogoniales.

First, consider a row of autospores, such as those in Figure 8.54, in which the mother wall's rupture site was not at the center of the cell, but instead displaced to one end. This change in the position of breakage of the autospore wall must have coincided with the appearance of strong apical/basal polarity in the evolving filament or its constituent cells. Cycles of cell division, exactly analogous to those in Figure 8.54, would have given rise to the wall morphologies set out in Figure 8.58. Now consider the evolution of H-shaped wall segments (Fig. 8.59) in this primitive filamentous form. The cross wall between daughter cells became a single integral wall as before (Fig. 8.55). If we can assume that a thin remnant of a new autospore wall was secreted within the parental autospore wall (as shown in Fig. 8.59) and that the wall rupture site is created anew in the newly secreted wall (as it must have been for the situation depicted in Fig. 8.54), then we have in fact achieved the wall morphology of the extant Oedogoniales, including the formation of the two types of caps whose position is related to the wall rupture site. These two assumptions are not only reasonable but almost certainly justified, since the appropriate phenomena are demonstrably present in the extant organism. I need not emphasize the invariable existence of a wall discontinuity in all cells of this order (Chapter 5). Furthermore, inspection of almost any suitably stained micrograph in this book of the wall in *Oedogonium* and *Bulbochaete* will reveal the pattern of secondary wall deposition indicated in Figure 8.59, associated with the formation of caps. A few of the most obvious examples are given: Figures 5.27 and 5.46 unmistakably show a layered wall secreted under the sequence of downward-facing caps, while Figures 5.45 and 5.135 show the same type of wall layering associated with a tiered cap. I have drawn the initial cell in Figure 8.59 in the form of a germling to emphasize its polarity. The cap morphology arising after the first division in such a germling (Fig. 8.59A) is clearly visible in Figure 5.89. In this light, the extant Oedogoniales represent the end result of some quite minor variations acting within a generalized and perhaps widespread system that could have evolved a filament from unicells. In many genera, the wall weakening is represented by the dislocation between two overlapping segments of wall which slide apart to allow cell expansion and the creation of new H-shaped segments (an obvious example is given by *Microspora*). However, the asymmetry of the H-shaped segments in the Oedogoniales may not allow them to overlap significantly (although the discontinuity does pass somewhat diagonally across the wall), and, therefore, the segments could not slide apart in an analogous fashion. Instead, the segments effectively abut one another, and the adaptation of a wound response or repair mechanism to cope with this intrinsically rather unsound structure has perhaps led to the unique method of cell division characteristic of the order (Chapter 5). Further experimentation with the wall rupture mechanism has probably resulted in the formation of the fertilization pore in the oogonium and (later?) in the evolution of hairs and branching in *Bulbochaete* (Chapter 5).

Elsewhere in filamentous Green Algae evidence of intermediate evolutionary stages or variations can be discerned; for example, short H-shaped wall segments have been reported in *Spirogyra* (Fritsch, 1935, pp. 320 et seq.). Neither is this form of wall secretion in phases associated with cell division peculiar to the Green Algae. A particularly striking example occurs in the genus *Bumilleria* (Xanthophyceae), which also forms H-shaped wall segments and whose wall is secreted in obvious layers. Germlings of the related *Tribonema* have a conspicuous wall weakening around the middle of the cell, where two separate wall segments overlap (Fig. 8.60). Upon elongation of this filament, the two segments slide apart as a new cylinder of wall is intercalated between them (Fig. 8.61). One can even draw analogies with the diatoms. Here, however, impregnation of the wall segments with silica has rendered them extremely rigid and consequently a steady reduction of size accompanies cell division in many of the daughter cells formed from an original parent (see Fritsch, 1935, pp. 609 et seq.). In filamentous diatoms, these siliceous walls are firmly stuck together, forming again a variant of the H-shaped segment.

Another aspect of wall formation, whose significance is uncertain, can be mentioned here. Some algae, notably the desmids and the Oedogoniales, secrete their wall in two distinct phases. Cell expansion is contained within an extensible wall layer (derived from the ring in the Oedogoniales), often clearly secreted by the

Golgi bodies, that may be analogous to the primary wall of higher plants. After cell expansion is essentially complete, a distinctly different thick and (largely) nonextensible wall is secreted within the "primary" wall layer, the latter being shed later in the case of desmids. This second phase of secretion may be equivalent to secondary wall deposition in higher plants and does not apparently involve the Golgi bodies. If distinctive phases of "primary" and "secondary" wall deposition do occur in other Green Algae, the layers may be difficult to distinguish morphologically, as indeed they often are in some secondarily thickened walls of higher plants.

Once the filamentous habit had been acquired, variation in the planes of cell division within the filament could obviously allow the development of more advanced organisms. It is doubtful whether the Chaetophorales are really distinctive enough, as described by Fritsch (1935) and others, to be set apart as a group from the simpler Ulotrichales. Genera such as *Ulothrix, Stigeoclonium, Chaetophra, Draparnaldia,* and *Fritschiella* show an obvious gradation of morphological properties between extremes in plant forms. Individual species in each of these genera display considerable morphological diversity often dependent upon growth conditions, and even *Ulothrix* may be branched (Floyd, Stewart, and Mattox, 1972a), parenchymatous, and have heterotrichous tendencies (Forest, 1956). These particular algae are all isogamous, although anisogamy has been reported in *Draparnaldia* (Forest, 1956). These organisms (*Ulothrix, Stigeoclonium,* et al., above) then seem to constitute a natural group. Furthermore, and most significantly in my opinion, Forest and others say that the oogamous forms of the Coleochaetaceae are "rather far removed" from these ulotrichalean algae (cf. Section 8.3c; Fig. 8.21). There are certainly no compelling reasons in any papers I have read to conclude that the ulotrichalean algae (including the Chaetophorales) are monophyletic or even all closely interrelated. Our suggestion of at least two distinct lines of progression in the Ulotrichales (Figs. 8.21, 8.31) seems compatible with much that has been previously written.

Indeed, it is increasingly clear that the Ulotrichales as presently defined comprise a heterogenous collection of genera, some included presumably because they cannot be better placed elsewhere. Stewart, Mattox, and Floyd (1973)

have begun attempts to reclassify this order, subdividing it initially into four (Ulotrichales, Chaetophorales, Microsporales, and Coleochaetales). They draw extensively on ultrastructural data—for example, on the spindle, cytokinetic apparatus, and structure of the pyrenoid. While it is too early to comment on this commendable reclassification, the need for which has become most pressing, two later papers by these authors (Mattox, Stewart, and Floyd, 1974; Birbeck, Stewart, and Mattox, 1974) show how useful ultrastructural data can be in taxonomy. They review the various attempts to classify the parenchymatous ulotrichalean alga *Shizomeris,* which many believe to have close affinities with the Ulvales (or Ulvaceae, if the reader prefers). They show that interphase cell morphology, cell division, and the structure of the zoospore strongly refute such an affinity, which lies nearer *Stigeoclonium;* some features, particularly the structure of the pyrenoid and the zoospore (Figs. 8.4–8.8), show that *Shizomeris* is also distinctly separable from *Stigeoclonium.* Such data and their other work are increasingly confirming their view that the pyrenoid's detailed structure has considerable taxonomic value. As they emphasize, the use of gross morphology and growth habit as a means of classifying such organisms has proved quite inadequate and misleading in many cases.

The two orders so difficult to place, the Oedogoniales and Conjugales, and also the Charales (or Charophyta, if one wants to stress their separateness from the other Green Algae), are generally agreed to have been derived long ago from ulotrichalean ancestors. The distinctive features of the Charales which set them far apart from the other Green Algae (Bold, 1967; Round, 1971) have already been sufficiently discussed in Chapter 7 and Section 8.3c. Prescott and some others (e.g., see in Fritsch, 1935, p. 361) consider the Conjugales also far enough removed from the other Green Algae to be placed in a subphylum by themselves, perhaps derived directly from the Volvocales (Prescott, 1968, p. 44). However, Fritsch (1935) probably represents the consensus when he states that there is not sufficient justification for this step. Klein and Cronquist (1967) say "most authors agree that the filamentous Zygnematales . . . were derived from the Chlorococcales or directly from the Tetrasporales," a consensus I myself have not noticed. Fritsch (1935, p. 362) does not hold with this view. Later, Klein and Cronquist mention that Chapman (1962) suggested an origin of the Conjugales (Zygnematales) from the Ulotrichales, but they do not like this latter proposition, using the relative "patterns of mitotic division" as one of their criteria to refute it. The work summarized in this book shows little relation in the patterns of cell division between the Chlorococcales and the Conjugales, and suggests instead affinities of the latter with *Klebsormidium* (p. 557). Regarding the most distinctive feature of the Conjugales, namely the lack of flagellate stages in their life cycle, Fritsch notes that sexual reproduction by ameboid gametes can also be found in certain species of *Chlamydomonas* and in the Chaetophoraceae. Prescott (1968, p. 124) has set out in considerable detail the possible phylogeny of the highly varied individual species in the Conjugales.

Once we come to the discussion of the origin of higher land plants, there is general agreement that the Ulotrichales (defined so as to include the Chaetophorales) constitute the logical group in which to seek their progenitors. Some (e.g., Church, 1919) have proposed origins from marine organisms such as large seaweeds, but these ideas have been generally refuted (Fritsch, 1916, 1921). Fritsch (1916) lists many reasons why the Chaetophorales should be predominant in such considerations, such as the great evolutionary potential of the group following the acquisition of the heterotrichous habit, the extreme adaptability of the group in occupying diverse habitats, the considerable experimentation that has taken place in the development of their form, and their terrestrial tendencies. Regarding the last attribute, it is remarkable how many orders of Green Algae contain species which are adapted for terrestrial existence. Examples are *Oedocladium* in the Oedogoniales, *Sirogonium* in the Conjugales, *Fritschiella* in the Ulotrichales, and *Chlorococcum* in the Chlorococcales. (For many more examples, see Fritsch, 1921, p. 168; Fritsch, 1922.) Fritsch (1916) singled out *Coleochaete* as an organism possibly related to the progenitors of the bryophytes, but in later papers, tends to favor organisms related to *Fritschiella* in this capacity, and Fott (1965) also says that the line of morphological advance in recent Green Algae end with this latter organism.

We can gain considerable insight into the evolution of higher plants by comparing the Green Algae with the Brown Algae (Fritsch, 1949). No one now seriously believes that the Brown Algae did give rise to higher plants, but we can see clearly several important parallels between evolution in the advanced members of the Brown Algae and higher plants themselves. Both Green and Brown Algae show the progression toward oogamy during evolutionary advance, and heterotrichy is highly developed among the Browns, often accompanied by virtual loss of the prostrate system (Fritsch, 1942a). We can follow how life cycles must have changed as well. Several Green Algae (e.g., *Ulva, Draparnaldiopsis, Cladophora,* and possibly *Fritschiella*) have apparently evolved from the haploid life cycle common to most Green Algae to the isomorphic alternation of generations (i.e., the haploid and diploid organisms are morphologically identical). No haploid life cycles are known in the Brown Algae (Fritsch, 1949), the simplest life cycle being that of the isomorphic Ectocarpaceae, which, not surprisingly, are also the simplest known members of that large class. The Brown Algae have gone on to develop a heteromorphic alternation of generations whose diploid phase has often become structurally highly differentiated both internally and externally. (Isomorphic and heteromorphic are used in the sense defined by Fritsch, 1942b.) Bower (1935) mentions the "stolid conservatism" of extant ulotrichalean algae such as *Ulothrix* and *Coleochaete* which has led to "evolutionary inertia," since they have not managed to postpone meiosis during germination of the zygospore, an innovation that would allow them a diploid phase in their life cycle. Incidentally, some algae, notably *Oedogonium* (Section 5.1d), and apparently some desmids (Section 6.4c), may suffer failure of meiotic divisions during germination, and larger diploid and apparently perfectly viable organisms are thereby formed, but this diversion into diploid seems relatively uncontrolled. Bower (1935) emphasizes the great importance of acquiring an alternation of generations in amphibial algae as one of the vital characteristics necessary for successful colonization of the land. It is, however, most difficult to spell out in detail why alternation of generations, and in particular why the development of a prolonged diploid phase of vegetative existence, should have been so vitally important for evolutionary advance. This matter has been discussed at length in an excellent paper by Raper and Flexer (1970). They emphasize how there is a "totally compelling correlation between somatic or vegetative diploidity and biological success." Diploidy seems to offer significant advantages in the evolution of larger and more complex organisms. An excellent example of this principle is apparent in the general progression of plant development from Green Algae through bryophytes and on to the most advanced land plants. During this progression, the diploid phase of the life cycle has become steadily more dominant in duration, size, and morphological complexity; in the flowering plants, the ultimate achievement in plant evolution, the haploid phase has become insignificant when compared to the vegetative diploid structure. All animals above the protozoa (and probably many protozoa as well) are also diploid, but it has not yet been possible to set out in a brief and compelling argument why this principle should hold so well (Raper and Flexer, 1970).

As an even more striking example of parallelism between the Brown Algae and green plants, we may note that the laminariales possess sieve tubes, complete with callose (Fritsch, 1945b, p. 232), and in many Brown Algae vegetative growth has become increasingly confined to meristematic regions, often being in part related to sequences of cell divisions in a tetrahedral apical cell, as happens in many archegoniate plants (cf. also the Charles). Fritsch (1949) concludes that forms somewhat analogous to those in the extant Brown Algae which must have bridged the gap between the Green Algae and higher plants evolved on the land and have long been extinct, so that all we have today in the Greens are assorted simpler forms that are aquatic or semiterrestrial.

One puzzling problem, however, concerns the fact that although oogamy has developed in a number of orders in the Green Algae, these all have haploid life cycles, whereas alternation of generations has appeared in a few isogamous species alone. Thus, a dichotomy exists which has given rise to separate phases of algal advancement that one might have expected to have progressed concurrently (i.e., we might expect that oogamy would develop along with alternation of generations; see Fritsch, 1949, pp. 113 et seq.).

Fritsch (1945a) summarizes many of the characteristics that were probably possessed by the first transmigrants onto the dry land. They were initially almost certainly heterotrichous, but may have soon abandoned this habit as the prostrate portion was lost. (Some extant advanced ulotrichalean algae have retained solely their erect portion of the plant body.) Initially isogamous, they developed isomorphic alternation of generations and not necessarily concurrently would have evolved toward oogamy and heteromorphism. The acquisition of a parenchymatous nature is important, leading to the possibility of developing large plant bodies (Bower, 1935; Fritsch, 1943). These plants would also have developed a cuticle resistant to desiccation, an increasing tendency toward apical growth, and eventual elaboration of conducting tissues, phloem, and xylem as they acquired the ability to transform metabolic wastes into the useful material lignin, so valuable in making cell walls waterproof, extremely strong, and durable. As mentioned many times, there are examples of remarkably close parallel development in groups of algae whose advanced members now possess combinations of these features. Fritsch repeatedly emphasizes the predominant position occupied by the Ulotrichales in these considerations, with *Fritschiella* being eminently suited as an example of an organism with many of the necessary characteristics. Bower (1935) notes that further massive development of the plant body and localization of growth in organisms derived from forms like *Fritschiella* can easily be visualized as leading to archegoniate plants. But all evidence indicates that *Fritschiella* contains a phycoplast instead of a phragmoplast. Is this really important? I personally suspect that it is, and the readers will now have to decide for themselves on this subject. Dittmer (1964) and Fritsch (1916) suggest that *Coleochaete* is related to the progenitors of the liverworts. (Dittmer also suggests that the Tracheophyta were derived from the siphonous lines, a view little supported elsewhere.) Wesley (1928) and Fritsch (1942a,b) note that *Coleochaete* alone forms more than four cells directly from germinating zygospores, the closest approach in the Green Algae to a sporophyte like that of the liverworts. (The Charales have a protonemal stage at germination; Chapter 7.) Also *Coleochaete* surrounds the oospore with cells that may function in its protection, as do the Charales; the prevalence of this form of covering (archegonium) in primitive land plants testifies to its importance in maintaining the viability of the vulnerable egg cell in a desiccating environment (Bower, 1935).

While extant members of the Ulotrichales may be more or less stabilized at their present phase of development, *Fritschiella* and various other terrestrial Green Algae could now be striving to accomplish the difficult and dangerous adaptation to terrestrial existence that their relatives successfully undertook many millions of years ago. Is *Fritschiella* stuck because (among other reasons) it lacks and cannot now evolve a phragmoplast? We may illuminate these and other interesting problems concerned with the evolution of land plants if we trace the lineage of the phragmoplast and phycoplast in different groups of Green Algae and particularly in the Ulotrichales. Other extremely diverse classes of Red and Brown Algae have also succeeded in evolving large and highly complex plant forms. Why should these classes have failed so completely to adapt to life out of water, when the Green Algae have been successful? Such questions should have fascinating answers, for the successful migration of Green Algae onto the dry land was one of the most important steps in the evolution of life, since it changed the face of the earth and thereby provided a suitable habitat for the development of all other terrestrial creatures and eventually man himself.

APPENDIX I
STEREO-ELECTRON MICROSCOPY

A useful technique in electron microscopy is the generation of three-dimensional images of fixed specimens. Such images can be obtained from sectioned material using transmission electron microscopy, particularly with thick sections and a high-voltage microscope (such as in Figs. 5.35–5.37); scanning microscopy provides easily and conveniently obtainable stereoviews of whole microscopic objects.

The principle of stereomicroscopy is easy to understand. Depth perception is based upon the fact that our eyes view the same scene simultaneously from slightly different angles, and the mind combines the two slightly dissimilar images into one. The perception of depth is generated in part from the parallax of objects with respect to one another when viewed from the two angles. To obtain stereomicrographs of small objects, we need only obtain two images of the object from slightly different angles, and present these two images, the right way around, one to each eye. In practice, the object is photographed in one position, and then it is tilted or rotated a small amount (e.g. about 12°) and photographed again. Care has to be taken to keep the magnification constant. (The detailed method used by the author for obtaining Figs. S.1–S.14 is given in Pickett-Heaps, 1973a.) Prints of the micrographs are then carefully mounted in register (the right way around!). When these pairs are viewed simultaneously, the mind generates the three-dimensional image of the object.

Small stereoviewers are readily available for viewing such pairs of micrographs. However, it is quite easy to learn the knack of visualizing the object without such a viewer. I suggest trying the following procedure:

1. If you are shortsighted, remove glasses.

2. Hold the pictures in front of the eyes, about 12 inches away.

3. Look over the picture and focus on some distant object.

4. Lower the eyes and look THROUGH rather than AT the pictures. One has to focus the image of the pictures at 12 inches while the eyes are diverging, looking at infinity.

5. Rotate the page slightly from side to side. Each eye forms an image of both pictures; when you can get the left image from the left eye into exact register with the right image from the right eye (i.e. while looking through the page), the combination will turn into a three-dimensional view of the object. Rocking the page in this manner is helpful (to me, anyway!) in getting the two images into such exact register.

S.1. *Pediastrum boryanum.*

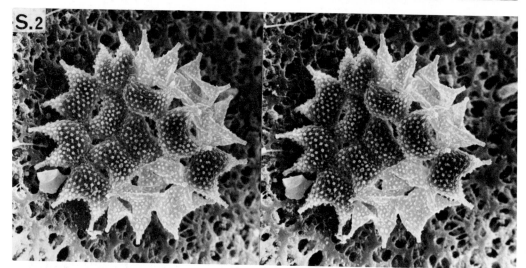

S.2. *P. boryanum;* several of the cells are empty, having formed daughter colonies.

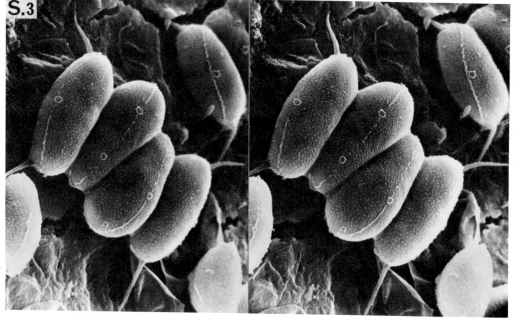

S.3. *Scenedesmus quadricauda* #276.

S.4. Zoospore of *Oedogonium cardiacum* (Fig. 5.67).

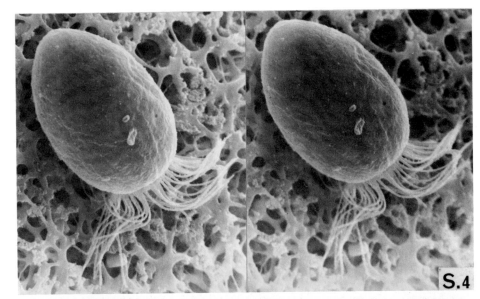

S.5. *Staurastrum manfeldtii.*

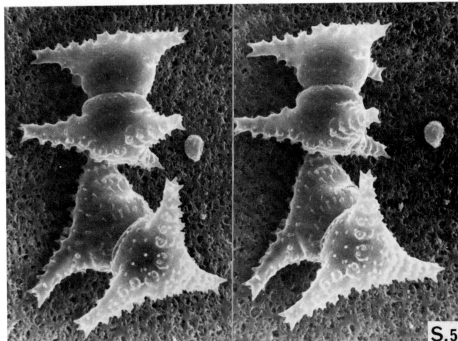

S.6. *St. pingue;* tetraradiate cell (Fig. 6.105).

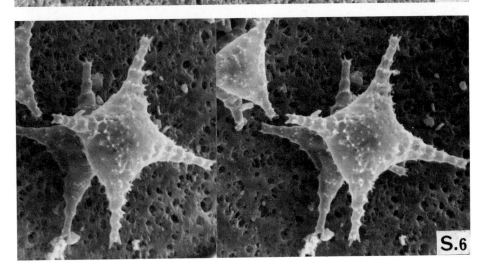

S.7. *St. furcigerum* (Fig. 6.169).

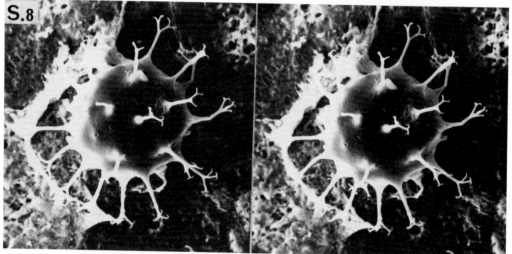

S.8. Zygote of *St. furcigerum* (Fig. 6.171).

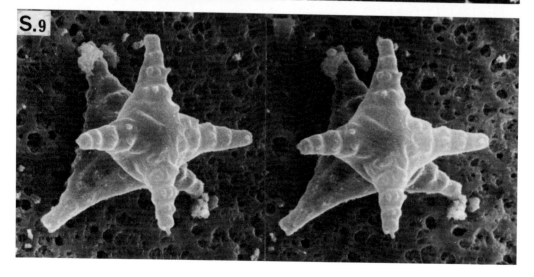

S.9. 'Janus' cell (tri/tetra-radiate) of *St. pingue* (Fig. 6.116).

S.10. *St. gracile*
(Fig. 6.106).

S.11. *Micrasterias radiata;*
recently divided daughter
cells (Fig. 6.109).

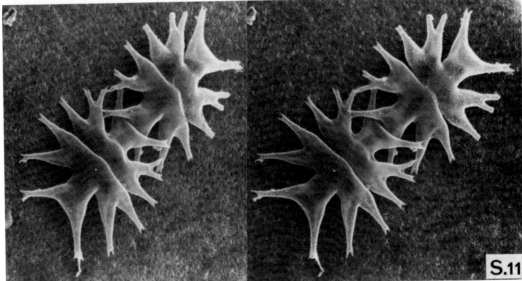

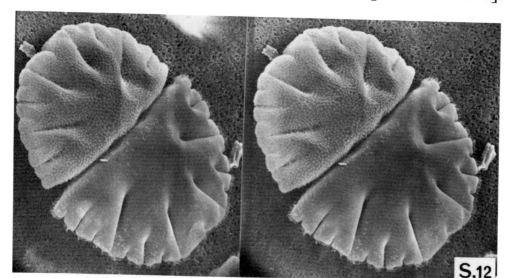

S.12. *M. angulosa.*

S.13. *M. thomasiana;*
triradiate, diploid variant
(Fig. 6.126 C).

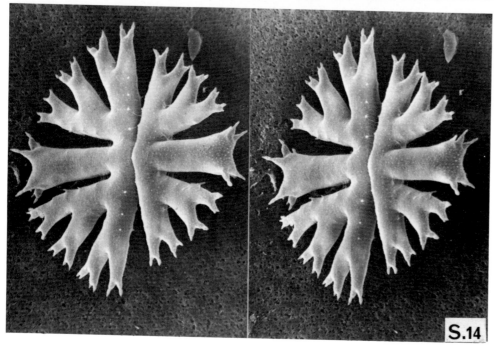

S.14. *M. sol.* var. *extensa*
(Fig. 6.108).

APPENDIX II
FILMS ON GREEN ALGAE

In my teaching experience, films concerned with aspects of the life cycles of algae have proved to be of greatest value for stimulating the interest and involvement of students. Unfortunately, there do not seem to be many such films widely available. I include here a list of films on Green Algae that I have seen and which are, or soon will be, obtainable on loan. This list is not comprehensive; indeed, I would be most grateful to hear from readers of other films that should be included in this list.

Available from the Indiana University Audio Visual Center, Bloomington, Indiana, 47401.

STARR, R. C., and FLATEN, C. M.: *Volvox:* Structure, Reproduction and Differentiation in *V. carteri,* strains HK9 and HK10. Cat. No. FSC 1257.

Available from the Institut für Wissenschaftlichen Film, 34 Göttingen, Nonnenstieg 72, Germany. Some of these are also held by the Pennsylvania State University Audio Visual Services, 6 Willard Building, University Park, Pa. 16802. The titles and subtitles are in German and I have translated them into English for this book.

SCHLOSSER, U.: *Chlamydomonas reinhardi* (Volvocales) - Asexual reproduction. Cat. No. E 1318. 10 min.

GRELL, K.-G. : *Gonium pectorale* (Phytomon-adina) - Asexual development. Cat. No. E 656. 4 1/2 min.

GRELL, K. -G. : *Pleodorina californica* (Phytomonadina) - Asexual development. Cat. No. E 657. 4 1/2 min.

KIERMAYER, O. : *Micrasterias denticulata* (Desmidiaceae) - Morphogenesis. Cat. No. E 868. 10 min.

KIERMAYER, O. : *Micrasterias denticulata* (Desmidiaceae) - Morphogenesis under reduced turgor. Cat. No. E 869. 11 min.

KIES, L. : Sexual reproduction in *Micrasterias papillifera* (Conjugatophyceae). Cat. No. C 1064. 10 min.

PIRSON, A., and KAISER, U. : Asexual reproduction in the green alga *Hydrodictyon reticulatum.* Cat. No. C 1042. 7 1/2 min.

PIRSON, A. and KAISER, U. : Sexual reproduction in the green alga *Hydrodictyon reticulatum.* Cat. No. C 1043. 12 min.

Available from BFA Educational Media, 2211 Michigan Ave., Santa Monica, Calif. 90404.

GREEN, P. B. : Growth and morphogenesis in Plant Cells. (Concerned with cell elongation in *Nitella.*) Cat. No. DB 505. 13 min.

MILLINGTON, W. F. : Colony Formation in *Pediastrum boryanum.* Cat. No. DB 514. 7 min.

APPENDIX III
ON THE USE OF
FIELD SPECIMENS

It may be as well to comment briefly on a technique we employ in our laboratory for ultrastructural research, that causes some misgivings in botanists, namely the use of specimens taken directly from the field and whose species identification may not be possible. There is often (but not always) one overriding advantage with such material—it gives excellent or even superb ultrastructural micrographs, for reasons which are totally obscure. So often, algae are quite difficult to grow in culture and it may be even more difficult to induce these cultures to differentiate (e.g., undergo zoosporogenesis or sexual reproduction) at will. One very serious problem we and many other workers have consistently encountered is the often poor quality of ultrastructural image obtained from cultured algae, even if the cultures are viable and very healthy by other criteria. Sometimes the results are good, but too often they vary from being marginally acceptable to being totally unusable, often because the cytoplasm is extremely dense. This is not just a problem in aesthetics. If the tissue suffers much physical damage (e.g., plasmolysis, etc.) from fixation, artifacts abound, particularly in membrane structures; if the cytoplasm is dense, many important features (e.g., microtubular and microfilamentous systems) are very difficult to discern or may be overlooked entirely.

Thus, the use of material gathered from the wild has proved of great value to research in our own group. The question can be raised: Are the observations obtained from such specimens repeatable by looking at other different species? We believe (and indeed on occasion, have shown) that this is generally the case. Many basic mitotic and morphogenetic phenomena are probably quite consistent within genera and sometimes within orders: for example, vegetative cell division and zoosporogenesis are quite obviously very similar in *Bulbochaete* and *Oedogonium*, but some minor variations have crept in (e.g., the slightly different structure of the zoospore vesicle in the two genera). We would also be very surprised, for example, if our observations on cell division in the unidentified species of *Spirogyra* (Fowke and Pickett-Heaps, 1969a) turned out to be atypical of the genus in general, a conclusion supported by the often remarkably close correlation of our electron micrographs with the drawings of earlier workers (e.g., McAllister, 1913). In contrast, less fundamentally important aspects of cell structure may show considerably more variation even between species in a given genus. An obvious extreme example is the diversity in structure of the scales formed by certain groups of algae, and the ornamentative features of *Scenedesmus* (such features may indeed serve to delineate species of Scenedesmus: see Komarek and Ludvik, 1972), and mutant variants may appear even in the clones of a given strain of an alga.

There are further, more subtle problems in working with some algal material. Attempts to work with "named" species of algae may create a false sense of security. The two species of *Scenedesmus* described in Chapter 4 have both been designated *S. quadricauda;* they are quite different, and Komarek and Ludvik (1972) have clearly shown that they are two different species, neither *S. quadricauda*! This discrepancy is not the fault of the Culture Collections (whose job is not to confirm identifications of cultures placed in their care) and perhaps not even of the workers who deposit the strains. Many green algae are remarkable for their morphological plasticity when grown under

585

different culture conditions, and key diagnostic features may not be discernable with the light microscope. The recent reclassification of *"Ulothrix subtilissima"* as *"Hormidium subtilissimum"* (Mattox and Bold, 1962) has turned out to be of considerably greater significance than could have been guessed at a few years ago. (The later renaming of *Hormidium* as *Klebsormidium* was due to a nomenclatural conflict.) Even the gross morphological features clearly visible with the light microscope and used in some classical phylogenies and taxonomies have occasionally proved very misleading, as has been emphasized in Chapter 8 (see also Stewart and Mattox, 1975).

In summary, we feel use of organisms growing in the wild often presents a source of superb experimental material, often neglected only because it is difficult and sometimes impossible to attach a specific name to it. This is a pity. There are certainly some serious problems to be considered when using such material. There are also problems in using organisms with deceptive labels and taxonomies attached to them, where one's ignorance of their true affinities is not so obvious! One answer of course is to provide living material of specimens studied to Culture Collections, for examination by others who may wish to confirm one's results. It is, however, distressing how often such organisms, blooming in rather smelly ditches and ponds, will quietly expire when one attempts to grow them in the pristine, vitamin-fortified environment of a research laboratory.

BIBLIOGRAPHY

AKINS, W. (1941). A cytological study of *Carteria crucifera.* *Bull. Torrey Bot. Club* 68, 429–445.

ALDRICH, H.C. (1969). The ultrastructure of mitosis in myxamoebae and plasmodia of *Physarum flavicomum. Am. J. Bot. 56,* 290–299.

ALLSOPP, A. (1969). Phylogenetic relationships of the procaryota and the origin of the eucaryotic cell. *New Phytol. 68,* 591–612.

ATKINSON, A.W., JR., GUNNING, B.E.S., and JOHN, P.C.L. (1972). Sporopollenin in the cell wall of *Chlorella* and other algae: Ultrastructure, chemistry, and incorporation of ^{14}C-acetate, studied in synchronous cultures. *Planta 107,* 1–32.

ATKINSON, A.W., JR., GUNNING, B.E.S., JOHN, P.C.L., and McCULLOUGH, W. (1971). Centrioles and microtubules in *Chlorella. Nature (Lond.) 234,* 24–25.

BANKS, H.P. (1970). Major evolutionary events and the geological record of plants. *Biol. Rev. 45,* 451–454.

BANKS, H.P., CHESTERS, K.I.M., HUGHES, N.F., JOHNSON, G.A.L., JOHNSON, H.M., and MOORE, L.R. (1967). Thallophyta I. In *The Fossil Record.* (Symposium published by the Geological Society of London.) Belfast: University Press, pp. 162–180.

BECH-HANSEN, C.W., and FOWKE, L.C. (1972). Mitosis in *Mougeotia* sp. *Can. J. Bot. 50,* 1811–1816.

BEHNKE, O., and FORER, A. (1967). Evidence for four classes of microtubules and pigment migration in the melanophores of *Fundulus heteroclitus* L. *Protoplasma 61,* 322–345.

BELAJEFF, J. (1894). Ueber Bau and Entwichelung der Spermatozoiden der Pflanzen. *Flora, (Jena)* 1–48.

BELCHER, J.H. (1968). The fine structure of *Furcilla stigmatophora* (Skuja) Korshikov. *Arch. Mikrobiol. 60,* 84–94.

BERKALOFF, C. (1966). Cytologie végétale: Observations sur l'organisation infrastructurale d'une volvocale. *C.*

R. Acad. Sci. (Paris) 262, 1232–1234.

BICUDO, C.E. de M., and SORMUS, L. (1972). Polymorphism in the desmid *Micrasterias laticeps* and its taxonomical implications. *J. Phycol. 8,* 237–242.

BIEBEL, P. (1964). The sexual cycle of *Netrium digitus. Am. J. Bot. 51,* 697–704.

BIRBECK, T.R., STEWART, K.D., and MATTOX, K.R. (1974). The cytology and classification of *Schizomeris leibleinii.* II. The structure of quadriflagellate zoospores. *Phycologia 13,* 71–79.

BISALPUTRA, T., ASHTON, F.M., and WEIER, T.E. (1966). Role of dictyosomes in wall formation during cell division in *Chlorella vulgaris. Am. J. Bot. 53,* 213–216.

BISALPUTRA, T., and STEIN, J.R. (1966). The development of cytoplasmic bridges in *Volvox aureus. Can. J. Bot. 44,* 1697–1702.

BISALPUTRA, T., and WEIER, T.E. (1963). The cell wall of *Scenedesmus quadricauda. Am. J. Bot. 50,* 1011–1019.

BISALPUTRA, T., WEIER, T.E., RISLEY, E.B., and ENGELBRECHT, A.H.P. (1964). The pectic layer of the cell wall of *Scenedesmus quadricauda. Am. J. Bot. 51,* 548–551.

BLACK, M., DOWNIE, C., ROSS, R., and SARJEANT, W.A.S. (1967). Thallophyta II. In *The Fossil Record.* (Symposium published by the Geological Society of London.) Belfast: University Press, pp. 181–209.

BOLD, H.C. (1967). *Morphology of Plants.* New York: Harper & Row.

BONNER, J.T. (1971). Morphogenetic Movement in Plants. In *Topics in the Study of Life.* New York: Harper & Row, pp. 176–188.

BOWER, F.O. (1935). *Primitive Land Plants.* London: Macmillan.

BRACKER, C.E. (1967). Ultrastructure of fungi. *Ann. Rev. Phytopathol. 5,* 343–374.

BRADLEY, M.O. (1973). Microfilaments and cytoplasmic streaming: Inhibition of streaming with cytochalasin.

J. Cell Sci. 12, 327–343.

BRANDHAM, P.E. (1965). Polyploidy in desmids. *Can. J. Bot. 43,* 405–417.

———— (1966). Time-lapse studies of conjugation in *Cosmarium botrytis.* I. Gamete fusion and spine formation. *Rev. Algol. 8,* 312–316.

———— (1967). Time-lapse studies of conjugation in *Cosmarium botrytis.* II. Pseudoconjugation and an anisogamous mating behavior involving chemotaxis. *Can. J. Bot. 45,* 483–493.

BRANDHAM, P.E., and GODWARD, M.B.E. (1964). The production and inheritance of the haploid triradiate form in *Cosmarium botrytis. Phycologia 4,* 75–83.

BRANTON, D., and PARK, R.B. (1967). Subunits in chloroplast lamellae. *J. Ultrastruct. Res. 19,* 283.

BRÅTEN, T. (1971). The ultrastructure of fertilization and zygote formation in the green alga *Ulva mutabilis* Føyn. *J. Cell Sci. 9,* 621–635.

———— (1973). Autoradiographic evidence for the rapid disintegration of one chloroplast in the zygote of the green alga *Ulva mutabilis. J. Cell Sci. 12,* 385–389.

BROWN, R. (1960). *Plant Physiology.* (Edited by F. C. Steward.) New York: Academic Press, Vol. IA, p. 3.

BROWN, R.M., JR., and ARNOTT, H.J. (1970). Structure and function of the algal pyrenoid. I. Ultrastructure and cytochemistry during zoosporogenesis of *Tetracystis excentrica. J. Phycol. 6,* 14–22.

BROWN, R.M., JR., JOHNSON, C., and BOLD, H.C. (1968). Electron and phase-contrast microscopy of sexual reproduction in *Chlamydomonas moewusii. J. Phycol. 4,* 100–120.

BROWN, R.M., JR., and McLEAN, R.J. (1969). New taxonomic criteria in classification of *Chlorococcum* species. II. Pyrenoid fine structure. *J. Phycol. 5,* 114–118.

BUFFALOE, N.D. (1958). A comparative cytological study of four species of *Chlamydomonas. Bull. Torrey Bot. Club 85,* 157–178.

BURNSIDE, B. (1971). Microtubules and microfilaments in newt neurulation. *Devel. Biol. 26,* 416–441.

BURTON, P.R. (1968). Effects of various treatments on microtubules and axial units of lung-fluke spermatozoa. *Z. Zellforsch. Mikrosk. Anat. 87,* 226–248.

BYERS, B., and ABRAMSON, D.H. (1968). Cytokinesis in Hela: Post-telophase delay and microtubule-associated motility. *Protoplasma 66,* 413–435.

CAIN, J.R., MATTOX, K.R., and STEWART, K.D. (1973a). The cytology of zoosporogenesis in the filamentous green algae genus *Klebsormidium. Trans. Am. Microsc. Soc. 92,* 398–404.

———— (1973b). Conditions of illumination and zoosporogenesis in *Klebsormidium flaccidum. J. Phycol. 10,* 134–137.

CAMPBELL, E.O., and SARAFIS, V. (1972). *Schizomeris:* A growth form of *Stigeoclonium tenue* (Chlorophyta: Chaetophoroceae). *J. Phycol. 8,* 276–282.

CAROTHERS, Z.B. (1973). Studies of spermatogenesis in the Hepaticae. IV. On the blepharoplast of *Blasia. Am. J. Bot. 60,* 819–828.

CAROTHERS, Z.B., and KREITNER, G.L. (1967). Studies of spermatogenesis in the Hepaticae. I. Ultrastructure of the Vierergruppe in *Marchantia. J. Cell Biol. 33,*

43–51.

———— (1968). Studies of spermatogenesis in the Hepaticae. II. Blepharoplast structure in the spermatid of *Marchantia. J. Cell Biol. 36,* 603–616.

CARTER, N. (1919a). Studies on the chloroplasts of desmids. I. *Ann. Bot. 33,* 215–254.

———— (1919b). Studies on the chloroplasts of desmids. II. *Ann. Bot. 33,* 295–304.

———— (1920a). Studies on the chloroplasts of desmids. III. The chloroplasts of *Cosmarium. Ann. Bot. 34,* 285–289.

———— (1920b). Studies on the chloroplasts of desmids. IV. *Ann. Bot. 34,* 301–319.

CAVALIER-SMITH, T. (1970). Electron microscope evidence for chloroplast fusion in zygotes of *Chlamydomonas reinhardii. Nature (Lond.) 228,* 333–335.

CAVE, M.S., and POCOCK, M.A. (1951). Karyological studies in the Volvocaceae. *Am. J. Bot. 38,* 800–811.

CHADEFAUD, M. (1960). Les végétaux nonvasculaires (cryptogamie). In *Traité de Botanique Systematique.* (Edited by M. Chadefaud and L. Emberger). Vol. 1, pp. 10108 et seq.

CHALONAR, W.G. (1970). The rise of the first land plants. *Biol. Rev. 45,* 353–377.

CHAPMAN, V.J. (1962). *The Algae.* London: Macmillan.

CHRISTENSEN, T. (1962). Alger. In *Botanik (Systematisk Botanik)* 1st edition. (Edited by T.W. Böcher, M. Lange, and T. Sørensen.)

———— (1966). Alger. In *Botanik (Systematisk Botanik),* 2nd edition, (Edited by T.W. Böcher, M. Lange, and T. Sørensen.) pp. 172 et seq.

CHURCH, A.H. (1919). Thalassiophyta and the subaerial transmigration. *Oxf. Bot. Mem. 3.*

CLEMÉNÇON, H., and FOTT, B. (1968). Submikroskopische Strukturen auf der Zellwand einer *Chlorella:* Ähnlichen Alge. *Acta Univ. Carol. 1967,* 185–188.

CLOWES, F.A.L., and JUNIPER, B.E. (1968). *Plant Cells.* Oxford: Blackwell.

COLEMAN, A.W. (1959). Sexual isolation in *Pandorina morum. J. Protozool. 6,* 249–264.

CONARD, A. (1947). Sur la division cellulaire chez *Oedogonium. Acad. Roy. Belg. Classe Sci. 21,* 1–87.

CONWAY, K., and TRAINOR, F.R. (1972). *Scenedesmus* morphology and flotation. *J. Phycol. 8,* 138–143.

COOK, P.W. (1962). Growth and reproduction of *Bulbochaete hiloensis* in unialgal culture. *Trans. Am. Microsc. Soc. 81,* 384–395.

COSS, R.A. (1974). Mitosis in *Chlamydomonas reinhardii.* Basal bodies and the mitotic apparatus. *J. Cell Biol. 63,* 325–329.

COSS, R.A., and PICKETT-HEAPS, J.D. (1973). Gametogenesis in the green alga *Oedogonium cardiacum.* I. The cell divisions leading to formation of spermatids and oogonia. *Protoplasma 78,* 21–39.

———— (1974a). Gametogenesis in the green alga *Oedogonium cardiacum:* II. Spermiogenesis. *Protoplasma 81,* 297–311.

———— (1974b). The effects of isopropyl N-phenyl carbamate on the green alga *Oedogonium cardiacum.* I. Cell division. *J. Cell Biol. 63,* 84–98.

DARDEN, W.H., JR. (1966). Sexual differentiation in *Volvox aureus. J. Protozool. 13*, 239–255.

DAVIS, J.S. (1964). Colony form in *Pediastrum. Bot. Gaz. 125*, 129–131.

_____ (1966a). Akinetes of *Tetraedron. Trans. Am. Microsc. Soc. 85*, 573–575.

_____ (1966b). Akinetes, reproduction and colony form in the green alga *Sorastrum. Trans. Ill. St. Acad. Sci. 59*, 275–280.

_____ (1967). The life cycle of *Pediastrum simplex. J. Phycol. 3*, 95–103.

DEASON, T.R., and DARDEN, W.H., JR. (1971). The Male Initial and Mitosis in *Volvox*. In *Contributions in Phycology*. (Edited by B.C. Parker and R.M. Brown.) Lawrence, Kansas: Allen Press, pp. 67–90.

DEASON, T.R., DARDEN, W.H., JR., and ELY, S. (1969). The development of sperm packets of the M5 strain of *Volvox aureus. J. Ultrastruct. Res. 26*, 85–94.

DE BARY, A. (1854). Ueber die Algengattungen *Oedogonium* und *Bulbochaete*. Abhandlung Senckenberg. *Naturf. Ges. Frankfurt 1*, 29–105.

_____ (1858). *Untersuchungen über die Familie der Conjugaten*. Leipzig.

DELANGE, R.J., CHANG, J.Y., SHAPER, J.H., MARTINEZ, R.J., KOMATSU, S.K., and GLAZER, A.N. (1973). On the amino-acid sequence of flagellin from *Bacillus subtilis* 168: Comparison with other bacterial flagellins. *Proc. Natl. Acad. Sci. U.S.A. 70*, 3428–3431.

DESIKACHARY, T.V., and SUNDARALINGAM, V.S. (1962). Affinities and interrelationships of the Characeae. *Phycologia 2*, 11–16.

DICKERSON, R.E. (1972). The structure and history of an ancient protein. *Sci. Am. 226*, 58–72.

DICKERSON, R.E., and GEIS, I. (1969). *The Structure and Action of Proteins*. New York: Harper & Row.

DIPPELL, R.V. (1968). The development of basal bodies in *Paramecium. Proc. Natl. Acad. Sci. U.S.A. 61*, 461–468.

DITTMER, H.J. (1964). *Phylogeny and Form in the Plant Kingdom*. New York: Van Nostrand Reinhold.

DIXON, P.S. (1970). A critique of the taxonomy of marine algae. *Ann. N.Y. Acad. Sci. 175*, 617–622.

DOLZMANN, R., and DOLZMANN, P. (1964). Untersuchungen über die Feinstruktur und die Funktion der Plasmodesmen von *Volvox aureus. Planta 61*, 332–345.

DUCKETT, J.G. (1973). An ultrastructural study of the differentiation of the spermatozoid of *Equisetum. J. Cell Sci. 12*, 95–129.

ELY, T.H., and DARDEN, W.H. (1972). Concentration and purification of the male-inducing substance from *Volvox aureus* M5. *Microbios 5*, 51–56.

ETTL, V.H. (1971). *Chlamydomonas*, a suitable model-organism for comparative cytomorphological studies. *Arch. Hydrobiol. Suppl. (Algological Studies 5) 39*, 259–300.

ETTL, H., and MANTON, I. (1964). Die feinere Struktur von *Pedinomonas minor* Korschikoff. *Nova Hedwigia 8*, 421–451.

EVANS, L.V., and CHRISTIE, A.O. (1970). Studies on the ship-fouling alga *Enteromorpha*. I. Aspects of the fine-structure and biochemistry of swimming and newly settled zoospores. *Ann. Bot. 34*, 451–466.

FAWCETT, D.W., ANDERSON, W.A., and PHILLIPS, D.M. (1971). Morphogenetic factors influencing the shape of the sperm head. *Devel. Biol. 26*, 220–251.

FAWCETT, D.W., and PORTER, K.R. (1954). A study of the fine structure of ciliated epithelia. *J. Morphol. 94*, 221.

FLOYD, G.L., STEWART, K.D., and MATTOX, K.R. (1971). Cytokinesis and plasmodesmata in *Ulothrix. J. Phycol. 7*, 306–309.

_____ (1972a). Comparative cytology of *Ulothrix* and *Stigeoclonium. J. Phycol. 8*, 68–81.

_____ (1972b). Cellular organization, mitosis and cytokinesis in the Ulotrichalean alga *Klebsormidium. J. Phycol. 8*, 176–184.

FOREST, H.S. (1956). Some aspects of evolution in the isogamous, filamentous Chlorophyceae and their relation to the classification of the Chlorophyceae. *Bull. Torrey Bot. Club 83*, 141–150.

FOTT, B. (1965). Evolutionary tendencies among algae and their position in the plant kingdom. *Preslia (Praha) 37*, 117–126.

_____ (1968). *Chodatella* stages in *Scenedesmus. Acta Univ. Carol. 1967*, 189–196.

FOWKE, L.C., and PICKETT-HEAPS, J.D. (1969a). Cell division in *Spirogyra*. I. Mitosis. *J. Phycol. 5*, 240–259.

_____ (1969b). Cell division in *Spirogyra*. II. Cytokinesis. *J. Phycol. 5*, 273–281.

_____ (1971). Conjugation in *Spirogyra. J. Phycol. 7*, 285–294.

FRASER, T.W., and GUNNING, B.E.S. (1969). The ultrastructure of plasmodesmata in the filamentous green alga, *Bulbochaete hiloensis* (Nordst.) Tiffany. *Planta 88*, 244–254.

_____ (1973). Ultrastructure of the hairs of the filamentous green alga *Bulbochaete hiloensis* (Nordst.) Tiffany: An apoplastidic plant cell with a well-developed Golgi apparatus. *Planta 113*, 1–19.

FREDERICK, S.E., GRUBER, P.J., and TOLBERT, N.E. (1973). The occurrence of glycolate dehydrogenase and glycolate oxidase in green plants: An evolutionary survey. *Plant Physiol. 52*, 318–323.

FREDERICK, S.E., and NEWCOMB, E.H. (1969). Cytochemical localization of catalase in leaf microbodies (peroxisomes). *J. Cell Biol. 43*, 343–353.

FREY-WYSSLING, A., and MUHLETHALER, K. (1965). *Ultrastructural Plant Cytology*. Amsterdam: Elsevier.

FRIEDBERG, I., GOLDBERG, I., and OHAD, I. (1971). A prolamellar body-like structure in *Chlamydomonas reinhardi. J. Cell Biol. 50*, 268–275.

FRIEDMANN, I., COLWIN, A.L., and COLWIN, L.H. (1968). Fine-structural aspects of fertilization in *Chlamydomonas reinhardi. J. Cell Sci. 3*, 115–128.

FRITSCH, F.E. (1902a). The structure and development of the young plants in *Oedogonium. Ann. Bot. 16*, 466–485.

_____ (1902b). The germination of the zoospores in *Oedogonium. Ann. Bot. 16*, 412–417.

_____ (1904). Algological notes. No. 5. Some points in the structure of a young *Oedogonium. Ann. Bot. 18*,

648–653.

——— (1916). The algal ancestry of the higher plants. *New Phytol.* 15, 233–250.

——— (1921). Thalassiophyta and the algal ancestry of the higher plants. *New Phytol.* 20, 165–178.

——— (1922). The terrestrial algae. *J. Ecol.* 10, 220–236.

——— (1929). Evolutionary sequence and affinities among Protophyta. *Biol. Rev.* 4, 103–151.

——— (1935). *The Structure and Reproduction of the Algae*, Vol. I. London: Cambridge University Press.

——— (1942a). Studies in the comparative morphology of the algae. I. Heterotrichy and juvenile stages. *Ann. Bot.* 6, 397–412.

——— (1942b). Studies in the comparitve morphology of the algae. II. The algal life-cycle. *Ann. Bot.* 6, 533–563.

——— (1943). Studies in the comparative morphology of the algae. III. Evolutionary tendencies and affinities among Phaeophyceae. *Ann. Bot.* 7, 63–87.

——— (1945a). Studies in the comparative morphology of the algae. IV. Algae and archegoniate plants. *Ann. Bot.* 9, 1–29.

——— (1945b). *The Structure and Reproduction of the Algae*, Vol. II. London: Cambridge University Press.

——— (1949). The lines of algal advance. *Biol. Rev.* 24, 94–124.

FULTON, C. (1971). Centrioles. In *Origin and Continuity of Cell Organelles*. New York: Springer-Verlag, pp. 170–221.

FULTON, C., and DINGLE, A.D. (1971). Basal bodies, but not centrioles, in *Naegleria*. *J. Cell Biol.* 51, 826–836.

GAWLIK, S.R., and MILLINGTON, W.F. (1969). Pattern formation and the fine structure of the developing cell wall in colonies of *Pediastrum boryanum*. *Am. J. Bot.* 56, 1084–1093.

GEITLER, L. (1960). Spontane Rotation und Oscillation des Chromatophores in den Haarzellen und Zoosporangien von *Coleochaete soluta*. *Planta* 55, 115–142.

——— (1961a). Spontaneous partial rotation and oscillation of the protoplasm in *Coleochaete* and other Chlorophyceae. *Am. J. Bot.* 48, 738–741.

——— (1961b). Über die Polarität der Zellen und Fäden von *Oedogonium*. *Österr. Bot. Z.* 108, 5–19.

GERISCH, G. (1959). Die Zelldifferenzierung bei *Pleodorina californica* Shaw und die Organisation der Phytomonadinenkolonien. *Arch. Protistenk.* 104, 292–361.

GIBBONS, I.R., and GRIMSTONE, A.V. (1960). On flagellar structure in certain flagellates. *J. Cell Biol.* 7, 697–715.

GIBBS, S.P. (1970). The comparative ultrastructure of the algal chloroplast. *Ann. N.Y. Acad. Sci.* 175, 454–473.

GODWARD, M.B.E. (editor) (1966). *The Chromosomes of the Algae*. London: Edward Arnold.

GOLDSTEIN, M. (1964). Speciation and mating behavior in *Eudorina*. *J. Protozool.* 11, 317–344.

——— (1967). Colony differentiation in *Eudorina*. *Can. J. Bot.* 45, 1591–1596.

GOODENOUGH, U.W. (1970). Chloroplast division and pyrenoid formation in *Chlamydomonas reinhardi*. *J. Phycol.* 6, 1–6.

GOODENOUGH, U.W., and STAEHELIN, L.A. (1971). Structural differentiation of stacked and unstacked chloroplast membranes. *J. Cell Biol.* 48, 594–619.

GRAMBAST, L.J., and LACEY, W.S. (1967). Bryophyta and Charophyta. In *The Fossil Record*. (Symposium published by the Geological Society of London.) Belfast: University Press, pp. 211–217.

GRANHOLM, N.H., and BAKER, J.R. (1970). Cytoplasmic microtubules and the mechanism of avian gastrulation. *Devel. Biol.* 23, 563–584.

GREEN, P.B. (1964). Cinematic observations on growth and division of chloroplasts in *Nitella*. *Am. J. Bot.* 51, 334–342.

——— (1965). Pathways of cellular morphogenesis: A diversity in *Nitella*. *J. Cell Biol.* 27, 343–363.

——— (1968). Growth physicis in *Nitella:* A method for continuous in vivo analysis of extensibility based on a micro-manometer technique for turgor pressure. *Plant Physiol.* 43, 1169–1184.

GREEN, P.B., ERICKSON, R.O., and RICHMOND, P.A. (1970). On the physical basis of cell morphogenesis. *Ann. N.Y. Acad. Sci.* 175, 712–731.

GREEN, P.B., and KING, A. (1966). A mechanism for the origin of specifically oriented textures in development with special reference to *Nitella* wall texture. *Aust. J. Biol. Sci.* 19, 421–437.

GREEN, P.B., and STANTON, F.W. (1967). Turgor pressure: Direct manometric measurement in single cells of *Nitella*. *Science (N.Y.)* 155, 1675–1676.

GREMLING, G. (1939). Sur la division cellulaire chez *Microspora amoena* (Kutz). *Bull. Soc. Roy. Bot. Belg.* 62, 49–62.

GRIFFITHS, D.J. (1970). The pyrenoid. *Bot. Rev.* 36, 29–58.

GROVE, S.N., and BRACKER, C.E. (1970). Protoplasmic organization of hyphal tips among fungi: Vesicles and spitzenkörper. *J. Bacteriol.* 104, 989–1009.

GUSSEWA, K. (1931). Über geschlechtliche und ungeschlechtliche Fortpflanzung von *Oedogonium capillare* Ktz. im Lichte der sie Bestimmenden Verhältnisse. *Planta* 12, 293–326.

HAMBURG, L.K. (1972). Geschlechtliche Fortpflanzung von *Micrasterias papillifera* (Conjugatophyceae). Cat. # C 1064 (16-mm film). Göttingen: Wissenschaftlichen Film.

HAWKINS, A.F., and LEEDALE, G.F. (1971). Zoospore structure and colony formation in *Pediastrum* spp. and *Hydrodiction reticulatum* (L.) Langerheim. *Ann. Bot.* 35, 201–211.

HEATH, I.B., and DARLEY, W.M. (1972). Observations on the ultrastructure of the male gametes of *Biddulphia levis* Ehr. *J. Phycol.* 8, 51–59.

HEATH, I.B., and GREENWOOD, A.D. (1968). Electron microscopic observations of dividing somatic nuclei in *Saprolegnia*. *J. Gen. Microbiol.* 53, 287–289.

HEJNOWICZ, Z., and SIEVERS, A. (1971). Mathematical model of geotropically bending *Chara* rhizoids. *Z. Pflanzenphysiol.* 66, 34–48.

HEPLER, P.K., and JACKSON, W.T. (1969). Isopropyl N-phenylcarbamate affects spindle microtubule orientation in dividing endosperm cells of *Haemanthus katherinae* Baker. *J. Cell Sci. 5,* 727–743.

HIBBERD, D.J., and LEEDALE, G.F. (1971). A new algal class: The Eustigmatophyceae. *Taxon 20,* 523–525.

HIGHAM, M.T., and BISALPUTRA, T. (1970). A further note on the surface structure of *Scenedesmus* coenobium. *Can. J. Bot. 48,* 1839–1841.

HILL, G.J.C., and MACHLIS, L. (1968). An ultrastructural study of vegetative cell division in *Oedogonium borisianum. J. Phycol. 4,* 261–271.

HOBBS, M.J. (1971). The fine structure of *Eudorina illinoiensis* (Kofoid) Paschen. *Br. Phycol. J. 6,* 81–103.

HOFFMAN, L.R. (1960). Chemotaxis of *Oedogonium* sperms. *Southwestern Naturalist 5,* 111–116.

_____ (1961). Studies on the Morphology, Cytology and Reproduction of *Oedogonium* and *Oedocladium.* Ph.D. dissertation, University of Texas, Austin.

_____ (1965). Cytological studies of *Oedogonium.* I. Oospore germination in *O. foveolatum. Am. J. Bot. 52,* 173–181.

_____ (1966). The fine structure of zoospore development in *Oedogonium cardiacum. J. Phycol. 2* (Suppl.), 5.

_____ (1967). Observations on the fine structure of *Oedogonium.* III. Microtubular elements in the chloroplasts of *O. cardiacum. J. Phycol. 3,* 212–221.

_____ (1970). Observations on the fine structure of *Oedogonium.* IV. The striated component of the compound flagellar "roots" of *O. cardiacum. Can. J. Bot. 48,* 189–196.

_____ (1971). Observations on the Fine Structure of *Oedogonium.* VII. The Oogonium Prior to Fertilization. In *Contributions in Phycology.* (Edited by B.C. Parker and R.M. Brown.) Lawrence, Kansas: Allen Press, pp. 93–106.

_____ (1973a). Fertilization in *Oedogonium.* I. Plasmogamy. *J. Phycol. 9,* 62–84.

_____ (1973b). Fertilization in *Oedogonium.* II. Polyspermy. *J. Phycol. 9,* 296–301.

HOFFMAN, L.R., and MANTON, I. (1962). Observations of the fine structure of the zoospore of *Oedogonium cardiacum* with special reference to the flagellar apparatus. *J. Exp. Bot. 13,* 443–449.

_____ (1963). Observations of the fine structure of *Oedogonium.* II. The spermatozoid of *O. cardiacum. Am. J. Bot. 50,* 455–463.

HOLDSWORTH, R.H. (1971). The isolation and partial characterization of the pyrenoid protein of *Eremosphaera viridis. J. Cell Biol. 51,* 499–513.

HOPKINS, J.M. (1970). Subsidiary components of the flagella of *Chlamydomonas reinhardii. J. Cell Sci. 7,* 823–839.

HOSHAW, R.W. (1968). Biology of the Filamentous Conjugating Algae. In *Algae, Man and the Environment.* (Edited by D.F. Jackson.) Syracuse: Syracuse University Press.

HOWARD, K.L., and MOORE, R.T. (1970). Ultrastructure of oogenesis in *Saprolegnia terrestris. Bot. Gaz. 131,* 311–336.

HUTT, W., and KOCHERT, G. (1971). Effects of some protein and nucleic acid synthesis inhibitors on fertilization in *Volvox carteri. J. Phycol. 7,* 316–320.

ICHIDA, A.A., and FULLER, M.S. (1968). Ultrastructure of mitosis in the aquatic fungus *Catenaria anguillulae. Mycologia 60,* 141–155.

ICHIMURA, T. (1971). Sexual cell division and conjugation-papilla formation in sexual reproduction of *Closterium strigosum.* In *Proceedings, Seventh International Seaweed Symposium,* Sapporo, Japan, pp. 208–214.

IKUSHIMA, N., and MARUYAMA, S. (1968). The protoplasmic connection in *Volvox. J. Protozool. 15,* 136–140.

INGOLD, C.T. (1973). Cell arrangement in coenobia of *Pediastrum. Ann. Bot. 37,* 389–394.

INOUÉ, S., and SATO, H. (1967). Cell motility by labile association of molecules. *J. Gen. Physiol. 50,* 259–292.

IYENGAR, M.O.P. (1932). *Fritschiella,* a new terrestrial member of the Chaetophoraceae. *New Phytol. 31,* 329–335.

JANE, F.W., and WOODHEAD, N. (1941). The formation of "H-pieces" in the walls of *Ulothrix* and *Hormidium. New Phytol. 40,* 183–188.

JAROSCH, R. (1958). Die Protoplasmafibrillen der Characeen. *Protoplasma 50,* 93–108.

JOHN, P.C.L., McCULLOUGH, W., ATKINSON, A.W., FORDE, B.G., and GUNNING, B.E.S. (1973). The Cell Cycle in *Chlorella.* In *The Cell Cycle in Development and Differentiation.* (Edited by M. Balls and F.S. Billet; Symposium of the British Society for Developmental Biology.) London: Cambridge University Press, pp. 61–76.

JOHNSON, U.G., and PORTER, K.R. (1968). Fine structure of cell division in *Chlamydomonas reinhardi. J. Cell Biol. 38,* 403–425.

JORDAN, E.G., and GODWARD, M.B.E. (1969). Some observations on the nucleolus in *Spirogyra. J. Cell Sci. 4,* 3–15.

KALLIO, P. (1959). The relationship between nuclear quantity and cytoplasmic units in *Micrasterias. Ann. Acad. Sci. Fenn. A IV Biol. 44,* 1–44.

_____ (1968). On the morphogenetic system of *Micrasterias sol. Ann. Acad. Sci. Fenn. A IV Biol. 124,* 5–22.

KALLIO, P., and HEIKKILÄ, H. (1969). UV-induced facies change in *Micrasterias torreyi. Österr. Bot. Z. 116,* 226–243.

KAMITSUBO, E. (1966a). Motile protoplasmic fibrils in cells of Characeae. II. Linear fibrillar structure and its bearing on protoplasmic streaming. *Proc. Jap. Acad. 42,* 640–643.

_____ (1966b). Motile protoplasmic fibrils in cells of Characeae. I. Movement of fibrillar loops. *Proc. Jap. Acad. 42,* 507–511.

_____ (1972). Motile protoplasmic fibrils in cells of the Characeae. *Protoplasma 74,* 53–70.

KAMIYA, N. (1968). Cytoplasmic Streaming. In *Cellular Dynamics.* (Edited by L.D. Peachey; Proceedings, First Interdisciplinary Conference on Cellular Dynamics.) New York: New York Academy of Science, pp. 127–158.

KARLING, J.S. (1926). Nuclear and cell division in *Nitella* and *Chara. Bull. Torrey Bot. Club 53,* 319–379.

KATER, J. McA. (1929). Morphology and division of *Chlamydomonas* with reference to phylogeny of the flagellate neuromotor system. *Univ. Calif. Publ. Zool. 33*, 125–168.

KELLAND, J.L. (1964). Inversion in *Volvox*. Ph.D. thesis, Princeton University.

KIERMAYER, O. (1965a). *Micrasterias denticulata* (desmidiaceae): Morphogenese. Cat. # E 868 (16-mm film). Göttingen: Wissenschaftlichen Film.

———. (1965b). *Micrasterias denticulata* (Desmidiaceae): Morphogenese bie reduziertem Turgor. Cat. # E 869 (16-mm film). Göttingen: Wissenschaftlichen Film.

——— (1966). Septumbildung und Cytomorphogenese von *Micrasterias denticulata* nach der Einwirkung von Äthanol. *Planta 71*, 305–313.

——— (1968). The distribution of microtubules in differentiating cells of *Micrasterias denticulata* breb. *Planta 83*, 223–236

——— (1970a). Causal aspects of cytomorphogenesis in *Micrasterias*. *Ann. N.Y. Acad. Sci. 175*, 686–701.

——— (1970b). Elektronenmikroskopische untersuchungen zum Problem der Cytomorphogenese von *Micrasterias denticulata* Breb. I. Allgemeiner Überblick. *Protoplasma 69*, 97–132.

KIERMAYER, O., and STAEHELIN, L.A. (1972). Feinstruktur von Zellwand und Plasmamembran bei *Micrasterias denticulata* breb. nach Gefrierätzung. *Protoplasma 74*, 227–237.

KIES, V.L. (1967). Über Zellteilung und Zygotenbildung bei *Roya obtusa* (Breb.) West et West. *Mitt. Staatsinst. Allg. Bot. Hamburg 12*, 35–42.

KIES, L. (1968). Uber die zygotenbildung bei *Micrasterias papillifera* Breb. *Flora (Jena) 157*, 301–313.

——— (1970a). Elektronenmikroskopische untersuchungen über Bildung und Struktur der Zygotenwand bei *Micrasterias papillifera* (Desmidiaceae). I. Das Exospor. *Protoplasma 70*, 21–47.

——— (1970b). Elektronenmikroskopische untersuchungen über Bildung und Struktur der Zygotenwand bei *Micrasterias papillifera* (Desmidiaceae). II. Die Struktur von Mesospor und Endospor. *Protoplasma 71*, 139–146.

KIRK, J.T., and TILNEY-BASSETT, R.A.E. (1967). *The Plastids; Their Chemistry, Structure, Growth and Inheritance*. San Francisco: Freeman.

KLEIN, R.M., and CRONQUIST, A. (1967). A consideration of the evolutionary and taxonomic significance of some biochemical, micromorphological, and physiological characters in the Thallophytes. *Q. Rev. Biol. 42*, 105–296.

KOCHERT, G. (1968). Differentiation of reproductive cells in *Volvox carteri*. *J. Protozool. 15*, 438–452.

KOCHERT, G., and OLSON. L.W. (1970a). Ultrastructure of *Volvox carteri*. I. The asexual colony. *Arch. Mikrobiol. 74*, 19–30.

——— (1970b). Endosymbiotic bacteria in *Volvox carteri*. *Trans. Am. Microsc. Soc. 89*, 475–478.

KOCHERT, G., and YATES, I. (1970). A UV-labile morphogenetic substance in *Volvox carteri*. *Devel. Biol. 23*, 128–135.

KOMAREK, V.J., and LUDVIK, J. (1972). Die Zellwandultrastruktur als taxonomisches Merkmal in der Gattung *Scenedesmus*. 2. Taxonomische Auswertung der untersuchen Arten. Algological Studies 6 *(Arch. Hydrobiol.* Suppl. *41)*, 11–47.

KORN, R.W. (1969). Chloroplast inheritance in *Cosmarium turpinii* Breb. *J. Phycol. 5*, 332–336.

KREITNER, G.L. (1970). The Ultrastructure of Spermatogenesis in the Liverwort, *Marchantia polymorpha*. Ph.D. dissertation, University of Illinois, Urbana.

KRETSCHMER, H. (1930). Beitrage zur Zytologie von *Oedogonium*. *Arch. Protistenk. 71*, 101.

KUBAI, D.F., and RIS, H. (1969). Division in the dinoflagellate *Gyrodinium cohnii* (Schiller): A new type of nuclear reproduction. *J. Cell Biol. 40*, 508–528.

KURODA, K. (1964). *Primitive Motile Systems in Cell Biology*. (Edited by R.D. Allen and N. Kamiya.) New York: Academic Press, p. 31.

LANG, N.J. (1963a). Electron microscopy of the Volvocaceae and Astrephomenaceae. *Am. J. Bot. 50*, 280–300.

——— (1963b). An additional ultrastructural component of flagella. *J. Cell Biol. 19*, 631–634.

LEADBEATER, B., and DODGE, J.D. (1967). An electron microscope study of nuclear and cell division in a dinoflagellate. *Arch. Mikrobiol. 57*, 239–254.

LEDBETTER, M.C., and PORTER, K.R. (1970). *Introduction to the Fine Structure of Plant Cells*. New York: Springer-Verlag.

LEEDALE, G.F. (1970). Phylogenetic aspects of nuclear cytology in the algae. *Ann. N.Y. Acad. Sci. 175*, 429–453.

LEMBI, C.A., and HERNDON, W.R. (1966). Fine structure of the pseudocilia of *Tetraspora*. *Can. J. Bot. 44*, 710–712.

LEMBI, C.A., and LANG, N.J. (1965). Electron microscopy of *Carteria* and *Chlamydomonas*. *Am. J. Bot. 52*, 464–477.

LEMBI, C.A., and WALNE, P.L. (1969). Interconnections between cytoplasmic microtubules and basal bodies of tetrasporalean pseudocilia. *J. Phycol. 5*, 202–205.

——— (1971). Ultrastructure of pseudocilia in *Tetraspora lubrica* (Roth) Ag. *J. Cell Sci. 9*, 569–579.

LENZENWEGER, R. von (1968a). Der Verlauf der Zygotenkeimung bei *Micrasterias rotata* (Grev.) Ralfs. *Arch. Protistenk. 111*, 1–11.

——— (1968b). Lebenszyklus und Zygotenkeimung bei der Zieralge *Micrasterias*. *Mikrokosmos 57*, 270–275.

LESSIE, P.E., and LOVETT, J.S. (1968). Ultrastructural changes during sporangium formation and zoospore differentiation in *Blastocladiella emersonii*. *Am. J. Bot. 55*, 220–236.

LEVINE, R.P., and EBERSOLD, W.T. (1960). The genetics and cytology of *Chlamydomonas*. *Ann. Rev. Microbiol. 14*, 197–216.

LING, H.U., and TYLER, P.A. (1972). The process and morphology of conjugation in desmids, especially the genus *Pleurotaenium*. *Br. Phycol. J. 7*, 65–79.

LIPPERT, B.E. (1967). Sexual reproduction in *Closterium moniliferum* and *Closterium ehrenbergii*. *J. Phycol. 3*, 182–

198.

LLOYD, F.E. (1928). Further observations on the behaviour of gametes during maturation and conjugation in *Spirogyra. Protoplasma 4,* 45–66.

LOTT, J.N.A., HARRIS, G.P., and TURNER, C.D. (1972). The cell wall of *Cosmarium botrytis. J. Phycol. 8,* 232–236.

LØVLIE, A. (1969a). Cell size, nucleic acids, and synthetic efficiency in the wild type and a growth mutant of the multicellular algae *Ulva mutabilis* Føyn. *Devel. Biol. 20,* 349–367.

_____ (1969b). Increase in function during growth of a single cell organelle. *J. Cell Biol. 43,* 389–395.

LØVLIE, A., and BRÅTEN, T. (1968). On the division of cytoplasm and chloroplast in the multicellular green alga *Ulva mutabilis* Føyn. *Exp. Cell Res. 51,* 211–220.

_____ (1970). On mitosis in the multicellular alga *Ulva mutabilis* Føyn. *J. Cell Sci. 6,* 109–129.

LOWY, J., and SPENCER, M. (1968). Structure and function of bacterial flagella. *Symp. Soc. Exp. Biol. 22,* 215–236.

LUDUENA, R.F., and WOODWARD, D.O. (1973). Isolation and partial characterization of a- and B-tubulin from outer doublets of sea-urchin sperm and microtubules of chick-embryo brain. *Proc. Natl. Acad. Sci. U.S.A. 70,* 3594–3598.

LUTMAN, B.F. (1910). The cell structure of *Closterium ebrenbergii* and *Closterium moniliferum. Bot. Gaz. 49,* 241–255.

_____ (1911). Cell and nuclear division in *Closterium. Bot. Gaz. 51,* 401–430.

MANTON, I. (1959). Electron microscopical observations on a very small flagellate: The problem of *Chromulina pusilla* Butcher. *J. Marine Biol. Assoc. U.K. 38,* 319–333.

_____ (1964a). Observations with the electron microscope on the division cycle in the flagellate *Prymnesium parvum* Carter. *J. Roy. Microsc. Soc. 83,* 317–325.

_____ (1964b). Observations on the fine structure of the zoospore and young germling of *Stigeoclonium. J. Exp. Bot. 15,* 399–411.

_____ (1965). Some phyletic implications of flagellar structure in plants. *Rec. Adv. Bot. Res. 2,* 1–21.

MANTON, I., CLARKE, B., and GREENWOOD, A.D. (1955). Observations with the electron microscope on biciliate and quadriciliate zoospores in green algae. *J. Exp. Bot. 6,* 126–128.

MANTON, I., KOWALLIK, K., and VON STOSCH, H.A. (1969). Observations on the fine structure and development of the spindle at mitosis and meiosis in a marine centric diatom *(Lithodesmium undulatum).* I. Preliminary survey of mitosis in spermatogonia. *J. Microsc. 89,* 295–320.

_____ (1970). Observations on the fine structure and development of the spindle at mitosis and meiosis in a marine centric diatom *(Lithodesmium undulatum).* I.V The second meiotic division and conclusion. *J. Cell Sci. 7,* 407–443.

MANTON, I., RAYNS, D. G., ETTL, H., and PARKE, M. (1965). Further observations on green flagellates with scaly flagella: The genus *Heteromastix* Korshikov. *J. Marine Biol. Assoc. U.K. 45,* 241–255.

MANTON, I., and VON STOSCH, H.A. (1966). Observations on the fine structure of the male gamete of the marine centric diatom *Lithodesmium undulatum. J. Roy. Microsc. Soc, 85,* 119–134.

MARCENKO, E. (1973). On the nature of bristles in *Scenedesmus. Arch. Mikrobiol. 88,* 153–161.

MARCHANT, H.J. (1974a). Mitosis, cytokinesis and colony formation in *Pediastrum boryanum. Ann. Bot.* In press.

_____ (1974b). Mitosis, cytokinesis and colony formation in the green alga *Sorastrum. J. Phycol. 10,* 107–120.

MARCHANT, H.J., and PICKETT-HEAPS, J.D. (1970). Ultrastructure and differentiation of *Hydrodictyon reticulatum.* I. Mitosis in the coenobium. *Aust. J. Biol. Sci. 23,* 1173–1186.

_____ (1971). Ultrastructure and differentiation of *Hydrodictyon reticulatum.* II. Formation of zooids within the coenobium. *Aust. J. Biol. Sci. 24,* 471–486.

_____ (1972a). Ultrastructure and differentiation of *Hydrodictyon reticulatum.* III. Formation of the vegetative daughter net. *Aust. J. Biol. Sci. 25,* 265–278.

_____ (1972b). Ultrastructure and differentiation of *Hydrodictyon reticulatum.* IV. Conjugation of gametes and the development of zygospores and azygospores. *Aust. J. Biol. Sci. 25,* 279–291.

_____ (1972c). Ultrastructure and differentiation of *Hydrodictyon reticulatum.* V. Development of polyhedra. *Aust. J. Biol. Sci. 25,* 1187–1197.

_____ (1972d). Ultrastructure and differentiation of *Hydrodictyon reticulatum.* VI. Formation of the germ net. *Aust. J. Biol. Sci. 25,* 1199–1213.

_____ (1973). Mitosis and cytokinesis in *Coleochaete scutata. J. Phycol. 9,* 461–471.

MARCHANT, H.J., PICKETT-HEAPS, J.D., and JACOBS, K. (1973). An ultrastructural study of zoosporogenesis and the mature zoospore of *Klebsormidium flaccidum. Cytobios 8,* 95–107.

MARGULIS, L. (1970). *Origin of Eukaryotic Cells.* New Haven: Yale University Press.

MASSALSKI, A., and TRAINOR, F.R. (1971). Capitate appendages on *Scenedesmus* culture 16 walls. *J. Phycol. 7,* 210–212.

MATTOX, K.R. (1971). Zoosporogenesis and Resistent Cell Formation in *Hormidium flaccidum.* In *Contributions in Phycology.* (Edited by B. C. Parker and R. M. Brown, Jr.) Lawrence, Kansas: Allen Press, pp. 137–144.

MATTOX, K.R., and BOLD, H.C. (1962). *Phycological Studies.* III. *The Taxonomy of Certain ulotrichalean Algae.* University of Texas Publication No. 6222. Austin: University of Texas.

MATTOX, K.R., and STEWART, K.D. (1973a). Observations on the zoospores of *Pseudendoclonium basiliense* and *Trichosarcina polymorpha. Can. J. Bot. 51,* 1425–1430.

_____ (1973b). Cytological characterization of the Ulvaceae. *J. Phycol. 9,* 6–7 (abstract).

_____ (1974). A comparative study of cell division in *Trichosarcina polymorpha* and *Pseudendoclonium basiliense* (Chlorophyceae). In preparation.

MATTOX, K.R., STEWART, K.D., and FLOYD, G.L. (1974). The cytology and classification of *Schizomeris leibleinii.* I. The vegetative thallus. *Phycologia 13,* 63–70.

McALLISTER, F. (1913). Nuclear division in *Tetraspora lubrica. Ann. Bot. 27,* 681–695.

——— (1931). The formation of the achromatic figure in *Spirogyra setiformis. Am. J. Bot. 18,* 838–853.

McBRIDE, G. E. (1967). Cytokinesis in the green alga *Fritschiella. Nature (Lond.) 216,* 939.

——— (1968). Ultrastructure of the *Coleochaete scutata* zoospore *J. Phycol. 4,* (Suppl.), 6 (abstract).

——— (1970). Cytokinesis and ultrastructure in *Fritschiella tuberosa* Iyengar. *Arch. Protistenk. 112,* 365–375.

——— (1971). The flagellar base in *Coleochaete* and its evolutionary significance. *J. Phycol. 7* (Suppl.), 13 (abstract).

McDONALD, K. (1972). The ultrastructure of mitosis in the marine red alga *Membranoptera platyphylla. J. Phycol. 8,* 156–166.

McLEAN, R.J., and PESSONEY, F. (1970). A large scale quasi-crystalline lamellar lattice in chloroplasts of the green alga *Zygnema. J. Cell Biol. 45,* 522–531.

——— (1971). Formation and Resistance of Akinetes of *Zygnema. In Contributions in Phycology.* (Edited by B. C. Parker and R. M. Brown, Jr.) Lawrence, Kansas: Allen Press, pp. 145–152.

McREYNOLDS, J.S. (1961). Growth determination in *Hydrodictyon. Bull. Torrey Bot. Club 88,* 397–403.

MILLINGTON, W.F., and GAWLIK, S.R. (1970). Ultrastructure and initiation of wall pattern in *Pediastrum boryanum. Am. J. Bot. 57,* 552–561.

MISHRA, N. C., and THRELKELD, S. F. H. (1968). Genetic studies in *Eudorina. Genet. Res. 11,* 21–31.

MIX, M. (1966). Licht und Elektronenmikropische untersuchungen an Desmidiaceen. XII. Zur Feinstruktur der Zellwande und Mikrofibrillen einiger Desmidiaceen vom *Cosmarium* Typ. *Arch. Mikrobiol. 55,* 116–133.

——— (1969). Zur Feinstruktur der Zellwande in der Gattung *Closterium* (Desmidiaceae) unter besonderer Berücksichtigung des Porensystems. *Arch. Mikrobiol. 68,* 306–325.

MIZUKAMI, I., and GALL, J. (1966). Centriole replication. II. Sperm formation in the fern, *Marsilea,* and the cycad, *Zamia. J. Cell Biol. 29,* 97–111.

MOESTRUP, Ø. (1970). The fine structure of mature spermatozoids of *Chara corallina,* with special reference to microtubles and scales. *Planta 93,* 295–308.

——— (1972). Observations on the fine structure of spermatozoids and vegetative cells of the green alga *Golekinia. Br. Phycol. J. 7,* 169–183.

MOSER, J.W. (1970). An Ultrastructural Study of Spermatogenesis in *Phaoceros laevis* subsp. *carolinianus.* Ph.D. thesis, University of Illinois, Urbana.

MOSER, J.W., and KREITNER, G.L. (1970). Centrosome structure in *Anthoceros laevis* and *Marchantia polymorpha. J. Cell Biol. 44,* 454–458.

MOTTIER, D.M. (1904). The development of the spermatozoid in *Chara. Ann. Bot. 18,* 245–254.

NAGAI, R., and REBHUN, L.I. (1966). Cytoplasmic microfilaments in streaming *Nitella* cells. *J. Ultrastruct. Res. 14,* 571–589.

NORRIS, R.E., and PEARSON, B.R. (1973). Ultrastructure of mitosis and cytokinesis in a marine species of *Pyramimonas. J. Phycol. 9* (Suppl.), 16.

NORSTOG, K.J. (1967). Fine structure of the spermatozoid of *Zamia* with special reference to the flagellar apparatus. *Am. J. Bot. 54,* 831–840.

——— (1968). Fine structure of the spermatozoid of *Zamia:* Observations on the microtubule system and related structures. *Phytomorphology 18,* 350–356.

O'BRIEN, T.P. (1972). The cytology of cell-wall formation in some eukaryotic cells. *Bot. Rev. 38,* 87–118.

OHASHI, H. (1930). Cytological study of *Oedogonium. Bot. Gaz. 90,* 177–197.

OJIMA, S., and TANAKA, K. (1970). Studies on the growth and development in *Spirogyra.* I. Diurnal movement of the filaments. *Sci. Rep. Hirosaki Univ. 17,* 15–26.

OLSON, L.W., and KOCHERT, G. (1970). Ultrastructure of *Volvox carteri.* II. The kinetosome. *Arch. Mikrobiol. 74,* 31–40.

OTT, D.W., and BROWN, R.M. (1972). Light and electron microscopical observations on mitosis in *Vaucheria litorea* Hofman *ex* C. Agardh. *Br. Phycol. J. 7,* 361–374.

——— (1974). Developmental cytology of the genus *Vaucheria.* I. Organization of the vegetative filament. *Br. Phycol. J. 9,* 111–126.

PAOLILLO, D.J., KREITNER, G.L., and REIGHARD, J.A. (1968a). Spermatogenesis in *Polytrichum juniperinum.* I. The origin of the apical body and the elongation of the nucleus. *Planta 78,* 226–247.

——— (1968b). Spermatogenesis in *Polytrichum juniperinum.* II. The mature sperm. *Planta 78,* 248–261.

PARKE, M., and MANTON, I. (1965). Preliminary observations on the fine structure of *Prasinocladus marinus. J. Marine Biol. Assoc. U.K. 45,* 525–536.

PERKINS, F.O. (1970). Formation of centriole and centriole-like structures during meiosis and mitosis in *Labyrinthula* sp. (Rhizopodea, Labyrinthulida). *J. Cell Sci. 6,* 629–653.

PICKETT-HEAPS, J.D. (1967a). The effects of colchicine on the ultrastructure of dividing plant cells, xylem wall differentiation and distribution of cytoplasmic microtubules. *Devel. Biol. 15,* 206–236.

——— (1967b). Ultrastructure and differentiation in *Chara* sp. I. Vegetative cells. *Aust. J. Biol. Sci. 20,* 539–551.

——— (1967c). Ultrastructure and differentiation in *Chara* sp. II. Mitosis. *Aust. J. Biol. Sci. 20,* 883–894.

——— (1967d). Preliminary attempts at ultrastructural polysaccharide localization in root tip cells. *J. Histochem. Cytochem. 15,* 442–455.

——— (1968a). Ultrastructure and differentiation in *Chara* sp. III. Formation of the antheridium. *Aust. J. Biol. Sci. 21,* 255–274.

——— (1968b). Ultrastructure and differentiation in *Chara* (fibrosa). IV. Spermatogenesis. *Aust. J. Biol. Sci. 21,* 655–690.

——— (1968c). Microtubule-like structure in the growing plastids of chloroplasts of two algae. *Planta 81,* 193–200.

_____ (1968d). Further ultrastructural observations on polysaccharide localization in plant cells. *J. Cell Sci.* 3, 55–64.

_____ (1969a). The evolution of the mitotic apparatus: An attempt at comparative ultrastructural cytology in dividing plant cells. *Cytobios 1* (3), 257–280.

_____ (1969b). Preprophase microtubules and stomatal differentiation: Some effects of centrifugation of symmetrical and asymmetrical division. *J. Ultrastruct. Res.* 27, 24–44.

_____ (1969c). Preprophase microtubule bands in some abnormal mitotic cells of wheat. *J. Cell Sci.* 4, 397–420.

_____ (1969d). Preprophase microtubules and stomatal differentiation in *Commelian cyanea*. *Aust. J. Biol. Sci.* 22, 375–391.

_____ (1970a). Some ultrastructural features of *Volvox*, with particular reference to the phenomenon of inversion. *Planta 90*, 174–190.

_____ (1970b). Mitosis and autospore formation in the green alga *Kirchneriella lunaris*. *Protoplasma 70*, 325–348.

_____ (1970c). The behavior of the nucleolus during mitosis in plants. *Cytobios 6*, 69–78.

_____ (1971a). The autonomy of the centriole: Fact or fallacy? *Cytobios 3*, 205–214.

_____ (1971b). Reproduction by zoospores in *Oedogonium*. I. Zoosporogenesis. *Protoplasma 72*, 275–314.

_____ (1972a). Variation in mitosis and cytokinesis in plant cells: Its significance in the phylogeny and evolution of ultrastructural systems. *Cytobios 5*, 59–77.

_____ (1972b). Cell division in *Tetraedron*. *Ann. Bot.* 36, 693–701.

_____ (1972c). Reproduction by zoospores in *Oedogonium*. II. Emergence of the zoospore and motile phase. *Protoplasma 74*, 149–168.

_____ (1972d). Reproduction by zoospores in *Oedogonium*. III. Differentiation of the germling. *Protoplasma 74*, 169–194.

_____ (1972e). Reproduction by zoospores in *Oedogonium*. IV. Cell division in the germling and evidence concerning the possible evolution of the wall rings. *Protoplasma 74*, 195–212.

_____ (1972f). A possible virus infection in the green alga *Oedogonium*. *J. Phycol.* 8, 44–47.

_____ (1972g). Cell division in *Cosmarium*. *J. Phycol.* 8, 343–360.

_____ (1972h). Cell division in *Klebsormidium subtilissimum* (formerly *Ulothrix subtilissima*) and its possible phylogenetic significance. *Cytobios 6*, 167–183.

_____ (1973a). Stereo scanning electron microscopy of desmids. *J. Micros.* 99, 109–116.

_____ (1973b). Cell division and wall structure in *Microspora*. *New Phytol.* 72, 347–355.

_____ (1973c). Cell division in *Tetraspora*. *Ann. Bot.* 37, 1017–1027.

_____ (1973d). Cell division in *Bulbochaete*. I. Divisions utilizing the wall ring. *J. Phycol.* 9, 408–420.

_____ (1974a). Scanning electron microscopy of some cultured desmids. *Trans. Am. Micros. Soc.* 93, 1–23.

_____ (1974b). Cell division in Stichococcus. *Br. Phycol. J.* 9, 63–73.

_____ (1974c). Structural and phylogenetic aspects of microtubular systems in gametes and zoospores of certain green algae. *Bot. J. Linbean. Soc.*

_____ (1974d). Plant Microtubules In *Dynamic Aspects of Ultrastructure.* (Edited by A. W. Robards.) New York: McGraw-Hill.

_____ (1974e). Cell division in *Bulbochaete*. II. Hair cell formation. *J. Phycol.* 10, 148–163.

_____ (1974f). The evolution of mitosis and the eucaryotic condition. *BioSystems.* 6, 37–48.

PICKETT-HEAPS, J.D., and FOWKE, L.C. (1969). Cell division in *Oedogonium*. I. Mitosis, cytokinesis and cell elongation. *Aust. J. Biol. Sci.* 22, 857–894.

_____ (1970a). Cell division in *Oedogonium*. II. Nuclear division in *O. cardiacum*. *Aust. J. Biol. Sci.* 23, 71–92.

_____ (1970b). Cell division in *Oedogonium*. III. Golgi bodies, wall structure and wall formation in *O. cardiacum*. *Aust. J. Biol. Sci.* 23, 93–113.

_____ (1970c). Mitosis, cytokinesis, and cell elongation in the desmid, *Closterium littorale*. *J. Phycol.* 6, 189–215.

_____ (1971). Conjugation in the desmid *Closterium littorale*. *J. Phycol.* 7, 37–50.

PICKETT-HEAPS, J.D., and MARCHANT, H.J. (1972). The phylogeny of the green algae: A new proposal. *Cytobios 6*, 255–264.

PICKETT-HEAPS, J.D., and NORTHCOTE, D.H. (1966a). Organization of microtubules and endoplasmic reticulum during mitosis and cytokinesis in wheat meristems. *J. Cell Sci.* 1, 109–120.

_____ (1966b). Cell division in the formation of the stomatal stomal complex in the young leaves of wheat. *J. Cell Sci.* 1, 121–128.

PICKETT-HEAPS, J.D., and OTT, D.W. (1974). Cell structure and division in *Pedinomonas*. *Cytobios*. In press.

PICKETT-HEAPS, J.D., and STAEHELIN, L.A. (1974). The ultrastructure of *Scenedesmus*. II. Cell division and colony formation. In preparation.

POCOCK, M.A (1960). *Hydrodictyon*: A comparative biological study. *J. S. Afr. Bot.* 26, 167–319.

PORTER, D. (1972). Cell division in the marine slime mold, *Labyrinthula* sp., and the role of the bothrosome in extracellular membrane production. *Protoplasma 74*, 427–448.

POWERS, J.H. (1908). Further studies in *Volvox*. *Trans. Am. Microsc. Soc.* 28, 141–175.

PRESCOTT, G.W. (1968). *The Algae: A Review.* Boston: Houghton Mifflin.

_____ (1970). *How To Know the Fresh Water Algae.* Dubuque, Iowa: Brown.

PRINGSHEIM, N. (1861). Ueber die Dauerschwarme des Wassernetzes. *Monatsb. Berl. Akad.* 775–794.

PROSKAUER, J. (1952). On the nuclear cytology of *Hydrodictyon*. *Hydrobiologia 4*, 399–408.

RANDALL, J. (1969). The flagellar apparatus as a model

organelle for the study of growth and morphopoiesis. *Proc. Roy. Soc., Ser. B 173*, 31–62.

RANDALL, J., WARR, J.R., HOPKINS, J.M., and McVILLTIE, A. (1964). A single-gene mutation of *Chlamydomonas reinhardii* affecting motility: A genetic and electron microscope study. *Nature (Lond.) 203*, 912–914.

RAPER, J. R., and FLEXER, A.S. (1970). The road to diploidy with emphasis on a detour. *Symp. Soc. Gen. Microbiol. 20*, 401–432.

RAWITSCHER-KUNKEL, E., and MACHLIS, L. (1962). The hormonal integration of sexual reproduction in *Oedogonium. Am. J. Bot. 49*, 177–183.

RAYNS, D.G., and GODWARD, M.B.E. (1965). A quantitative study of mitosis in *Eudorina elegans* in culture. *J. Exp. Bot. 16*, 569–580.

RETALLACK, B., and BUTLER, R.D. (1970a). The development and structure of pyrenoids in *Bulbochaete hiloensis. J. Cell Sci. 6*, 229–241.

_____ (1970b). The development and structure of the zoospore vesicles in *Bulbochaete hiloensis. Arch. Mikrobiol. 72*, 223–237.

_____ (1972). Reproduction in *Bulbochaete hiloensis* (Nordst.) Tiffany. I. Structure of the zoospore. *Arch. Mikrobiol. 86*, 265–280.

_____ (1973). Reproduction in *Bulbochaete hiloensis* (Nordst.) Tiffany. II. Sexual reproduction. *Arch. Mikrobiol. 90*, 343–364.

RETALLACK, E.T., and von MALTZAHN, K.E. (1968). Some observations on zoosporogenesis in the female strain of *Oedogonium cardiacum. Can. J. Bot. 46*, 767–771.

REYNOLDS, N. (1940). Seasonal variations in *Staurastrum paradoxum* Meyen. *New Phytol. 39*, 86–89.

RHODES, R.G., and STOFAN, P.E. (1967). *Tetraspora, Chlorosocus*, and *Phaeosphaera*, a unique example of parallel evolution in the algae. *J. Phycol. 3*, 87–89.

RICE, H.V., and LAETSCH, W.M. (1967). Observations on the morphology and physiology of *Marsilea* sperm. *Am. J. Bot. 54*, 856–866.

RINGO, D.L. (1967). Flagellar motion and fine structure of the flagellar apparatus in *Chlamydomonas. J. Cell Biol. 33*, 543–571.

RIS, H., and PLAUT, W. (1962). Ultrastructure of DNA-containing areas in the chloroplast of *Chlamydomonas. J. Cell Biol. 13*, 383–391.

ROBERTS, K., GURNEY-SMITH, M., and HILLS, G.J. (1972). Structure, composition and morphogenesis of the cell wall of *Chlamydomonas reinhardi. J. Ultrastruct. Res. 40*, 599–613.

ROSENBAUM, J.I., MOULDER, J.E., and RINGO, D.L. (1969). Flagellar elongation and shortening in *Chlamydomonas. J. Cell Biol. 41*, 600–619.

ROSOWSKI, J.R., and HOSHAW, R.W. (1971). Results of an attempt to obtain pyrenoids of *Zygnema* by bulk-isolation methods. *J. Phycol. 7*, 312–316.

ROSS, I.K. (1968). Nuclear membrane behavior during mitosis in normal and heteroploid myxomycetes. *Protoplasma 66*, 173–184.

ROUND, F.E. (1963). The taxonomy of the Chlorophyta. *Br. Phycol. J. 2*, 224–235.

_____ (1971). The taxonomy of the Chlorophyta. II. *Br. Phycol. J. 6*, 235–264.

SCAGEL, R.F., BRANDONI, R.J., ROUSE, G.E., SCHOFIELD, W.B., STEIN, J.R., and TAYLOR, T.M.C. (1965). *An Evolutionary Survey of the Plant Kingdom.* Belmont, California: Wadsworth.

SCHLÖSSER, U. (1966). *Chlamydomonas reinhardi* (Volvocales): Asexuelle Vermehrung. Cat. # E 1318, (16-mm., silent film). Göttingen: Wissenschaftlichen Film.

SCHOPF, J.W. (1970). Precambrian micro-organisms and evolutionary events prior to the origin of vascular plants. *Biol. Rev. 45*, 319–352.

SCHÖTZ, F., BATHELT, H., ARNOLD, C., and SCHIMMER, O. (1972). Die Architektur und Organisation der *Chlamydomonas*: Zelle Ergebnisse der Elektronenmikroskopie von Serienschnitten und der daraus resultierenden dreidimensionalen Rekonstruktion. *Protoplasma 75*, 229–254.

SCHROEDER, T.E. (1970). The contractile ring. I. Fine structure of dividing mammalian (HeLa) cells and the effects of cytochalasin B. *Z. Zellforsch. Mikrosk. Anat. 109*, 431–449.

SELMAN, G.G. (1966). Experimental evidence for the nuclear control of differentiation in *Micrasterias. J. Embryol. Exp. Morphol. 16*, 469–485.

SHELANSKI, M.L., and TAYLOR, E.W. (1968). Properties of the protein subunit of central-pair and outer-doublet microtubules of sea urchin flagella. *J. Cell Biol. 38*, 304–315.

SIEVERS, A. (1965). Elektronenmikroskopische untersuchunger zur geotropischen Reaktion. I. Über besonderheiten im feinbau der Rhizoide von *Chara foetida. Z. Pflanzenphysiol. 53*, 193–213.

_____ (1967a). Elektronenmikroskopische untersuchungen zur geotropischen Reaktion. II. Die polare Organisation des normal wachsenden Rhizoids von *Chara foetida. Protoplasma 64*, 225–253.

_____ (1967b). Elektronenmikroskopische untersuchungen zur geotropischen Reaktion. III. Die transversale Polarisierung der Rhizoidspitze von *Chara foetida* nach 5 bis 10 minuten Horizontallage. *Z. Pflanzenphysiol. 57*, 462–473.

SIEVERS, A., and SCHRÖTER, K. (1971). Versuch einer, Kausalanalyse der geotropischen Reaktionskette im *Chara*-rhizoid. *Planta 96*, 339–353.

SILVA, P.C., MATTOX, K.R., and BLACKWELL, W.H. (1972). The generic name *Hormidium* as applied to green algae. *Taxon 21*, 639–645.

SINGH, R.N. (1941). On some phases in the life history of the terrestrial alga, *Fritschiella tuberosa* Iyeng., and its autecology. *New Phytol. 40*, 170–182.

_____ (1947). *Fritschiella tuberosa* Iyeng. *Ann. Bot. 11*, 159–164.

SMITH, G.M. (1914). The cell structure and colony formation in *Scenedesmus. Arch. Protistenk. 32*, 278–297.

_____ (1916). Cytological studies in the Protococcales. II. Cell structure and zoospore formation in *Pediastrum boryanum* (Turp.), Menagh. *Ann. Bot. 30*, 467–479.

_____ (1918). Cytological studies in the Protococcales. III. Cell structure and autospore formation in *Tetraedron minimum* (A. Br.), Hansg. *Ann. Bot. 32*,

459–464.

——————— (1950). *The Fresh-Water Algae of the United States*, 2nd edition. New York: McGraw-Hill.

SOEDER, C.J. (1965). Elektronmikroskopische untersuchungen die Protoplastenteileung bei *Chlorella fusca* Shihira et Krauss. *Arch. Mikrobiol. 50*, 368–377.

SOYER, M.O. (1971). Structure du noyau des *Blastodinium* (dinoflagelles parasites): Division et condensation chromatique. *Chromosoma 33*, 70–114.

SPOONER, B.S., and WESSELLS, N.K. (1972). An analysis of salivary gland morphogenesis: Role of cytoplasmic microfilaments and microtubules. *Devel. Biol. 27*, 38–54.

STAEHELIN, L.A., and KIERMAYER, O. (1970). Membrane differentiation in the Golgi complex of *Micrasterias denticulata* breb. visualized by freeze-etching. *J. Cell Sci. 7*, 787–792.

STAEHELIN, L.A., and PICKETT-HEAPS, J.D. (1974). The Ultrastructure of *Scenedesmus* (Chlorophyceae). I. Species with the "reticulate" or "warty" type of ornamental layer. In preparation.

STARR, R.C. (1949). Zygospore germination in *Cosmarium botrytis* var. *subtumidum. Proc. Natl. Acad. Sci. U.S.A. 35*, 453–456.

——————— (1954a). Heterothallism in *Cosmarium botrytis* var. *subtumidum. Am. J. Bot. 41*, 601–607.

——————— (1954b). Reproduction by zoospores in *Tetraedron bitridens. Am. J. Bot. 41*, 17–20.

——————— (1955a). Zygospore germination in *Cosmarium botrytis* var. *subtumidum. Am. J. Bot. 42*, 577–581.

——————— (1955b). Isolation of sexual strains of placoderm desmids. *Bull. Torrey Bot. Club 82*, 261–265.

——————— (1958). The production and inheritance of the triradiate form in *Cosmarium turpinii. Am. J. Bot. 45*, 243–248.

——————— (1959). Sexual reproduction in certain species of *Cosmarium. Arch. Protistenk. 104*, 155–164.

——————— (1964). The culture collection of algae at Indiana University. *Am. J. Bot. 51*, 1013–1044.

——————— (1969). Structure, reproduction and differentiation in *Volvox carteri* f. *nagariensis* Iyengar, strains HK 9 and 10. *Arch. Protistenk. 111*, 204–222.

——————— (1970a). Control of differentiation in *Volvox. Devel. Biol. Suppl. 4*, 59–100.

——————— (1970b). *Volvox pocockiae*, a new species with dwarf males. *J. Phycol. 6*, 234–239.

——————— (1971). Sexual Reproduction in *Volvox africanus.* In *Contributions in Phycology.* (Edited by B. C. Parker and R. M. Brown.) Lawrence, Kansas: Allen Press, pp. 59–66.

STARR, R.C., and FLATEN, C.M. (1970). *Volvox:* Structure, reproduction and differentiation in *V. carteri,* strains HK9 and HK10. Cat. # FSC 1257 (16-mm film). Bloomington: Indiana University Audio-Visual Center.

STEINECKE, F. (1929). Hemizellulosen bei *Oedogonium. Bot. Arch. 24*, 391–403.

STEVENSON, I., and LLOYD, F.P. (1971a). Ultrastructure of nuclear division in *Paramecium aurelia.* I. Mitosis in the micronucleus. *Aust. J. Biol. Sci. 24*, 963–975.

——————— (1971b). Ultrastructure of nuclear division in *Paramecium aurelia.* II. Amitosis of the macronucleus. *Aust. J. Biol. Sci. 24*, 977–987.

STEWART, K.D., and MATTOX, K.R. (1975). Comparative cytology, evolution and classification of the green algae with some consideration of the origin of other organisms with Chlorophylls *a* and *b. Bot. Rev.* (in press).

STEWART, K.D., FLOYD, G.L., MATTOX, K.R., and DAVIS, M.E. (1972). Cytochemical demonstration of a single peroxisome in a filamentous green alga. *J. Cell Biol. 54*, 431–434.

STEWART, K.D., MATTOX, K.R., and CHANDLER, C.D. (1974). Mitosis and cytokinesis in *Platymonas subcordiformis,* a scaly green flagellate. *J. Phycol. 10*, 65–80.

STEWART, K.D., MATTOX, K.R., and FLOYD, G.L. (1973). Mitosis, cytokinesis, the distribution of plasmodesmata and the cytological characteristics in the Ulotrichales, Ulvales and Chaetophorales: Phylogenetic and taxonomic considerations. *J. Phycol. 9*, 128–140.

STRASBURGER, E. (1880). *Zellbildung und Zellteilung,* 3rd edition. Jena.

SZOLLOSI, D., COLARCO, P., and DONAHUE, R.P. (1972). Absence of centrioles in the first and second meiotic spindles of mouse oocytes. *J. Cell Sci. 11*, 521–541.

TEWS, L.L. (1969). Dimorphism in *Cosmarium botrytis* var. *depressum. J. Phycol. 5*, 270–271.

TIFFANY, L.H. (1928). The algal genus *Bulbochaete. Trans. Am. Microsc. Soc. 47*, 121–177.

——————— (1957). The Oedogoniaceae. III. *Bot. Rev. 23*, 47–63.

TILDEN, J.E. (1968). *The Algae and Their Life Relations.* New York: Hafner.

TILNEY, L.G., and GIBBONS, J.R. (1968). Differential effects of antimitotic agents on the stability and behavior of cytoplasmic and ciliary microtubules. *Protoplasma 65*, 167–179.

TIPPIT, D.H., and PICKETT-HEAPS, J.D. (1974). Experimental investigations into morphogenesis in *Micrasterias. Protoplasma. 81*, 271–296.

TRAINOR, F.R. (1963). Zoospores in *Scenedesmus obliquus. Science (N.Y.) 142*, 1673–1674.

——————— (1964). The effect of composition of the medium on morphology in *Scenedesmus obliquus. Can. J. Bot. 42*, 515–518.

——————— (1966). Phototaxis in *Scenedesmus. Can. J. Bot. 44*, 1427–1429.

——————— (1969). *Scenedesmus* morphogenesis: Trace elements and spine formation. *J. Phycol. 5*, 185–190.

——————— (1971). Development of Form in *Scenedesmus.* In *Contributions in Phycology.* (Edited by B. C. Parker and R. M. Brown.) Lawrence, Kansas: Allen Press, pp. 81–92.

TRAINOR, F.R., and BURG, C.A. (1965a). *Scenedesmus obliquus* sexuality. *Science (N.Y.) 148*, 1094–1095.

——————— (1965b). Detection of bristles in *Scenedesmus* species. *J. Phycol. 1*, 139–144.

TRAINOR, F.R., and MASSALSKI, A. (1971). Ultrastructure of *Scenedesmus* strain 614 bristles. *Can. J. Bot. 49*, 1273–1276.

TRAINOR, F.R., and ROSKOSKY, F.G. (1967). Control of unicell formation in a soil *Scenedesmus. Can. J. Bot. 45*, 1657–1664.

TRAINOR, F.R., and ROWLAND, H.L. (1968). Control of colony and unicell formation in a synchronized *Scenedesmus. J. Phycol. 4*, 310–317.

TSCHERMAK, E. (1943). Vergleichende und experimentelle Cytologische untersuchungen an der Gattung *Oedogonium. Chromosoma 5*, 493–518.

TURNER, F.R. (1968). An ultrastructural study of plant spermatogenesis: Spermatogenesis in *Nitella. J. Cell Biol. 37*, 370–393.

————— (1970). The effects of colchicine on spermatogenesis in *Nitella. J. Cell Biol. 46*, 220–234.

TUTTLE, A.H. (1910). Mitosis in *Oedogonium. J. Exp. Zool. 9*, 143–157.

UNDERBRINK, A.G., and SPARROW, A.G. (1968). The fine structure of the alga *Brachiomonas submarina. Bot. Gaz. 129*, 259–266.

VAN WISSELINGH, C. (1909). Zur Physiologie der Spirogyrazelle. *Beih. Bot. Centralbl. 24*, 133–210.

VARMA, A.K., and MITRA, A.K. (1964). Observations on the life history of *Fritschiella tuberosa* Iyengar. *Bull. Jap. Soc. Phycol. 12*, 44–47.

VIRGIN, H.I. (1968). Light and chloroplast movements. *Symp. Soc. Exp. Biol. 22*, 329–352.

WADDINGTON, C.H. (1963). Ultrastructure aspects of cellular differentiation. *Symp. Soc. Exp. Biol. 17*, 85–97.

WAGNER, G., HAUPT, W., and LAUX, A. (1972). Reversible inhibition of chloroplast movement by cytochalasin B in green alga *Mougeotia. Science (N.Y.) 176*, 808–809.

WALNE, P.L. (1967). The effects of colchicine on cellular organization in *Chlamydomonas*. II. Ultrastructure. *Am. J. Bot. 54*, 564–577.

WALNE, P.L., and ARNOTT, H.J. (1967). The comparative ultrastructure and possible function of eyespots: *Euglena granulata* and *Chlamydomonas eugametos. Planta 77*, 325–353.

WANKA, F. (1965). The use of colchicine in investigation of the life cycle of *Chlorella. Arch. Mikrobiol. 52*, 305–318.

————— (1968). Ultrastructural changes during normal and colchicine inhibited cell division of *Chlorella. Protoplasma 66*, 105–130.

WANKA, F., and MULDERS, P.F.M. (1967). The effect of light on DNA synthesis and related processes in synchronous cultures of *Chlorella. Arch. Mikrobiol. 58*, 257–269.

WARIS, H., and KALLIO, P. (1964). Morphogenesis in *Micrasterias. Adv. Morphog. 4*, 45–82.

WARR, J.R., McVITTIE, A., RANDALL, J., and HOPKINS, J.M. (1966). Genetic control of flagellar structure in *Chlamydomonas reinhardii. Genet. Res. 7*, 335–351.

WATSON, M.W., and ARNOTT, H.J. (1973). Ultrastructural morphology of *Microthamnion* zoospores. *J. Phycol. 9*, 15–27.

WESLEY, O.C. (1928). Asexual reproduction in *Coleochaete. Bot. Gaz. 86*, 1–31.

————— (1930). Spermatogenesis in *Coleochaete scutata. Bot. Gaz. 89*, 180–191.

WESSELS, N.K., SPOONER, B.S., ASH, J.F., BRADLEY, M.O., LUDUENA, M.A., TAYLOR, E.L., WRENN, J.T. and YAMADA, K.M. (1971). Microfilaments in cellular and developmental processes. *Science (N.Y.) 171*, 135–143.

WEST, G.S., and FRITSCH, F.E. (1926). *A Treatise on the British Fresh Water Algae,* 2nd edition. London: Cambridge University Press.

WEST, K.R., and PITMAN, M.G. (1967). Ionic relations and ultrastructure in *Ulva lactuca. Aust. J. Biol. Sci. 20*, 901–914.

WIESE, L. (1969). Algae. In *Fertilization*. (Edited by C. B. Metz and A. Monroy.) New York: Academic Press, Vol. II. pp. 135–188.

WILSON, H.J., WANKA, F., and LINSKENS, H.F. (1973). The relationship between centrioles, microtubules and cell plate initiation in *Chlorella pyrenoidosa. Planta 109*, 259–267.

WINTER, P.A., and BIEBEL, P. (1967). Conjugation in a heterothallic *Staurastrum. Proc. Pennsylvania Acad. Sci. 40*, 76–79.

WITMAN, G.B., CARLSON, K., BERLINER, J., and ROSENBAUM, J.L. (1972). *Chlamydomonas* flagella. I. Isolation and electrophoretic analysis of microtubules, matrix, membranes, and mastigonemes. *J. Cell Biol. 54*, 507–539.

WITMAN, G.B., CARLSON, K., and ROSENBAUM, J.I. (1972). *Chlamydomonas* flagella. II. The distribution of tubulins 1 and 2 in the outer doublet microtubules. *J. Cell Biol. 54*, 540–555.

WOOD, R. D. (1952). The Characeae. *Bot. Rev. 18*, 317–323.

WOODCOCK, C.L.F., and MILLER, G.J. (1973). Ultrastructural features of the life cycle of *Acetabularia mediterranea*. I. Gametogenesis. *Protoplasma 77*, 313–329.

ZICKLER, D. (1970). Division spindle and centrosomal plaques during mitosis and meiosis in some Ascomycetes. *Chromosoma 30*, 287–304.

ACKNOWLEDGMENTS

It is a pleasure to acknowledge those who have contributed in diverse ways to the writing of this book. In particular, my colleagues, Drs. Larry Fowke, Harvey Marchant, Donald Ott, Kent McDonald, Ron Coss and David Tippit, have provided endless stimulation in the laboratory, besides doing much of the hard work for the research described. I have been doubly fortunate in working in two departments whose chairmen, besides being exemplary scientists, have always been sympathetic, most encouraging and ever ready to give support to my group: Dr. Denis Carr in Canberra and Dr. Keith Porter in Boulder. Many research scientists in the field have also allowed me to use their beautiful micrographs; they are acknowledged in the figure legends. Without such help, a book such as this is almost impossible to produce. Ms. Elizabeth Marchant, Judy Andreozzi and Peggy Lloyd put in many long hours printing micrographs, and typing and editing the manuscript, with astonishing good will. I would also like to use this opportunity to acknowledge the grant support given by our research program from both the National Science Foundation, and the National Institutes of Health, Department of Health, Education and Welfare; without their support, our own efforts would indeed have been impossible. Finally, I am grateful to the editors of the following journals for permission to reproduce micrographs: The American Journal of Botany, Annals of Botany, Archiv fur Protistenkunde, The Australian Journal of Biological Sciences, The British Phycological Journal, "Contributions in Phycology", Cytobios, Developmental Biology, Osterrische Botanische Zeitschrift, The Journal of Cell Biology, The Journal of Cell Science, The Journal of Microscopy, The Journal of Protozoology, The Journal of Phycology, Microkosmos, Nature, The New Phytologist, Planta, Protoplasma. Transactions of the American Microscopical Society, and Zeitschrift fur Pflanzenphysiologie. Reproductions from the Canadian Journal of Botany are by permission of the National Research Council of Canada.

INDEX

Acetabularia, 432, 536
acetolysis, 80, 102, 112, 148
actinomycin D, 432, 436
akinetes, 18, 90, 150
  formation, 382 *et seq.*
alternation of generations, 210, 218, 574 *et seq.*
  isomorphic, 210, 218, 574
  heteromorphic, 212, 574
amitosis, 470, 516
amyloplasts, 510
androgonidia, 60 *et seq.*
androspore, 268, 298
  germination, 270, 298
angiosperms, 524–526
anisogamy, 14, 170, 378, 382, 454, 560
Ankistrodesmus, 72, Sect. 3.4
antheridia, 188, 272 *et seq.*, 298, 300, 470, 478, Sect. 7.4, 504, 506, 508, 510
Anthoceros, pyrenoid, 24
apical cap (gamete), 130 *et seq.*
aplanospores, 186
ascomycetes, 518, 524, 568
autospores, 42, 70–82, 86, 90, 94–98, 108, 110–112, 526, 544–546, 570–572
  scales, 562
axillary shoot, 468, 482
azygospores (azygotes), 70, 90, 114–116, 130 *et seq.*, 142

bacteria, endosymbiotic, 63
basal body (see centriole), 26–30, 88, 244, 262, 496–498, 514 *et seq.*, 540, 563, Sect. 8.4a
  complex, 36, 44–46, 50, 66–67, 498
  extra, non-functional, 28, 63, 528, 563
  origin de novo, 72, 246, 526
  replication, 44, 48, 66, 564
basal particles, 250, 262, 276
basidiomycetes, 516, 524, 568
Biddulphia (flagella), 30
Binuclearia, 570

Blasia, 534
Blastocladiella, 569
Brachiomonas, 42
bristles, *Pediastrum*, 140
  Scenedesmus, 98, 104
Brown algae, 170, 212, 574–76
bryophytes, 467, 496, 522–526, 534 *et seq.*, Sect. 8.3c, 565, 574–575
  centrioles, 88
  pyrenoids, 24
  sperm, 194, 496, 504, 532 *et seq.*, 556
Bulbochaete, 220, 224, 232, 270, 278–80, Sect. 5.2, 572
  hair cell formation, 278–280
Bumilleria, 572

caffeine, 438
calcification of walls, 474, 508, 564
capitulum (primary & secondary), 490 *et seq.*, 510
caps, division (Oedogoniales), 232 *et seq.*, 284 *et seq.*, 294 *et seq.*, 302 *et seq.*, 572
carotenoids, 82
Carteria, 22, 514
caryogamy, 58, 132, 216, 366, 380, 410
Catenaria, 569
cell division, asymmetrical (unequal)
  Charales, 468, 478
  Coleochaete, 194
  Oedogoniales, 272, 276, 286 *et seq.*, 298, 302, 306, 310, 314
  Volvocales, 12, 59–62
cell plate, 54, 78, 170, 190, 196, 202–204, 218, 274, 374, 386, 438, 480–482, 494, 508, 520, 548–550, 554, 570
centriole (basal body), 26, 30, 44, 76, 84, 88, 92, 106, 116, 144, 172–174, 178, 182–184, 188, 196–204, 214, 480, 490, 494–496, 502, 510, 514 *et seq.*, 540–544, 558, 571
  assembly, 28, 66
  complex, 36, 52, 84–86, 94, 120

loss, 78
origin de novo, 2, 72, 84, 88, 126, 138, 246, 274, 302, 492, 526, Sect. 8.4a
replication, 88, 198
Chaetophora, 573
Chaetophorales, 167–168, 204, 467, 556, 562, 573–574
Chara, 114, 182, 186, 218, 402, 468 *et seq.,* Sects. 7.1–7.6, 522–524, 532, 556
flagellar apparatus, 32, 186
manchette microtubules, 34, Sect. 7.5
Charales, 186–188, 194, Chap. 7, 522, 534, 558–560, 565, 573–575
fossil, 565
Charophyta, (see Charales), 467
chemotaxis, 56, 60, 128, 268, 278, 280, 454
Chlamydomonadineae, 8, 20, 568
Chlamydomonas, Chap. 2, 69–70, 74–76, 108, 120, 130, 168, 206, 216–218, 496, 514, 520, 528–530, 540, 544, 560, 563–565, 568–571, 574
mating types, 14, 56
Chlorella, 42, 70–72, Sect. 3.2, 102, 106, 110, 148, 520, 542 *et seq.,* 562
Chlorellaceae, 72
Chlorococcaceae, 72
Chlorococcales, Chap. 3, 186, 198, 532, 542 *et seq.,* 558 *et seq.,* 565, Sect. 8.4c, 573–574
derivation from Volvocales, 42, 66, 69 *et seq.,* 74, 540
Chlorococcum, 70–72, Sect. 3.1, 78, 542 *et seq.,* 574
Chlorococcus, 568
Chlorodendrineae, 568
Chlorogonium, 16, 54
chloroplast, 24 *et seq.,* 74, 90, 116, 138, 144, 148, 154, 176, 184, 188–192, 198, 204, 218–220, 246, 288–290, 358–360, 368, 382–384, 388, 410, 414, 444, 460, 470–472, 476, 490, 510, 571
division, 40, 48, 76, 86, 172, 178, 188, 196, 214, 230, 286, 360, 368, 382, 390 *et seq.,* 400, 412, 416, 450, 564
fusion during sexual reproduction, 59, 218
inheritance, 218, 462
membrane structure, 26
rotation, 384
chloroplast DNA, 24
recombination, 59
choanoid body, 58
Chodatella,, 98
Chromophyta, 565
chromoplasts, 490
chromosomes, polytene, 42
Chromulina, 28
Chrysophyceae, 568
Cladophora, 574
Cladophorales, 546, 569
Closterium, 360–362, 366, 374, 382, 386, Sect. 6.3a, 414, 424, 444–448, 452–454, 460, 554
Coelastraceae, 72, 96
coenobia, 8 *et seq.,* 59 *et seq.,* 96 *et seq.,* 112 *et seq.*
coenocytic condition, 92, 112 *et seq.,* 134, 140, 168,

468, 516, 542 *et seq.,* 550–552, 569
colchicine, 30, 65, 78, 128, 146, 192, 240, 362, 384, 400–402, 442, 476, 492, 504, 510, 550
effects on colony formation 128, 146, 154, 550
—flagellar regeneration, 38
—inversion, 64
—nuclear migration, 362, 394, 442
—spermatogenesis, 502 *et seq.*
Coleochaetaceae, 186, 560, 573
Coleochaete, 170, 174, 182, 186, Sect. 4.3, 218, 482, 504, 532. 556 *et seq.,* 564, 574–575
Conjugales, Chap. 6, 524, 550, 554, 556 *et seq.,* 573–574
conjugation, 362 *et seq.,* 368, 376 *et seq.,* 382–384, 404 *et seq.,* 452 *et seq.*
lateral, 376
scalariform, 376
conjugation tube, 362–366, 376 *et seq.,* 382, 410–412, 454, 458–60
contractile vacuoles, 22, 34, 118–120, 184, 206, 256, 280, 362, 380, 454, 526, 565
replication, 40
coronet, 508 *et seq.*
cortical cells *(Chara),* 468–470, 478
Cosmarium, 18, 236, 374, 386, 394, 414 *et seq.,* 424–426, 440–442, Sect. 6.4b, 452–454, 460, 464
Ctenocladus, 554
cuticle, 575
cycads, 526, 532
cycloheximide, 436
effects on flagellar regeneration, 36
Cylindrocapsa, 552, 564, 570–571
Cylindrocystis, 358
cytochalasin B, 64–65, 384, 438, 476
cytoplasmic bridges (Volvocales), 14, 59 *et seq.*

Dasycladales, 560, 564
Desmidioideae (see also desmids, placoderm), 258, 556
desmids, 572–574
filamentous, 416
placoderm, 358–360, 374, 382, 386, 460
—constricted, Sect. 6.4, 452
—unconstricted, Sect. 6.3
saccoderm, 358, Sect. 6.1, 374, 382, 406
variability in size, ornamentation, symmetry etc. 418, Sect. 6.4a, 464
diatoms, 88, 522, 567–568, 572
flagella, 30
Dictyosphaericeae, 72
dinoflagellates, 516–518, 564, 567
diploidy, 282, 426, 430, 464, 575
—and genetic buffering, 78
division caps, (Oedogoniales)—see caps, division
division ring, (Oedogoniales)—see ring, division
Draparnaldia, 573
Draparnaldiopsis, 574
Dunaliella, 562
dwarf male, 268–270, 298 *et seq.*

Ectocarpaceae, 574

Elodea, 562
endoplasmic reticulum, 80, 106, 112, 220, 244, 248–254, 264, 288, 372, 478, 492, 496, 500, 510, 569
Enteromorpha, 206
epiphyte, 186
Equisetum, 534, 562
Eremosphaera, 24
Eremosphaeraceae, 72
Euastropsis, 142
Euconjugatae, 358, 556
Eudorina, 10–18, 61
euglenids (and *Euglena*), 516–518, 564–568
eyespots, 10, 24, 40, 59–60, 63, 66, 176, 206, 254, 262, 296, 560

fenestra, polar, 46, 65–66, 76, 84, 94, 106, 118, 196, 204, 214, 516–518, 552, 563
ferns, 532, 569
fertilization, 58 *et seq.*, 130 *et seq.*, 142, 214 *et seq.*, 270, 276 *et seq.*, 280, 376–378, 408 *et seq.*, 454, 460, 506–508
fertilization tubule, 58, 130
fibrous ring, 248, 262, 274, 530
filamentous form, evolution of, 570 *et seq.*
flagella, 254, 357, 502, 514 *et seq.*
    fusion (during mating), 57, 216
    hairs, 28
    heterokont, 496, 563–565
    isokont, 563–565
    motion, 26
    mutants, 30
    regeneration experiments, 36 *et seq.*
    retraction (gametes, zoospores), 124–126, 184, 206, 216
    scales, 194, 500, 564
    shedding (gametes, zoospores), 256–258, 262, 280
flagella apparatus, 26, 88, 184–186, 194, 206, 246, 262, 274, 280, 296, 302, 526 *et seq.*, 563, 565, 571
    basal cartwheel, 28, 44, 63—degeneration, 498
    phylogeny, Sect. 8.4a
fossils (algal), 474, 508, 564–565
Fritschiella, Sect. 4.8, 548, 556, 573–576

gametes, 14–18, 54 *et seq.*, 60 *et seq.*, 100, 116–118, 126 *et seq.*, 142, 196, 200, 204, 208, 216–218, 357, 366, 404, 412, 452, 456, 460, 514, 574
    phylogenetic aspects of structure, Sect. 8.2
Geminella, 570
genetic buffering, 78
geotrophism, 482 *et seq.*
germ net *(Hydrodictyon)*, 116, 138 *et seq.*
ghosts of cells, 102, 148
Ginkgo, 526
glycolate dehydrogenase, 562
glycolate oxidase, 562
Golenkinia, flagella, 30
Golgi bodies, 108, 112, 172, 178, 192, 196, 202, 212, 218–220, 224, 230, 234–236, 244, 250–254, 262, 274, 284, 288–294, 302, 308, 378–

380, 384, 410, 440–442, 446, 474, 482 *et seq.*, 492, 500, 564, 569, 573
    membranes, 442
gone, 366, 412, 462, 466
gonidia, 12, 59 *et seq.*
Gonium, 12, 536
gymnosperms, 526
Gymnozyga, 416

Haemanthus, **synctium,** 546
hair cells, 186–188, Sect. 4.3b, 194, 278, 282–284, 302
    formation, 288 *et seq.*, 312–314
heterothallism, 14
    advantages, 16
heterotrichy, 168–170, 186, 204, 218, 573–575
high voltage electron microscopy, 230, 240
higher plants—see land plants
holdfast, 168, 176, 196, 208, 220, 236–238, 242, 258–260, 264, 282
homothallism, 14
    advantages, 16
Hormidium (see *Klebsormidium*), 176
hormones (inducers), 61–62, 268, 404
hyaline layer and vesicle, 244 *et seq.*, 254–256, 276, 296–298, 302, 366, 412, 454
Hyalotheca, 386, 400
Hydra, 74
Hydrodictyaceae, 72, 90, 140
Hydrodictyon, 56, 70–72, 90, 98, 110, Sect. 3.7, 140–150, 154, 206, 216, 520, 528, 544–546, 550
hydroxyproline, 22

inducers—see hormones
internodes, 468 *et seq.*, 474–478
inversion, 12, 18, 60, 63 *et seq.*, 536, 542
    evolution of, 18–20, 540
ion transport, 114, 212, 474
isogamy, 14, 56, 70, 98, 170, 176, 196, 200, 204, 216–218, 454–456, 536, 560, 564, 575
isopropyl-N-phenyl carbamate (IPC), 240–242, 442
isthmus, 414, 424, 442–446, 450, 454–456

janus cell, 418, 426

kinetochores, 200, 226, 230, 240, 286, 386
    fibres, 226, 230
Kirchneriella, 70–74, Sect. 3.3, 88, 90, 96, 154, 526, 544, 570
Klebsormidium, 52, 172 Sect. 4.2, 188, 192–196, 202, 214, 218, 504, 516, 532, 548 *et seq.*, 556 *et seq.*, 570–571, 574

Labyrinthula, 88
land plants, 220, 370–374, 400–402, 438, 478–482, 490, 520–522, 526, 532, 573
    progenitors, 168, 174, 218, 534, 548, 554, *et seq.*, Sect. 8.4d
Leptosira, 536, 563
lichens, 74
life cycle, haploid, 78, 458, 508, 574

lignin, 575
Lithodesmium, 88
  flagella, 30, 88
liverworts, 562, 569, 575

macrandrous reproduction, 268 *et seq.,* 300
manchette, 498 *et seq.,* 534 *et seq.*
  adjunct, 498 *et seq.*
manubrium, 490 *et seq.*
Marchantia, 532
Marsilea, 534
mastigonemes, 28
mating types, 14, 56, 216, 376, 404, 454
meiosis, 59–61, 116, 134, 196, 282, 357, 366, 380, 412, 462–464, 508, 574
Membranopetra, 516, 566
meristem (apical), 468–470, 575
Mesotaenioideae (see also desmids, saccoderm), 358
metaphase band (microtubules), 46, 76
Micrasterias, 374, 414 *et seq.,* Sect. 6.4a, 444, 450–452, 458, 462
microfilaments, 65, 472, 476, 520
  in inversion, 64
Microspora, 88, 198, Sect. 4.5, 520, 526, 562
  evolution of wall, 570 *et seq.*
Microthamnion, 530
microtubules, 172, 184–186, 190–192, 218, 246, 250, 276, 288, 308, 362, 370–374, 392 *et seq.,* 442–444, 476–478, 498 et seq., 518, 528 *et seq.,* Sect. 8.4a
  chemistry, 30, 504
  involvement in colony formation, 110, 126 et seq., 138, 144–146, 154
  —forming vacuolar envelope, 120
  —inversion, 64
  —shape changes, 64, 110, 126–128, 146, 154, 498 *et seq.*
  persistent (after telophase), 180, 190, 373–374, 392 *et seq.,* 442–448, 536 *et seq.,* 554
  —and evolution of phragmoplast, 554 *et seq.*
  wall, 176–178, 182, 220, 264, 288, 292, 370, 378, 386–388, 398 *et seq.,* 406, 442, 446, 476, 486, 490
microtubule-like structures,
  in chloroplast, 218, 510
  in cytoplasm, 92, 492, 510
  in nuclei, 474
microtubule-organizing-center—see MTOC
mitochondria, 172, 176, 494 *et seq.,* 532, 569
  single, in *Chlorella,* 74
mosses, 562, 569
Mougeotia, 368, Sect. 6.2c, 554
Mougeotiaceae, 384
MTOC (microtubule-organizing-center), 34, 88, 106, 172, 242, 248, 274, 394, 404, 498, 518, 563, Sect. 8.4a
mucilage, 102, 236, 260, 270, 274, 278, 358–360, 370, 378–380, 388, 406, 412–414, 444, 454, 492, 520, 570
  basal, 242 *et seq.,* 254, 296
multilayered structure, 184–186, 194, 532 *et seq.*

nannandrous reproduction, 268 *et seq.,* 298 *et seq.*
Netrium, 358, Sect. 6.1a, 368, 382, 386, 400, 466, 562
Nitella, 114, 472–474, 482, 494, 498, 500, 504, 508, 532, 562
  flagellar apparatus, 32
node, 468 *et seq.,* 476, 488, 504
nuclear migration (post mitotic), 362, 368, 382, 390 *et seq.,* 416, 442, 448–450
nucleolus, 132, 288, 474, 496
  during mitosis, 44, 76, 84, 94, 172, 178, 196, 202, 214, 226, 286, 362, 370–374, 386, 390, 444, 480

Oedocladium, 220, 574
Oedogoniales, 170, Chap. 5, 358, 538, 558, 573–574
  evolution of wall structure, 570 *et seq.*
Oedogonium, 186, 198, 206–208, 218, Sect. 5.1, 284, 294–296, 298, 302–304, 520, 524, 532, 550, 572–574
  contractile vacuoles in zoospores, 22, 256
  flagellar apparatus, 32, 532
  rootlet microtubules, 34
Oocystaceae, 72
oogamy, 14, 70, 170, 186, 266, 470, 530, 536, 552, 560, 564, 573–575
oogenesis, 276 *et seq.,* 298 *et seq.,* 302 *et seq.,* Sect. 7.6
oogonial pore—see pore, fertilization
oogonium, 16, 188, 266 *et seq.,* 276 *et seq.,* 302 *et seq.,* 470, 474, 478, 482, 488, 492, 504 *et seq.*
  fossil, 565
oospores, 276 *et seq.,* 302 *et seq.*

palmella stages, 8, 65, 72, 568
Pandorina, 10–14, 536
papillae (see conjugation tube), 378, 406 *et seq.*
Paramecium, 28, 74
parthenospores, 18, 282, 312
pectin, 100–102, 236
Pediastrum, 70–72, 82, 90, 98, 110, 124, 128, 136–138, Sect. 3.8, 150, 154, 528, 544, 550
Pedinomonas, 540, 563 *et seq.,* 571
Penium, 386
perinuclear envelope, 76, 80, 84, 94, 106, 118, 144, 152, 542, 546
peroxisome, 172, 176–178, 182–184, 188, 196
Phaeoceros, 534
Phaeosphaera, 568
phialopore, 60, 63
phloem, 575
phragmoplast, 52, 190, 218, 374, 386, 480, 494, 510, 520–522, 536, Sect. 8.3c, Sect. 8.3d, 575–6
  evolution, 374
phycomycetes, 518, 569
phycoplast, 52, 67, 70, 78, 86, 90, 94–96, 106, 120, 134, 144, 152, 170, 174, 180, 196–198, 202–204, 214, 218–219, 226, 230, 240, 272–274, 286, 290–292, 302, 306, 520 *et seq.,* 532, 544, Sect. 8.3c, Sect. 8.3d, 575–576
  function, 50, 120, 550
  phylogenetic significance, 50 *et seq.,* 67, 575 *et seq.*

plakea, 18, 60, 536

plasmalemma, 22, 58, 64, 110, 124, 232, 244, 256, 402, 424, 442, 448–450, 474–476, 518

plasmodesmata, 63, 170, 176, 188, 196–198, 202, 218–220, 232, 244, 284, 296, 370, 474, 496

plasmogamy—see fertilization

plastids, 494 *et seq.*

Platydorina, 10, 536
    inversion, 20

Platymonas, 562–563

Pleodorina, 10–16

Pleurastrum, 563

Pleurotaenium, 412

pollen, 526

polyhedra, 90–92, 114–116, 130 *et seq.,* 142, 148, 544

polyploidy, 42, 418, 426, 440, 464–466

polyspermy, 282

Polytrichum, 534

pore, fertilization, 61, 270, 276, 280, 294, 308, 314
    formation, 278, 308, 312

pore, mucilage, 358, 388, 414, 420, 442–444
    formation, 450

Prasinophyceae, 562, Sect. 8.3d, 568

procaryotic cells, 518, 566 *et seq.*

prolamellar bodies, 26

protonema, 470, 508, 569, 575

Prototheca, 82

protozoa, 516, 522, 565–567, 575

Prymnesium, 28

Pseudendoclonium, 552–554, 562–564, 571

pseudocilia, 66

Psilotum, 562

Pyramimonas, 563 *et seq.,* 571

pyrenoid, 24, 59, 86, 90, 96, 108, 116–118, 126, 140, 172, 176, 196–198, 204, 208, 214, 220, 360, 368, 382, 388, 414, 444, 460, 464, 476
    division, 40, 48, 76, 178, 382, 450, 470
    taxonomic significance, 24, 38, 72, 573

Pyrrophyta, 568

Radiophyllum, 570

Raphidonema, 174, 556–558

Red algae, 170, 516, 566–567, 576

rhizoids, 168–170, 218, 258, 262–264, 468–470, 482 *et seq.*

rhizoplast, 530, 564

ribosomes, 254, 264, 408, 500

ring, division (Oedogoniales), 220 *et seq.,* 266, 272, 276, 284, 298–302, 306
    abnormal, 286

rootlet (root) microtubules, 32 *et seq.,* 206, 262, 274, 526–528, 563, 566–568

rootlet templates, 34, 94, 246, 274, 504, 526, 530, 568

rotation of protoplast
    during division, 40, 44, 67, 72, 108
    hair cells, 190 *et seq.*

Roya, 366

Saprolegnia, 569

scales (see flagella scales), 194, 564
    taxonomic significance, 562

Scenedesmus, 42, 70–72, 82, 90, Sect. 3.6, 148, 544

Schizomeris, 168, 530, 569, 573

Selaginella, 562

Selenastraceae, 72, 82

semicells, 388, 414
    morphogenesis, 400–402, 416 *et seq.*
    —after germination, 462 *et seq.*

septum, 362, 374–376, 390–392, 398, 416, 422–424, 430, 444–446
    results of perforated septum, 402 *et seq.,* 424, 428, 452

shield cells, 488 *et seq.*

sieve tubes, 575

silver hexamine (polysaccharide localization), 230–232, 244, 262–264

Siphonales, 546, 560, 569

Siphonoclades, 560

Siphonous evolution, 72, 513, Sect. 8.3b, Sect. 8.4c, 575

Sirogonium, 574

slime molds, 516, 522

Sorastrum, 72, 110, 146, Sect. 3.9, 528

sperm, 188, 194, 198, 250, 266 *et seq.,* 296–298, 302, 488, 508

sperm packets, 16–18, 60, 64–65

spermatogenesis, 186, 268 *et seq.,* 302, 490, Sect. 7.5
    bryophytes, 88

Sphaerocarpos, 534

Sphaeroplea, 168

spherosomes, 492

spindle fibres, 240

spindles, multipolar, 240

spines, *Scenedesmus,* 96 *et seq.*
    zygospores, 454, 458–460

Spirogyra, 202, 357, 366–368, Sect. 6.2a, 382, 386, 406, 410, 416, 480–482, 520, 554, 560–562, 572
    replicate end walls, 416

spitzenkörper, 292, 482

Spondylomoraceae, 540

spores, 16

sporophyte, 188, 575

sporopollenin (see trilaminar sheath), 80–82, 102, 148

statoliths, 484 *et seq.*

Staurastrum, 414, 418, 452, 460, 464

sterile cells (surrounding oogonia, antiheridia), 470, 488, 508 *et seq.,* 560, 575

Stichococcus, Sect. 4.1, 176, 182, 516, 548, 556–558

Stigeoclonium, 168, 182, Sect. 4.6, 216–218, 520, 528–530, 548, 556 *et seq.,* 564, 573

stigma (see eyespots), 24, 176

streaming, cytoplasmic, 360, 390, 438, 470 *et seq.,* 476–478

striated fibres, 32, 44, 206, 250, 262, 274, 530, 564

suffultory (supporting) cell, 270, 276–278, 302–4, 310–2

swarming vesicle, 136, 140, 144, 150, 154–156

terrestial algae, 218, 574

Tetraedron, 70–74, 88, Sect. 3.5, 150–152, 526, 544, 570

Tetraspora, 20, 35, Sect. 2.3, 74–76, 214, 540, 548, 568

Tetrasporineae (Tetrasporaceae, Tetrasporales), 8, 20, 66, 72, 168, 532, 536 *et seq.,* 546–548, 560, 568–569, 573

Tolypella, 508

tonoplast, 120, 124, 136, 546

Tribonema, 572

Trichosarcina, 552–554, 562, 571

trilaminar sheath, 80, 102, 110–112

turgor pressure, 112, 118, 234–236, 380, 408, 424, 438, 458, 474, 486

Ulothrix, 167, 174–6, 182, Sect. 4.4, 200–204, 214, 218, 394, 548, 556, 570–571, 573–4

U. zonata, 554

Ulotrichalean evolution—see Ulotrichales

Ulotrichales, Chap. 4, 358, 522, 532, 536, 542, Sect. 8.3c, 569, Sect. 8.4d

Ulva, 52, 168, Sect. 4.7, 552, 569, 574

Ulvaceae, 168, 208, 212, 554, 573

Uronema, 174

vacuolar envelope, 118 *et seq.,* 136, 144, 546

vacuole, 114, 136, 176–8, 182–184, 192, 214, 234, 258, 372, 380, 388, 406, 450, 474, 478, 510

Vaucheria, 518, 569

Vaucheriaceae, 569

vierergruppe, 534

vinblastine, 442

virus, 264

Volvocales, Chap. 2, 170, 186, 536 *et seq.,* 544, 558 *et seq.;* 565, Sect. 8.4b, Sect. 8.4c, 569, 573

polarity in colonial forms, 10

structure of colonies, 18

Volvocine evolution, 8 *et seq.,* Sect. 8.3a, 565, Sect. 8.4b

Volvox, 10–16, 20, 50, Sect. 2.2, 214, 404, 528, 540, 565

extra basal bodies, 28

wall, H-shaped segments, 198 *et seq.*

H-shaped segments and evolution, 570 *et seq.*

primary, 224, 236, 294, 416, 424, 446, 450–452, 458, 572

secondary, 224, 236, 294, 298, 306–308, 414, 442, 450, 458, 572

secretion, 80, 84, 148, 154, 200, 224, 236, 260, 290, 308, 360, 378, 390, 406, 410, 424, 438–440, 444–446, 450, 454, 482 *et seq.,* 572 *et seq.*

wound response, 236, 260, 266, 294, 572

Xanthophyceae, 568–569, 572

xylem, 575

wall thickenings, 490–492

Zamia, 534

zoospore, 61, 70–74, 82, 88, 90–92, 96–98, 114–126, 134–150, 154, 176, 182–86, 192 *et seq.,* 204–208, 218, 224, 242 *et seq.,* 268, 310, 542, Sect. 8.3c

"bryophytan" and "Chlamydomonad," 528 *et seq.*

germination, 168, 184, 194, 206–208, 242 *et seq.,* 258 *et seq.,* 470, 570–571

phylogenetic aspects of structure, 514, Sect. 8.2

zoosporogenesis, 118–120, 182, 242 *et seq.,* 258, 296 et seq., 550, 571

Zygnema, 24, 360, 368, Sect. 6.2b, 400

Zygnematales—see Conjugales

Zygnemoideae, 358–360, Sect. 6.2, 376

zygote (zygospores), 58–59, 90, 100, 116, 130 *et seq.,* 142, 196, 216, 276 *et seq.,* 357, 366, 376 *et seq.,* 382, 404 *et seq.,* 410, 454 *et seq.,* 575

germination, 59, 61, 100, 114, 134, 142, 276 *et seq.,* 310, 366, 380, 410, 460 *et seq.,* 574

ABOUT THE BOOK

The text of this book is set in Video Garamond, a modified version of a typeface originally designed in 1540 by the French typographer Claude Garamond. Video typefaces are set on photosensitive paper by a computer-driven cathode ray tube. The display type is Microgramma Bold Extended. The book was designed by Joseph Vesely, printed at Murray Printing Company and bound by Haddon Craftsmen, Inc. The cover was printed by Lehigh Press.